# Partial Differential Equations for Scientists and Engineers

# Partial Differential Equations for Scientists and Engineers

Third Edition


Tyn Myint-U, Ph.D.

Professor of Mathematics
Department of Mathematics and Computer Science
Manhattan College
New York

with

Lokenath Debnath, Ph.D.

Professor and Chairman
Department of Mathematics
University of Central Florida
Orlando, Florida



P T R Prentice Hall
Englewood Cliffs, New Jersey 07632

**Library of Congress Cataloging-in-Publication Data**

Tyn Myint U.
Partial differential equations for scientists and engineers.

Bibliography: p.
Includes index.
1. Differential equations, Partial. 2. Science —Mathematics. 3. Engineering mathematics. I. Debnath, Lokenath. II. Title.
QA377.T96 1987 515.3'53 87-15620


A Simon & Schuster Company
Englewood Cliffs, New Jersey 07632



Printed in the United States of America
10 9 8 7 6 5 4 3

ISBN 0-13-051665-1

Prentice-Hall International (UK) Limited, *London*
Prentice-Hall of Australia Pty. Limited, *Sydney*
Prentice-Hall Canada Inc., *Toronto*
Prentice-Hall Hispanoamericana, S. A., *Mexico*
Prentice-Hall of India Private Limited, *New Delhi*
Prentice-Hall of Japan, Inc., *Tokyo*
Simon & Schuster Asia Pte. Ltd., *Singapore*
Editora Prentice-Hall do Brasil, Ltda., *Rio de Janeiro*

To
U and Mrs. Hla Din
U and Mrs. Thant

To the memory of Jogesh Chandra Debnath

# Contents

# Preface

The theory of partial differential equations has long been one of the most important fields in mathematics. This is essentially due to the frequent occurrence and the wide range of applications of partial differential equations in many branches of physics, engineering, and other sciences. With much interest and great demand for theory and applications in diverse areas of science and engineering, several excellent books on PDEs have been published. This book is written for the purpose of presenting an approach based mainly on the mathematics, physics, and engineering problems and their solutions, and also to construct a course appropriate for all students of mathematical, physical and engineering sciences. Our primary objective, therefore, is not concerned with an elegant exposition of general theory, but to provide students with the fundamental concepts, the underlying principles, a wide range of applications, and various methods of solution of partial differential equations.

This book, a revised and expanded version of the second edition published in 1980, was written for a one-semester course in the theory and applications of partial differential equations. It has been used by advanced undergraduate or beginning graduate students in applied mathematics, physics, engineering, and other applied sciences. The prerequisite for its study is a standard calculus sequence with elementary ordinary differential equations. This revised edition is in part based on lectures given by Tyn Myint-U at Manhattan College and by Lokenath Debnath at the University of Central Florida. This revision preserves the basic content and style of the earlier editions, which were written by Tyn Myint-U alone. However, the authors have made some major additions and changes in this third edition in order to modernize the contents and to improve clarity. Two new chapters added are on nonlinear PDEs, and on numerical and approximation methods. New materials emphasizing applications have been inserted. New examples and exercises have been provided. Many physical interpretations of mathematical solutions have been added. Also, the authors have improved the exposition by reorganizing some material and by making examples, exercises, and applications more prominent in the text. These additions and changes have been made with the student uppermost in mind.

The first chapter gives an introduction to partial differential equations. The second chapter deals with the mathematical models representing physical and engineering problems that yield the three basic types of PDEs. Included are only important equations of most common interest in physics and engineering. The third chapter constitutes an account of the classification of linear PDEs of second order in two independent variables into hyperbolic, parabolic and elliptic types and, in addition, illustrates the determination of the general solution for a class of relatively simple equations.

Cauchy's problem, the Goursat problem, and the initial-boundary value problems involving hyperbolic equations of the second order are presented in Chapter 4. Special attention is given to the physical significance of solutions and the methods of solution of the wave equation in Cartesian, spherical polar, and cylindrical polar coordinates. The fifth chapter contains a fuller treatment of Fourier Series and integrals essential for the study of PDEs. Also included are proofs of several important theorems concerning Fourier series and integrals.

Separation of variables is one of the simplest methods, and the most widely used method, for solving PDEs. The basic concept and separability conditions necessary for its application are discussed in the sixth chapter. This is followed by some well-known problems of applied mathematics, mathematical physics, and engineering sciences along with a detailed analysis of each problem. Special emphasis is also given to the existence and uniqueness of the solutions and to the fundamental similarities and differences in the properties of the solutions to the various PDEs. In Chapter 7, self-adjoint eigenvalue problems are treated in depth, building on their introduction in the preceding chapter. In addition, Green's function and its applications to eigenvalue problems and boundary value problems for ordinary differential equations are presented. Following the general theory of eigenvalues and eigenfunctions, the most common special functions including the Bessel, Legendre, and Hermite functions are discussed as examples of the major role of special functions in physical and engineering sciences. Application to heat conduction problems and the Schrödinger equation for the linear harmonic oscillator are also included.

Boundary value problems and the maximum principle are described in Chapter 8, and emphasis is placed on the existence, uniqueness and well-posedness of solutions. Higher dimensional boundary value problems and the method of eigenfunction expansion are treated in the ninth chapter which also includes several applications to the vibrating membrane, waves in three dimensions, heat conduction in a rectangular volume, the three-dimensional Schrödinger equation in a central field of force, and the hydrogen atom. Chapter 10 deals with the basic concepts and construction of Green's function and its application to boundary value problems.

Chapter 11 provides an introduction to the use of integral transform methods and their applications to numerous problems in applied mathematics, mathematical physics, and engineering sciences. The fundamental properties and the techniques of Fourier, Laplace, Hankel, and Mellin transforms are discussed in some detail. Applications to problems concerning heat flows, fluid flows, elastic waves, current and potential electric transmission lines are included in this chapter.

Chapters 12 and 13 are entirely new. First order and second order nonlinear PDEs are covered in Chapter 12. Most of the contents of this chapter have been developed during the last twenty-five years. Several new nonlinear PDEs including the one-dimensional nonlinear wave equation, Whitham's equation, Burgers' equation, the Korteweg de Vries equation, and the nonlinear Schrödinger equation are solved. The solutions of these equations are then discussed with physical significance. Special emphasis is given to the fundamental similarities and differences in the properties of the solutions to the corresponding linear and nonlinear equations under consideration.

The final chapter is devoted to the major numerical and approximation methods for finding solutions of PDEs. A fairly detailed treatment of the explicit and implicit finite difference methods is given with applications. The variational method and the Euler–Lagrange equations are described with many applications. Also included are the Rayleigh–Ritz, the Galerkin, and the Kantorovich methods of approximation with many illustrations and applications.

This new edition contains almost four hundred examples and exercises which are either directly associated with applications or phrased in terms of the physical and engineering contexts in which they arise. The exercises truly complement the text, and answers to most exercises are provided at the end of the book. The Appendix has been expanded to include some basic properties of the Gamma function and the tables of Fourier, Laplace, and Hankel transforms. For students wishing to know more about the subject or to have further insight into the subject matter, important references are listed in the Bibliography.

The chapters on the mathematical models, Fourier series and integrals, and eigenvalue problems are self-contained, so these chapters can be omitted for those students who have prior knowledge of the subject.

An attempt has been made to present a clear and concise exposition of the mathematics used in analyzing a variety of problems. With this in mind, the chapters are carefully organized to enable students to view the material in an orderly perspective. For example, the results and theorems in the chapters on Fourier series and integrals, and on eigenvalue problems are explicitly mentioned, whenever necessary, to avoid confusion with their use in the development of PDEs. A wide range of problems subject to various boundary conditions has been included to improve the student's understanding.

In this third edition, specific changes and additions include the following:

1. Chapter 2 on mathematical models has been revised by adding a list of the most common linear PDEs in applied mathematics, mathematical physics, and engineering science.
2. The chapter on the Cauchy problem has been expanded by including the wave equations in spherical and cylindrical polar coordinates. Examples and exercises on these wave equations and the energy equation have been added.
3. Eigenvalue problems have been revised with an emphasis on Green's functions and applications. A section on the Schrödinger equation for the linear harmonic oscillator has been added. Higher dimensional boundary value problems with an emphasis on applications, and a section on the hydrogen atom and on the three-dimensional Schrödinger equation in a central field of force have been added to Chapter 9.
4. Chapter 11 has been extensively reorganized and revised in order to include Hankel and Mellin transforms and their applications, and has added new sections on the asymptotic approximation method and the finite Hankel transform with applications. Many new examples and exercises, some new materials with applications, and physical interpretations of mathematical solutions have also been included.
5. A new chapter on nonlinear PDEs of current interest and their applications has been added with considerable emphasis on the fundamental similarities and the distinguishing differences in the properties of the solutions to the nonlinear and corresponding linear equations.
6. Chapter 13 is also new. It contains a fairly detailed treatment of explicit and implicit finite difference methods with their stability analysis. A large section on the variational methods and the Euler–Lagrange equations has been included with many applications. Also included are the Rayleigh–Ritz, the Galerkin, and the Kantorovich methods of approximation with illustrations and applications.

7. Many new applications, examples, and exercises have been added to deepen understanding. Expanded versions of the tables of Fourier, Laplace, and Hankel transforms are included. The bibliography has been updated with more recent and important references.

As a text on partial differential equations for students in applied mathematics, physics, engineering, and applied sciences, this edition provides the student with the art of combining mathematics with intuitive and physical thinking to develop the most effective approach to solving problems of interest.

In preparing this edition, the authors wish to express their sincere thanks to those who have read the manuscript and offered many valuable suggestions and comments. The authors also wish to express their thanks to the editor and the staff of Elsevier-North Holland, Inc. for their kind help and cooperation.

Tyn Myint-U
Lokenath Debnath

CHAPTER ONE

# Introduction

## 1.1 BASIC CONCEPTS AND DEFINITIONS

A differential equation that contains, in addition to the dependent variable and the independent variables, one or more partial derivatives of the dependent variable is called a *partial differential equation*. In general, it may be written in the form

$$f(x, y, \ldots, u, u_x, u_y, \ldots, u_{xx}, u_{xy}, \ldots) = 0 \tag{1.1.1}$$

involving several independent variables $x, y, \ldots$, an unknown function $u$ of these variables, and the partial derivatives[1] $u_x, u_y, \ldots, u_{xx}, u_{xy}, \ldots$ of the function. Here Eq. (1.1.1) is considered in a suitable domain $D$ of the $n$-dimensional space $R^n$ in the independent variables $x, y, \ldots$. We seek functions $u = u(x, y, \ldots)$ which satisfy Eq. (1.1.1) identically in $D$. Such functions, if they exist, are called *solutions* of Eq. (1.1.1). From these many possible solutions we attempt to select a particular one by introducing suitable additional conditions.

For instance,

$$\begin{gathered} uu_{xy} + u_x = y \\ u_{xx} + 2yu_{xy} + 3xu_{yy} = 4 \sin x \\ (u_x)^2 + (u_y)^2 = 1 \\ u_{xx} - u_{yy} = 0 \end{gathered} \tag{1.1.2}$$

[1] Subscripts on dependent variables denote differentiations, e.g.,

$$u_x = \partial u / \partial x \qquad u_{xy} = \partial^2 u / \partial y\, \partial x$$

are partial differential equations. The functions

$$u(x, y) = (x + y)^3$$
$$u(x, y) = \sin (x - y)$$

are solutions of the last equation of (1.1.2), as can easily be verified.

The *order* of a partial differential equation is the order of the highest-ordered partial derivative appearing in the equation. For example,

$$u_{xx} + 2xu_{xy} + u_{yy} = e^y$$

is a second-order partial differential equation, and

$$u_{xxy} + xu_{yy} + 8u = 7y$$

is a third-order partial differential equation.

A partial differential equation is said to be *linear* if it is linear in the unknown function and all its derivatives with coefficients depending only on the independent variables; it is said to be *quasilinear* if it is linear in the highest-ordered derivative of the unknown function. For example, the equation

$$yu_{xx} + 2xyu_{yy} + u = 1$$

is a second-order linear partial differential equation, whereas

$$u_x u_{xx} + xuu_y = \sin y$$

is a second-order quasilinear partial differential equation. The equation which is not linear is called a *nonlinear* equation.

We shall be primarily concerned with linear second-order partial differential equations, which frequently arise in problems of mathematical physics. The most general second-order linear partial differential equation in $n$ independent variables has the form

$$\sum_{i,j=1}^{n} A_{ij} u_{x_i x_j} + \sum_{i=1}^{n} B_i u_{x_i} + Fu = G \tag{1.1.3}$$

where we assume without loss of generality that $A_{ij} = A_{ji}$. We also assume that $B_i$, $F$, and $G$ are functions of the $n$ independent variables $x_i$.

If $G$ is identically zero, the equation is said to be *homogeneous*; otherwise it is *nonhomogeneous*.

The general solution of a linear ordinary differential equation of $n$th order is a family of functions depending on $n$ independent arbitrary constants. In the case of partial differential equations, the general solution depends on arbitrary functions rather than on arbitrary constants. To illustrate this, consider the equation

$$u_{xy} = 0$$

If we integrate this equation with respect to $y$, we obtain

$$u_x(x, y) = f(x)$$

A second integration with respect to $x$ yields

$$u(x, y) = g(x) + h(y)$$

where $g(x)$ and $h(y)$ are arbitrary functions.

Suppose $u$ is a function of three variables, $x$, $y$, and $z$. Then for the equation

$$u_{yy} = 2$$

one finds the general solution

$$u(x, y, z) = y^2 + yf(x, z) + g(x, z)$$

where $f$ and $g$ are arbitrary functions of two variables $x$ and $z$.

We recall that in the case of ordinary differential equations, the first task is to find the general solution, and then a particular solution is determined by finding the values of arbitrary constants from the prescribed conditions. But, for partial differential equations, selecting a particular solution satisfying the supplementary conditions from the general solution of a partial differential equation may be as difficult as, or even more difficult than, the problem of finding the general solution itself. This is so because the general solution of a partial differential equation involves arbitrary functions; the specialization of such a solution to the particular form which satisfies supplementary conditions requires the determination of these arbitrary functions, rather than merely the determination of constants.

For linear homogeneous ordinary differential equations of order $n$, a linear combination of $n$ linearly independent solutions is a solution. Unfortunately, this is not true, in general, in the case of partial differential equations. This is due to the fact that the solution space of every homogeneous linear partial differential equation is infinite dimensional. For example, the partial differential equation

$$u_x - u_y = 0 \tag{1.1.4}$$

can be transformed into the equation

$$2u_\eta = 0$$

by the transformation of variables

$$\xi = x + y$$
$$\eta = x - y$$

The general solution is

$$u(x, y) = f(x + y)$$

where $f(x+y)$ is an arbitrary function. Thus, we see that each of the functions

$$(x+y)^n$$
$$\sin n(x+y)$$
$$\cos n(x+y)$$
$$\exp n(x+y), \qquad n = 1, 2, 3, \ldots$$

is a solution of Eq. (1.1.4). The fact that a simple equation such as (1.1.4) yields infinitely many solutions is an indication of an added difficulty which must be overcome in the study of partial differential equations. Thus, we generally prefer to directly determine the particular solution of a partial differential equation satisfying prescribed supplementary conditions.

## 1.2 MATHEMATICAL PROBLEMS

A problem consists of finding an unknown function of a partial differential equation satisfying appropriate supplementary conditions. These conditions may be initial and/or boundary conditions. For example

| | | | |
|---|---|---|---|
| *P.D.E.* | $u_t - u_{xx} = 0$ | $0 < x < l,$ | $t > 0$ |
| *I.C.* | $u(x, 0) = \sin x$ | $0 \leqslant x \leqslant l$ | |
| *B.C.* | $u(0, t) = 0$ | | $t \geqslant 0$ |
| *B.C.* | $u(l, t) = 0$ | | $t \geqslant 0$ |

is a problem which consists of a partial differential equation and three supplementary conditions. The equation describes the heat conduction in a rod of length $l$. The last two conditions are called the *boundary conditions* which describe the function at two prescribed boundary points. The first condition is known as the *initial condition* which prescribes the unknown function $u(x, t)$ throughout the given region at some initial time $t$, in this case $t = 0$. This problem is known as the *initial-boundary value problem*. Mathematically speaking, the time and the space coordinates are regarded as independent variables. In this respect, the initial condition is merely a point prescribed on the $t$-axis and the boundary conditions are prescribed, in this case, as two points on the $x$-axis. Initial conditions are usually prescribed at a certain time $t = t_0$ or $t = 0$, but it is not customary to consider the other end point of a given time interval.

In many cases, in addition to prescribing the unknown function, other conditions such as their derivatives are specified on the boundary.

In considering the problem of unbounded domain, the solution can be determined uniquely by prescribing initial conditions only. The problem is called the *initial value problem*.[2] The solution of such a problem may be interpreted physically as the solution unaffected by the boundary conditions at infinity. For problems affected by the boundary at infinity, boundedness conditions on the behavior of solutions at infinity must be prescribed.

A mathematical problem is said to be *properly posed* if it satisfies the following requirements:

1. Existence: There is at least one solution.
2. Uniqueness: There is at most one solution.
3. Continuity: The solution depends continuously on the data.

The first requirement is an obvious logical condition, but we must keep in mind that we cannot simply state that the mathematical problem has a solution just because the physical problem has a solution. We may well be erroneously developing a mathematical model, say, consisting of a partial differential equation whose solution may not exist at all. The same can be said about the uniqueness requirement. In order to really reflect the physical problem that has a unique solution, the mathematical problem must have a unique solution.

For physical problems, it is not sufficient to know that the problem has a unique solution. Hence the last requirement is not only useful but also essential. If the solution is to have physical significance, a small change in the initial data must produce a small change in the solution. The data in a physical problem are normally obtained from experiment, and are approximated in order to solve the problem by numerical or approximate methods. It is essential to know that the process of making an approximation to the data produces only a small change in the solution.

## 1.3 LINEAR OPERATORS

An operator is a mathematical rule which, when applied to a function, produces another function. For example, in the expressions

$$L[u] = \frac{\partial^2 u}{\partial x^2} + \frac{\partial^2 u}{\partial y^2}$$

$$M[u] = \frac{\partial^2 u}{\partial x^2} - \frac{\partial u}{\partial x} + x\frac{\partial u}{\partial y}$$

$L = \partial^2/\partial x^2 + \partial^2/\partial y^2$ and $M = \partial^2/\partial x^2 - \partial/\partial x + x(\partial/\partial y)$ are called the *differential operators*.

[2] The mathematical definition is given in Chapter 4.

An operator is said to be *linear* if it satisfies the following:

1. A constant $c$ may be taken outside the operator:

$$L[cu] = cL[u] \tag{1.3.1}$$

2. The operator operating on the sum of two functions gives the sum of the operator operating on the individual functions:

$$L[u_1 + u_2] = L[u_1] + L[u_2] \tag{1.3.2}$$

We may combine (1.3.1) and (1.3.2) as

$$L[c_1u_1 + c_2u_2] = c_1L[u_1] + c_2L[u_2] \tag{1.3.3}$$

where $c_1$ and $c_2$ are constants. This can be extended to a finite number of functions. If $u_1, u_2, \ldots, u_k$ are $k$ functions and $c_1, c_2, \ldots, c_k$ are $k$ constants, then by repeated application of Eq. (1.3.3)

$$L\left[\sum_{j=1}^{k} c_ju_j\right] = \sum_{j=1}^{k} c_jL[u_j] \tag{1.3.4}$$

We may now define the sum of two linear differential operators formally. If $L$ and $M$ are two linear operators, then the sum of $L$ and $M$ is defined as

$$(L + M)[u] = L[u] + M[u] \tag{1.3.5}$$

where $u$ is a sufficiently differentiable function. It can be readily shown that $L + M$ is also a linear operator.

The product of two linear differential operators $L$ and $M$ is the operator which produces the same result as is obtained by the successive operations of the operators $L$ and $M$ on $u$, that is,

$$LM[u] = L(M[u]) \tag{1.3.6}$$

in which we assume that $M[u]$ and $L(M[u])$ are defined. It can be readily shown that $LM$ is also a linear operator.

Linear differential operators satisfy the following:

1. $L + M = M + L$ (commutative) (1.3.7)
2. $(L + M) + N = L + (M + N)$ (associative) (1.3.8)
3. $(LM)N = L(MN)$ (associative) (1.3.9)
4. $L(M + N) = LM + LN$ (distributive) (1.3.10)

For linear differential operators with constant coefficients,

5. $LM = ML$ (commutative) (1.3.11)

EXAMPLE 1.3.1. Let $L = \dfrac{\partial^2}{\partial x^2} + x\dfrac{\partial}{\partial y}$ and $M = \dfrac{\partial^2}{\partial y^2} - y\dfrac{\partial}{\partial y}$

$$LM[u] = \left(\frac{\partial^2}{\partial x^2} + x\frac{\partial}{\partial y}\right)\left(\frac{\partial^2 u}{\partial y^2} - y\frac{\partial u}{\partial y}\right)$$
$$= \frac{\partial^4 u}{\partial x^2\,\partial y^2} - y\frac{\partial^3 u}{\partial x^2\,\partial y} + x\frac{\partial^3 u}{\partial y^3} - xy\frac{\partial^2 u}{\partial y^2} - x\frac{\partial u}{\partial y}$$
$$ML[u] = \left(\frac{\partial^2}{\partial y^2} - y\frac{\partial}{\partial y}\right)\left(\frac{\partial^2 u}{\partial x^2} + x\frac{\partial u}{\partial y}\right)$$
$$= \frac{\partial^4 u}{\partial y^2\,\partial x^2} + x\frac{\partial^3 u}{\partial y^3} - y\frac{\partial^3 u}{\partial y\,\partial x^2} - xy\frac{\partial^2 u}{\partial y^2}$$

which shows that $LM \neq ML$.

Now let us consider a linear second-order partial differential equation. In the case of two independent variables, such an equation takes the form

$$A(x, y)u_{xx} + B(x, y)u_{xy} + C(x, y)u_{yy} + D(x, y)u_x + E(x, y)u_y + F(x, y)u = G(x, y) \quad (1.3.12)$$

where $A$, $B$, $C$, $D$, $E$, and $F$ are the coefficients, and $G$ is the nonhomogeneous term.

If we denote

$$L = A\frac{\partial^2}{\partial x^2} + B\frac{\partial^2}{\partial x\,\partial y} + C\frac{\partial^2}{\partial y^2} + D\frac{\partial}{\partial x} + E\frac{\partial}{\partial y} + F$$

then Eq. (1.3.12) may be written in the form

$$L[u] = G \quad (1.3.13)$$

Very often the square bracket is omitted and one simply writes

$$Lu = G$$

Let $v_1, v_2, \ldots, v_n$ be $n$ functions which satisfy

$$L[v_j] = G_j, \quad j = 1, 2, \ldots, n$$

and let $w_1, w_2, \ldots, w_n$ be $n$ functions which satisfy

$$L[w_j] = 0, \quad j = 1, 2, \ldots, n.$$

If we let

$$u_j = v_j + w_j$$

then the function

$$u = \sum_{j=1}^{n} u_j$$

satisfies

$$L[u] = \sum_{j=1}^{n} G_j$$

This is called the *principle of superposition.*

In particular, if $v$ is a particular solution of Eq. (1.3.13), that is, $L[v] = G$, and $w$ is a solution of the associated homogeneous equation, that is, $L[w] = 0$, then $u = v + w$ is a solution of $L[u] = G$.

## 1.4 SUPERPOSITION

We may express supplementary conditions using the operator notation. For instance, the initial boundary value problem

$$\begin{aligned}
u_{tt} - c^2 u_{xx} &= G(x, t) && 0 < x < l, && t > 0 \\
u(x, 0) &= g_1(x) && 0 \leqslant x \leqslant l && \\
u_t(x, 0) &= g_2(x) && 0 \leqslant x \leqslant l && \\
u(0, t) &= g_3(t) && && t \geqslant 0 \\
u(l, t) &= g_4(t) && && t \geqslant 0
\end{aligned} \tag{1.4.1}$$

may be written in the form

$$\begin{aligned}
L[u] &= G \\
M_1[u] &= g_1 \\
M_2[u] &= g_2 \\
M_3[u] &= g_3 \\
M_4[u] &= g_4
\end{aligned} \tag{1.4.2}$$

where $g_i$ are the prescribed functions and the subscripts on operators are assigned arbitrarily.

Now let us consider the problem

$$\begin{aligned}
L[u] &= G \\
M_1[u] &= g_1 \\
M_2[u] &= g_2 \\
&\vdots \\
M_n[u] &= g_n
\end{aligned} \tag{1.4.3}$$

By virtue of the linearity of the equation and the supplementary conditions, we may divide problem (1.4.3) into a series of problems as follows:

$$\begin{array}{ccccc}
L[u_1] = G & L[u_2] = 0 & & L[u_n] = 0 \\
M_1[u_1] = 0 & M_1[u_2] = g_1 & & M_1[u_n] = 0 \\
M_2[u_1] = 0 & M_2[u_2] = 0 & \cdots & M_2[u_n] = 0 \\
\vdots & \vdots & & \vdots \\
M_n[u_1] = 0 & M_n[u_2] = 0 & \cdots & M_n[u_n] = g_n \\
(1.4.4) & (1.4.5) & & (1.4.6)
\end{array}$$

Then the solution of problem (1.4.3) is given by

$$u = \sum_{i=1}^{n} u_i \tag{1.4.7}$$

Let us consider one of the subproblems, say, (1.4.5). Suppose we find a sequence of functions $\phi_1, \phi_2, \ldots,$ which may be finite or infinite, satisfying the homogeneous system

$$\begin{aligned}
L[\phi_i] &= 0 \\
M_2[\phi_i] &= 0 \\
&\vdots \\
M_n[\phi_i] &= 0, \qquad i = 1, 2, 3, \ldots
\end{aligned} \tag{1.4.8}$$

and suppose we can express $g_1$ in terms of the series

$$g_1 = c_1 M_1[\phi_1] + c_2 M_1[\phi_2] + \ldots \tag{1.4.9}$$

Then the linear combination

$$u_1 = c_1 \phi_1 + c_2 \phi_2 + \ldots \tag{1.4.10}$$

is the solution of problem (1.4.5). In the case of an infinite number of terms in the linear combination (1.4.10), we require that the infinite series be uniformly convergent and sufficiently differentiable, and that all the series $N_k(u_i)$ where $N_0 = L$, $N_j = M_j$ for $j = 1, 2, \ldots, n$ converge uniformly.

## 1.5 EXERCISES

**1.** For each of the following, state whether the partial differential equation is linear, quasilinear or nonlinear. If it is linear, state whether it is homogeneous or nonhomogencous, and give its order.

a. $u_{xx} + xu_y = y$

b. $uu_x - 2xyu_y = 0$

c. $u_x^2 + uu_y = 1$

d. $u_{xxxx} + 2u_{xxyy} + u_{yyyy} = 0$

e. $u_{xx} + 2u_{xy} + u_{yy} = \sin x$

f. $u_{xxx} + u_{xyy} + \log u = 0$

g. $u_{xx}^2 + u_x^2 + \sin u = e^y$

**2.** Verify that the functions

$$u(x, y) = x^2 - y^2$$
$$u(x, y) = e^x \sin y$$

are solutions of the equation

$$u_{xx} + u_{yy} = 0$$

**3.** Show that $u = f(xy)$ where $f$ is an arbitrary differentiable function satisfies

$$xu_x - yu_y = 0$$

and verify that the functions $\sin(xy)$, $\cos(xy)$, $\log(xy)$, $e^{xy}$, and $(xy)^3$ are solutions.

**4.** Show that $u = f(x)g(y)$ where $f$ and $g$ are arbitrary twice differentiable functions satisfies

$$uu_{xy} - u_x u_y = 0$$

**5.** Determine the general solution of the differential equation

$$u_{yy} + u = 0$$

**6.** Find the general solution of

$$u_{xy} + u_x = 0$$

by setting $u_x = v$.

**7.** Find the general solution of

$$u_{xx} - 4u_{xy} + 3u_{yy} = 0$$

by assuming the solution to be in the form $u(x, y) = f(\lambda x + y)$ where $\lambda$ is an unknown parameter.

**8.** Find the general solution of

$$u_{xx} - u_{yy} = 0$$

**9.** Show that the general solution of

$$\frac{\partial^2 u}{\partial t^2} - c^2 \frac{\partial^2 u}{\partial x^2} = 0$$

is $u(x, t) = f(x - ct) + g(x + ct)$ where $f$ and $g$ are arbitrary twice differentiable functions.

**10.** Verify that the function

$$u = \phi(xy) + x\psi\left(\frac{y}{x}\right)$$

is the general solution of the equation

$$x^2 u_{xx} - y^2 u_{yy} = 0$$

**11.** If $u_x = v_y$ and $v_x = -u_y$, show that both $u$ and $v$ satisfy the Laplace equation

$$\nabla^2 \binom{u}{v} = 0$$

**12.** If $u$ is a homogeneous function of degree $n$, show that $u$ satisfies the first order equation

$$xu_x + yu_y = nu$$

CHAPTER TWO

# Mathematical Models

## 2.1 THE CLASSICAL EQUATIONS

The three basic types of second-order partial differential equations are:

a. The wave equation

$$u_{tt} - c^2(u_{xx} + u_{yy} + u_{zz}) = 0 \tag{2.1.1}$$

b. The heat equation

$$u_t - k(u_{xx} + u_{yy} + u_{zz}) = 0 \tag{2.1.2}$$

c. The Laplace equation

$$u_{xx} + u_{yy} + u_{zz} = 0 \tag{2.1.3}$$

In this section, we list a few more common linear partial differential equations of importance in applied mathematics, mathematical physics, and engineering science. Such a list naturally cannot ever be complete. Included are only equations of most common interest:

d. The Poisson equation

$$\nabla^2 u = f(x, y, z) \tag{2.1.4}$$

e. The Helmholtz equation

$$\nabla^2 u + \lambda u = 0 \tag{2.1.5}$$

f. The biharmonic equation

$$\nabla^4 u = \nabla^2(\nabla^2 u) = 0 \tag{2.1.6}$$

g. The biharmonic wave equation

$$u_{tt} + c^2 \nabla^4 u = 0 \tag{2.1.7}$$

h. The telegraph equation

$$u_{tt} + au_t + bu = c^2 u_{xx} \tag{2.1.8}$$

i. The Schrödinger equations in Quantum Physics

$$i\hbar\psi_t = \left[\left(-\frac{\hbar^2}{2m}\right)\nabla^2 + V(x, y, z)\right]\psi \tag{2.1.9}$$

$$\nabla^2\Psi + \frac{2m}{\hbar^2}[E - V(x, y, z)]\Psi = 0 \tag{2.1.10}$$

j. The Klein–Gordon equation

$$\Box u + \lambda^2 u = 0 \tag{2.1.11}$$

where

$$\nabla^2 \equiv \frac{\partial^2}{\partial x^2} + \frac{\partial^2}{\partial y^2} + \frac{\partial^2}{\partial z^2} \tag{2.1.12}$$

is the Laplace operator in rectangular Cartesian coordinates $(x, y, z)$,

$$\Box \equiv \nabla^2 - \frac{1}{c^2}\frac{\partial^2}{\partial t^2} \tag{2.1.13}$$

is the D'Alembertian, and in all equations $\lambda$, $a$, $b$, $c$, $m$, $E$ are constants and $h = 2\pi\hbar$ is the Planck constant.

Many problems in mathematical physics reduce to the solving of partial differential equations, in particular, the partial differential equations listed above. We will begin our study of these equations by first examining in detail the mathematical models representing physical problems.

## 2.2 THE VIBRATING STRING

One of the most important problems in mathematical physics is the vibration of a stretched string. Simplicity and frequent occurrence in many branches of mathematical physics make it a classic example in the theory of partial differential equations.

Let us consider a stretched string of length $l$ fixed at the end points. The problem here is to determine the equation of motion which characterizes the position $u(x, t)$ of the string at time $t$ after an initial disturbance is given.

In order to obtain a simple equation we make the following assumptions:

1. The string is flexible and elastic, that is, the string cannot resist bending moment and thus the tension in the string is always in the direction of the tangent to the existing profile of the string.

2. There is no elongation of a single segment of the string and hence, by Hooke's law, the tension is constant.
3. The weight of the string is small compared with the tension in the string.
4. The deflection is small compared with the length of the string.
5. The slope of the displaced string at any point is small compared with unity.
6. There is only pure transverse vibration.

We consider a differential element of the string. Let $T$ be the tension at the end points as shown in Fig. 2.2.1. The forces acting on the element of the string in the vertical direction are

$$T \sin \beta - T \sin \alpha$$

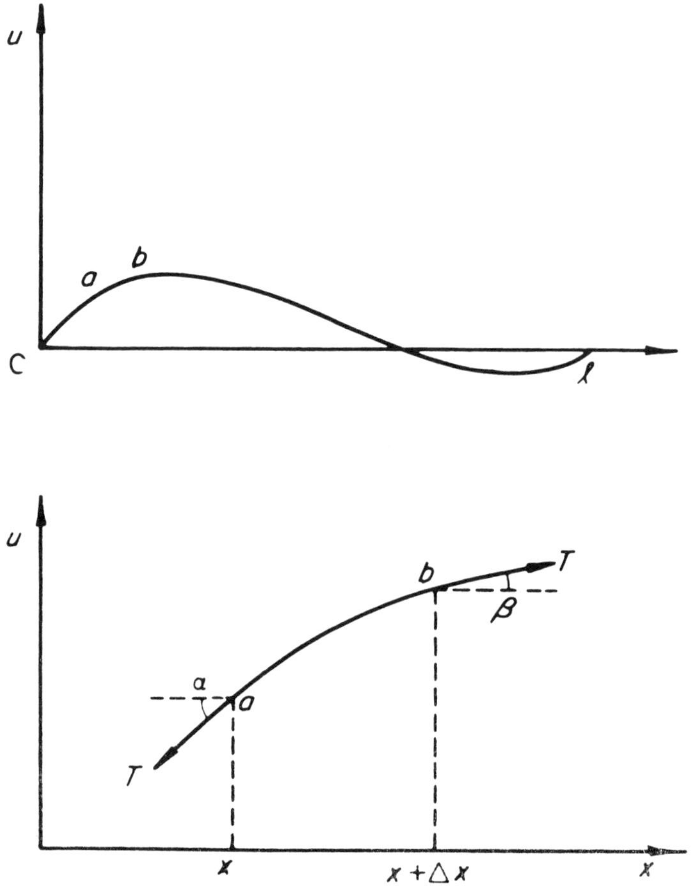


Figure 2.2.1

By Newton's second law of motion, the resultant force is equal to the mass times the acceleration. Hence,

$$T \sin \beta - T \sin \alpha = \rho \, \Delta s u_{tt} \tag{2.2.1}$$

where $\rho$ is the line density and $\Delta s$ is the small arc length of the string. Since the slope of the displaced string is small, we have

$$\Delta s \simeq \Delta x$$

Since the angles $\alpha$ and $\beta$ are small

$$\sin \alpha \simeq \tan \alpha \qquad \sin \beta \simeq \tan \beta$$

Thus Eq. (2.2.1) becomes

$$\tan \beta - \tan \alpha = \frac{\rho \, \Delta x}{T} u_{tt} \tag{2.2.2}$$

But, from calculus we know that

$$\tan \alpha = (u_x)_x$$

and

$$\tan \beta = (u_x)_{x+\Delta x}$$

at time $t$. Equation (2.2.2) may thus be written as

$$\frac{1}{\Delta x} [(u_x)_{x+\Delta x} - (u_x)_x] = \frac{\rho}{T} u_{tt}$$

In the limit as $\Delta x$ approaches zero, we find

$$u_{tt} = c^2 u_{xx} \tag{2.2.3}$$

where $c^2 = T/\rho$. This is called the *one-dimensional wave equation.*

If there is an external force $f$ per unit length acting on the string, Eq. (2.2.3) assumes the form

$$u_{tt} = c^2 u_{xx} + f^*, \qquad f^* = f/\rho \tag{2.2.4}$$

where $f$ may be pressure, gravitation, resistance, and so on.

## 2.3 THE VIBRATING MEMBRANE

The equation of the vibrating membrane occurs in a great number of problems in applied mathematics and mathematical physics. Before we derive the equation for the vibrating membrane we make certain simplifying assumptions as in the case of the vibrating string:

1. The membrane is flexible and elastic, that is, the membrane cannot resist bending moment and the tension in the membrane is always in the direction of the tangent to the existing profile of the membrane.

2. There is no elongation of a single element of the membrane and hence, by Hooke's law, the tension is constant.
3. The weight of the membrane is small compared with the tension in the membrane.
4. The deflection is small compared with the minimal diameter of the membrane.
5. The slope of the displayed membrane at any point is small compared with unity.
6. There is only pure transverse vibration.

We consider a small element of the membrane. Since the deflection and slope are small, the area of the element is approximately equal to $\Delta x \Delta y$. If $T$ is the tensile force per unit length, then the forces acting on the sides of the element are $T\,\Delta x$ and $T\,\Delta y$, as shown in Fig. 2.3.1.

The forces acting on the element of the membrane in the vertical direction are

$$T\,\Delta x \sin\beta - T\,\Delta x \sin\alpha + T\,\Delta y \sin\delta - T\,\Delta y \sin\gamma$$

Since the slopes are small, sines of the angles are approximately equal to their tangents. Thus the resultant force becomes

$$T\,\Delta x(\tan\beta - \tan\alpha) + T\,\Delta y(\tan\delta - \tan\gamma)$$

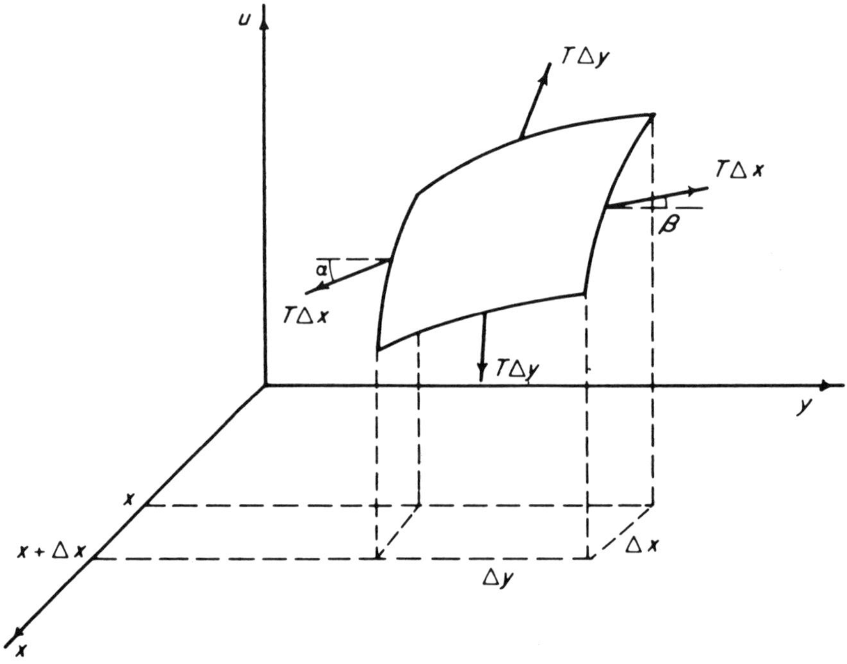


Figure 2.3.1

By Newton's second law of motion, the resultant force is equal to the mass times the acceleration. Hence

$$T\,\Delta x(\tan\beta - \tan\alpha) + T\,\Delta y(\tan\delta - \tan\gamma) = \rho\,\Delta A u_{tt} \quad (2.3.1)$$

where $\rho$ is the mass per unit area, $\Delta A \simeq \Delta x \Delta y$ is the area of this element, and $u_{tt}$ is computed at some point in the region under consideration. But from calculus, we have

$$\tan\alpha = u_y(x_1, y)$$
$$\tan\beta = u_y(x_2, y + \Delta y)$$
$$\tan\gamma = u_x(x, y_1)$$
$$\tan\delta = u_x(x + \Delta x, y_2)$$

where $x_1$ and $x_2$ are the values of $x$ between $x$ and $x + \Delta x$, and $y_1$ and $y_2$ are the values of $y$ between $y$ and $y + \Delta y$. Substituting these values in (2.3.1), we obtain

$$T\,\Delta x[u_y(x_2, y + \Delta y) - u_y(x_1, y)] + T\,\Delta y[u_x(x + \Delta x, y_2) - u_x(x, y_1)] = \rho\,\Delta x\,\Delta y u_{tt}$$

Division by $\rho\,\Delta x\,\Delta y$ yields

$$\frac{T}{\rho}\left[\frac{u_y(x_2, y + \Delta y) - u_y(x_1, y)}{\Delta y} + \frac{u_x(x + \Delta x, y_2) - u_x(x, y_1)}{\Delta x}\right] = u_{tt} \quad (2.3.2)$$

In the limit as $\Delta x$ approaches zero and $\Delta y$ approaches zero, we obtain

$$u_{tt} = c^2(u_{xx} + u_{yy}) \quad (2.3.3)$$

where $c^2 = T/\rho$. This equation is called the *two-dimensional wave equation*.

If there is an external force $f$ per unit area acting on the membrane, Eq. (2.3.3) takes the form

$$u_{tt} = c^2(u_{xx} + u_{yy}) + f^* \quad (2.3.4)$$

where $f^* = f/\rho$

## 2.4 WAVES IN AN ELASTIC MEDIUM

If a small disturbance is originated at a point in an elastic medium, neighboring particles are set into motion, and the medium is put under a state of strain. We consider such states of motion to extend in all directions. We assume that the displacements of the medium are small and that we are not concerned with translation or rotation of the medium as a whole.

Let the body under investigation be homogeneous and isotropic. Let

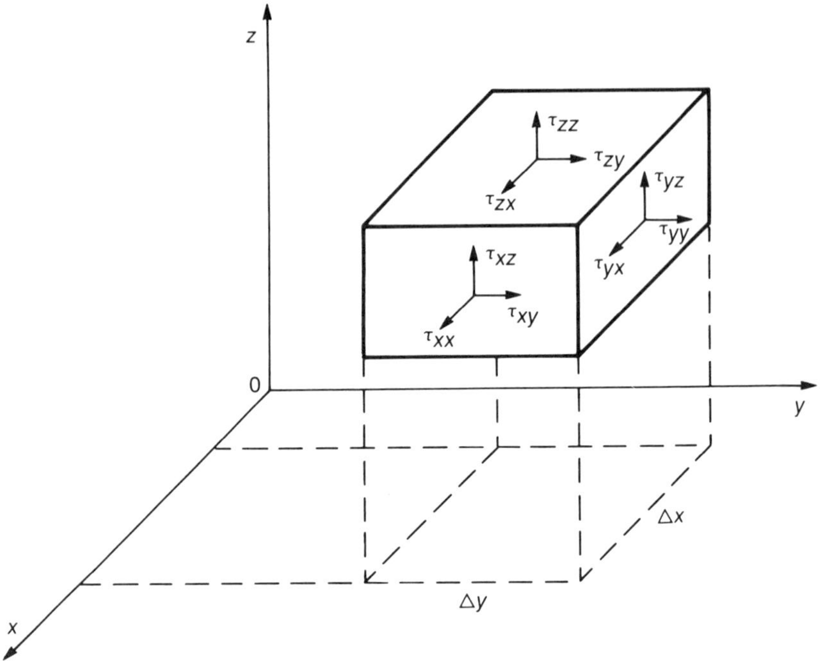


Figure 2.4.1

$\Delta V$ be a differential volume of the body, and let the stresses acting on the faces of the volume be $\tau_{xx}$, $\tau_{yy}$, $\tau_{zz}$, $\tau_{xy}$, $\tau_{xz}$, $\tau_{yx}$, $\tau_{yz}$, $\tau_{zx}$, $\tau_{zy}$. The first three stresses are called the *normal stresses* and the rest are called the *shear stresses.* (See Fig. 2.4.1)

We shall assume that the stress tensor $\tau_{ij}$ is symmetric,[3] that is,

$$\tau_{ij} = \tau_{ji} \quad i \neq j \quad i, j = x, y, z \tag{2.4.1}$$

Neglecting the body forces, the sum of all the forces acting on the volume element in the $x$-direction is

$$[(\tau_{xx})_{x+\Delta x} - (\tau_{xx})_x]\,\Delta y\,\Delta z + [(\tau_{xy})_{y+\Delta y} - (\tau_{xy})_y]\,\Delta z\,\Delta x + [(\tau_{xz})_{z+\Delta z} - (\tau_{xz})_z]\,\Delta x\,\Delta y$$

By Newton's law of motion this resultant force is equal to the mass times the acceleration. Thus we obtain

$$[(\tau_{xx})_{x+\Delta x} - (\tau_{xx})_x]\,\Delta y\,\Delta z + [(\tau_{xy})_{y+\Delta y} - (\tau_{xy})_y]\,\Delta z\,\Delta x + [(\tau_{xz})_{z+\Delta z} - (\tau_{xz})_z]\,\Delta x\,\Delta y = \rho\Delta x\,\Delta y\,\Delta z u_{tt} \tag{2.4.2}$$

[3] The condition of the rotational equilibrium of the volume element.

where $\rho$ is the density of the body and $u$ is the displacement component in the $x$-direction. Hence, in the limit as $\Delta V$ approaches zero, we obtain

$$\frac{\partial \tau_{xx}}{\partial x}+\frac{\partial \tau_{xy}}{\partial y}+\frac{\partial \tau_{xz}}{\partial z}=\rho\frac{\partial^2 u}{\partial t^2} \tag{2.4.3}$$

Similarly, the following two equations corresponding to $y$ and $z$ directions are obtained:

$$\frac{\partial \tau_{yx}}{\partial x}+\frac{\partial \tau_{yy}}{\partial y}+\frac{\partial \tau_{yz}}{\partial z}=\rho\frac{\partial^2 v}{\partial t^2} \tag{2.4.4}$$

$$\frac{\partial \tau_{zx}}{\partial x}+\frac{\partial \tau_{zy}}{\partial y}+\frac{\partial \tau_{zz}}{\partial z}=\rho\frac{\partial^2 w}{\partial t^2} \tag{2.4.5}$$

where $v$ and $w$ are the displacement components in the $y$ and $z$ directions respectively.

We may now define linear strains [40] as

$$\begin{aligned}
\varepsilon_{xx}&=\frac{\partial u}{\partial x} &\qquad \varepsilon_{yz}&=\frac{\partial w}{\partial y}+\frac{\partial v}{\partial z}\\
\varepsilon_{yy}&=\frac{\partial v}{\partial y} &\qquad \varepsilon_{zx}&=\frac{\partial u}{\partial z}+\frac{\partial w}{\partial x}\\
\varepsilon_{zz}&=\frac{\partial w}{\partial z} &\qquad \varepsilon_{xy}&=\frac{\partial v}{\partial x}+\frac{\partial u}{\partial y}
\end{aligned} \tag{2.4.6}$$

in which $\varepsilon_{xx}$, $\varepsilon_{yy}$, $\varepsilon_{zz}$ represent unit elongations and $\varepsilon_{yz}$, $\varepsilon_{zx}$, $\varepsilon_{xy}$ represent unit shearing strains.

In the case of an isotropic body, generalized Hooke's law takes the form

$$\begin{aligned}
\tau_{xx}&=\lambda\theta+2\mu\varepsilon_{xx} &\qquad \tau_{yz}&=\mu\varepsilon_{yz}\\
\tau_{yy}&=\lambda\theta+2\mu\varepsilon_{yy} &\qquad \tau_{zx}&=\mu\varepsilon_{zx}\\
\tau_{zz}&=\lambda\theta+2\mu\varepsilon_{zz} &\qquad \tau_{xy}&=\mu\varepsilon_{xy}
\end{aligned} \tag{2.4.7}$$

where $\theta=\varepsilon_{xx}+\varepsilon_{yy}+\varepsilon_{zz}$ is called the dilatation and $\lambda$ and $\mu$ are Lame's constants.

Expressing stresses in terms of displacements we have

$$\begin{aligned}
\tau_{xx}&=\lambda\theta+2\mu\frac{\partial u}{\partial x}\\
\tau_{xy}&=\mu\left(\frac{\partial v}{\partial x}+\frac{\partial u}{\partial y}\right)\\
\tau_{xz}&=\mu\left(\frac{\partial w}{\partial x}+\frac{\partial u}{\partial z}\right)
\end{aligned} \tag{2.4.8}$$

By differentiating Eqs. (2.4.8), we obtain

$$\begin{aligned}\frac{\partial \tau_{xx}}{\partial x} &= \lambda \frac{\partial \theta}{\partial x} + 2\mu \frac{\partial^2 u}{\partial x^2} \\ \frac{\partial \tau_{xy}}{\partial y} &= \mu \frac{\partial^2 v}{\partial x\, \partial y} + \mu \frac{\partial^2 u}{\partial y^2} \\ \frac{\partial \tau_{xz}}{\partial z} &= \mu \frac{\partial^2 w}{\partial x\, \partial z} + \mu \frac{\partial^2 u}{\partial z^2}\end{aligned} \tag{2.4.9}$$

Substitution of Eqs. (2.4.9) into Eq. (2.4.3) yields

$$\lambda \frac{\partial \theta}{\partial x} + \mu\left(\frac{\partial^2 u}{\partial x^2} + \frac{\partial^2 v}{\partial x\, \partial y} + \frac{\partial^2 w}{\partial x\, \partial z}\right) + \mu\left(\frac{\partial^2 u}{\partial x^2} + \frac{\partial^2 u}{\partial y^2} + \frac{\partial^2 u}{\partial z^2}\right) = \rho \frac{\partial^2 u}{\partial t^2} \tag{2.4.10}$$

We note that

$$\frac{\partial^2 u}{\partial x^2} + \frac{\partial^2 v}{\partial x\, \partial y} + \frac{\partial^2 w}{\partial x\, \partial z} = \frac{\partial}{\partial x}\left(\frac{\partial u}{\partial x} + \frac{\partial v}{\partial y} + \frac{\partial w}{\partial z}\right) = \frac{\partial \theta}{\partial x}$$

and introduce the notation

$$\Delta = \nabla^2 = \frac{\partial^2}{\partial x^2} + \frac{\partial^2}{\partial y^2} + \frac{\partial^2}{\partial z^2}$$

The symbol $\Delta$ or $\nabla^2$ is called the *Laplace operator*. Hence, Eq. (2.4.10) becomes

$$(\lambda + \mu) \frac{\partial \theta}{\partial x} + \mu \nabla^2 u = \rho \frac{\partial^2 u}{\partial t^2} \tag{2.4.11}$$

In a similar manner, we obtain the other two equations which are

$$(\lambda + \mu) \frac{\partial \theta}{\partial y} + \mu \nabla^2 v = \rho \frac{\partial^2 v}{\partial t^2} \tag{2.4.12}$$

$$(\lambda + \mu) \frac{\partial \theta}{\partial z} + \mu \nabla^2 w = \rho \frac{\partial^2 w}{\partial t^2} \tag{2.4.13}$$

In vector form, the equations of motion assume the form

$$(\lambda + \mu)\, \text{grad div}\, \mathbf{u} + \mu \nabla^2 \mathbf{u} = \rho \mathbf{u}_{tt} \tag{2.4.14}$$

where $\mathbf{u} = u\mathbf{i} + v\mathbf{j} + w\mathbf{k}$ and $\theta = \text{div}\, \mathbf{u}$.

(i) If div $\mathbf{u} = 0$, the general equation becomes

$$\mu \nabla^2 \mathbf{u} = \rho \mathbf{u}_{tt}$$

or

$$\mathbf{u}_{tt} = c^2 \nabla^2 \mathbf{u} \tag{2.4.15}$$

where the velocity $c$ of a propagated wave is

$$c = \sqrt{\mu/\rho}$$

This is the case of an equivoluminal wave propagation, since the volume expansion $\theta$ is zero for waves moving with this velocity. Sometimes these waves are called *waves of distortion* because the velocity of propagation depends on $\mu$ and $\rho$; the shear modulus $\mu$ characterizes the distortion and rotation of the volume element.

(ii) When curl $\mathbf{u} = \mathbf{0}$, the identity

$$\text{curl curl } \mathbf{u} = \text{grad div } \mathbf{u} - \nabla^2 \mathbf{u}$$

gives

$$\text{grad div } \mathbf{u} = \nabla^2 \mathbf{u}$$

Then the general equation becomes

$$(\lambda + 2\mu)\nabla^2 \mathbf{u} = \rho \mathbf{u}_{tt}$$

or

$$\mathbf{u}_{tt} = c^2 \nabla^2 \mathbf{u} \tag{2.4.16}$$

where the velocity of propagation is

$$c = \sqrt{\frac{\lambda + 2\mu}{\rho}}$$

This is the case of an *irrotational* or *dilatational* wave propagation, since curl $\mathbf{u} = \mathbf{0}$ describes irrotational motion. Equations (2.4.15) and (2.4.16) are called the *three-dimensional wave equations*.

In general the wave equation may be written as

$$u_{tt} = c^2 \nabla^2 u \tag{2.4.17}$$

where the Laplace operator may be one, two, or three dimensional. The importance of the wave equation stems from the fact that this type of equation arises in many physical problems; for example, sound waves in space, electrical vibration in a conductor, torsional oscillation of a rod, shallow water waves, linearized supersonic flow in a gas, waves in an electric transmission line, waves in magnetohydrodynamics, and longitudinal vibrations of a bar.

## 2.5 CONDUCTION OF HEAT IN SOLIDS

We consider a domain $D^*$ bounded by a closed surface $B^*$. Let $u(x, y, z, t)$ be the temperature at a point $(x, y, z)$ at time $t$. If the temperature is not constant, heat flows from places of higher temperature to places of lower temperature. Fourier's law states that the rate of flow is proportional to the gradient of the temperature. Thus the velocity of the heat flow in an isotropic body is

$$\mathbf{v} = -K \operatorname{grad} u \tag{2.5.1}$$

where $K$ is a constant, called the thermal conductivity of the body.

Let $D$ be an arbitrary domain bounded by a closed surface $B$ in $D^*$. Then the amount of heat leaving $D$ per unit time is

$$\iint_B v_n \, ds$$

where $v_n = \mathbf{v} \cdot \mathbf{n}$ is the component of $\mathbf{v}$ in the direction of the outer unit normal $\mathbf{n}$ of $B$. Thus by Gauss' theorem (Divergence theorem)

$$\begin{aligned}\iint_B v_n \, ds &= \iiint_D \operatorname{div}(-K \operatorname{grad} u)\, dx\, dy\, dz \\ &= -K \iiint_D \nabla^2 u \, dx\, dy\, dz \end{aligned} \tag{2.5.2}$$

But the amount of heat in $D$ is given by

$$\iiint_D \sigma\rho u \, dx\, dy\, dz \tag{2.5.3}$$

where $\rho$ is the density of the material of the body and $\sigma$ is its specific heat. Assuming that integration and differentiation are interchangeable, the rate of decrease of heat in $D$ is

$$-\iiint_D \sigma\rho \frac{\partial u}{\partial t}\, dx\, dy\, dz \tag{2.5.4}$$

Since the rate of decrease of heat in $D$ must be equal to the amount of heat leaving $D$ per unit time, we have

$$-\iiint_D \sigma\rho u_t \, dx\, dy\, dz = -K \iiint_D \nabla^2 u \, dx\, dy\, dz$$

or

$$\iiint_D [\sigma\rho u_t - K\nabla^2 u]\, dx\, dy\, dz = 0 \tag{2.5.5}$$

for an arbitrary $D$ in $D^*$. We assume that the integrand is continuous. If we suppose that the integrand is not zero at a point $(x_0, y_0, z_0)$ in $D$, then by continuity, the integrand is not zero in the small region surrounding the point $(x_0, y_0, z_0)$. Continuing in this fashion we extend the region encompassing $D$. Hence the integral must be nonzero. This contradicts (2.5.5). Thus, the integrand is zero everywhere, that is,

$$u_t = k\nabla^2 u \tag{2.5.6}$$

where $k = K/\sigma\rho$. This is known as the *heat equation*.

This type of equation appears in a great variety of problems in mathematical physics, for example the concentration of diffusing material, the motion of a tidal wave in a long channel, transmission in electrical cables, and unsteady boundary layers in viscous fluid flows.

## 2.6 THE GRAVITATIONAL POTENTIAL

In this section we shall derive one of the most well-known equations in the theory of partial differential equations, the Laplace equation.

We consider two particles of masses $m$ and $M$, at $P$ and $Q$ as shown in Fig. 2.6.1. Let $r$ be the distance between them. Then according to Newton's law of gravitation, a force proportional to the product of their masses, and inversely proportional to the square of the distance between them, is given in the form

$$F = G\frac{mM}{r^2} \tag{2.6.1}$$

where $G$ is the gravitational constant.

It is customary in potential theory to choose the unit of force so that $G = 1$. Thus $F$ becomes

$$F = \frac{mM}{r^2} \tag{2.6.2}$$

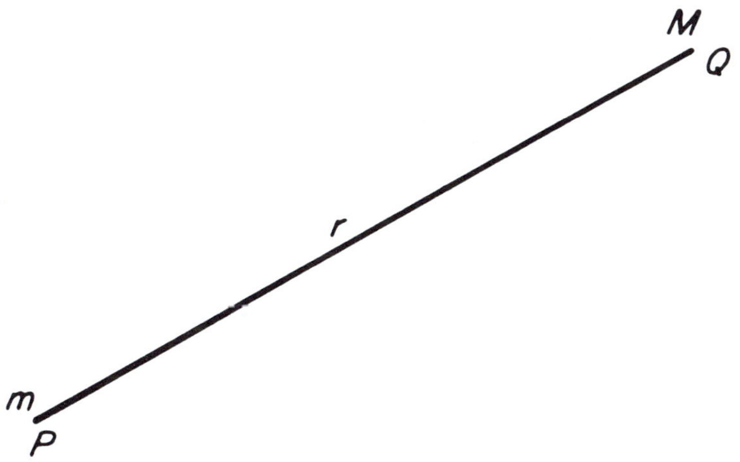


Figure 2.6.1

If $\mathbf{r}$ represents the vector $PQ$, the force per unit mass at $Q$ due to the mass at $P$ may be written as

$$\mathbf{F} = \frac{-m\mathbf{r}}{r^3} = \nabla\left(\frac{m}{r}\right) \tag{2.6.3}$$

which is called the intensity of the gravitational field of force.

We suppose that a particle of unit mass moves under the attraction of the particle of mass $m$ at $P$ from infinity up to $Q$. The work done by the force $\mathbf{F}$ is

$$\int_{\infty}^{r} \mathbf{F}\, d\mathbf{r} = \int_{\infty}^{r} \nabla\left(\frac{m}{r}\right) d\mathbf{r} = \frac{m}{r} \tag{2.6.4}$$

This is called the *potential* at $Q$ due to the particle at $P$. We denote this by

$$V = -\frac{m}{r} \tag{2.6.5}$$

Hence the intensity of force at $P$ is

$$\mathbf{F} = \nabla\left(\frac{m}{r}\right) = -\nabla V \tag{2.6.6}$$

We shall now consider a number of masses $m_1, m_2, \ldots, m_n$, whose distances from $Q$ are $r_1, r_2, \ldots, r_n$, respectively. Then the force of attraction per unit mass at $Q$ due to the system is

$$\mathbf{F} = \sum_{k=1}^{n} \nabla \frac{m_k}{r_k} = \nabla \sum_{k=1}^{n} \frac{m_k}{r_k} \tag{2.6.7}$$

The work done by the forces acting on a particle of unit mass is

$$\int_{\infty}^{r} \mathbf{F} \cdot d\mathbf{r} = \sum_{k=1}^{n} \frac{m_k}{r_k} = -V \tag{2.6.8}$$

The potential satisfies

$$\nabla^2 V = -\nabla^2 \sum_{k=1}^{n} \frac{m_k}{r_k} = -\sum_{k=1}^{n} \nabla^2 \frac{m_k}{r_k} = 0,\ r_k \neq 0 \tag{2.6.9}$$

In the case of a continuous distribution of mass in some volume $R$, we have, as in Fig. 2.6.2

$$V(x, y, z) = \iiint_R \frac{\rho(\xi, \eta, \zeta)}{r}\, dR \tag{2.6.10}$$

where $r = \sqrt{(x-\xi)^2 + (y-\eta)^2 + (z-\zeta)^2}$ and $Q$ is outside the body. It

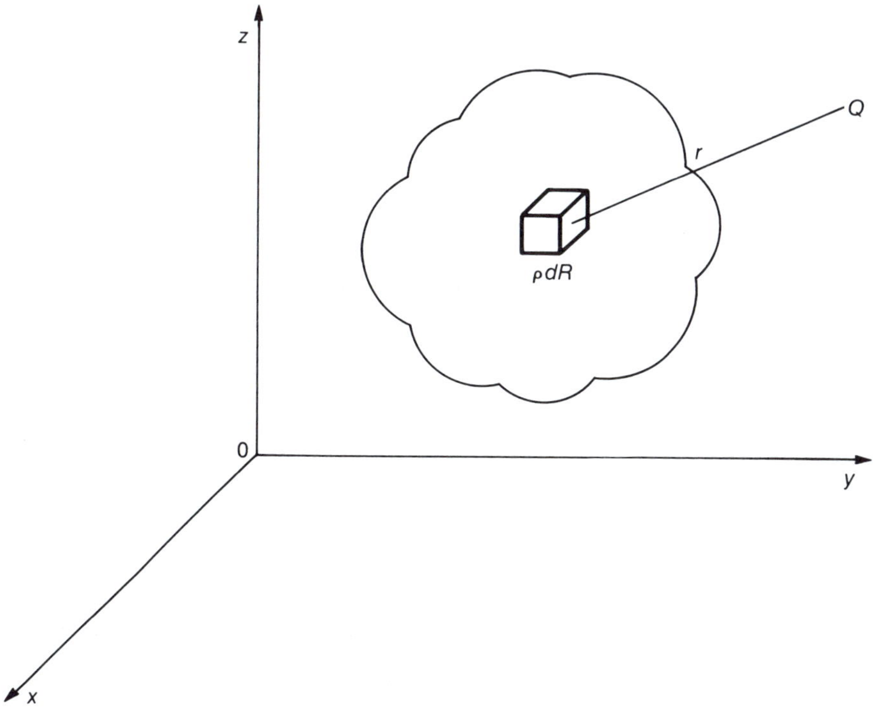


Figure 2.6.2

immediately follows that

$$\nabla^2 V = 0 \tag{2.6.11}$$

This equation is called the *Laplace equation,* also known as the *potential equation.* It appears in many physical problems, such as those of electrostatic potentials, potentials in hydrodynamics, and harmonic potentials in the theory of elasticity. We observe that the Laplace equation can be viewed as the special case of the heat and the wave equations when the dependent variables involved are independent of time.

## 2.7 EXERCISES

**1.** Show that the equation of motion of a long string is

$$u_{tt} = c^2 u_{xx} - g$$

where $g$ is the gravitational acceleration.

**2.** Derive the *damped wave equation* of a string

$$u_{tt} + au_t = c^2 u_{xx}$$

where the damping force is proportional to the velocity and $a$ is a constant. Considering a restoring force proportional to the displacement of a string, show that the resulting equation is

$$u_{tt} + au_t + bu = c^2 u_{xx}$$

where $b$ is a constant. This equation is called the *telegraph equation.*

**3.** Consider the transverse vibration of a uniform beam. Adopting Euler's beam theory, the moment $M$ at a point can be written as

$$M = -EIu_{xx}$$

where $EI$ is called the flexural rigidity, $E$ is the elastic modulus, and $I$ is the moment of inertia of the cross section of the beam. Show that the transverse motion of the beam may be described by

$$u_{tt} + c^2 u_{xxxx} = 0$$

where $c^2 = EI/\rho A$, $\rho$ is the density, and $A$ is the cross-sectional area of the beam.

**4.** Derive the deflection equation of a thin elastic plate

$$\nabla^4 u = q/D$$

where $q$ is the uniform load per unit area, $D$ is the flexural rigidity of the plate, and

$$\nabla^4 u = u_{xxxx} + 2u_{xxyy} + u_{yyyy}$$

**5.** Derive the one dimensional heat equation

$$u_t = ku_{xx}$$

Assuming that heat is also lost by radioactive exponential decay of the material in the bar, show that the above equation becomes

$$u_t = ku_{xx} + he^{-\alpha x}$$

where $h$ and $\alpha$ are constants.

**6.** Starting from *Maxwell's equations* in electrodynamics, show that in a conducting medium electric intensity $\mathbf{E}$, magnetic intensity $\mathbf{H}$, and current density $\mathbf{J}$ satisfy

$$\nabla^2 \mathbf{X} = \mu\varepsilon \mathbf{X}_{tt} + \mu\sigma \mathbf{X}_t$$

where $\mathbf{X}$ represents $\mathbf{E}$, $\mathbf{H}$, and $\mathbf{J}$, $\mu$ is the magnetic inductive capacity, $\varepsilon$ is the electric inductive capacity, and $\sigma$ is the electrical conductivity.

**7.** Derive the *continuity equation*

$$\rho_t + \operatorname{div}(\rho\mathbf{u}) = 0$$

and *Euler's equation of motion*

$$\rho[\mathbf{u}_t + (\mathbf{u}\cdot\operatorname{grad})\mathbf{u}] + \operatorname{grad} p = 0$$

in fluid dynamics.

**8.** In the derivation of the Laplace equation (2.6.11) the potential at $Q$ which is outside the body is ascertained. Now determine the potential at $Q$ when it is inside the body, and show that it satisfies the *Poisson equation*

$$\nabla^2 u = -4\pi\rho$$

where $\rho$ is the density of the body.

**9.** Setting $U = e^{ikt}u$ in the wave equation $U_{tt} = \nabla^2 U$ and setting $U = e^{-k^2 t}u$ in the heat equation $U_t = \nabla^2 U$, show that $u(x, y, z)$ satisfies the *Helmholtz equation*

$$\nabla^2 u + k^2 u = 0$$

**10.** The Maxwell equations in vacuum are

$$\nabla\times\mathbf{E} = -\frac{\partial\mathbf{B}}{\partial t}, \qquad \nabla\times\mathbf{B} = \mu\varepsilon\frac{\partial\mathbf{E}}{\partial t}$$

$$\nabla\cdot\mathbf{E} = 0, \qquad \nabla\cdot\mathbf{B} = 0$$

where $\mu$ and $\varepsilon$ are universal constants. Show that the magnetic field $\mathbf{B} = (0, B_y(x, t), 0)$ and the electric field $\mathbf{E} = (0, 0, E_z(x, t))$ satisfy the wave equation

$$\frac{\partial^2 u}{\partial t^2} = c^2\frac{\partial^2 u}{\partial x^2}$$

where $u = B_y$ or $E_z$ and $c = (\mu\varepsilon)^{-\frac{1}{2}}$ is the speed of light.

**11.** The equations of gas dynamics are linearized for small perturbations about a constant state $u = 0$, $\rho = \rho_0$, and $p = p(\rho_0)$ with $c_0^2 = p'(\rho_0)$. In terms of velocity potential $\phi$ defined by $\mathbf{u} = -\nabla\phi$, the perturbation equations are

$$\rho_t + \rho_0 \operatorname{div}\mathbf{u} = 0$$

$$p - p_0 = -\rho_0\phi_t$$

$$\rho - \rho_0 - -\frac{\rho_0}{c_0^2}\phi_t$$

Show that $\phi$ satisfies the three dimensional wave equation

$$\phi_{tt} = c_0^2\nabla^2\phi$$

where

$$\nabla^2 \equiv \frac{\partial^2}{\partial x^2} + \frac{\partial^2}{\partial y^2} + \frac{\partial^2}{\partial z^2}$$

**12.** Consider a slender body moving in a gas with arbitrary constant velocity $U$, and suppose $(x_1, x_2, x_3)$ represents the frame of reference in which the motion of the gas is small and described by the equations of problem 11. The body moves in the negative $x_1$ direction, and $(x, y, z)$ denotes the coordinates fixed with respect to the body so that the coordinate transformation is $(x, y, z) = (x_1 + Ut, x_2, x_3)$. Show that the wave equation $\phi_{tt} = c_0^2\nabla^2\phi$ reduces to the form

$$(M^2 - 1)\Phi_{xx} = \Phi_{yy} + \Phi_{zz}$$

where $M \equiv U/c_0$ is the Mach number and $\Phi$ is the potential in the new frame of reference $(x, y, z)$.

**13.** Consider the motion of a gas in a taper tube of cross section $A(x)$. Show that the equation of continuity and the equation of motion are

$$\rho = \rho_0\left(1 - \frac{\partial \xi}{\partial x} - \frac{\xi}{A}\frac{\partial A}{\partial x}\right) = \rho_0\left[1 - \frac{1}{A}\frac{\partial}{\partial x}(A\xi)\right]$$

and

$$\rho_0 \frac{\partial^2 \xi}{\partial t^2} = -\frac{\partial p}{\partial x}$$

where $x$ is the distance along the length of the tube, $\xi(x)$ is the displacement function, $p = p(\rho)$ is the pressure-density relation, $\rho_0$ is the average density, and $\rho$ is the local density of the gas.

Hence derive the equation of motion

$$\xi_{tt} = c^2 \frac{\partial}{\partial x}\left[\frac{1}{A}\frac{\partial}{\partial x}(A\xi)\right], \qquad c^2 = \frac{\partial p}{\partial \rho}$$

Find the equation of motion when $A$ is constant. If $A(x) = a_0 \exp(2\alpha x)$ where $a_0$ and $\alpha$ are constants, show that the above equation takes the form

$$\xi_{tt} = c^2(\xi_{xx} + 2\alpha\xi_x)$$

**14.** Consider the current $I(x, t)$ and the potential $V(x, t)$ at a point $x$ and time $t$ of a uniform electric transmission line with resistance $R$, inductance $L$, capacity $C$, and leakage conductance $G$ per unit length. Show that both $I$ and $V$ satisfy the system of equations

$$LI_t + RI = -V_x$$
$$CV_t + GV = -I_x$$

Derive the equations for $I$ and $V$ in the following cases:

(i) Lossless transmission line $(R = G = 0)$

(ii) Ideal submarine cable $(L = G = 0)$

(iii) Heaviside's distortionless line $(R/L = G/C = \text{constant} = k)$

**15.** The Fermi–Pasta–Ulam model is used to describe waves in an anharmonic lattice of length $l$ consisting of a row of $n$ identical masses $m$, each connected to the next by nonlinear springs of constant $\kappa$. The masses are at a distance $h = l/n$ apart, and the springs when extended or compressed by an amount $d$ exert a force $F = \kappa(d + \alpha d^2)$ where $\alpha$ measures the strength of nonlinearity. The equation of motion of the $i$th mass is

$$m\ddot{y}_i = \kappa[(y_{i+1} - y_i) - (y_i - y_{i-1}) + \alpha\{(y_{i+1} - y_i)^2 - (y_i - y_{i-1})^2\}]$$

where $i = 1, 2, 3 \ldots n$, $y_i$ is the displacement of the $i$th mass from its equilibrium position, and $\kappa$, $\alpha$ are constants with $y_0 = y_n = 0$.

Assume a continuum approximation of this discrete system so that the Taylor expansions

$$y_{i+1} - y_i = hy_x + \frac{h^2}{2!}y_{xx} + \frac{h^3}{3!}y_{xxx} + \frac{h^4}{4!}y_{xxxx} + 0(h^5)$$

$$y_i - y_{i-1} = hy_x - \frac{h^2}{2!}y_{xx} + \frac{h^3}{3!}y_{xxx} - \frac{h^4}{4!}y_{xxxx} + 0(h^5)$$

can be used to derive the nonlinear differential equation

$$y_{tt} = c^2[1 + 2\alpha hy_x]y_{xx} + 0(h^4)$$

$$y_{tt} = c^2[1 + 2\alpha hy_x]y_{xx} + \frac{c^2h^2}{12}y_{xxxx} + 0(h^5)$$

where

$$c^2 = \frac{\kappa h^2}{m}$$

Using a change of variable $\xi = x - ct$, $\tau = c\alpha ht$, shows that $u = y_\xi$ satisfies the $K\,dV$ equation

$$u_\tau + uu_\xi + \beta u_{\xi\xi\xi} = 0(\varepsilon^2), \qquad \varepsilon = \alpha h, \qquad \beta = \frac{h}{24\alpha}.$$

CHAPTER THREE

# Classification of Second-order Equations

## 3.1 SECOND-ORDER EQUATIONS IN TWO INDEPENDENT VARIABLES

The general linear second-order partial differential equation in one dependent variable $u$ may be written as

$$\sum_{i,j}^{n} A_{ij} u_{x_i x_j} + \sum_{i}^{n} B_i u_{x_i} + Fu = G \tag{3.1.1}$$

in which we assume $A_{ij} = A_{ji}$ and $A_{ij}$, $B_i$, $F$, and $G$ are real-valued functions defined in some region of the space $(x_1, x_2, \ldots, x_n)$.

Here we shall be concerned with second order equations in the dependent variable $u$ and the independent variables $x$, $y$. Hence Eq. (3.1.1) can be put in the form

$$Au_{xx} + Bu_{xy} + Cu_{yy} + Du_x + Eu_y + Fu = G \tag{3.1.2}$$

where the coefficients are functions of $x$ and $y$ and do not vanish simultaneously. We shall assume that the function $u$ and the coefficients are twice continuously differentiable in some domain $R$.

The classification of second-order equations[4] is based upon the possibility of reducing Eq. (3.1.2) by a coordinate transformation to *canonical* or *standard* form at a point. An equation is said to be

[4] The classification of partial differential equations is suggested by the classification of the quadratic equation of conic sections in analytic geometry. The equation

$$Ax^2 + Bxy + Cy^2 + Dx + Ey + F = 0$$

is hyperbolic, parabolic, or elliptic accordingly as $B^2 - 4AC$ is positive, zero, or negative.

*hyperbolic, parabolic,* or *elliptic* at a point $(x_0, y_0)$ accordingly as

$$B^2(x_0, y_0) - 4A(x_0, y_0)C(x_0, y_0) \tag{3.1.3}$$

is positive, zero, or negative. If this is true at all points, then the equation is said to be hyperbolic, parabolic, or elliptic in a domain. In the case of two independent variables, a transformation can always be found to reduce the given equation to the canonical form in a given domain. However, in the case of several independent variables, it is not, in general, possible to find such a transformation.

To transform Eq. (3.1.2) to a canonical form we make a change of independent variables. Let the new variables be

$$\begin{aligned} \xi &= \xi(x, y) \\ \eta &= \eta(x, y) \end{aligned} \tag{3.1.4}$$

Assuming that $\xi$ and $\eta$ are twice continuously differentiable and that the Jacobian

$$J = \begin{vmatrix} \xi_x & \xi_y \\ \eta_x & \eta_y \end{vmatrix} \tag{3.1.5}$$

is nonzero in the region under consideration, then $x$ and $y$ can be determined uniquely from the system (3.1.4). Let $x$ and $y$ be twice continuously differentiable functions of $\xi$ and $\eta$. Then we have

$$\begin{aligned} u_x &= u_\xi \xi_x + u_\eta \eta_x \\ u_y &= u_\xi \xi_y + u_\eta \eta_y \\ u_{xx} &= u_{\xi\xi} \xi_x^2 + 2u_{\xi\eta} \xi_x \eta_x + u_{\eta\eta} \eta_x^2 + u_\xi \xi_{xx} + u_\eta \eta_{xx} \\ u_{xy} &= u_{\xi\xi} \xi_x \xi_y + u_{\xi\eta}(\xi_x \eta_y + \xi_y \eta_x) + u_{\eta\eta} \eta_x \eta_y + u_\xi \xi_{xy} + u_\eta \eta_{xy} \\ u_{yy} &= u_{\xi\xi} \xi_y^2 + 2u_{\xi\eta} \xi_y \eta_y + u_{\eta\eta} \eta_y^2 + u_\xi \xi_{yy} + u_\eta \eta_{yy} \end{aligned} \tag{3.1.6}$$

Substituting these values in Eq. (3.1.2) we obtain

$$A^* u_{\xi\xi} + B^* u_{\xi\eta} + C^* u_{\eta\eta} + D^* u_\xi + E^* u_\eta + F^* u = G^* \tag{3.1.7}$$

where

$$\begin{aligned} A^* &= A\xi_x^2 + B\xi_x \xi_y + C\xi_y^2 \\ B^* &= 2A\xi_x \eta_x + B(\xi_x \eta_y + \xi_y \eta_x) + 2C\xi_y \eta_y \\ C^* &= A\eta_x^2 + B\eta_x \eta_y + C\eta_y^2 \\ D^* &= A\xi_{xx} + B\xi_{xy} + C\xi_{yy} + D\xi_x + E\xi_y \\ E^* &= A\eta_{xx} + B\eta_{xy} + C\eta_{yy} + D\eta_x + E\eta_y \\ F^* &= F \\ G^* &= G \end{aligned} \tag{3.1.8}$$

The resulting equation (3.1.7) is in the same from as the original equation (3.1.2) under the general transformation (3.1.4). The nature of the equation remains invariant under such a transformation if the Jacobian does not vanish. This can be seen from the fact that the sign of the discriminant does not alter under the transformation, that is,

$$B^{*2} - 4A^*C^* = J^2(B^2 - 4AC) \tag{3.1.9}$$

which can be easily verified. It should be noted here that the equation can be of a different type at different points of the domain, but for our purpose we shall assume that the equation under consideration is of the single type in a given domain.

The classification of Eq. (3.1.2) depends on the coefficients $A(x, y)$, $B(x, y)$, and $C(x, y)$ at a given point $(x, y)$. We shall, therefore, rewrite Eq. (3.1.2) as

$$Au_{xx} + Bu_{xy} + Cu_{yy} = H \tag{3.1.10}$$

where

$$H = H(x, y, u, u_x, u_y)$$

and Eq. (3.1.7) as

$$A^*u_{\xi\xi} + B^*u_{\xi\eta} + C^*u_{\eta\eta} = H^* \tag{3.1.11}$$

where

$$H^* = H^*(\xi, \eta, u, u_\xi, u_\eta)$$

## 3.2 CANONICAL FORMS

In this section we shall consider the problem of reducing Eq. (3.1.10) to canonical form.

We suppose first that none of $A$, $B$, $C$, is zero. Let $\xi$ and $\eta$ be the new variables such that the coefficients $A^*$ and $C^*$ in Eq. (3.1.11) vanish. Thus, from (3.1.8), we have

$$A^* = A\xi_x^2 + B\xi_x\xi_y + C\xi_y^2 = 0$$
$$C^* = A\eta_x^2 + B\eta_x\eta_y + C\eta_y^2 = 0$$

These two equations are of the same type and hence we may write them in the form

$$A\zeta_x^2 + B\zeta_x\zeta_y + C\zeta_y^2 = 0 \tag{3.2.1}$$

in which $\zeta$ stands for either of the functions $\xi$ or $\eta$. Dividing through by $\zeta_y^2$, Eq. (3.2.1) becomes

$$A\left(\frac{\zeta_x}{\zeta_y}\right)^2 + B\left(\frac{\zeta_x}{\zeta_y}\right) + C = 0 \tag{3.2.2}$$

Along the curve $\zeta = \text{constant}$, we have

$$d\zeta = \zeta_x\, dx + \zeta_y\, dy = 0$$

Thus,

$$\frac{dy}{dx} = -\frac{\zeta_x}{\zeta_y} \tag{3.2.3}$$

and therefore, Eq. (3.2.2) may be written in the form

$$A\left(\frac{dy}{dx}\right)^2 - B\left(\frac{dy}{dx}\right) + C = 0 \tag{3.2.4}$$

the roots of which are

$$\frac{dy}{dx} = (B + \sqrt{B^2 - 4AC})/2A \tag{3.2.5}$$

$$\frac{dy}{dx} = (B - \sqrt{B^2 - 4AC})/2A \tag{3.2.6}$$

These equations, which are known as the *characteristic equations*, are the ordinary differential equations for families of curves in the $xy$-plane along which $\xi = \text{constant}$ and $\eta = \text{constant}$. The integrals of Eqs. (3.2.5) and (3.2.6) are called the *characteristic curves*. Since the equations are first order ordinary differential equations, the solutions may be written as

$$\phi_1(x, y) = c_1, \qquad c_1 = \text{constant}$$
$$\phi_2(x, y) = c_2, \qquad c_2 = \text{constant}$$

Hence the transformation

$$\xi = \phi_1(x, y)$$
$$\eta = \phi_2(x, y)$$

will transform Eq. (3.1.10) to a canonical form.

## (A) Hyperbolic Type

If $B^2 - 4AC > 0$, then integration of Eqs. (3.2.5) and (3.2.6) yield two real and distinct families of characteristics. Eq. (3.1.11) reduces to

$$u_{\xi\eta} = H_1 \tag{3.2.7}$$

where $H_1 = H^*/B^*$. It can be easily shown that $B^* \neq 0$. This form is called the *first canonical form of the hyperbolic equation.*

Now if the new independent variables

$$\alpha = \xi + \eta$$
$$\beta = \xi - \eta \tag{3.2.8}$$

are introduced then Eq. (3.2.7) is transformed into

$$u_{\alpha\alpha} - u_{\beta\beta} = H_2(\alpha, \beta, u, u_\alpha, u_\beta) \tag{3.2.9}$$

This form is called the *second canonical form of the hyperbolic equation.*

## (B) Parabolic Type

In this case, we have $B^2 - 4AC = 0$, and Eqs. (3.2.5) and (3.2.6) coincide. Thus, there exists one real family of characteristics, and we obtain only a single integral $\xi = \text{constant}$ (or $\eta = \text{constant}$).

Since $B^2 = 4AC$ and $A^* = 0$, we find that

$$A^* = A\xi_x^2 + B\xi_x\xi_y + C\xi_y^2 = (\sqrt{A}\xi_x + \sqrt{C}\xi_y)^2 = 0$$

From this it follows that

$$\begin{aligned} B^* &= 2A\xi_x\eta_x + B(\xi_x\eta_y + \xi_y\eta_x) + 2C\xi_y\eta_y \\ &= 2(\sqrt{A}\xi_x + \sqrt{C}\xi_y)(\sqrt{A}\eta_x + \sqrt{C}\eta_y) \\ &= 0 \end{aligned}$$

for arbitrary values of $\eta(x, y)$ which is functionally independent of $\xi(x, y)$; for instance, if $\eta = y$, the Jacobian does not vanish in the domain of parabolicity.

Division of Eq. (3.1.11) by $C^*$ yields

$$u_{\eta\eta} = H_3(\xi, \eta, u, u_\xi, u_\eta) \qquad C^* \neq 0 \tag{3.2.10}$$

This is called the *canonical form of the parabolic equation.*

Eq. (3.1.11) may also assume the form

$$u_{\xi\xi} = H_3^*(\xi, \eta, u, u_\xi, u_\eta) \tag{3.2.11}$$

if we choose $\eta = \text{constant}$ as the integral of Eq. (3.2.5)

## (C) Elliptic Type

For an equation of the elliptic type, we have $B^2 - 4AC < 0$. Consequently the quadratic equation (3.2.4) has no real solutions, but it has two complex conjugate solutions which are continuous complex-valued functions of the real variables $x$ and $y$. Thus, in this case there are no real characteristic curves. However, if the coefficients $A$, $B$ and $C$ are analytic functions[5] of $x$ and $y$, then one can consider Eq. (3.2.4) for complex $x$ and $y$.

---

[5] A function of two real variables $x$, $y$ is said to be analytic in a certain domain if in some neighborhood of every point $(x_0, y_0)$ of this domain, the function can be represented as a Taylor series in the variables $(x - x_0)$ and $(y - y_0)$.

Since $\xi$ and $\eta$ are complex, we introduce the new real variables

$$\alpha = \frac{1}{2}(\xi + \eta)$$

$$\beta = \frac{1}{2i}(\xi - \eta) \tag{3.2.12}$$

so that

$$\xi = \alpha + i\beta$$

$$\eta = \alpha - i\beta \tag{3.2.13}$$

First, we transform Eq. (3.1.10). We then have

$$A^{**}(\alpha, \beta)u_{\alpha\alpha} + B^{**}(\alpha, \beta)u_{\alpha\beta} + C^{**}(\alpha, \beta)u_{\beta\beta} = H_4(\alpha, \beta, u, u_\alpha, u_\beta) \tag{3.2.14}$$

in which the coefficients assume the same form as the coefficients in Eq. (3.1.11). With the use of transformation (3.2.13), the equations $A^* = C^* = 0$ become

$$(A\alpha_x^2 + B\alpha_x\alpha_y + C\alpha_y^2) - (A\beta_x^2 + B\beta_x\beta_y + C\beta_y^2)$$
$$+ i[2A\alpha_x\beta_x + B(\alpha_x\beta_y + \alpha_y\beta_x) + 2C\alpha_y\beta_y] = 0$$
$$(A\alpha_x^2 + B\alpha_x\alpha_y + C\alpha_y^2) - (A\beta_x^2 + B\beta_x\beta_y + C\beta_y^2)$$
$$- i[2A\alpha_x\beta_x + B(\alpha_x\beta_y + \alpha_y\beta_x) + 2C\alpha_y\beta_y] = 0$$

or

$$(A^{**} - C^{**}) + iB^{**} = 0$$
$$(A^{**} - C^{**}) - iB^{**} = 0$$

These equations are satisfied if and only if

$$A^{**} = C^{**} \quad \text{and} \quad B^{**} = 0$$

Hence, Eq. (3.2.14) transforms into

$$A^{**}u_{\alpha\alpha} + A^{**}u_{\beta\beta} = H_4(\alpha, \beta, u, u_\alpha, u_\beta)$$

Dividing through by $A^{**}$, we obtain

$$u_{\alpha\alpha} + u_{\beta\beta} = H_5(\alpha, \beta, u, u_\alpha, u_\beta) \tag{3.2.15}$$

where $H_5 = H_4/A^{**}$. This is called the *canonical form of the elliptic equation.*

EXAMPLE 3.2.1. Consider the equation

$$y^2u_{xx} - x^2u_{yy} = 0$$

Here

$$A = y^2 \qquad B = 0 \qquad C = -x^2$$

Thus,

$$B^2 - 4AC = 4x^2y^2 > 0$$

The equation is hyperbolic everywhere except on the coordinate axes $x = 0$ and $y = 0$. From the characteristic equations (3.2.5) and (3.2.6), we have

$$\frac{dy}{dx} = \frac{x}{y}$$

$$\frac{dy}{dx} = -\frac{x}{y}$$

After integration of these equations, we obtain

$$\frac{1}{2}y^2 - \frac{1}{2}x^2 = c_1$$

$$\frac{1}{2}y^2 + \frac{1}{2}x^2 = c_2$$

The first of these curves is a family of hyperbolas

$$\frac{1}{2}y^2 - \frac{1}{2}x^2 = c_1$$

and the second is a family of circles

$$\frac{1}{2}y^2 + \frac{1}{2}x^2 = c_2$$

To transform the given equation to the canonical form, we consider

$$\xi = \frac{1}{2}y^2 - \frac{1}{2}x^2$$

$$\eta = \frac{1}{2}y^2 + \frac{1}{2}x^2$$

From the relations (3.1.6), we have

$$\begin{aligned}
u_x &= u_\xi \xi_x + u_\eta \eta_x \\
&= -xu_\xi + xu_\eta \\
u_y &= u_\xi \xi_y + u_\eta \eta_y \\
&= yu_\xi + yu_\eta \\
u_{xx} &= u_{\xi\xi}\xi_x^2 + 2u_{\xi\eta}\xi_x\eta_x + u_{\eta\eta}\eta_x^2 + u_\xi \xi_{xx} + u_\eta \eta_{xx} \\
&= x^2u_{\xi\xi} - 2x^2u_{\xi\eta} + x^2u_{\eta\eta} - u_\xi + u_\eta \\
u_{yy} &= u_{\xi\xi}\xi_y^2 + 2u_{\xi\eta}\xi_y\eta_y + u_{\eta\eta}\eta_y^2 + u_\xi \xi_{yy} + u_\eta \eta_{yy} \\
&= y^2u_{\xi\xi} + 2y^2u_{\xi\eta} + y^2u_{\eta\eta} + u_\xi + u_\eta
\end{aligned}$$

Thus, the given equation assumes the canonical form

$$u_{\xi\eta} = \frac{\eta}{2(\xi^2 - \eta^2)} u_\xi - \frac{\xi}{2(\xi^2 - \eta^2)} u_\eta$$

EXAMPLE 3.2.2. Consider the partial differential equation

$$x^2 u_{xx} + 2xy u_{xy} + y^2 u_{yy} = 0$$

In this case, the discriminant is

$$B^2 - 4AC = 4x^2y^2 - 4x^2y^2 = 0$$

The equation is therefore parabolic everywhere. The characteristic equation is

$$\frac{dy}{dx} = \frac{y}{x}$$

and hence the characteristics are

$$\frac{y}{x} = c$$

which is the equation of a family of straight lines.

Consider the transformation

$$\xi = \frac{y}{x}$$

$$\eta = y$$

where $\eta$ is chosen arbitrarily. The given equation is then reduced to the canonical form

$$y^2 u_{\eta\eta} = 0$$

Thus,

$$u_{\eta\eta} = 0 \quad \text{for} \quad y \neq 0$$

EXAMPLE 3.2.3. The equation

$$u_{xx} + x^2 u_{yy} = 0$$

is elliptic everywhere except on the coordinate axis $x = 0$ because

$$B^2 - 4AC = -4x^2 < 0, \; x \neq 0$$

The characteristic equations are

$$\frac{dy}{dx} = ix$$

$$\frac{dy}{dx} = -ix$$

Integration yields

$$2y - ix^2 = c_1$$
$$2y + ix^2 = c_2$$

Thus, if we write

$$\xi = 2y - ix^2$$
$$\eta = 2y + ix^2$$

and hence

$$\alpha = \frac{1}{2}(\xi + \eta) = 2y$$
$$\beta = \frac{1}{2i}(\xi - \eta) = -x^2$$

we obtain the canonical form

$$u_{\alpha\alpha} + u_{\beta\beta} = -\frac{1}{2\beta}u_\beta$$

It should be remarked here that a given partial differential equation may be of a different type in a different domain. Thus, for example, *Tricomi's equation*

$$u_{xx} + xu_{yy} = 0 \tag{3.2.16}$$

is elliptic for $x > 0$ and hyperbolic for $x < 0$, since $B^2 - 4AC = -4x$. For a detailed treatment, see Hellwig [19].

## 3.3 EQUATIONS WITH CONSTANT COEFFICIENTS

In the case of an equation with real constant coefficients, the equation is of the single type at all points in the domain. This is because the discriminant $B^2 - 4AC$ is a constant.

From the characteristic equation

$$\frac{dy}{dx} = (B + \sqrt{B^2 - 4AC})/2A$$

and (3.3.1)

$$\frac{dy}{dx} = (B - \sqrt{B^2 - 4AC})/2A,$$

we can see that the characteristics

$$y = \left(\frac{B + \sqrt{B^2 - 4AC}}{2A}\right)x + c_1$$
$$y = \left(\frac{B - \sqrt{B^2 - 4AC}}{2A}\right)x + c_2 \tag{3.3.2}$$

are two families of straight lines. Consequently, the characteristic coordinates take the form

$$\xi = y - \lambda_1 x$$
$$\eta = y - \lambda_2 x \tag{3.3.3}$$

where

$$\lambda_{1,2} = \frac{B \pm \sqrt{B^2 - 4AC}}{2A} \tag{3.3.4}$$

The linear second-order partial differential equation with constant coefficients may be written as

$$Au_{xx} + Bu_{xy} + Cu_{yy} + Du_x + Eu_y + Fu = G(x, y) \tag{3.3.5}$$

## (A) Hyperbolic Equation

When $B^2 - 4AC > 0$, the equation is of hyperbolic type, in which case the characteristics form two distinct families.

Using (3.3.3), Eq. (3.3.5) becomes

$$u_{\xi\eta} = D_1 u_\xi + E_1 u_\eta + F_1 u + G_1(\xi, \eta) \tag{3.3.6}$$

where $D_1$, $E_1$, and $F_1$ are constants. Here, since the coefficients are constants, the lower order terms are expressed explicitly.

When $A = 0$, Eq. (3.3.1) does not hold. In this case, the characteristic equation may be put in the form

$$-B(dx/dy) + C(dx/dy)^2 = 0$$

which may again be rewritten as

$$dx/dy = 0$$
$$-B + C(dx/dy) = 0$$

Integration gives

$$x = c_1$$
$$x = (B/C)y + c_2$$

where $c_1$ and $c_2$ are integration constants. Thus, the characteristic coordinates are

$$\xi = x$$
$$\eta = x - (B/C)y \tag{3.3.7}$$

Under this transformation, Eq. (3.3.5) reduces to the canonical form

$$u_{\xi\eta} = D_1^* u_\xi + E_1^* u_\eta + F_1^* u + G_1^*(\xi, \eta) \tag{3.3.8}$$

where $D_1^*$, $E_1^*$ and $F_1^*$ are constants.

## (B) Parabolic Equation

When $B^2 - 4AC = 0$, the equation is of parabolic type, in which case only one real family of characteristics exists. From Eq. (3.3.4), we find that

$$\lambda_1 = \lambda_2 = (B/2A)$$

so that the single family of characteristics is given by

$$y = (B/2A)x + c_1$$

where $c_1$ is an integration constant. Thus, we have

$$\begin{aligned} \xi &= y - (B/2A)x \\ \eta &= hy + kx \end{aligned} \tag{3.3.9}$$

where $\eta$ is chosen arbitrarily such that the Jacobian of the transformation is not zero, and $h$ and $k$ are constants.

With the proper choice of the constants $h$ and $k$ in the transformation (3.3.9), Eq. (3.3.5) reduces to

$$u_{\eta\eta} = D_2 u_\xi + E_2 u_\eta + F_2 u + G_2(\xi, \eta) \tag{3.3.10}$$

where $D_2$, $E_2$, and $F_2$ are constants.

If $B = 0$, we can see at once from the relation

$$B^2 - 4AC = 0$$

that $C$ or $A$ vanishes. The given equation is then already in the canonical form. Similarly, in the other cases when $A$ or $C$ vanishes, $B$ vanishes. The given equation is then also in canonical form.

## (C) Elliptic Equation

When $B^2 - 4AC < 0$, the equation is of elliptic type. In this case, the characteristics are complex conjugates.

The characteristic equations yield

$$\begin{aligned} y &= \lambda_1 x + c_1 \\ y &= \lambda_2 x + c_2 \end{aligned} \tag{3.3.11}$$

where $\lambda_1$ and $\lambda_2$ are complex numbers. Accordingly, $c_1$ and $c_2$ are allowed to take on complex values. Thus,

$$\begin{aligned} \xi &= y - (a + ib)x \\ \eta &= y - (a - ib)x \end{aligned} \tag{3.3.12}$$

where $\lambda_{1,2} = a \pm ib$ in which $a$ and $b$ are real constants, and $a = B/2A$ and $b = \sqrt{4AC - B^2}/2A$.

Introduce the new variables

$$\alpha = \frac{1}{2}(\xi + \eta) = y - ax$$
$$\beta = \frac{1}{2i}(\xi - \eta) = -bx \tag{3.3.13}$$

Application of this transformation readily reduces Eq. (3.3.5) to the canonical form

$$u_{\alpha\alpha} + u_{\beta\beta} = D_3 u_\alpha + E_3 u_\beta + F_3 u + G_3(\alpha, \beta) \tag{3.3.14}$$

where $D_3$, $E_3$ and $F_3$ are constants.

We note that $B^2 - AC < 0$, so neither $A$ nor $C$ is zero.

EXAMPLE 3.3.1. Consider the equation

$$4u_{xx} + 5u_{xy} + u_{yy} + u_x + u_y = 2$$

Since $A = 4$, $B = 5$, $C = 1$, and $B^2 - 4AC = 9 > 0$, the equation is hyperbolic. Thus the characteristic equations take the form

$$\frac{dy}{dx} = 1$$
$$\frac{dy}{dx} = \frac{1}{4}$$

and hence the characteristics are

$$y = x + c_1$$
$$y = (x/4) + c_2$$

The transformation

$$\xi = y - x$$
$$\eta = y - (x/4)$$

therefore reduces the given equation to the canonical form

$$u_{\xi\eta} = \frac{1}{3}u_\eta - \frac{8}{9}$$

This is the first canonical form.

The second canonical form may be obtained by the transformation

$$\alpha = \xi + \eta$$
$$\beta = \xi - \eta$$

as

$$u_{\alpha\alpha} - u_{\beta\beta} = \frac{1}{3}u_\alpha - \frac{1}{3}u_\beta - \frac{8}{9}$$

EXAMPLE 3.3.2. The equation

$$u_{xx} - 4u_{xy} + 4u_{yy} = e^y$$

is parabolic since $A = 1$, $B = -4$, $C = 4$ and $B^2 - 4AC = 0$. Thus, we have from Eq. (3.3.9)

$$\xi = y + 2x$$
$$\eta = y$$

in which $\eta$ is chosen arbitrarily. By means of this mapping, the equation transforms into

$$u_{\eta\eta} = \frac{1}{4} e^{\eta}$$

EXAMPLE 3.3.3. Consider the equation

$$u_{xx} + u_{xy} + u_{yy} + u_x = 0$$

Since $A = 1$, $B = 1$, $C = 1$, and $B^2 - 4AC = -3 < 0$, the equation is elliptic.

We have

$$\lambda_{1,2} = \frac{B \pm \sqrt{B^2 - 4AC}}{2A} = \frac{1}{2} \pm i\frac{\sqrt{3}}{2}$$

and hence,

$$\xi = y - \left(\frac{1}{2} + i\frac{\sqrt{3}}{2}\right)x$$
$$\eta = y - \left(\frac{1}{2} - i\frac{\sqrt{3}}{2}\right)x$$

Introducing the new variables

$$\alpha = \frac{1}{2}(\xi + \eta) = y - \frac{1}{2}x$$
$$\beta = \frac{1}{2i}(\xi - \eta) = -\frac{\sqrt{3}}{2}x$$

the given equation is then transformed into canonical form

$$u_{\alpha\alpha} + u_{\beta\beta} = \frac{2}{3}u_{\alpha} + \frac{2}{\sqrt{3}}u_{\beta}$$

EXAMPLE 3.3.4. Consider the wave equation

$$u_{tt} - c^2 u_{xx} = 0, \quad c \text{ is constant}$$

Since $A = -c^2$, $B = 0$, $C = 1$, and $B^2 - 4AC = 4c^2 > 0$, the wave equation is hyperbolic everywhere. According to (3.2.4), the equation of characteristics is

$$-c^2\left(\frac{dt}{dx}\right)^2 + 1 = 0$$

or

$$dx^2 - c^2\,dt^2 = 0$$

Therefore

$$x + ct = \xi = \text{constant}$$
$$x - ct = \eta = \text{constant}$$

Thus the characteristics are straight lines, which are shown in Fig. 3.3.1. The characteristics form a natural set of coordinates for a hyperbolic equation.

In terms of new coordinates $\xi$ and $\eta$ defined above, we obtain

$$u_{xx} = u_{\xi\xi} + 2u_{\xi\eta} + u_{\eta\eta}$$
$$u_{tt} = c^2(u_{\xi\xi} - 2u_{\xi\eta} + u_{\eta\eta})$$

so that the wave equation becomes

$$-4c^2 u_{\xi\eta} = 0$$

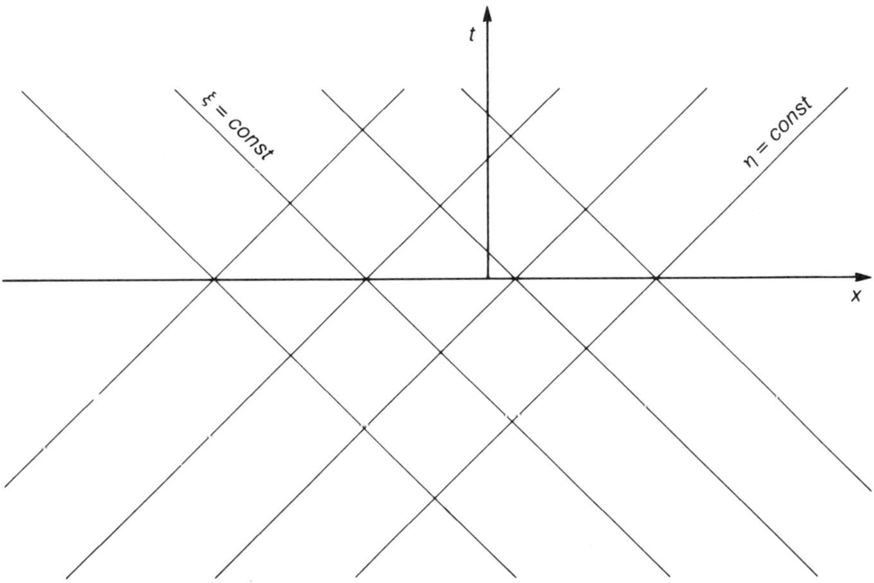


Figure 3.3.1. Characteristics for the wave equation.

Since $c \neq 0$, we have

$$u_{\xi\eta} = 0$$

Integrating with respect to $\xi$, we obtain

$$u_\eta = \psi_1(\eta)$$

where $\psi_1$ is the arbitrary function of $\eta$. Integrating again with respect to $\eta$, we obtain

$$u(\xi, \eta) = \int \psi_1(\eta)\, d\eta + \phi(\xi)$$

If we set $\psi(\eta) = \int \psi_1(\eta)\, d\eta$, the solution is

$$u(\xi, \eta) = \phi(\xi) + \psi(\eta)$$

which is, in terms of the original variables $x$ and $t$,

$$u(x, t) = \phi(x + ct) + \psi(x - ct)$$

provided $\phi$ and $\psi$ are arbitrary but twice differentiable functions.

Note that $\phi$ is constant on 'wavefronts' $x = -ct + \xi$ that travel toward decreasing $x$, as $t$ increases, whereas $\psi$ is constant on wavefronts $x = ct + \eta$ that travel toward increasing $x$ as $t$ increases. Thus, any general solution can be expressed as the sum of two waves, one traveling to the right with constant velocity $c$ and the other traveling to the left with the same velocity $c$.

## 3.4 GENERAL SOLUTION

In general it is not so simple to determine the general solution of a given equation. Sometimes further simplification of the canonical form of an equation may yield the general solution. If the canonical form of the equation is simple, then the solution can be immediately ascertained.

EXAMPLE 3.4.1. Find the general solution of

$$x^2 u_{xx} + 2xy u_{xy} + y^2 u_{yy} = 0$$

In Example 3.2.2, using the transformation $\xi = y/x$, $\eta = y$, this equation was reduced to the canonical form

$$u_{\eta\eta} = 0 \quad \text{for} \quad y \neq 0$$

Integrating twice with respect to $\eta$, we obtain

$$u(\xi, \eta) = \eta f(\xi) + g(\xi)$$

where $f(\xi)$ and $g(\xi)$ are arbitrary functions. In terms of the independent variables $x$ and $y$, we have

$$u(x, y) = yf\left(\frac{y}{x}\right) + g\left(\frac{y}{x}\right)$$

EXAMPLE 3.4.2. Determine the general solution of

$$4u_{xx} + 5u_{xy} + u_{yy} + u_x + u_y = 2$$

Using the transformation $\xi = y - x$, $\eta = y - (x/4)$, the canonical form of this equation is (see Example 3.3.1)

$$u_{\xi\eta} = \frac{1}{3}u_\eta - \frac{8}{9}$$

By means of the substitution $v = u_\eta$, the preceding equation reduces to

$$v_\xi = \frac{1}{3}v - \frac{8}{9}$$

This can be easily integrated by separating the variables. Integrating with respect to $\xi$, we have

$$v = \frac{8}{3} + \frac{1}{3}e^{(\xi/3)}F(\eta)$$

Integrating again with respect to $\eta$, we obtain

$$u(\xi, \eta) = \frac{8}{3}\eta + \frac{1}{3}g(\eta)e^{\xi/3} + f(\xi)$$

where $f(\xi)$ and $g(\eta)$ are arbitrary functions. The general solution of the given equation is therefore

$$u(x, y) = \frac{8}{3}\left(y - \frac{x}{4}\right) + \frac{1}{3}g\left(y - \frac{x}{4}\right)e^{\frac{1}{3}(y-x)} + f(y - x)$$

EXAMPLE 3.4.3. Obtain the general solution of

$$3u_{xx} + 10u_{xy} + 3u_{yy} = 0$$

Since $B^2 - 4AC = 64 > 0$, the equation is hyperbolic. Thus, from Eqs. (3.3.2), the characteristics are

$$y = 3x + c_1$$

$$y = \frac{1}{3}x + c_2$$

Using the transformation

$$\xi = y - 3x$$
$$\eta = y - \frac{1}{3}x$$

the given equation is reduced to

$$\frac{64}{3}u_{\xi\eta} = 0$$

Hence, we obtain

$$u_{\xi\eta} = 0$$

Integration yields

$$u(\xi, \eta) = f(\xi) + g(\eta)$$

In terms of the original variables, the general solution is

$$u(x, y) = f(y - 3x) + g\left(y - \frac{x}{3}\right)$$

## 3.5 SUMMARY AND FURTHER SIMPLIFICATION

We summarize the classification of the linear second-order partial differential equations with constant coefficients in two independent variables.

$$\text{hyperbolic:} \quad u_{rs} = a_1 u_r + a_2 u_s + a_3 u + f_1 \tag{3.5.1}$$
$$u_{rr} - u_{ss} = a_1^* u_r + a_2^* u_s + a_3^* u + f_1^* \tag{3.5.2}$$
$$\text{parabolic:} \quad u_{ss} = b_1 u_r + b_2 u_s + b_3 u + f_2 \tag{3.5.3}$$
$$\text{elliptic:} \quad u_{rr} + u_{ss} = c_1 u_r + c_2 u_s + c_3 u + f_3 \tag{3.5.4}$$

where $r$ and $s$ represent the new independent variables in the linear transformation

$$r = r(x, y)$$
$$s = s(x, y) \tag{3.5.5}$$

and the Jacobian $J \neq 0$.

To simplify Eq. (3.5.1) further, we introduce the new dependent variable

$$v = ue^{-(ar+bs)} \tag{3.5.6}$$

where $a$ and $b$ are undetermined coefficients. Finding the derivatives, we

obtain

$$
\begin{aligned}
u_r &= (v_r + av)e^{ar+bs} \\
u_s &= (v_s + bv)e^{ar+bs} \\
u_{rr} &= (v_{rr} + 2av_r + a^2v)e^{ar+bs} \\
u_{rs} &= (v_{rs} + av_s + bv_r + abv)e^{ar+bs} \\
u_{ss} &= (v_{ss} + 2bv_s + b^2v)e^{ar+bs}
\end{aligned}
$$

Substitution of these in Eq. (3.5.1) yields

$$v_{rs} + (b - a_1)v_r + (a - a_2)v_s + (ab - a_1a - a_2b - a_3)v = f_1e^{-(ar+bs)}$$

In order that the first derivatives vanish, we set

$$a = a_2 \quad \text{and} \quad b = a_1$$

Thus, the above equation becomes

$$v_{rs} = (a_1a_2 + a_3)v + g_1$$

where $g_1 = f_1e^{-(a_2r+a_1s)}$. In a similar manner, we can transform Eqs. (3.5.2)–(3.5.4). Thus, we have the following transformed equations corresponding to Eqs. (3.5.1)–(3.5.4):

$$
\begin{aligned}
&\text{hyperbolic:} && v_{rs} = h_1v + g_1 \\
& && v_{rr} - v_{ss} = h_1^*v + g_1^* \qquad (3.5.7) \\
&\text{parabolic:} && v_{ss} = h_2v + g_2 \\
&\text{elliptic:} && v_{rr} + v_{ss} = h_3v + g_3
\end{aligned}
$$

In the case of partial differential equations in several independent variables or in higher order, the classification is considerably more complex. For further reading, see Courant and Hilbert [7, 62].

## 3.6 EXERCISES

**1.** Determine the region in which the given equation is hyperbolic, parabolic, or elliptic, and transform the equation in the respective region to the canonical form.

a. $xu_{xx} + u_{yy} = x^2$

b. $u_{xx} + y^2u_{yy} = y$

c. $u_{xx} + xyu_{yy} = 0$

d. $x^2u_{xx} - 2xyu_{xy} + y^2u_{yy} = e^x$

e. $u_{xx} + u_{xy} - xu_{yy} = 0$

f. $e^xu_{xx} + e^yu_{yy} = u$

g. $\sin^2 x\, u_{xx} + \sin 2x\, u_{xy} + \cos^2 x\, u_{yy} = x$

h. $u_{xx} - y u_{xy} + x u_x + y u_y + u = 0$

**2.** Obtain the general solution of

i. $x^2 u_{xx} + 2xy u_{xy} + y^2 u_{yy} + xy u_x + y^2 u_y = 0$

ii. $r u_{tt} - c^2 r u_{rr} - 2c^2 u_r = 0$, $c =$ constant

Check the solutions by substitution. [Hint: Let $v = ru$]

**3.** Find the characteristics and characteristic coordinates, and reduce the following equations to canonical form.

a. $u_{xx} + 2u_{xy} + 3u_{yy} + 4u_x + 5u_y + u = e^x$

b. $2u_{xx} - 4u_{xy} + 2u_{yy} + 3u = 0$

c. $u_{xx} + 5u_{xy} + 4u_{yy} + 7u_y = \sin x$

d. $u_{xx} + u_{yy} + 2u_x + 8u_y + u = 0$

e. $u_{xy} + 2u_{yy} + 9u_x + u_y = 2$

f. $6u_{xx} - u_{xy} + u = y^2$

g. $u_{xy} + u_x + u_y = 3x$

h. $u_{yy} - 9u_x + 7u_y = \cos y$

**4.** Determine the general solutions of

i. $u_{xx} - \dfrac{1}{c^2} u_{yy} = 0$, $\quad c =$ constant

ii. $u_{xx} + u_{yy} = 0$ [Hint: Let $c = i = \sqrt{-1}$ in $(i)$]

iii. $u_{xxxx} + 2u_{xxyy} + u_{yyyy} = 0$ [Hint: Let $z = x + iy$]

iv. $u_{xx} - 3u_{xy} + 2u_{yy} = 0$

v. $u_{xx} + u_{xy} = 0$

vi. $u_{xx} + 10u_{xy} + 9u_{yy} = y$

**5.** Transform the following equations to the form

$$v_{\xi\eta} = cv, \qquad c = \text{constant}$$

by introducing the new variables

$$v = u e^{-(a\xi + b\eta)}$$

where $a$ and $b$ are undetermined coefficients.

i. $u_{xx} - u_{yy} + 3u_x - 2u_y + u = 0$

ii. $3u_{xx} + 7u_{xy} + 2u_{yy} + u_y + u = 0$

**6.** Given the parabolic equation

$$u_{xx} = au_t + bu_x + cu + f$$

where the coefficients are constants, by the substitution

$$u = ve^{\frac{1}{2}bx}$$

for the case $c = -(b^2/4)$ show that the given equation is reduced to the heat equation

$$v_{xx} = av_t + g$$

where

$$g = fe^{-bx/2}$$

**7.** Reduce the Tricomi equation

$$u_{xx} + xu_{yy} = 0$$

into the canonical form

i. $u_{\xi\eta} - [6(\xi - \eta)]^{-1}(u_\xi - u_\eta) = 0, \qquad \text{for } x < 0$

ii. $u_{\alpha\alpha} + u_{\beta\beta} + \dfrac{1}{3\beta} u_\beta = 0, \qquad x > 0 \qquad \left(\alpha = \dfrac{3y}{2},\ \beta = -x^{3/2}\right)$

Show that the characteristic curves for $x < 0$ are cubic parabolas.

CHAPTER FOUR

# The Cauchy Problem

## 4.1 THE CAUCHY PROBLEM

In the theory of ordinary differential equations, by the initial value problem we mean the problem of finding the solution of a given differential equation with the appropriate number of initial conditions prescribed at an initial point. For example, the second-order ordinary differential equation

$$\frac{d^2u}{dt^2} = f\left(t, u, \frac{du}{dt}\right)$$

and the initial conditions

$$u(t_0) = \alpha$$

$$\frac{du}{dt}(t_0) = \beta$$

constitute the initial value problem.

An analogous problem can be defined in the case of partial differential equations. Here we shall state the problem involving second-order partial differential equations in two independent variables.

We consider a second-order partial differential equation for the function $u$ in the independent variables $x$ and $y$, and suppose that this equation can be solved explicitly for $u_{yy}$, and hence can be represented in the form

$$u_{yy} = F(x, y, u, u_x, u_y, u_{xx}, u_{xy}) \tag{4.1.1}$$

For some value $y = y_0$, we prescribe the initial values of the unknown function and of the derivative with respect to $y$

$$\begin{aligned} u(x, y_0) &= f(x) \\ u_y(x, y_0) &= g(x) \end{aligned} \tag{4.1.2}$$

The problem of determining the solution of Eq. (4.1.1) satisfying the initial conditions (4.1.2) is known as the *initial-value problem*. For instance, the initial-value problem of a vibrating string is the problem of finding the solution of the equation

$$u_{tt} = c^2 u_{xx}$$

satisfying the initial conditions

$$\begin{aligned} u(x, t_0) &= u_0(x) \\ u_t(x, t_0) &= v_0(x) \end{aligned}$$

where $u_0(x)$ is the initial displacement and $v_0(x)$ is the initial velocity.

In initial-value problems, the initial values usually refer to the data assigned at $y = y_0$. It is not essential that these values be given along the line $y = y_0$; they may very well be prescribed along some curve $L_0$ in the $xy$ plane. In such a context, the problem is called the *Cauchy problem* instead of the initial value problem, although the two names are actually synonymous.

Let us consider the equation

$$Au_{xx} + Bu_{xy} + Cu_{yy} = F(x, y, u, u_x, u_y) \tag{4.1.3}$$

where $A$, $B$, $C$ are functions of $x$ and $y$. Let $(x_0, y_0)$ denote points on a smooth curve $L_0$ in the $xy$ plane. Also let the parametric equations of this curve $L_0$ be

$$x_0 = x_0(\lambda), \qquad y_0 = y_0(\lambda) \tag{4.1.4}$$

where $\lambda$ is a parameter.

We suppose that two functions $f(\lambda)$ and $g(\lambda)$ are prescribed along the curve $L_0$. The Cauchy problem is now one of determining the solution $u(x, y)$ of Eq. (4.1.3) in the neighborhood of the curve $L_0$ satisfying the Cauchy conditions

$$u = f(\lambda) \tag{4.1.5a}$$

$$\frac{\partial u}{\partial n} = g(\lambda) \tag{4.1.5b}$$

on the curve $L_0$. $n$ is the direction of the normal to $L_0$ which lies to the left of $L_0$ in the counterclockwise direction of increasing arc length. The functions $f(\lambda)$ and $g(\lambda)$ are called the *Cauchy data*.

For every point on $L_0$, the value of $u$ is specified by Eq. (4.1.5a). Thus, the curve $L_0$ represented by Eq. (4.1.4) with the condition (4.1.5a) yields a twisted curve $L$ in $(x, y, u)$ space whose projection on the $xy$ plane is the curve $L_0$. Thus, the solution of the Cauchy problem is a surface, called an integral surface, in the $(x, y, u)$ space passing through $L$ and satisfying the condition (4.1.5b), which represents a tangent plane to the integral surface along $L$.

If the function $f(\lambda)$ is differentiable, then along the curve $L_0$, we have

$$\frac{du}{d\lambda} = \frac{\partial u}{\partial x}\frac{dx}{d\lambda} + \frac{\partial u}{\partial y}\frac{dy}{d\lambda} = \frac{df}{d\lambda} \tag{4.1.6}$$

and

$$\frac{\partial u}{\partial n} = \frac{\partial u}{\partial x}\frac{dx}{dn} + \frac{\partial u}{\partial y}\frac{dy}{dn} = g \tag{4.1.7}$$

but

$$\frac{dx}{dn} = -\frac{dy}{ds} \quad \text{and} \quad \frac{dy}{dn} = \frac{dx}{ds} \tag{4.1.8}$$

Eq. (4.1.7) may be written as

$$\frac{\partial u}{\partial n} = -\frac{\partial u}{\partial x}\frac{dy}{ds} + \frac{\partial u}{\partial y}\frac{dx}{ds} = g \tag{4.1.9}$$

Since

$$\begin{vmatrix} \dfrac{dx}{d\lambda} & \dfrac{dy}{d\lambda} \\ \dfrac{-dy}{ds} & \dfrac{dx}{ds} \end{vmatrix} = \frac{(dx)^2 + (dy)^2}{ds\, d\lambda} \neq 0 \tag{4.1.10}$$

it is possible to find $u_x$ and $u_y$ on $L_0$ from the system of Eqs. (4.1.6) and (4.1.9). Now that $u_x$ and $u_y$ are known on $L_0$, we find the higher derivatives by first differentiating $u_x$ and $u_y$ with respect to $\lambda$. Thus, we have

$$\frac{\partial^2 u}{\partial x^2}\frac{dx}{d\lambda} + \frac{\partial^2 u}{\partial x\,\partial y}\frac{dy}{d\lambda} = \frac{d}{d\lambda}\left(\frac{\partial u}{\partial x}\right) \tag{4.1.11}$$

$$\frac{\partial^2 u}{\partial x\,\partial y}\frac{dx}{d\lambda} + \frac{\partial^2 u}{\partial y^2}\frac{dy}{d\lambda} = \frac{d}{d\lambda}\left(\frac{\partial u}{\partial y}\right) \tag{4.1.12}$$

From Eq. (4.1.3), we have

$$A\frac{\partial^2 u}{\partial x^2} + B\frac{\partial^2 u}{\partial x\,\partial y} + C\frac{\partial^2 u}{\partial y^2} = F \tag{4.1.13}$$

where $F$ is known since $u_x$ and $u_y$ have been found. The system of equations can be solved for $u_{xx}$, $u_{xy}$, and $u_{yy}$, if

$$\begin{vmatrix} \frac{dx}{d\lambda} & \frac{dy}{d\lambda} & 0 \\ 0 & \frac{dx}{d\lambda} & \frac{dy}{d\lambda} \\ A & B & C \end{vmatrix} = C\left(\frac{dx}{d\lambda}\right)^2 - B\frac{dx}{d\lambda}\frac{dy}{d\lambda} + A\left(\frac{dy}{d\lambda}\right)^2 \neq 0 \qquad (4.1.14)$$

The equation

$$A\left(\frac{dy}{dx}\right)^2 - B\frac{dy}{dx} + C = 0 \qquad (4.1.15)$$

is the characteristic equation. It is then evident that the necessary condition for obtaining the second derivatives is that the curve $L_0$ must not be a characteristic curve.

If the coefficients of Eq. (4.1.3) and the function (4.1.5) are analytic, then all the derivatives of higher orders can be computed by the above process. The solution can then be represented in the form of a Taylor series:

$$u(x, y) = \sum_{n=0}^{\infty} \sum_{k=0}^{n} \frac{1}{k!\,(n-k)!} \frac{\partial^n u_0}{\partial x_0^k\, \partial y_0^{n-k}} (x - x_0)^k (y - y_0)^{n-k} \qquad (4.1.16)$$

which can be shown to converge in the neighborhood of the curve $L_0$. Thus we may state the famous Cauchy–Kowalewsky theorem.

## 4.2 CAUCHY–KOWALEWSKY THEOREM

Let the partial differential equation be given in the form

$$u_{yy} = F(y, x_1, x_2, \ldots, x_n, u, u_y, u_{x_1}, u_{x_2}, \ldots, u_{x_n}, u_{x_1y}, u_{x_2y}, \ldots, u_{x_ny}, u_{x_1x_1}, u_{x_2x_2}, \ldots, u_{x_nx_n}) \qquad (4.2.1)$$

and let the initial conditions

$$u = f(x_1, x_2, \ldots, x_n) \qquad (4.2.2)$$

$$u_y = g(x_1, x_2, \ldots, x_n) \qquad (4.2.3)$$

be given on the noncharacteristic manifold $y = y_0$.

If the function $F$ is analytic in some neighborhood of the point $(y^0, x_1^0, x_2^0, \ldots, x_n^0, u^0, u_y^0, \ldots)$ and if the functions $f$ and $g$ are analytic in some neighborhood of the point $(x_1^0, x_2^0, \ldots, x_n^0)$, then the Cauchy problem has a unique analytic solution in some neighborhood of the point $(y^0, x_1^0, x_2^0, \ldots, x_n^0)$.

For the proof, see Petrovsky [30].

The preceding statement seems equally applicable to hyperbolic, parabolic, or elliptic equations. However, we shall see that difficulties arise in formulating the Cauchy problem for nonhyperbolic equations. Consider, for instance, the Hadamard example.

The problem consists of the elliptic equation

$$u_{xx} + u_{yy} = 0$$

and the initial conditions

$$u(x, 0) = 0$$
$$u_y(x, 0) = n^{-1} \sin nx$$

The solution of this problem is

$$u(x, y) = n^{-2} \sinh ny \sin nx,$$

which can be easily verified.

It can be seen that, when $n$ tends to infinity, the function $n^{-1} \sin nx$ tends uniformly to zero. But, the solution $n^{-2} \sinh ny \sin nx$ does not become small, as $n$ increases for any nonzero $y$. It is obvious then that the solution does not depend continuously on the data. Thus, it is not a properly posed problem.

In addition to existence and uniqueness, the question of continuous dependence of solution on the initial data arises in connection with the Cauchy–Kowalewsky theorem. It is well known that any continuous function can accurately be approximated by polynomials. We can apply the Cauchy–Kowalewsky theorem with continuous data by using polynomial approximations only if a small variation in the analytic data leads to a small change in the solution.

## 4.3 HOMOGENEOUS WAVE EQUATION

To study the Cauchy problems for hyperbolic partial differential equations, it is quite natural to begin investigating the simplest and yet most important equation, the one dimensional wave equation, by the method of characteristics. The essential characteristic of the solution of the general wave equation is preserved in this simplified case.

We shall consider the following Cauchy problem of an infinite string with the initial condition

$$u_{tt} - c^2 u_{xx} = 0 \tag{4.3.1}$$

$$u(x, 0) = f(x) \tag{4.3.2}$$

$$u_t(x, 0) = g(x) \tag{4.3.3}$$

By the method of characteristics described in Chapter 3, the characteristic equation according to Eq. (3.2.4) is

$$dx^2 - c^2\, dt^2 = 0$$

which reduces to

$$dx + c\,dt = 0$$
$$dx - c\,dt = 0$$

The integrals are the straight lines

$$x + ct = c_1$$
$$x - ct = c_2$$

Introducing the characteristic coordinates

$$\xi = x + ct$$
$$\eta = x - ct$$

we obtain

$$u_{xx} = u_{\xi\xi} + 2u_{\xi\eta} + u_{\eta\eta}$$
$$u_{tt} = c^2(u_{\xi\xi} - 2u_{\xi\eta} + u_{\eta\eta})$$

Substitution of these in Eq. (4.3.1) yields

$$-4c^2 u_{\xi\eta} = 0$$

Since $c \neq 0$, we have

$$u_{\xi\eta} = 0$$

Integrating with respect to $\xi$, we obtain

$$u_\eta = \psi^*(\eta)$$

where $\psi^*(\eta)$ is the arbitrary function of $\eta$. Integrating again with respect to $\eta$, we obtain

$$u(\xi, \eta) = \int \psi^*(\eta)\,d\eta + \phi(\xi)$$

If we set $\psi(\eta) = \int \psi^*(\eta)\,d\eta$, we have

$$u(\xi, \eta) = \phi(\xi) + \psi(\eta)$$

where $\phi$ and $\psi$ are arbitrary functions. Transforming to the original variables $x$ and $t$, we find the general solution of the wave equation

$$u(x, t) = \phi(x + ct) + \psi(x - ct) \qquad (4.3.4)$$

provided $\phi$ and $\psi$ are twice differentiable functions.

Now applying the initial conditions (4.3.2) and (4.3.3), we obtain

$$u(x, 0) = f(x) = \phi(x) + \psi(x) \qquad (4.3.5)$$
$$u_t(x, 0) = g(x) = c\phi'(x) - c\psi'(x) \qquad (4.3.6)$$

Integration of Eq. (4.3.6) gives

$$\phi(x) - \psi(x) = \frac{1}{c}\int_{x_0}^{x} g(\tau)\, d\tau + K \tag{4.3.7}$$

where $x_0$ and $K$ are arbitrary constants. Solving for $\phi$ and $\psi$ from Eqs. (4.3.5) and (4.3.7), we obtain

$$\phi(x) = \frac{1}{2} f(x) + \frac{1}{2c}\int_{x_0}^{x} g(\tau)\, d\tau + \frac{K}{2}$$

$$\psi(x) = \frac{1}{2} f(x) - \frac{1}{2c}\int_{x_0}^{x} g(\tau)\, d\tau - \frac{K}{2}$$

The solution is thus given by

$$u(x, t) = \frac{1}{2}[f(x+ct) + f(x-ct)] + \frac{1}{2c}\left[\int_{x_0}^{x+ct} g(\tau)\, d\tau - \int_{x_0}^{x-ct} g(\tau)\, d\tau\right]$$

$$= \frac{1}{2}[f(x+ct) + f(x-ct)] + \frac{1}{2c}\int_{x-ct}^{x+ct} g(\tau)\, d\tau \tag{4.3.8}$$

This is called the *d'Alembert solution* of the Cauchy problem for the one dimensional wave equation.

It is easy to verify by direct substitution that $u(x, t)$, represented by (4.3.8), is the unique solution of the wave equation (4.3.1) provided $f(x)$ is twice continuously differentiable and $g(x)$ is continuously differentiable. This essentially proves the existence of the d'Alembert solution. By direct substitution, it can also be shown that the solution (4.3.8) is uniquely determined by the initial conditions (4.3.2) and (4.3.3). It is important to note that the solution $u(x, t)$ depends only on the initial values at points between $x - ct$ and $x + ct$ and not at all on initial values outside this interval on the line $t = 0$. This interval is called the *domain of dependence* of the variables $(x, t)$. Moreover, the solution depends continuously on the initial data, that is, the problem is well posed. Mathematically, this can be stated as follows:

For every $\varepsilon > 0$ and for each time interval $0 \leqslant t \leqslant t_0$, there exists a number $\delta(\varepsilon, t_0)$ such that

$$|u(x, t) - u^*(x, t)| < \varepsilon$$

whenever

$$|f(x) - f^*(x)| < \delta, \qquad |g(x) - g^*(x)| < \delta$$

The proof follows immediately from Eq. (4.3.8). We have

$$|u(x, t) - u^*(x, t)| \leqslant \frac{1}{2}|f(x+ct) - f^*(x+ct)|$$
$$+\frac{1}{2}|f(x-ct) - f^*(x-ct)|$$
$$+\frac{1}{2c}\int_{x-ct}^{x+ct} |g(\tau) - g^*(\tau)|\, d\tau < \varepsilon$$

where $\varepsilon = \delta(1 + t_0)$.

For any finite time interval $0 < t < t_0$, a small change in the initial data only produces a small change in the solution. This shows that the problem is well posed.

EXAMPLE 4.3.1. Find the solution of the initial value problem

$$u_{tt} = c^2 u_{xx}$$
$$u(x, 0) = \sin x$$
$$u_t(x, 0) = \cos x$$

From (4.3.8), we have

$$u(x, t) = \frac{1}{2}[\sin(x+ct) + \sin(x-ct)] + \frac{1}{2c}\int_{x-ct}^{x+ct} \cos\tau\, d\tau$$
$$= \sin x \cos ct + \frac{1}{2c}[\sin(x+ct) - \sin(x-ct)]$$
$$= \sin x \cos ct + \frac{1}{c}\cos x \sin ct$$

It follows from the d'Alembert solution that, if an initial displacement or an initial velocity is located in a small neighborhood of some point $(x_0, t_0)$, it can influence only the area $t > t_0$ bounded by two characteristics $x - ct = \text{constant}$ and $x + ct = \text{constant}$ with slope $\pm 1/c$ passing through the point $(x_0, t_0)$, as shown in Fig. 4.3.1. This means that the initial displacement propagates at the speed $c$, whereas the effect of the initial velocity propagates at all speeds up to $c$. This infinite sector is called the *domain of influence* of the point $(x_0, t_0)$.

According to Eq. (4.3.8), the value of $u(x_0, t_0)$ depends on the initial data $f$ and $g$ in the interval $[x_0 - ct_0, x_0 + ct_0]$ which is cut out of the initial line by the two characteristics $x - ct = \text{constant}$ and $x + ct = \text{constant}$ with slope $\pm 1/c$ passing through the point $(x_0, t_0)$. The interval $[x_0 - ct_0, x_0 + ct_0]$ on the line $t = 0$ is called the *domain of dependence*, as shown in Fig. 4.3.2.

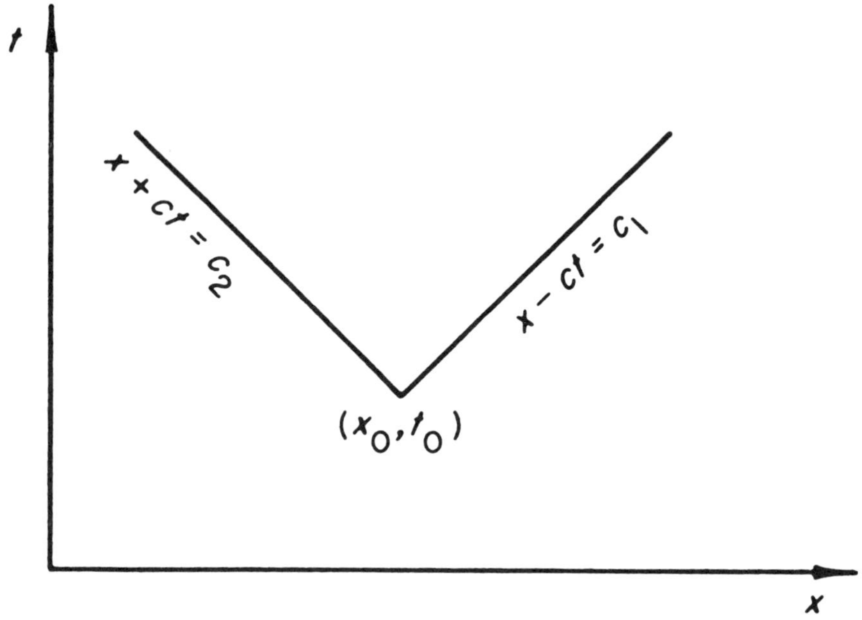


Figure 4.3.1

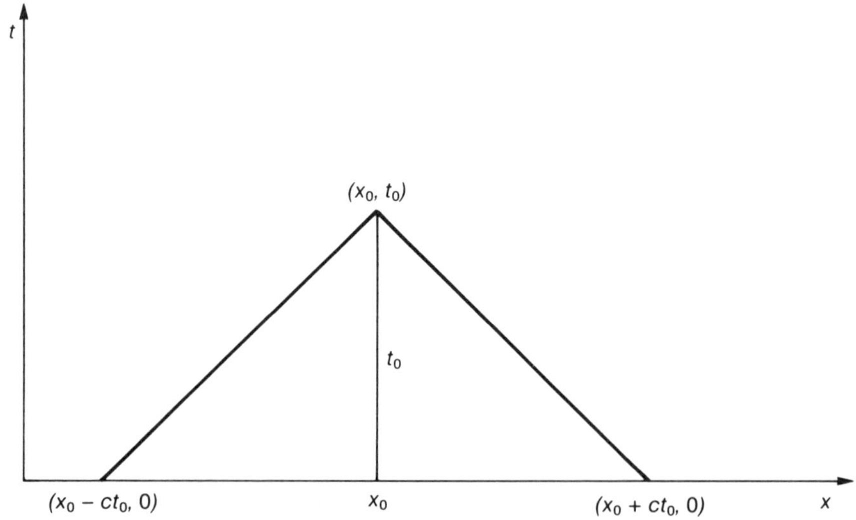


Figure 4.3.2

We will now investigate the physical significance of the d'Alembert solution (4.3.8) in greater detail. We rewrite the solution in the form

$$u(x, t) = \frac{1}{2} f(x + ct) + \frac{1}{2c} \int_0^{x+ct} g(\tau)\, d\tau + \frac{1}{2} f(x - ct) - \frac{1}{2c} \int_0^{x-ct} g(\tau)\, d\tau \tag{4.3.9}$$

or

$$u(x, t) = \phi(x + ct) + \psi(x - ct) \tag{4.3.10}$$

where

$$\phi(\xi) = \frac{1}{2} f(\xi) + \frac{1}{2c} \int_0^{\xi} g(\tau)\, d\tau \tag{4.3.11}$$

$$\psi(\eta) = \frac{1}{2} f(\eta) - \frac{1}{2c} \int_0^{\eta} g(\tau)\, d\tau \tag{4.3.12}$$

Evidently, $\phi(x + ct)$ represents a progressive wave travelling in the negative $x$-direction with speed $c$ without change of shape. Similarly, $\psi(x - ct)$ is also a progressive wave propagating in the positive $x$-direction with the same speed $c$ without change of shape. We shall examine this point in greater detail. Treat $\psi(x - ct)$ as a function of $x$ for a sequence of times $t$. At $t = 0$, the shape of this function is $u = \psi(x)$. At a subsequent time its shape is given by $u = \psi(x - ct)$ or $u = \psi(\xi)$ where $\xi = x - ct$ is the new coordinate obtained by translating the origin a distance $ct$ to the right. Thus the shape of the curve remains the same as time progresses, but moves to the right with velocity $c$ as shown in Fig. 4.3.3. This shows that $\psi(x - ct)$ represents a progressive wave travelling

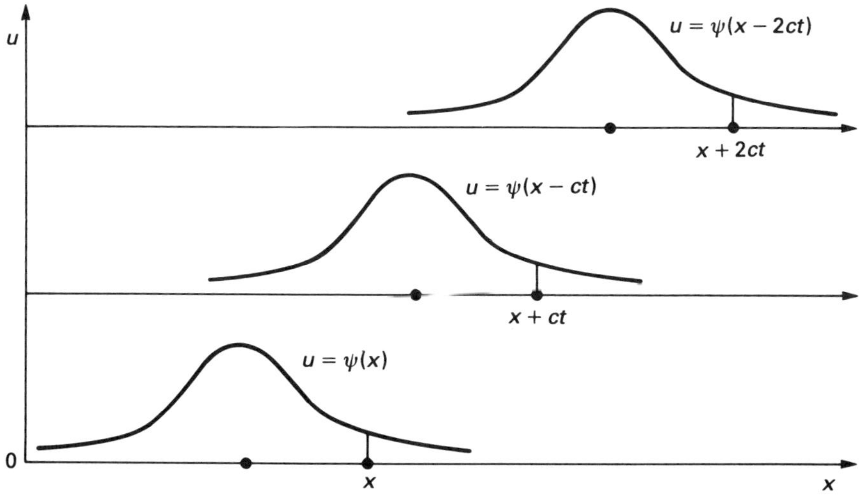


Figure 4.3.3

in the positive $x$-direction with velocity $c$ without change of shape. Similarly, $\phi(x+ct)$ is also a progressive wave propagating in the negative $x$-direction with the same speed $c$ without change of shape. For instance,

$$u(x, t) = \sin (x \mp ct) \tag{4.3.13}$$

represent sinusoidal waves travelling with speed $c$ in the positive and negative directions respectively without change of shape. The propagation of waves without change of shape is common to all linear wave equations.

To interpret the d'Alembert formula we consider two cases:

*Case 1.* We first consider the case when the initial velocity is zero, that is,

$$g(x) = 0$$

Then the d'Alembert solution has the form

$$u(x, t) = \frac{1}{2}[f(x+ct) + f(x-ct)]$$

Now suppose the initial displacement $f(x)$ is different from zero in an interval $(-b, b)$. Then in this case the forward and the backward waves are represented by

$$u = \frac{1}{2} f(x)$$

The waves are first initially superimposed, and then they separate and travel in opposite directions.

We consider $f(x)$ which has the form of a triangle. We draw a triangle with the ordinate $x = 0$ one-half that of the given function at that point, as shown in Fig. 4.3.4. If we displace these graphs and then take the sum of the ordinates of the displaced graphs, we obtain the shape of the string at any time $t$.

As can be seen from the figure, the waves travel in opposite directions away from each other. After both waves have passed the region of initial disturbance, the string returns to its rest position.

*Case 2.* We consider the case when the initial displacement is zero, that is,

$$f(x) = 0$$

and the d'Alembert formula assumes the form

$$u(x, t) = \frac{1}{2c}\int_{x-ct}^{x+ct} g(\tau)\, d\tau = \frac{1}{2}[G(x+ct) - G(x-ct)]$$

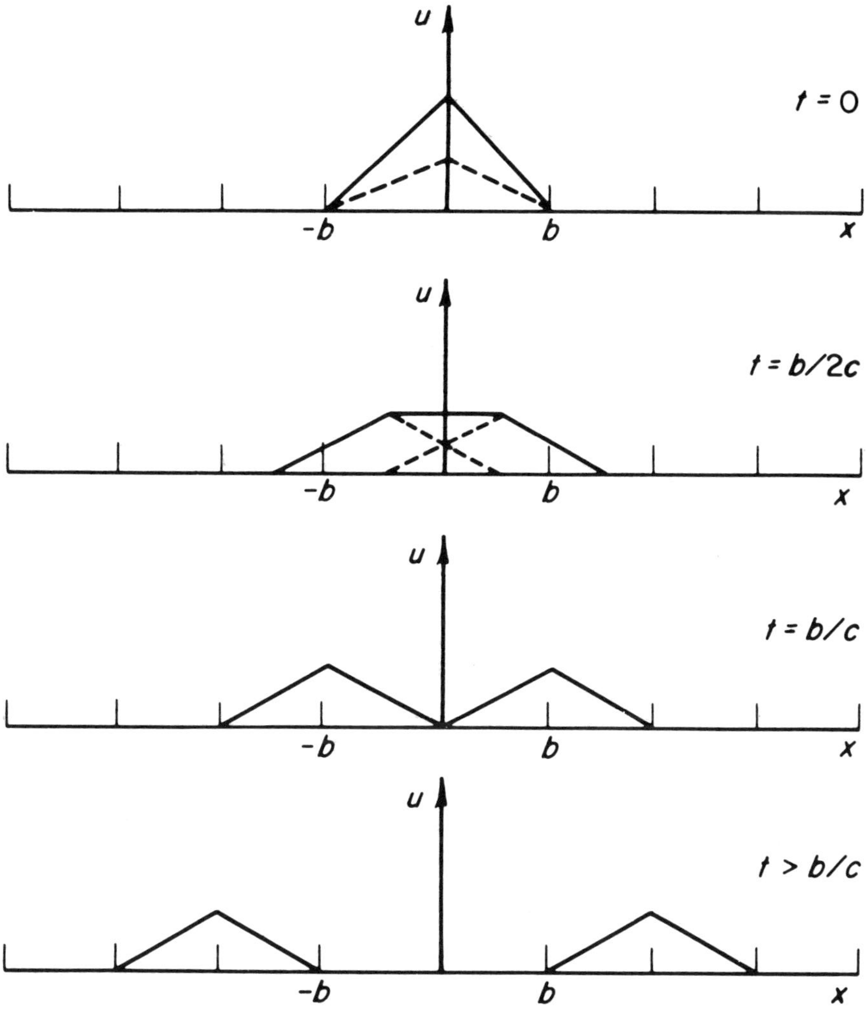


Figure 4.3.4

where

$$G(x) = \frac{1}{c}\int_{x_0}^{x} g(\tau)\, d\tau$$

If we take for the initial velocity

$$g(x) = \begin{cases} 0 & |x| > b \\ g_0 & |x| \leqslant b \end{cases}$$

then the function $G(x)$ is equal to zero for values of $x$ in the interval

$x \leqslant -b$, and

$$G(x) = \begin{cases} \dfrac{1}{c}\displaystyle\int_{-b}^{x} g_0\, d\tau = \dfrac{g_0}{c}(x+b) & -b \leqslant x \leqslant b \\ \dfrac{1}{c}\displaystyle\int_{-b}^{x} g_0\, d\tau = \dfrac{2bg_0}{c} & x > b \end{cases}$$

As in the previous case, the two waves which differ in sign travel in opposite directions on the $x$-axis. After some time $t$ the two functions $(1/2)G(x)$ and $-(1/2)G(x)$ move a distance $ct$. Thus, the graph of $u$ at time $t$ is obtained by summing the ordinates of the displaced graphs as shown in Fig. 4.3.5. As $t$ approaches infinity, the string will reach a state of rest, but it will not, in general, assume its original position. This displacement is known as the *residual displacement.*

In the preceding examples, we note that $f(x)$ is continuous but not continuously differentiable and $g(x)$ is discontinuous. To these initial data, there corresponds a generalized solution. By generalized solution we mean the following:

Let us suppose that the function $u(x, t)$ satisfies the initial conditions (4.3.2) and (4.3.3). Let $u(x, t)$ be the limit of a uniformly convergent sequence of solutions $u_n(x, t)$ which satisfies the wave equation (4.3.1) and the initial conditions

$$u_n(x, 0) = f_n(x)$$

$$\frac{\partial u_n}{\partial t}(x, 0) = g_n(x)$$

Let $f_n(x)$ be a continuously differentiable function and converge uniformly to $f(x)$, and let $g_n(x)$ be a continuously differentiable function and $\int_{x_0}^{x} g_n(\tau)\, d\tau$ approach uniformly to $\int_{x_0}^{x} g(\tau)\, d\tau$. Then the function $u(x, t)$ is called the *generalized solution* of the problem (4.3.1)–(4.3.3).

In general, it is interesting to discuss the effect of discontinuity of the function $f(x)$ at a point $x = x_0$, assuming $g(x)$ is a smooth function. Clearly it follows from (4.3.8) that $u(x, t)$ will be discontinuous at each point $(x, t)$ such that $x + ct = x_0$ or $x - ct = x_0$, that is, at each point of the two characteristic lines intersecting at the point $(x_0, 0)$. This means that discontinuities are propagated along the characteristic lines. At each point of the characteristic lines the partial derivatives of the function $u(x, t)$ fail to exist, and hence $u$ can no longer be a solution of the Cauchy problem in the usual sense. However, such a function may be called a *generalized solution* of the Cauchy problem. Similarly, if $f(x)$ is continuous but either $f'(x)$ or $f''(x)$ has a discontinuity at some point $x = x_0$, the first- or second-order partial derivatives of the solution $u(x, t)$ will be discontinuous along the characteristic lines through $(x_0, 0)$. Finally, a discontinuity in $g(x)$ at $x = x_0$ would lead to a discontinuity in the first- and second-order partial derivatives of $u$ along the characteristic

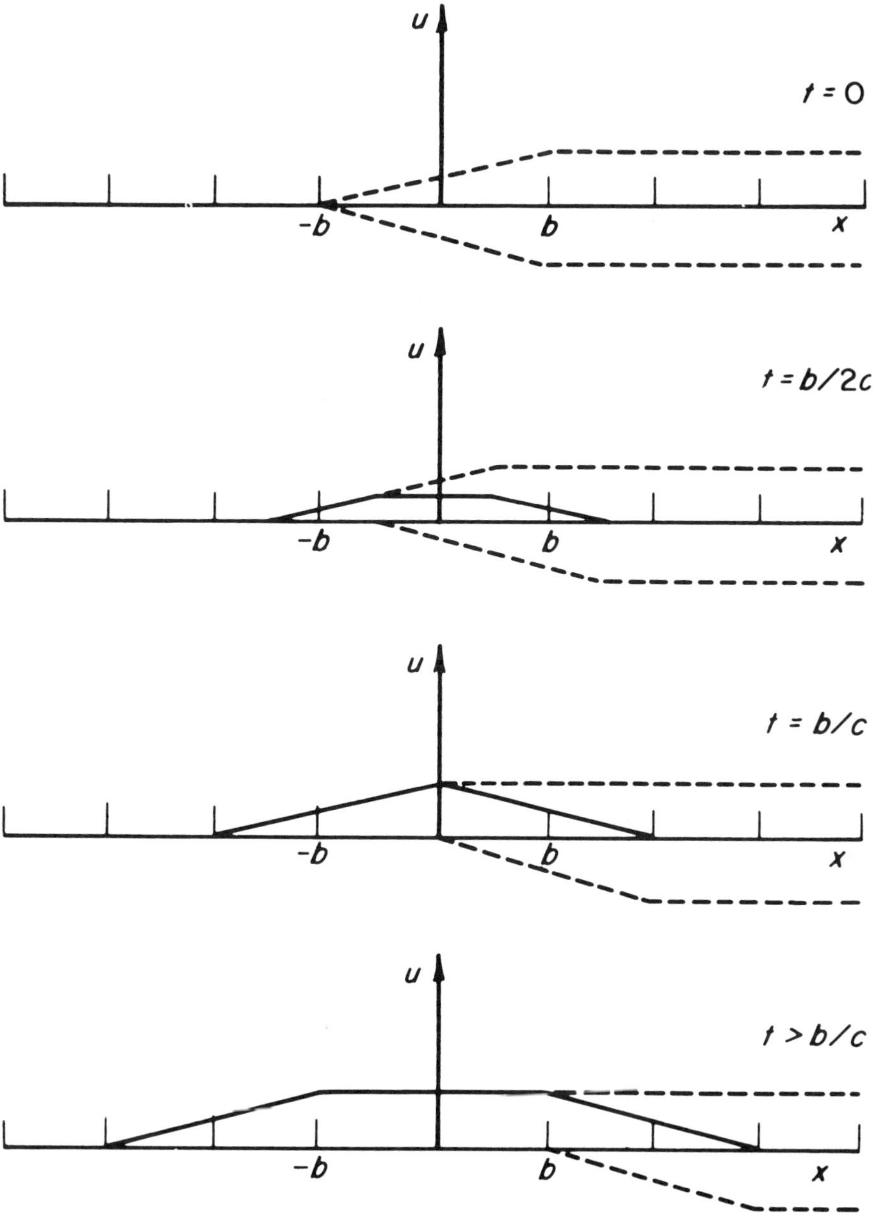


Figure 4.3.5

lines through $(x_0, 0)$, and a discontinuity in $g'(x)$ at $x_0$ will imply a discontinuity in the second-order partial derivatives of $u$ along the characteristic through $(x_0, 0)$. The solution given by (4.3.8) with $f$, $f'$, $f''$, $g$, and $g'$ piecewise continuous on $-\infty < x < \infty$ is usually called the *generalized solution* of the Cauchy problem.

## 4.4 INITIAL-BOUNDARY VALUE PROBLEMS

We have just determined the solution of the initial value problem for the infinite vibrating string. We will now study the effect of a boundary on the solution.

### (A) Semi-infinite String with a Fixed End

Let us first consider a semi-infinite vibrating string with a fixed end, that is,

$$\begin{aligned} u_{tt} &= c^2 u_{xx}, && 0 < x < \infty, \quad t > 0 \\ u(x, 0) &= f(x), && 0 \leq x < \infty \\ u_t(x, 0) &= g(x), && 0 \leq x < \infty \\ u(0, t) &= 0, && 0 \leq t < \infty \end{aligned} \tag{4.4.1}$$

It is evident here that the boundary condition at $x = 0$ produces a wave moving to the right with the velocity $c$. Thus, for $x > ct$, the solution is the same as that of the infinite string, and the displacement is influenced only by the initial data on the interval $[x - ct, x + ct]$, as shown in Fig. 4.4.1.

When $x < ct$, the interval $[x - ct, x + ct]$ extends onto the negative $x$-axis where $f$ and $g$ are not prescribed.

But from the d'Alembert formula

$$u(x, t) = \phi(x + ct) + \psi(x - ct) \tag{4.4.2}$$

where

$$\phi(\xi) = \frac{1}{2} f(\xi) + \frac{1}{2c} \int_0^{\xi} g(\tau)\, d\tau + \frac{K}{2} \tag{4.4.3}$$

$$\psi(\eta) = \frac{1}{2} f(\eta) - \frac{1}{2c} \int_0^{\eta} g(\tau)\, d\tau - \frac{K}{2} \tag{4.4.4}$$

we see that

$$u(0, t) = \phi(ct) + \psi(-ct) = 0$$

Hence

$$\psi(-ct) = -\phi(ct)$$

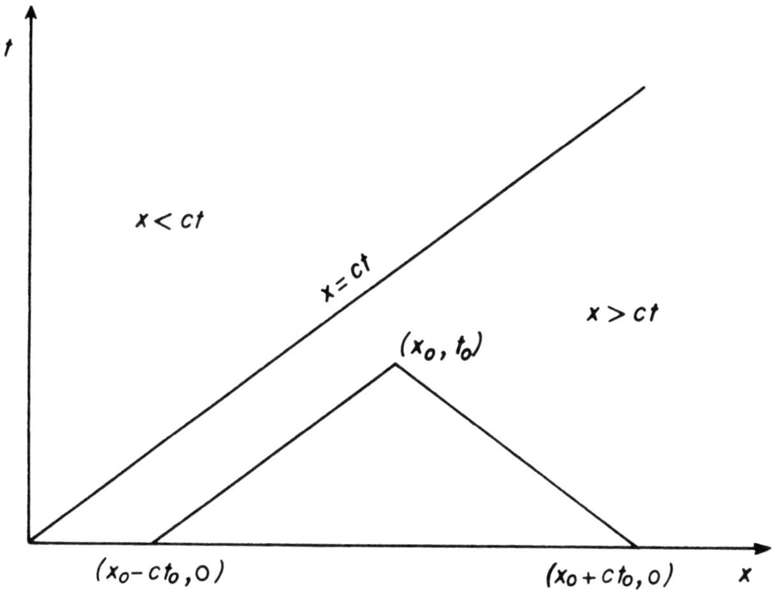


Figure 4.4.1

If we let $\alpha = -ct$, then

$$\psi(\alpha) = -\phi(-\alpha)$$

Replacing $\alpha$ by $x - ct$, we obtain for $x < ct$

$$\psi(x - ct) = -\phi(ct - x)$$

and hence

$$\psi(x - ct) = -\frac{1}{2} f(ct - x) - \frac{1}{2c} \int_0^{ct-x} g(\tau)\, d\tau - \frac{K}{2}$$

The solution of the initial-boundary value problem, therefore, is given by

$$u(x, t) = \frac{1}{2}[f(x + ct) + f(x - ct)] + \frac{1}{2c} \int_{x-ct}^{x+ct} g(\tau)\, d\tau \quad \text{for} \quad x > ct \quad (4.4.5)$$

$$u(x, t) = \frac{1}{2}[f(x + ct) - f(ct - x)] + \frac{1}{2c} \int_{ct-x}^{x+ct} g(\tau)\, d\tau \quad \text{for} \quad x < ct \quad (4.4.6)$$

In order for this solution to exist, $f$ must be twice continuously differentiable and $g$ must be continuously differentiable, and in addition

$$f(0) = f''(0) = g(0) = 0$$

Formula (4.4.6) has an interesting physical interpretation. If we draw the characteristics through the point $(x_0, t_0)$ in the region $x > ct$, we see, as pointed out earlier, that the displacement at $(x_0, t_0)$ is determined by the initial values on $[x_0 - ct_0, x_0 + ct_0]$.

If the point $(x_0, t_0)$ lies in the region $x < ct$ as shown in Fig. 4.4.1 we see that the characteristic $x + ct = x_0 + ct_0$ intersects the $x$-axis at $(x_0 + ct_0, 0)$. However, the characteristic $x - ct = x_0 - ct_0$ intersects the $t$-axis at $(0, t_0 - x_0/c)$, and the characteristic $x + ct = ct_0 - x_0$ through this point intersects the $x$-axis at $(ct_0 - x_0, 0)$. Thus, the disturbance at $(ct_0 - x_0, 0)$ travels along the backward characteristic $x + ct = ct_0 - x_0$, and is reflected at $(0, t_0 - x_0/c)$ as a forward wave $-\phi(ct_0 - x_0)$.

EXAMPLE 4.4.1. Determine the solution of the initial-boundary value problem

$$\begin{aligned} u_{tt} &= 4u_{xx}, && x > 0, \quad t > 0 \\ u(x, 0) &= |\sin x|, && x > 0 \\ u_t(x, 0) &= 0, && x \geqslant 0 \\ u(0, t) &= 0, && t \geqslant 0 \end{aligned}$$

For $x > 2t$

$$\begin{aligned} u(x, t) &= \frac{1}{2}[f(x + 2t) + f(x - 2t)] \\ &= \frac{1}{2}[|\sin (x + 2t)| + |\sin (x - 2t)|] \end{aligned}$$

and $x < 2t$

$$\begin{aligned} u(x, t) &= \frac{1}{2}[f(x + 2t) - f(2t - x)] \\ &= \frac{1}{2}[|\sin (x + 2t)| - |\sin (2t - x)|] \end{aligned}$$

Notice that $u(0, t) = 0$ is satisfied by $u(x, t)$ for $x < 2t$ (that is, $t > 0$).

## (B) Semi-infinite String with a Free End

Let us consider a semi-infinite string with a free end at $x = 0$. Thus, we will determine the solution of

$$\begin{aligned} u_{tt} &= c^2 u_{xx}, && 0 < x < \infty, \quad t > 0 \\ u(x, 0) &= f(x), && 0 \leqslant x < \infty \\ u_t(x, 0) &= g(x), && 0 \leqslant x < \infty \\ u_x(0, t) &= 0, && 0 \leqslant t < \infty \end{aligned} \tag{4.4.7}$$

As in the case of the fixed end, for $x > ct$ the solution is the same as that of the infinite string. For $x < ct$, from the d'Alembert formula

$$u(x, t) = \phi(x + ct) + \psi(x - ct)$$

we have

$$u_x(x, t) = \phi'(x + ct) + \psi'(x - ct)$$

Thus

$$u_x(0, t) = \phi'(ct) + \psi'(-ct) = 0$$

Integration yields

$$\phi(ct) - \psi(-ct) = K$$

where $K$ is a constant. Now, if we let $\alpha = -ct$, we obtain

$$\psi(\alpha) = \phi(-\alpha) - K$$

Replacing $\alpha$ by $x - ct$, we have

$$\psi(x - ct) = \phi(ct - x) - K$$

and hence

$$\psi(x - ct) = \frac{1}{2} f(ct - x) + \frac{1}{2c} \int_0^{ct-x} g(\tau)\, d\tau - \frac{K}{2}$$

The solution of the initial-boundary value problem, therefore, is given by

$$u(x, t) = \frac{1}{2}[f(x + ct) + f(x - ct)] + \frac{1}{2c} \int_{x-ct}^{x+ct} g(\tau)\, d\tau \quad \text{for} \quad x > ct \qquad (4.4.8)$$

$$u(x, t) = \frac{1}{2}[f(x + ct) + f(ct - x)] + \frac{1}{2c} \left[ \int_0^{x+ct} g(\tau)\, d\tau + \int_0^{ct-x} g(\tau)\, d\tau \right] \quad \text{for} \quad x < ct \qquad (4.4.9)$$

We note here that for this solution to exist, $f$ must be twice continuously differentiable and $g$ must be continuously differentiable, and in addition,

$$f'(0) = g'(0) = 0.$$

EXAMPLE 4.4.2. Find the solution of the initial boundary value problem

$$\begin{aligned} &u_{tt} = u_{xx}, && 0 < x < \infty, && t > 0 \\ &u(x, 0) = \cos \frac{\pi x}{2}, && 0 \leq x < \infty && \\ &u_t(x, 0) = 0, && 0 \leq x < \infty && \\ &u_x(0, t) = 0, && && t \geq 0 \end{aligned}$$

For $x > t$

$$u(x, t) = \frac{1}{2}\left[\cos\frac{\pi}{2}(x+t) + \cos\frac{\pi}{2}(x-t)\right]$$

$$= \cos\frac{\pi}{2}x \cos\frac{\pi}{2}t$$

and for $x < t$

$$u(x, t) = \frac{1}{2}\left[\cos\frac{\pi}{2}(x+t) + \cos\frac{\pi}{2}(t-x)\right]$$

$$= \cos\frac{\pi}{2}x \cos\frac{\pi}{2}t$$

## 4.5 NONHOMOGENEOUS BOUNDARY CONDITIONS

In the case of initial-boundary value problems with nonhomogeneous boundary condition, such as

$$\begin{aligned} u_{tt} &= c^2 u_{xx}, && x > 0, && t > 0 \\ u(x, 0) &= f(x), && x \geqslant 0 && \\ u_t(x, 0) &= g(x), && x \geqslant 0 && \\ u(0, t) &= p(t), && && t \geqslant 0 \end{aligned} \tag{4.5.1}$$

we proceed in a similar manner as in the case of homogeneous boundary conditions. Using Eq. (4.4.2), we apply the boundary condition to obtain

$$u(0, t) = \phi(ct) + \psi(-ct) = p(t)$$

If we let $\alpha = -ct$, we have

$$\psi(\alpha) = p\left(-\frac{\alpha}{c}\right) - \phi(-\alpha)$$

Replacing $\alpha$ by $x - ct$, the preceding relation becomes

$$\psi(x - ct) = p\left(t - \frac{x}{c}\right) - \phi(ct - x)$$

Thus, for $0 \leqslant x < ct$

$$\begin{aligned} u(x, t) &= p\left(t - \frac{x}{c}\right) + \frac{1}{2}[f(x+ct) - f(ct-x)] \\ &\qquad + \frac{1}{2c}\int_{ct-x}^{x+ct} g(\tau)\, d\tau \\ &= p\left(t - \frac{x}{c}\right) + \phi(x+ct) - \psi(ct-x) \end{aligned} \tag{4.5.2}$$

where $\phi(\xi)$ is given by (4.3.11) and $\psi(\eta)$ is

$$\psi(\eta)=\frac{1}{2}f(\eta)+\frac{1}{2c}\int_0^{\eta} g(\tau)\,d\tau$$

The solution for $x>ct$ is given by the solution of the infinite string (4.4.5).

In this case, in addition to the differentiability conditions satisfied by $f$ and $g$, as in the case of the problem with the homogeneous boundary condition, $p$ must be twice continuously differentiable in $t$ and

$$p(0)=f(0), \qquad p'(0)=g(0), \qquad p''(0)=c^2f''(0)$$

Let us next consider the initial-boundary value problem

$$\begin{aligned} u_{tt}&=c^2u_{xx}, && x>0, \quad t>0 \\ u(x,0)&=f(x), && x\geqslant 0 \\ u_t(x,0)&=g(x), && x\geqslant 0 \\ u_x(0,t)&=q(t), && t\geqslant 0 \end{aligned}$$

Using (4.4.2), we apply the boundary condition to obtain

$$u_x(0,t)=\phi'(ct)+\psi'(-ct)=q(t)$$

Then integration yields

$$\phi(ct)-\psi(-ct)=c\int_0^t q(\tau)\,d\tau+K$$

If we let $\alpha=-ct$, then

$$\psi(\alpha)=\phi(-\alpha)-c\int_0^{-\alpha/c} q(\tau)\,d\tau-K$$

Replacing $\alpha$ by $x-ct$, we obtain

$$\psi(x-ct)=\phi(ct-x)-c\int_0^{t-x/c} q(\tau)\,d\tau-K$$

The solution of the initial-boundary value problem for $x<ct$, therefore, is given by

$$u(x,t)=\frac{1}{2}[f(x+ct)+f(ct-x)]+\frac{1}{2c}\left[\int_0^{x+ct} g(\tau)\,d\tau+\int_0^{ct-x} g(\tau)\,d\tau\right] \\ -c\int_0^{t-x/c} q(\tau)\,d\tau \qquad (4.5.3)$$

Here $f$ and $g$ must satisfy the differentiability conditions, as in the case of the problem with the homogeneous boundary condition. In addition

$$f'(0) = q(0), \qquad g'(0) = q'(0)$$

The solution for the initial-boundary value problem involving the elastic boundary condition

$$u_x(0, t) + hu(0, t) = 0, \qquad h = \text{constant}$$

can also be constructed in a similar manner from the d'Alembert formula.

## 4.6 FINITE STRING WITH FIXED ENDS

The problem of the finite string is more complicated than that of the infinite string due to the repeated reflection of waves from the boundaries.

Let us consider the vibration of the string of length $l$ fixed at both ends. The problem is that of finding the solution of

$$\begin{aligned} u_{tt} &= c^2 u_{xx}, && 0 < x < l, \quad t > 0 \\ u(x, 0) &= f(x), && 0 \leq x \leq l \\ u_t(x, 0) &= g(x), && 0 \leq x \leq l \\ u(0, t) &= 0, && t \geq 0 \\ u(l, t) &= 0, && t \geq 0 \end{aligned} \tag{4.6.1}$$

From the previous results, we know that the solution of the wave equation is

$$u(x, t) = \phi(x + ct) + \psi(x - ct)$$

Applying the initial conditions, we have

$$\begin{aligned} u(x, 0) &= \phi(x) + \psi(x) = f(x), && 0 \leq x \leq l \\ u_t(x, 0) &= c\phi'(x) - c\psi'(x) = g(x), && 0 \leq x \leq l \end{aligned}$$

Solving for $\phi$ and $\psi$, we find

$$\phi(\xi) = \frac{1}{2} f(\xi) + \frac{1}{2c} \int_0^{\xi} g(\tau)\, d\tau + \frac{K}{2}, \qquad 0 \leq \xi \leq l \tag{4.6.2}$$

$$\psi(\eta) = \frac{1}{2} f(\eta) - \frac{1}{2c} \int_0^{\eta} g(\tau)\, d\tau - \frac{K}{2}, \qquad 0 \leq \eta \leq l \tag{4.6.3}$$

Hence

$$u(x, t) = \frac{1}{2} [f(x + ct) + f(x - ct)] + \frac{1}{2c} \int_{x-ct}^{x+ct} g(\tau)\, d\tau \tag{4.6.4}$$

for $0 \leq x + ct \leq l$ and $0 \leq x - ct \leq l$. The solution is thus uniquely determined by the initial data in the region

$$t \leqslant \frac{x}{c}, \qquad t \leqslant \frac{l-x}{c}, \qquad t \geqslant 0$$

For larger times, the solution depends on the boundary conditions. Applying the boundary conditions, we obtain

$$u(0, t) = \phi(ct) + \psi(-ct) = 0, \qquad t \geqslant 0 \tag{4.6.5}$$

$$u(l, t) = \phi(l + ct) + \psi(l - ct) = 0, \qquad t \geqslant 0 \tag{4.6.6}$$

If we set $\alpha = -ct$, Eq. (4.6.5) becomes

$$\psi(\alpha) = -\phi(-\alpha), \qquad \alpha \leqslant 0 \tag{4.6.7}$$

and if we set $\alpha = l + ct$, Eq. (4.6.6) takes the form

$$\phi(\alpha) = -\psi(2l - \alpha), \qquad \alpha \geqslant l \tag{4.6.8}$$

With $\xi = -\eta$ we may write Eq. (4.6.2) as

$$\phi(-\eta) = \frac{1}{2} f(-\eta) + \frac{1}{2c} \int_0^{-\eta} g(\tau)\, d\tau + \frac{K}{2}, \qquad 0 \leqslant -\eta \leqslant l \tag{4.6.9}$$

Thus from (4.6.7) and (4.6.9), we have

$$\psi(\eta) = -\frac{1}{2} f(-\eta) - \frac{1}{2c} \int_0^{-\eta} g(\tau)\, d\tau - \frac{K}{2}, \qquad -l \leqslant \eta \leqslant 0 \tag{4.6.10}$$

We see that the range of $\psi(\eta)$ is extended to $-l \leqslant \eta \leqslant l$.

If we put $\alpha = \xi$ in Eq. (4.6.8), we obtain

$$\phi(\xi) = -\psi(2l - \xi), \qquad \xi \geqslant l \tag{4.6.11}$$

Then by putting $\eta = 2l - \xi$ in Eq. (4.6.3), we obtain

$$\psi(2l - \xi) = \frac{1}{2} f(2l - \xi) - \frac{1}{2c} \int_0^{2l-\xi} g(\tau)\, d\tau - \frac{K}{2}, \qquad 0 \leqslant 2l - \xi \leqslant l \tag{4.6.12}$$

Substitution of this in Eq. (4.6.11) yields

$$\phi(\xi) = -\frac{1}{2} f(2l - \xi) + \frac{1}{2c} \int_0^{2l-\xi} g(\tau)\, d\tau + \frac{K}{2}, \qquad l \leqslant \xi \leqslant 2l \tag{4.6.13}$$

The range of $\phi(\xi)$ is thus extended to $0 \leqslant \xi \leqslant 2l$. Continuing in this manner, we obtain $\phi(\xi)$ for all $\xi \geqslant 0$ and $\psi(\eta)$ for all $\eta \leqslant l$. Hence, the solution is determined for all $0 \leqslant x \leqslant l$ and $t \geqslant 0$.

In order to observe the effect of the boundaries on the propagation of waves, the characteristics are drawn through the end points until they meet the boundaries and then continue inward as shown in Fig. 4.6.1. It can be seen from the diagram that only direct waves propagate in region

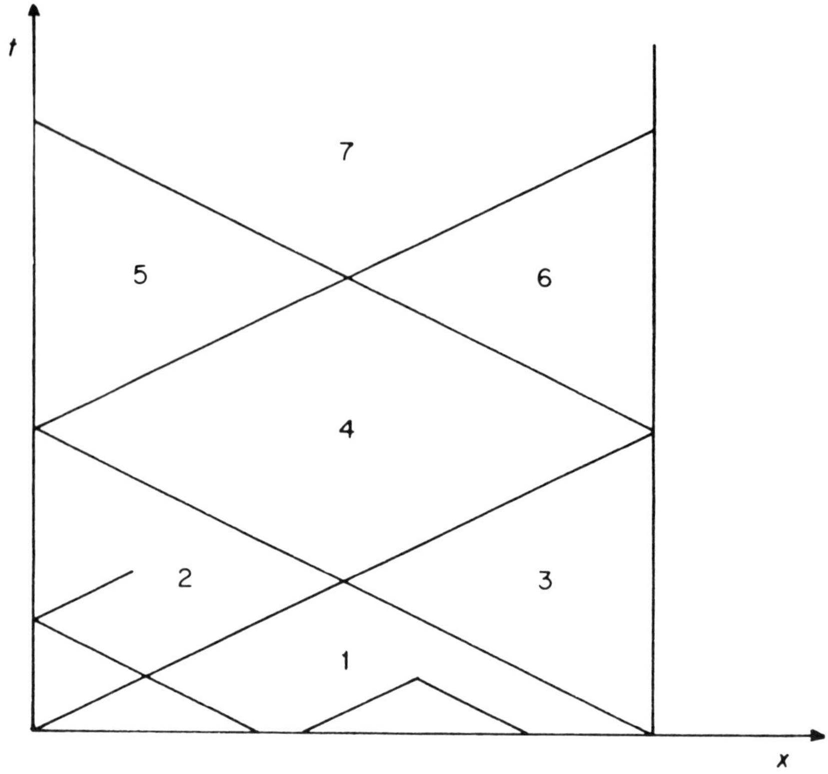


Figure 4.6.1

1. In regions 2 and 3, both direct and reflected waves propagate. In regions 4, 5, 6, . . . , several waves propagate along the characteristics reflected from both of the boundaries $x=0$ and $x=l$.

EXAMPLE 4.6.1. Determine the solution of

$$
\begin{aligned}
&u_{tt}=c^2u_{xx}, && 0<x<l, && t>0\\
&u(x,0)=\sin(\pi x/l), && 0\leqslant x\leqslant l\\
&u_t(x,0)=0, && 0\leqslant x\leqslant l\\
&u(0,t)=0, && && t\geqslant 0\\
&u(l,t)=0, && && t\geqslant 0
\end{aligned}
$$

From Eqs. (4.6.2) and (4.6.3), we have

$$\phi(\xi)=\frac{1}{2}\sin\frac{\pi\xi}{l}+\frac{K}{2}, \qquad 0\leqslant\xi\leqslant l$$

$$\psi(\eta)=\frac{1}{2}\sin\frac{\pi\eta}{l}-\frac{K}{2}, \qquad 0\leqslant\eta\leqslant l$$

Using Eq. (4.6.10), we obtain

$$\psi(\eta) = -\frac{1}{2}\sin\left(-\frac{\pi\eta}{l}\right) - \frac{K}{2}, \qquad -l \leqslant \eta \leqslant 0$$

$$= \frac{1}{2}\sin\frac{\pi\eta}{l} - \frac{K}{2}$$

From Eq. (4.6.13), we find

$$\phi(\xi) = -\frac{1}{2}\sin\frac{\pi}{l}(2l - \xi) + \frac{K}{2}, \qquad l \leqslant \xi \leqslant 2l$$

Again by Eq. (4.6.7) and from the preceding $\phi(\xi)$, we have

$$\psi(\eta) = \frac{1}{2}\sin\frac{\pi\eta}{l} - \frac{K}{2}, \qquad -2l \leqslant \eta \leqslant -l$$

Proceeding in this manner, we determine the solution

$$u(x, t) = \phi(\xi) + \psi(\eta)$$

$$= \frac{1}{2}\left[\sin\frac{\pi}{l}(x + ct) + \sin\frac{\pi}{l}(x - ct)\right]$$

for all $x$ in $(0, l)$ and for all $t > 0$.

The solution of the finite initial-boundary value problem

$$\begin{aligned} u_{tt} &= c^2 u_{xx}, && 0 < x < l, && t > 0 \\ u(x, 0) &= f(x), && 0 \leqslant x \leqslant l && \\ u_t(x, 0) &= g(x), && 0 \leqslant x \leqslant l && \\ u(0, t) &= p(t), && && t \geqslant 0 \\ u(l, t) &= q(t), && && t \geqslant 0 \end{aligned}$$

can be determined by the same method.

## 4.7 NONHOMOGENEOUS WAVE EQUATION

We shall consider next the Cauchy problem for the nonhomogeneous wave equation

$$u_{tt} = c^2 u_{xx} + h^*(x, t) \tag{4.7.1}$$

with the initial conditions

$$\begin{aligned} u(x, 0) &= f(x) \\ u_t(x, 0) &= g^*(x) \end{aligned} \tag{4.7.2}$$

By the coordinate transformation

$$y = ct \tag{4.7.3}$$

the problem is reduced to

$$u_{xx} - u_{yy} = h(x, y) \tag{4.7.4}$$

$$u(x, 0) = f(x) \tag{4.7.5}$$

$$u_y(x, 0) = g(x) \tag{4.7.6}$$

where $h(x, y) = -h^*/c^2$ and $g(x) = g^*/c$.

Let $P_0(x_0, y_0)$ be a point of the plane, and let $Q_0$ be the point $(x_0, 0)$ on the initial line $y = 0$. Then the characteristics $x \pm y = \text{constant}$ of Eq. (4.7.4) are two straight lines drawn through the point $P_0$ with slopes $\pm 1$. Obviously, they intersect the $x$-axis at the points $P_1(x_0 - y_0, 0)$ and $P_2(x_0 + y_0, 0)$, as shown in Fig. 4.7.1. Let the sides of the triangle $P_0P_1P_2$ be designated by $B_0$, $B_1$, and $B_2$, and let $R$ be the region representing the interior of the triangle and its boundaries $B$. Integrating both sides of Eq. (4.7.4), we obtain

$$\iint_R (u_{xx} - u_{yy})\, dR = \iint_R h(x, y)\, dR \tag{4.7.7}$$

Now we apply Green's theorem, and obtain

$$\iint_R (u_{xx} - u_{yy})\, dR = \oint_B (u_x\, dy + u_y\, dx) \tag{4.7.8}$$

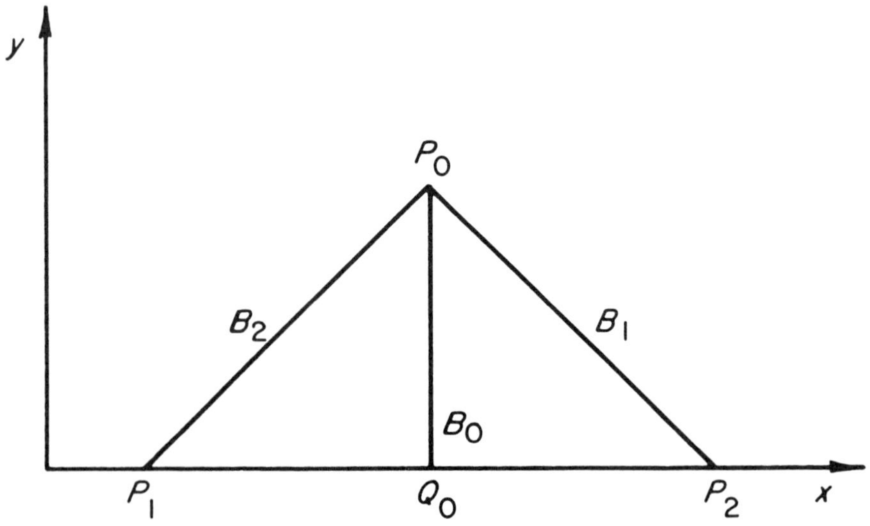


Figure 4.7.1

Since $B$ is composed of $B_0$, $B_1$, and $B_2$, we note that

$$\int_{B_0} (u_x\,dy + u_y\,dx) = \int_{x_0-y_0}^{x_0+y_0} u_y\,dx$$

$$\begin{aligned}\int_{B_1} (u_x\,dy + u_y\,dx) &= \int_{B_1} (-u_x\,dx - u_y\,dy)\\ &= u(x_0+y_0, 0) - u(x_0, y_0)\end{aligned}$$

$$\begin{aligned}\int_{B_2} (u_x\,dy + u_y\,dx) &= \int_{B_2} (u_x\,dx + u_y\,dy)\\ &= u(x_0-y_0, 0) - u(x_0, y_0)\end{aligned}$$

Hence,

$$\begin{aligned}\oint_B (u_x\,dy + u_y\,dx) &= -2u(x_0, y_0) + u(x_0-y_0, 0)\\ &\quad + u(x_0+y_0, 0) + \int_{x_0-y_0}^{x_0+y_0} u_y\,dx \end{aligned} \tag{4.7.9}$$

Combining Eqs. (4.7.7), (4.7.8) and (4.7.9), we obtain

$$\begin{aligned}u(x_0, y_0) &= \frac{1}{2}[u(x_0+y_0, 0) + u(x_0-y_0, 0)]\\ &\quad + \frac{1}{2}\int_{x_0-y_0}^{x_0+y_0} u_y\,dx - \frac{1}{2}\iint_R h(x, y)\,dR \end{aligned} \tag{4.7.10}$$

We have chosen $x_0$, $y_0$ arbitrarily, and as a consequence, we replace $x_0$ by $x$ and $y_0$ by $y$. Equation (4.7.10) thus becomes

$$u(x, y) = \frac{1}{2}[f(x+y) + f(x-y)] + \frac{1}{2}\int_{x-y}^{x+y} g(\tau)\,d\tau - \frac{1}{2}\iint_R h(x, y)\,dR$$

In terms of the original variables

$$u(x, t) = \frac{1}{2}[f(x+ct) + f(x-ct)] + \frac{1}{2c}\int_{x-ct}^{x+ct} g^*(\tau)\,d\tau - \frac{1}{2}\iint_R h(x, t)\,dR \tag{4.7.11}$$

EXAMPLE 4.7.1. Determine the solution of

$$u_{xx} - u_{yy} = 1$$
$$u(x, 0) = \sin x$$
$$u_y(x, 0) = x$$

It is easy to see that the characteristics are $x + y = \text{constant} = x_0 + y_0$

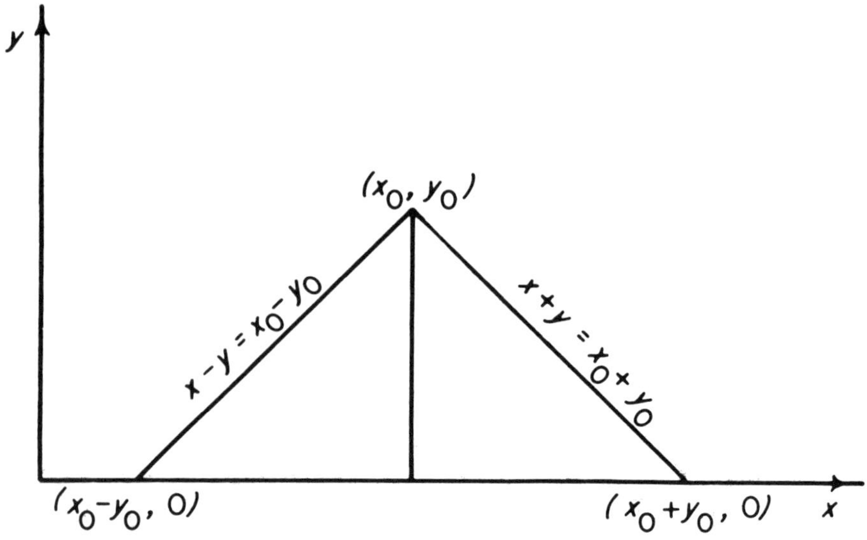


Figure 4.7.2

and $x - y = \text{constant} = x_0 - y_0$, as shown in Fig. 4.7.2. Thus,

$$u(x_0, y_0) = \frac{1}{2}[\sin(x_0 + y_0) + \sin(x_0 - y_0)]$$

$$+ \frac{1}{2}\int_{x_0-y_0}^{x_0+y_0} \tau\, d\tau - \frac{1}{2}\int_0^{y_0}\int_{y+x_0-y_0}^{-y+x_0+y_0} dxdy$$

$$= \frac{1}{2}[\sin(x_0 + y_0) + \sin(x_0 - y_0)] + x_0 y_0 - \frac{1}{2}y_0^2$$

Now dropping the subscript zero, we obtain the solution

$$u(x, y) = \frac{1}{2}[\sin(x + y) + \sin(x - y)] + xy - \frac{1}{2}y^2$$

## 4.8 RIEMANN METHOD

We shall discuss Riemann's method of integrating the linear hyperbolic equation

$$L[u] \equiv u_{xy} + au_x + bu_y + cu = f(x, y) \tag{4.8.1}$$

$L$ denotes the linear operator, and $a(x, y)$, $b(x, y)$, $c(x, y)$ and $f(x, y)$ are differentiable functions in some domain $D^*$. The method consists essentially of the derivation of an integral formula which represents the solution of the Cauchy problem.

Let $v(x, y)$ be a function having continuous second-order partial derivatives. Then we may write

$$\begin{aligned} vu_{xy} - uv_{xy} &= (vu_x)_y - (uv_y)_x \\ vau_x &= (avu)_x - u(av)_x \\ vbu_y &= (bvu)_y - u(bv)_y \end{aligned} \tag{4.8.2}$$

so that

$$vL[u] - uM[v] = U_x + V_y \tag{4.8.3}$$

where $M$ is the operator represented by

$$M[v] = v_{xy} - (av)_x - (bv)_y + cv \tag{4.8.4}$$

and

$$\begin{aligned} U &= auv - uv_y \\ V &= buv + vu_x \end{aligned} \tag{4.8.5}$$

The operator $M$ is called the *adjoint operator* of $L$. If $M = L$, then the operator $L$ is said to be *self-adjoint*. Now applying Green's theorem, we have

$$\iint_D (U_x + V_y)\, dx\, dy = \oint_C (U\, dy - V\, dx) \tag{4.8.6}$$

where $C$ is the closed curve bounding the region of integration $D$ which is in $D^*$.

Let $\Lambda$ be a smooth initial curve which is continuous, shown in Fig. 4.8.1. Since Eq. (4.8.1) is in first canonical form, $x$ and $y$ are the characteristic coordinates. We assume that the tangent to $\Lambda$ is nowhere parallel to $x$ or $y$ axes. Let $P(\alpha, \beta)$ be a point at which the solution to the Cauchy problem is sought. Line $PQ$ parallel to the $x$ axis intersects the initial curve $\Lambda$ at $Q$ and line $PR$ parallel to the $y$ axis intersects the curve $\Lambda$ at $R$. We suppose that $u$ and $u_x$ or $u_y$ are prescribed along $\Lambda$.

Let $C$ be the closed contour $PQRP$ bounding $D$. Since $dy = 0$ on $PQ$ and $dx = 0$ on $PR$, it follows immediately from Eqs. (4.8.3) and (4.8.6) that

$$\iint_D (vL[u] - uM[v])\, dx\, dy = \int_Q^R (U\, dy - V\, dx) + \int_R^P U\, dy - \int_P^Q V\, dx \tag{4.8.7}$$

From Eq. (4.8.5), we find

$$\int_P^Q V\, dx = \int_P^Q bvu\, dx + \int_P^Q vu_x\, dx$$

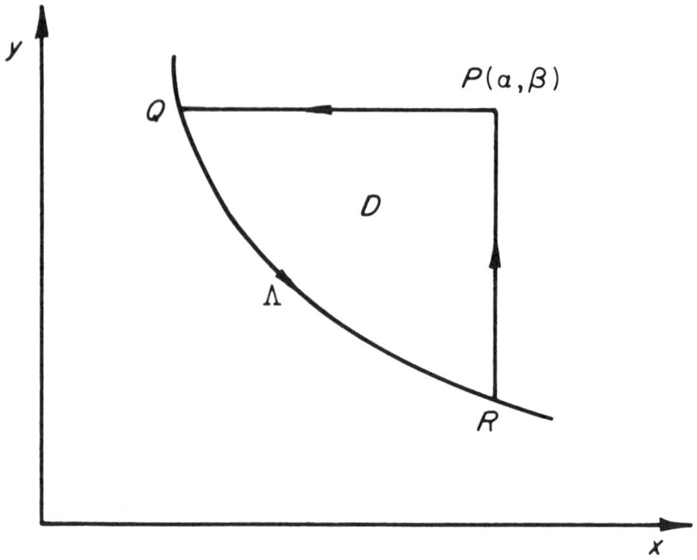


Figure 4.8.1

Integrating by parts, we obtain

$$\int_P^Q vu_x\,dx = [uv]_P^Q - \int_P^Q uv_x\,dx$$

Hence, we may write

$$\int_P^Q V\,dx = [uv]_P^Q + \int_P^Q u(bv - v_x)\,dx$$

Substitution of this integral in Eq. (4.8.7) yields

$$[uv]_P = [uv]_Q + \int_P^Q u(bv - v_x)\,dx - \int_R^P u(av - v_y)\,dy - \int_Q^R (U\,dy - V\,dx)$$

$$+ \iint_D (vL[u] - uM[v])\,dx\,dy \tag{4.8.8}$$

Suppose we can choose the function $v(x, y; \alpha, \beta)$ to be the solution of the adjoint equation

$$M[v] = 0 \tag{4.8.9}$$

satisfying the conditions

$$\begin{aligned} v_x &= bv \quad \text{when} \quad y = \beta \\ v_y &= av \quad \text{when} \quad x = \alpha \\ v &= 1 \quad \text{when} \quad x = \alpha \quad \text{and} \quad y = \beta \end{aligned} \tag{4.8.10}$$

The function $v(x, y; \alpha, \beta)$ is called the *Riemann function*. Since $L[u]=f$, Eq. (4.8.8) reduces to

$$[u]_P = [uv]_Q - \int_Q^R uv(a\,dy - b\,dx) + \int_Q^R (uv_y\,dy + vu_x\,dx) + \iint_D vf\,dx\,dy \tag{4.8.11}$$

This gives us the value of $u$ at the point $P$ when $u$ and $u_x$ are prescribed along the curve $\Lambda$. When $u$ and $u_y$ are prescribed, the identity

$$[uv]_R - [uv]_Q = \int_Q^R [(uv)_x\,dx + (uv)_y\,dy]$$

may be used to put Eq. (4.8.8) in the form

$$[u]_P = [uv]_R - \int_Q^R uv(a\,dy - b\,dx) - \int_Q^R (uv_x\,dx + vu_y\,dy) + \iint_D vf\,dx\,dy \tag{4.8.12}$$

By adding Eqs. (4.8.11) and (4.8.12), the value of $u$ at $P$ is given by

$$[u]_P = \frac{1}{2}([uv]_Q + [uv]_R) - \int_Q^R uv(a\,dy - b\,dx) - \frac{1}{2}\int_Q^R u(v_x\,dx - v_y\,dy)$$
$$+\frac{1}{2}\int_Q^R v(u_x\,dx - u_y\,dy) + \iint_D vf\,dx\,dy \tag{4.8.13}$$

which is the solution of the Cauchy problem in terms of the Cauchy data given along the curve $\Lambda$. It is easy to see that the solution at the point $(\alpha, \beta)$ depends only on the Cauchy data along the arc $QR$ on $\Lambda$. If the initial data were to change outside this arc $QR$, the solution would change only outside the triangle $PQR$. Thus, from Fig. 4.8.2, we can see that each characteristic separates the region in which the solution remains unchanged from the region in which it varies. Because of this fact, the unique continuation of the solution across any characteristic is not possible. This is evident from Fig. 4.8.2. The solution on the right of the characteristic $P_1R_1$ is determined by the initial data given on $Q_1R_2$, whereas the solution on the left is determined by the initial data given on $Q_1R_1$. If the initial data on $R_1R_2$ were changed, the solution on the right of $P_1R_1$ only will be affected.

It should be remarked here that the initial curve can intersect each characteristic at only one point. Suppose, for example, the initial curve $\Lambda$ intersects the characteristic at two points, as shown in Fig. 4.8.3. Then, the solution at $P$ obtained from the initial data on $QR$ will be different from the solution obtained from the initial data on $RS$. Hence, the Cauchy problem, in this case, is not solvable.

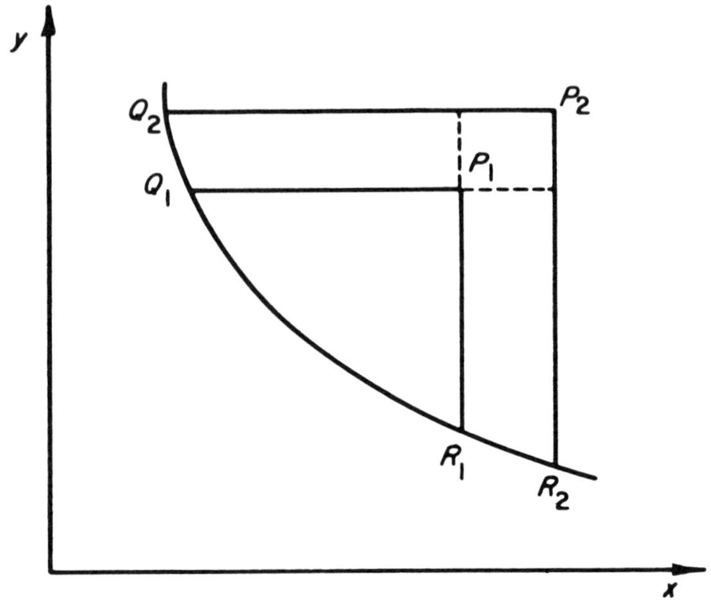


Figure 4.8.2

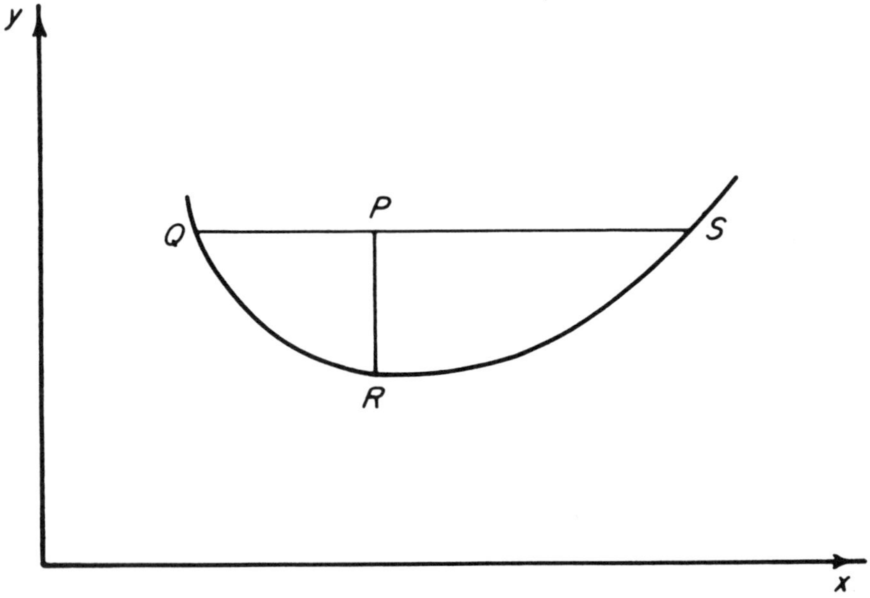


Figure 4.8.3

EXAMPLE 4.8.1. The telegraph equation

$$w_{tt} + a^* w_t + b^* w = c^2 w_{xx}$$

may be transformed into canonical form

$$L[u] = u_{\xi\eta} + ku = 0$$

by the successive transformations

$$w = u\, e^{-a^* t/2}$$

and

$$\xi = x + ct \qquad \eta = x - ct$$

where $k = (a^{*2} - 4b^*)/16c^2$.

We apply Riemann's method to determine the solution satisfying the initial conditions

$$u(x, 0) = f(x)$$
$$u_t(x, 0) = g(x)$$

Since

$$t = \frac{1}{2c}(\xi - \eta)$$

the line $t = 0$ corresponds to the straight line $\xi = \eta$ in the $\xi$–$\eta$ plane. The initial conditions may thus be transformed into

$$[u]_{\xi=\eta} = f(\xi) \tag{4.8.14}$$

$$[u_\xi - u_\eta]_{\xi=\eta} = c^{-1} g(\xi) \tag{4.8.15}$$

We shall now determine the Riemann function $v(\xi, \eta; \alpha, \beta)$ which satisfies

$$v_{\xi\eta} + kv = 0 \tag{4.8.16}$$

$$v_\xi(\xi, \beta; \alpha, \beta) = 0 \tag{4.8.17}$$

$$v_\eta(\alpha, \eta; \alpha, \beta) = 0 \tag{4.8.18}$$

$$v(\alpha, \beta; \alpha, \beta) = 1 \tag{4.8.19}$$

The differential equation (4.8.16) is self-adjoint, that is,

$$L[v] = M[v] = v_{\xi\eta} + kv$$

We assume that the Riemann function is of the form

$$v(\xi, \eta; \alpha, \beta) = F(s)$$

with the argument $s = (\xi - \alpha)(\eta - \beta)$. Substituting this value in Eq.

(4.8.16), we obtain

$$sF_{ss} + F_s + kF = 0$$

If we let $\lambda = \sqrt{4ks}$, the above equation becomes

$$F''(\lambda) + \frac{1}{\lambda} F'(\lambda) + F(\lambda) = 0$$

This is the Bessel equation of order zero, and the solution is

$$F(\lambda) = J_0(\lambda)$$

disregarding $Y_0(\lambda)$ which is unbounded at $\lambda = 0$. Thus, the Riemann function is

$$v(\xi, \eta; \alpha, \beta) = J_0(\sqrt{4k(\xi - \alpha)(\eta - \beta)})$$

which satisfies Eq. (4.8.16) and is equal to one on the characteristics $\xi = \alpha$ and $\eta = \beta$. Since $J_0'(0) = 0$, Eqs. (4.8.17) and (4.8.18) are satisfied. From this, it immediately follows that

$$[v_\xi]_{\xi=\eta} = \frac{\sqrt{k}\,(\xi - \beta)}{\sqrt{(\xi - \alpha)(\xi - \beta)}} [J_0'(\lambda)]_{\xi=\eta}$$

$$[v_\eta]_{\xi=\eta} = \frac{\sqrt{k}\,(\xi - \alpha)}{\sqrt{(\xi - \alpha)(\xi - \beta)}} [J_0'(\lambda)]_{\xi=\eta}$$

Thus, we have

$$[v_\xi - v_\eta]_{\xi=\eta} = \frac{\sqrt{k}\,(\alpha - \beta)}{\sqrt{(\xi - \alpha)(\xi - \beta)}} [J_0'(\lambda)]_{\xi=\eta} \tag{4.8.20}$$

From the initial condition

$$u(Q) = f(\beta) \quad \text{and} \quad u(R) = f(\alpha) \tag{4.8.21}$$

and substituting Eqs. (4.8.15), (4.8.19), and (4.8.20) into Eq. (4.8.13), we obtain

$$\begin{aligned} u(\alpha, \beta) = {} & \frac{1}{2}[f(\alpha) + f(\beta)] \\ & - \frac{1}{2}\int_\beta^\alpha \frac{\sqrt{k}\,(\alpha - \beta)}{\sqrt{(\tau - \alpha)(\tau - \beta)}} J_0'[\sqrt{4k(\tau - \alpha)(\tau - \beta)}] f(\tau)\, d\tau \\ & + \frac{1}{2c}\int_\beta^\alpha J_0[\sqrt{4k(\tau - \alpha)(\tau - \beta)}\,] g(\tau)\, d\tau \end{aligned} \tag{4.8.22}$$

Replacing $\alpha$ and $\beta$ by $\xi$ and $\eta$, and substituting the original variables $x$ and $t$, we obtain

$$u(x, t) = \frac{1}{2}[f(x + ct) + f(x - ct)] + \frac{1}{2}\int_{x-ct}^{x+ct} G(x, t, \tau)\, d\tau \tag{4.8.23}$$

where

$$G(x, t, \tau) = \{-2\sqrt{k}\, ctf(\tau)J_0[\sqrt{4k[(\tau-x)^2 - c^2t^2]}\,]\}/\sqrt{(\tau-x)^2 - c^2t^2}$$
$$+ c^{-1}g(\tau)J_0\{\sqrt{4k[(\tau-x)^2 - c^2t^2]}\,\}$$

If we set $k = 0$, we arrive at the d'Alembert formula for the wave equation

$$u(x, t) = \frac{1}{2}[f(x+ct) + f(x-ct)] + \frac{1}{2c}\int_{x-ct}^{x+ct} g(\tau)\, d\tau$$

## 4.9 GOURSAT PROBLEM

The Goursat problem is that of finding the solution of a linear hyperbolic equation

$$u_{xy} = a_1(x, y)u_x + a_2(x, y)u_y + a_3(x, y)u + h(x, y) \tag{4.9.1}$$

satisfying the prescribed conditions

$$u(x, y) = f(x) \tag{4.9.2}$$

on a characteristic, say, $y = 0$, and

$$u(x, y) = g(x) \tag{4.9.3}$$

on a monotonic increasing curve $y = y(x)$ which, for simplicity, is assumed to intersect the characteristic at the origin.

The solution in the region between the $x$-axis and the monotonic curve in the first quadrant can be determined by the method of successive approximations. The proof is given in Garabedian [15].

EXAMPLE 4.9.1. Determine the solution of the Goursat problem

$$u_{tt} = c^2 u_{xx} \tag{4.9.4}$$
$$u(x, t) = f(x) \quad \text{on} \quad x - ct = 0 \tag{4.9.5}$$
$$u(x, t) = g(x) \quad \text{on} \quad t = t(x) \tag{4.9.6}$$

where $f(0) = g(0)$.

The general solution of the wave equation is

$$u(x, t) = \phi(x+ct) + \psi(x-ct)$$

Applying the prescribed conditions, we obtain

$$f(x) = \phi(2x) + \psi(0) \tag{4.9.7}$$
$$g(x) = \phi(x + ct(x)) + \psi(x - ct(x)) \tag{4.9.8}$$

It is evident that

$$f(0) = \phi(0) + \psi(0) = g(0)$$

Now, if $s = x - ct(x)$, the inverse of it is $x = \alpha(s)$. Thus, Eq. (4.9.8) may be written as

$$g(\alpha(s)) = \phi(x + ct(x)) + \psi(s) \tag{4.9.9}$$

Replacing $x$ by $(x + ct(x))/2$ in Eq. (4.9.7), we obtain

$$f\left(\frac{x + ct(x)}{2}\right) = \phi(x + ct(x)) + \psi(0) \tag{4.9.10}$$

Thus, using (4.9.10), (4.9.9) becomes

$$\psi(s) = g(\alpha(s)) - f\left(\frac{\alpha(s) + ct(\alpha(s))}{2}\right) + \psi(0)$$

Replacing $s$ by $x - ct$, we have

$$\psi(x - ct) = g(\alpha(x - ct)) - f\left(\frac{\alpha(x - ct) + ct(\alpha(x - ct))}{2}\right) + \psi(0)$$

Hence, the solution is given by

$$u(x, t) = f\left(\frac{x + ct}{2}\right) - f\left(\frac{\alpha(x - ct) + ct(\alpha(x - ct))}{2}\right) + g(\alpha(x - ct)) \tag{4.9.11}$$

Let us consider a special case when the curve $t = t(x)$ is a straight line represented by $t - kx = 0$ with a constant $k > 0$. Then $s = x - ckx$ and hence $x = s/(1 - ck)$. Using these values in (4.9.11), we obtain

$$u(x, t) = f\left(\frac{x + ct}{2}\right) - f\left(\frac{(1 + ck)(x - ct)}{2(1 - ck)}\right) + g\left(\frac{x - ct}{1 - ck}\right) \tag{4.9.12}$$

When the values of $u$ are prescribed on both the characteristics, the problem of finding $u$ of a linear hyperbolic equation is called a *characteristic initial value problem*. This is the degenerate case of the Goursat problem.

Consider the characteristic initial value problem

$$u_{xy} = h(x, y) \tag{4.9.13}$$

$$u(x, 0) = f(x) \tag{4.9.14}$$

$$u(0, y) = g(y) \tag{4.9.15}$$

where $f$ and $g$ are continuously differentiable and $f(0) = g(0)$.

Integrating Eq. (4.9.13), we obtain

$$u(x, y) = \int_0^x \int_0^y h(\xi, \eta)\, d\eta\, d\xi + \phi(x) + \psi(y) \tag{4.9.16}$$

where $\phi$ and $\psi$ are arbitrary functions. Applying the prescribed conditions (4.9.14) and (4.9.15), we have

$$u(x, 0) = \phi(x) + \psi(0) = f(x) \tag{4.9.17}$$

$$u(0, y) = \phi(0) + \psi(y) = g(y) \tag{4.9.18}$$

Thus

$$\phi(x) + \psi(y) = f(x) + g(y) - \phi(0) - \psi(0) \tag{4.9.19}$$

But from (4.9.17), we have

$$\phi(0) + \psi(0) = f(0) \tag{4.9.20}$$

Hence, from (4.9.16), (4.9.19), and (4.9.20), we obtain

$$u(x, y) = f(x) + g(y) - f(0) + \int_0^x \int_0^y h(\xi, \eta)\, d\eta\, d\xi \tag{4.9.21}$$

EXAMPLE 4.9.2. Determine the solution of the characteristic initial value problem

$$u_{tt} = c^2 u_{xx}$$

$$u(x, t) = f(x) \quad \text{on} \quad x + ct = 0$$

$$u(x, t) = g(x) \quad \text{on} \quad x - ct = 0$$

where $f(0) = g(0)$.

Here it is not necessary to reduce the given equation into canonical form. The general solution of the wave equation is

$$u(x, t) = \phi(x + ct) + \psi(x - ct)$$

The characteristics are

$$x + ct = 0$$

$$x - ct = 0$$

Applying the prescribed conditions, we have

$$u(x, t) = \phi(2x) + \psi(0) = f(x) \quad \text{on} \quad x + ct = 0 \tag{4.9.22}$$

$$u(x, t) = \phi(0) + \psi(2x) = g(x) \quad \text{on} \quad x - ct = 0 \tag{4.9.23}$$

We observe that these equations are compatible, since $f(0) = g(0)$.

Now, replacing $x$ by $(x + ct)/2$ in Eq. (4.9.22) and replacing $x$ by $(x - ct)/2$ in Eq. (4.9.23), we have

$$\phi(x + ct) = f\left(\frac{x + ct}{2}\right) - \psi(0)$$

$$\psi(x - ct) = g\left(\frac{x - ct}{2}\right) - \phi(0)$$

Hence the solution is given by

$$u(x, t) = f\left(\frac{x+ct}{2}\right) + g\left(\frac{x-ct}{2}\right) - f(0) \tag{4.9.24}$$

We note that this solution can be obtained by substituting $k = -1/c$ into (4.9.12).

EXAMPLE 4.9.3. Find the solution of the characteristic initial value problem

$$y^3 u_{xx} - y u_{yy} + u_y = 0 \tag{4.9.25}$$

$$u(x, y) = f(x) \quad \text{on} \quad x + \frac{y^2}{2} = 4 \quad \text{for} \quad 2 \leqslant x \leqslant 4$$

$$u(x, y) = g(x) \quad \text{on} \quad x - \frac{y^2}{2} = 0 \quad \text{for} \quad 0 \leqslant x \leqslant 2$$

with $f(2) = g(2)$.

Since the equation is hyperbolic except for $y = 0$, we reduce it to the canonical form

$$u_{\xi\eta} = 0$$

where $\xi = x + y^2/2$ and $\eta = x - y^2/2$. Thus, the general solution is

$$u(x, y) = \phi\left(x + \frac{y^2}{2}\right) + \psi\left(x - \frac{y^2}{2}\right) \tag{4.9.26}$$

Applying the prescribed conditions, we have

$$f(x) = \phi(4) + \psi(2x - 4) \tag{4.9.27}$$

$$g(x) = \phi(2x) + \psi(0) \tag{4.9.28}$$

Now, if we replace $(2x - 4)$ by $(x - y^2/2)$ in (4.9.27) and $(2x)$ by $(x + y^2/2)$ in (4.9.28), we obtain

$$\psi\left(x - \frac{y^2}{2}\right) = f\left(\frac{x}{2} - \frac{y^2}{4} + 2\right) - \phi(4)$$

$$\phi\left(x + \frac{y^2}{2}\right) = g\left(\frac{x}{2} + \frac{y^2}{4}\right) - \psi(0)$$

Thus

$$u(x, y) = f\left(\frac{x}{2} - \frac{y^2}{4} + 2\right) + g\left(\frac{x}{2} + \frac{y^2}{4}\right) - \phi(4) - \psi(0)$$

But from (4.9.27) and (4.9.28), we see that

$$f(2) = \phi(4) + \psi(0) = g(2)$$

Hence

$$u(x, y) = f\left(\frac{x}{2} - \frac{y^2}{4} + 2\right) + g\left(\frac{x}{2} + \frac{y^2}{4}\right) - f(2)$$

## 4.10. SPHERICAL WAVE EQUATION

In spherical polar coordinates $(r, \theta, \phi)$, the wave equation (2.1.1) takes the form

$$\frac{1}{r^2}\frac{\partial}{\partial r}\left(r^2\frac{\partial u}{\partial r}\right) + \frac{1}{r^2 \sin\theta}\frac{\partial}{\partial\theta}\left(\sin\theta\frac{\partial u}{\partial\theta}\right) + \frac{1}{r^2\sin^2\theta}\frac{\partial^2 u}{\partial\phi^2} = \frac{1}{c^2}\frac{\partial^2 u}{\partial t^2} \quad (4.10.1)$$

Solutions of this equation are called *spherical symmetric waves* if $u$ depends on $r$ and $t$. Thus the solution $u = u(r, t)$ which satisfies the wave equation with spherical symmetry in three dimensional space is

$$\frac{1}{r^2}\frac{\partial}{\partial r}\left(r^2\frac{\partial u}{\partial r}\right) = \frac{1}{c^2}\frac{\partial^2 u}{\partial t^2} \quad (4.10.2)$$

Introducing a new dependent variable $U = ru(r, t)$, this equation reduces to a simple form

$$U_{tt} = c^2 U_{rr} \quad (4.10.3)$$

This is identical with the one dimensional wave equation (4.3.1) and has the general solution in the form

$$U(r, t) = \phi(r + ct) + \psi(r - ct) \quad (4.10.4)$$

or

$$u(r, t) = \frac{1}{r}[\phi(r + ct) + \psi(r - ct)] \quad (4.10.5)$$

This solution consists of two progressive spherical waves traveling with constant velocity $c$. The terms involving $\phi$ and $\psi$ represent the incoming waves to the origin and the outgoing waves from the origin respectively.

Physically, the solution for only outgoing waves generated by a source is of most interest, and has the form

$$u(r, t) = \frac{1}{r}\psi(r - ct) \quad (4.10.6)$$

where the explicit form of $\psi$ is to be determined from the properties of the source.

In the context of fluid flows, $u$ represents the velocity potential so that the limiting total flux through a sphere of center at the origin and radius $r$ is

$$Q(t) = \lim_{r \to 0} 4\pi r^2 u_r = -4\pi\psi(-ct) \quad (4.10.7)$$

In physical terms, we say that there is a simple (or monopole) point source of strength $Q(t)$ located at the origin. Thus the solution (4.10.6) can be expressed in terms of $Q$ as

$$u(r, t) = -\frac{1}{4\pi r} Q\left(t - \frac{r}{c}\right) \tag{4.10.8}$$

This represents the velocity potential of the point source, and $u_r$ is called the *radial velocity*. In fluid flows, the difference between the pressure at any time $t$ and the equilibrium value is given by

$$p - p_0 = \rho u_t = -\frac{\rho}{4\pi r} \dot{Q}\left(t - \frac{r}{c}\right) \tag{4.10.9}$$

where $\rho$ is the density of the fluid.

Following an analysis similar to § 4.3, the solution of the initial value problem with the initial data

$$u(r, 0) = f(r), \qquad u_t(r, 0) = g(r), \qquad r \geqslant 0 \tag{4.10.10ab}$$

where $f$ and $g$ are continuously differentiable, is given by

$$u(r, t) = \frac{1}{2r}\left[(r + ct)f(r + ct) + (r - ct)f(r - ct) + \frac{1}{c}\int_{r-ct}^{r+ct} \tau g(\tau)\, d\tau\right] \tag{4.10.11}$$

provided $r \geqslant ct$. However, when $r < ct$ this solution fails because $f$ and $g$ are not defined for $r < 0$. This initial data at $t = 0$, $r \geqslant 0$ determine the solution $u(r, t)$ only up to the characteristic $r = ct$ in the $r - t$ plane. To find $u$ for $r < ct$, we require $u$ to be finite at $r = 0$ for all $t \geqslant 0$, that is, $U = 0$ at $r = 0$. Thus the solution for $U$ is

$$U(r, t) = \frac{1}{2}\left[(r + ct)f(r + ct) + (r - ct)f(r - ct) + \frac{1}{c}\int_{r-ct}^{r+ct} \tau g(\tau)\, d\tau\right] \tag{4.10.12}$$

provided $r \geqslant ct \geqslant 0$, and

$$U(r, t) = \frac{1}{2}[\phi(ct + r) + \psi(ct - r)], \qquad ct \geqslant r \geqslant 0 \tag{4.10.13}$$

where

$$\phi(ct) + \psi(ct) = 0 \quad \text{for} \quad ct \geqslant 0 \tag{4.10.14}$$

In view of the fact that $U_r + \frac{1}{c} U_t$ is constant on each characteristic $r + ct = \text{constant}$, it turns out that

$$\phi'(ct + r) = (r + ct)f'(r + ct) + f(r + ct) + \frac{1}{c}(r + ct)g(r + ct)$$

or

$$\phi'(ct) = ctf'(ct) + f(ct) + tg(ct)$$

Integration gives

$$\phi(t) = tf(t) + \frac{1}{c}\int_0^t \tau g(\tau)\, d\tau + \phi(0)$$

so that

$$\psi(t) = -tf(t) - \frac{1}{c}\int_0^t \tau g(\tau)\, d\tau - \phi(0)$$

Substituting these values into (4.10.13) and using $U(r, t) = ru(r, t)$, we obtain, for $ct > r$,

$$u(r, t) = \frac{1}{2r}\left[(ct + r)f(ct + r) - (ct - r)f(ct - r) + \frac{1}{c}\int_{ct-r}^{ct+r} \tau g(\tau)\, d\tau\right] \tag{4.10.15}$$

## 4.11 CYLINDRICAL WAVE EQUATION

In cylindrical polar coordinates $(R, \theta, z)$, the wave equation (2.1.1) assumes the form

$$u_{RR} + \frac{1}{R}u_R + \frac{1}{R^2}u_{\theta\theta} + u_{zz} = \frac{1}{c^2}u_{tt} \tag{4.11.1}$$

If $u$ depends only on $R$ and $t$, this equation becomes

$$u_{RR} + \frac{1}{R}u_R = \frac{1}{c^2}u_{tt} \tag{4.11.2}$$

Solutions of (4.11.2) are called *cylindrical waves.*

In general, it is not easy to find the solution of (4.11.1). However, we shall solve this equation by using the method of separation of variables in Chapter 6. Here we derive the solution for outgoing cylindrical waves from the spherical wave solution (4.10.8). We assume that sources of constant strength $Q(t)$ per unit length are distributed uniformly on the $z$-axis. The solution for the cylindrical waves produced by the line source is given by the total disturbance

$$u(R, t) = -\frac{1}{4\pi}\int_{-\infty}^{\infty} \frac{1}{r}Q\left(t - \frac{r}{c}\right) dz = -\frac{1}{2\pi}\int_0^{\infty} \frac{1}{r}Q\left(t - \frac{r}{c}\right) dz \tag{4.11.3}$$

where $R$ is the distance from the $z$-axis so that $R^2 = (r^2 - z^2)$.

Substitution of $z = R \sinh \xi$ and $r = R \cosh \xi$ in (4.11.3) gives

$$u(R, t) = -\frac{1}{2\pi}\int_0^\infty Q\left(t - \frac{R}{c}\cosh \xi\right) d\xi \qquad (4.11.4)$$

This is usually considered as the cylindrical wave function due to a source of strength $Q(t)$ at $R = 0$. It follows from (4.11.4) that

$$u_{tt} = -\frac{1}{2\pi}\int_0^\infty Q''\left(t - \frac{R}{c}\cosh \xi\right) d\xi \qquad (4.11.5)$$

$$u_R = \frac{1}{2\pi c}\int_0^\infty \cosh \xi \, Q'\left(t - \frac{R}{c}\cosh \xi\right) d\xi \qquad (4.11.6)$$

$$u_{RR} = -\frac{1}{2\pi c^2}\int_0^\infty \cosh^2 \xi \, Q''\left(t - \frac{R}{c}\cosh \xi\right) d\xi \qquad (4.11.7)$$

which give

$$c^2\left(u_{RR} + \frac{1}{R}u_R\right) - u_{tt} = \frac{1}{2\pi}\int_0^\infty \frac{d}{d\xi}\left[\frac{c}{R}Q'\left(t - \frac{R}{c}\cosh \xi\right)\sinh \xi\right] d\xi$$

$$= \lim_{\xi \to \infty}\left[\frac{c}{2\pi R}Q'\left(t - \frac{R}{c}\cosh \xi\right)\sinh \xi\right] = 0$$

provided the differentiation under the sign of integration is justified and the above limit is zero. This means that $u(R, t)$ satisfies the cylindrical wave equation (4.11.2).

In order to find the asymptotic behavior of the solution as $R \to 0$, we substitute $\cosh \xi = \dfrac{c(t - \zeta)}{R}$ into (4.11.4) and (4.11.6) to obtain

$$u = -\frac{1}{2\pi}\int_{-\infty}^{t - R/c} \frac{Q(\zeta)\, d\zeta}{\left[(t - \zeta)^2 - \dfrac{R^2}{c^2}\right]^{\frac{1}{2}}} \qquad (4.11.8)$$

$$u_R = \frac{1}{2\pi}\int_{-\infty}^{t - R/c} \left(\frac{t - \zeta}{R}\right) \frac{Q'(\zeta)\, d\zeta}{\left[(t - \zeta)^2 - \dfrac{R^2}{c^2}\right]^{\frac{1}{2}}} \qquad (4.11.9)$$

which, in the limit $R \to 0$, give

$$u_R \sim \frac{1}{2\pi R}\int_{-\infty}^{t} Q'(\zeta)\, d\zeta = \frac{1}{2\pi R}Q(t) \qquad (4.11.10)$$

This leads to the result

$$\lim_{R \to 0} 2\pi R\, u_R = Q(t) \qquad (4.11.11)$$

or

$$u(R, t) \sim \frac{1}{2\pi} Q(t) \log R \quad \text{as} \quad R \to 0. \tag{4.11.12}$$

We next investigate the nature of the cylindrical wave solution near the wavefront ($R = ct$) and in the far field ($R \to \infty$). We assume $Q(t) = 0$ for $t < 0$ so that the lower limit of integration in (4.11.8) may be taken to be zero, and the solution is non-zero for $\tau = t - \frac{R}{c} > 0$ where $\tau$ is the time passed after the arrival of the wavefront. Consequently, (4.11.8) becomes

$$u(R, t) = -\frac{1}{2\pi} \int_0^{\tau} \frac{Q(\zeta)\, d\zeta}{\left[(\tau - \zeta)\left(\tau - \zeta + \frac{2R}{c}\right)\right]^{\frac{1}{2}}} \tag{4.11.13}$$

Since $0 < \zeta < \tau$, $\frac{2R}{c} > \frac{R}{c} > \tau > \tau - \zeta > 0$, so that the second factor under the radical is approximately equal to $\frac{2R}{c}$ when $R \gg c\tau$, and hence

$$u(R, t) \sim -\frac{1}{2\pi} \left(\frac{c}{2R}\right)^{\frac{1}{2}} \int_0^{\tau} \frac{Q(\zeta)\, d\zeta}{(\tau - \zeta)^{\frac{1}{2}}} = -\left(\frac{c}{2R}\right)^{\frac{1}{2}} q(\tau)$$
$$= -\left(\frac{c}{2R}\right)^{\frac{1}{2}} q\left(t - \frac{R}{c}\right), \qquad R \gg \frac{ct}{2} \tag{4.11.14}$$

where

$$q(\tau) = \frac{1}{2\pi} \int_0^{\tau} \frac{Q(\zeta)\, d\zeta}{\sqrt{\tau - \zeta}} \tag{4.11.15}$$

Evidently, the amplitude involved in the solution (4.11.14) decays like $R^{-\frac{1}{2}}$ for large $R$ ($R \to \infty$).

EXAMPLE 4.11.1. Determine the asymptotic form of the solution (4.11.4) for a harmonically oscillating source of frequency $\omega$.

We take the source in the form $Q(t) = q_0 \exp[-i(\omega + i\varepsilon)t]$ where $\varepsilon$ is positive and small so that $Q(t) \to 0$ as $t \to -\infty$. The small imaginary part $\varepsilon$ of $\omega$ will make insignificant contributions to the solution at finite times as $\varepsilon \to 0$. Thus the solution (4.11.4) becomes

$$u(R, t) = -\frac{q_0}{2\pi} e^{-i\omega t} \int_0^{\infty} \exp\left[\frac{i\omega R}{c} \cosh \xi\right] d\xi$$
$$= -\frac{iq_0}{4} e^{-i\omega t} H_0^{(1)}\left(\frac{\omega R}{c}\right) \tag{4.11.16}$$

where $H_0^{(1)}(z)$ is the Hankel function given by

$$H_0^{(1)}(z) = \frac{2}{\pi i}\int_0^\infty \exp(iz\cosh\xi)\,d\xi \tag{4.11.17}$$

In view of the asymptotic expansion of $H_0^{(1)}(z)$ in the form

$$H_0^{(1)}(z) \sim \left(\frac{2}{\pi z}\right)^{\frac{1}{2}} \exp\left[i\left(z - \frac{\pi}{4}\right)\right], \qquad z \to \infty \tag{4.11.18}$$

the asymptotic solution for $u(R, t)$ in the limit $\frac{\omega R}{c} \to \infty$ is

$$u(R, t) \sim -\frac{iq_0}{4}\left(\frac{2c}{\pi\omega R}\right)^{\frac{1}{2}} \exp\left[-i\left\{\omega t - \frac{\omega R}{c} - \frac{\pi}{4}\right\}\right]$$

This represents the cylindrical wave propagating with constant velocity $c$. The amplitude of the wave decays like $R^{-\frac{1}{2}}$ as $R \to \infty$.

EXAMPLE 4.11.2. For a supersonic flow ($M > 1$) past a solid body of revolution, the perturbation potential $\Phi$ satisfies the cylindrical wave equation

$$\Phi_{RR} + \frac{1}{R}\Phi_R = N^2\Phi_{xx}, \qquad N^2 = M^2 - 1$$

where $R$ is the distance from the path of the moving body and $x$ is the distance from the nose of the body.

It follows from problem 12 in Exercises 2.7 that $\Phi$ satisfies the equation

$$\Phi_{yy} + \Phi_{zz} = N^2\Phi_{xx}$$

This represents a two-dimensional wave equation with $x \leftrightarrow t$ and $N^2 \leftrightarrow \frac{1}{c^2}$. For a body of revolution with $(y, z) \leftrightarrow (R, \theta)$, $\frac{\partial}{\partial\theta} \equiv 0$, the above equation becomes

$$\Phi_{RR} + \frac{1}{R}\Phi_R = N^2\Phi_{xx}$$

## 4.12 EXERCISES

**1.** Determine the solution of each of the following initial-value problems:

a. $u_{tt} - c^2u_{xx} = 0 \qquad u(x, 0) = 0 \qquad u_t(x, 0) = 1$

b. $u_{tt} - c^2u_{xx} = 0 \qquad u(x, 0) = \sin x \qquad u_t(x, 0) = x^2$

c. $u_{tt} - c^2u_{xx} = 0 \qquad u(x, 0) = x^3 \qquad u_t(x, 0) = x$

d. $u_{tt} - c^2u_{xx} = 0 \qquad u(x, 0) = \cos x \qquad u_t(x, 0) = e^{-1}$

e. $u_{tt} - c^2 u_{xx} = 0 \qquad u(x, 0) = \log(1 + x^2) \qquad u_t(x, 0) = 2$

f. $u_{tt} - c^2 u_{xx} = 0 \qquad u(x, 0) = x \qquad u_t(x, 0) = \sin x$

**2.** Determine the solution of each of the following initial-value problems:

a. $u_{tt} - c^2 u_{xx} = x \qquad u(x, 0) = 0 \qquad u_t(x, 0) = 3$

b. $u_{tt} - c^2 u_{xx} = x + ct \qquad u(x, 0) = x \qquad u_t(x, 0) = \sin x$

c. $u_{tt} - c^2 u_{xx} = e^x \qquad u(x, 0) = 5 \qquad u_t(x, 0) = x^2$

d. $u_{tt} - c^2 u_{xx} = \sin x \qquad u(x, 0) = \cos x \qquad u_t(x, 0) = 1 + x$

e. $u_{tt} - c^2 u_{xx} = xe^t \qquad u(x, 0) = \sin x \qquad u_t(x, 0) = 0$

f. $u_{tt} - c^2 u_{xx} = 2 \qquad u(x, 0) = x^2 \qquad u_t(x, 0) = \cos x$

**3.** A gas which is contained in a sphere of radius $R$ is at rest initially, and the initial condensation is given by $s_0$ inside the sphere and zero outside the sphere. The condensation is related to the velocity potential by

$$s = (1/c^2)u_t$$

at all times, and the velocity potential satisfies

$$u_{tt} = \nabla^2 u$$

Determine the condensation $s$ for all $t > 0$.

**4.** Solve the initial value problem

$$u_{xx} + 2u_{xy} - 3u_{yy} = 0$$
$$u(x, 0) = \sin x$$
$$u_y(x, 0) = x$$

**5.** Find the longitudinal oscillation of a rod subject to the initial conditions

$$u(x, 0) = \sin x$$
$$u_t(x, 0) = x$$

**6.** By the Riemann method, solve

a. $\sin^2 \mu \, \phi_{xx} - \cos^2 \mu \, \phi_{yy} - (\lambda^2 \sin^2 \mu \cos^2 \mu)\phi = 0$

$\phi(0, y) = f_1(y) \qquad \phi(x, 0) = g_1(x)$

$\phi_x(0, y) = f_2(y) \qquad \phi_y(x, 0) = g_2(x)$

b. $x^2 u_{xx} - t^2 u_{tt} = 0$

$u(x, t_1) = f(x)$

$u_t(x, t_2) = g(x)$

**7.** Determine the solution of the initial-boundary value problem

$$\begin{aligned} u_{tt} &= 4u_{xx}, && 0<x<\infty, && t>0 \\ u(x,0) &= x^4, && 0\leq x<\infty && \\ u_t(x,0) &= 0, && 0\leq x<\infty && \\ u(0,t) &= 0, && && t\geq 0 \end{aligned}$$

**8.** Determine the solution of the initial-boundary value problem

$$\begin{aligned} u_{tt} &= 9u_{xx}, && 0<x<\infty, && t>0 \\ u(x,0) &= 0, && 0\leq x<\infty && \\ u_t(x,0) &= x^3, && 0\leq x<\infty && \\ u_x(0,t) &= 0, && && t\geq 0 \end{aligned}$$

**9.** Determine the solution of the initial-boundary value problem

$$\begin{aligned} u_{tt} &= 16u_{xx}, && 0<x<\infty, && t>0 \\ u(x,0) &= \sin x, && 0\leq x<\infty && \\ u_t(x,0) &= x^2, && 0\leq x<\infty && \\ u(0,t) &= 0, && && t\geq 0 \end{aligned}$$

**10.** In the initial-boundary value problem

$$\begin{aligned} u_{tt} &= c^2u_{xx}, && 0<x<l, && t>0 \\ u(x,0) &= f(x), && 0\leq x\leq l && \\ u_t(x,0) &= g(x), && 0\leq x\leq l && \\ u(0,t) &= 0, && && t\geq 0 \end{aligned}$$

if $f$ and $g$ are extended as the odd functions, show that $u(x, t)$ is given by (4.4.5) for $x > ct$ and (4.4.6) for $x < ct$.

**11.** In the initial-boundary value problem

$$\begin{aligned} u_{tt} &= c^2u_{xx}, && 0<x<l, && t>0 \\ u(x,0) &= f(x), && 0\leq x\leq l && \\ u_t(x,0) &= g(x), && 0\leq x\leq l && \\ u_x(0,t) &= 0, && && t\geq 0 \end{aligned}$$

if $f$ and $g$ are extended as the even functions, show that $u(x, t)$ is given by (4.4.8) for $x > ct$ and (4.4.9) for $x < ct$.

**12.** Determine the solution of the initial-boundary value problem

$$\begin{aligned} &u_{tt} = c^2 u_{xx}, && 0 < x < \infty, && t > 0 \\ &u(x, 0) = f(x), && 0 \leqslant x < \infty \\ &u_t(x, 0) = 0, && 0 \leqslant x < \infty \\ &u_x(0, t) + hu(0, t) = 0, && && t \geqslant 0,\ h = \text{constant} \end{aligned}$$

State the compatibility condition of $f$.

**13.** Find the solution of the problem

$$\begin{aligned} &u_{tt} = c^2 u_{xx}, && at < x < \infty, && t > 0 \\ &u(x, 0) = f(x), && 0 < x < \infty \\ &u_t(x, 0) = 0, && 0 < x < \infty \\ &u(at, t) = 0, && && t > 0 \end{aligned}$$

where $f(0) = 0$ and $a$ is constant.

**14.** Find the solution of the initial-boundary value problem

$$\begin{aligned} &u_{tt} = u_{xx}, && 0 < x < 2, && t > 0 \\ &u(x, 0) = \sin \pi x/2, && 0 \leqslant x \leqslant 2 \\ &u_t(x, 0) = 0, && 0 \leqslant x \leqslant 2 \\ &u(0, t) = 0, && && t \geqslant 0 \\ &u(2, t) = 0, && && t \geqslant 0. \end{aligned}$$

**15.** Find the solution of the initial-boundary value problem

$$\begin{aligned} &u_{tt} = 4u_{xx}, && 0 < x < 1, && t > 0 \\ &u(x, 0) = 0, && 0 \leqslant x \leqslant 1 \\ &u_t(x, 0) = x(1 - x), && 0 \leqslant x \leqslant 1 \\ &u(0, t) = 0, && && t \geqslant 0 \\ &u(1, t) = 0, && && t \geqslant 0 \end{aligned}$$

**16.** Determine the solution of the initial-boundary value problem

$$\begin{aligned} &u_{tt} = c^2 u_{xx}, && 0 < x < l, && t > 0 \\ &u(x, 0) = f(x), && 0 \leqslant x \leqslant l \\ &u_t(x, 0) = g(x), && 0 \leqslant x \leqslant l \\ &u_x(0, t) = 0, && && t \geqslant 0 \\ &u_x(l, t) = 0, && && t \geqslant 0 \end{aligned}$$

by extending $f$ and $g$ as the even functions about $x = 0$ and $x = l$.

**17.** Determine the solution of the initial-boundary value problem

$$\begin{aligned} u_{tt} &= c^2 u_{xx}, & 0<x<l, \quad & t>0 \\ u(x,0) &= f(x), & 0\leqslant x\leqslant l & \\ u_t(x,0) &= g(x), & 0\leqslant x\leqslant l & \\ u(0,t) &= p(t), & & t\geqslant 0 \\ u(l,t) &= q(t), & & t\geqslant 0 \end{aligned}$$

**18.** Determine the solution of the initial-boundary value problem

$$\begin{aligned} u_{tt} &= c^2 u_{xx}, & 0<x<l, \quad & t>0 \\ u(x,0) &= f(x), & 0\leqslant x\leqslant l & \\ u_t(x,0) &= g(x), & 0\leqslant x\leqslant l & \\ u_x(0,t) &= p(t), & & t\geqslant 0 \\ u_x(l,t) &= q(t), & & t\geqslant 0 \end{aligned}$$

**19.** Solve the characteristic initial value problem

$$xy^3 u_{xx} - x^3 y u_{yy} - y^3 u_x + x^3 u_y = 0$$

$$u(x,y) = f(x) \quad \text{on} \quad y^2 - x^2 = 8 \quad \text{for} \quad 0\leqslant x\leqslant 2$$

$$u(x,y) = g(x) \quad \text{on} \quad y^2 + x^2 = 16 \quad \text{for} \quad 2\leqslant x\leqslant 4$$

with $f(2) = g(2)$.

**20.** Solve the Goursat problem

$$xy^3 u_{xx} - x^3 y u_{yy} - y^3 u_x + x^3 u_y = 0$$

$$u(x,y) = f(x) \quad \text{on} \quad y^2 + x^2 = 16 \quad \text{for} \quad 0\leqslant x\leqslant 4$$

$$u(x,y) = f(y) \quad \text{on} \quad x = 0 \quad \text{for} \quad 0\leqslant y\leqslant 4$$

where $f(0) = g(4)$.

**21.** Solve

$$u_{tt} = c^2 u_{xx}$$

$$u(x,t) = f(x) \quad \text{on} \quad t = t(x)$$

$$u(x,t) = g(x) \quad \text{on} \quad x + ct = 0$$

where $f(0) = g(0)$.

**22.** Solve the characteristic initial value problem

$$x u_{xx} - x^3 u_{yy} - u_x = 0, \quad x \neq 0$$

$$u(x,y) = f(y) \quad \text{on} \quad y - \frac{x^2}{2} = 0 \quad \text{for} \quad 0\leqslant y\leqslant 2$$

$$u(x,y) = g(y) \quad \text{on} \quad y + \frac{x^2}{2} = 4 \quad \text{for} \quad 2\leqslant y\leqslant 4$$

where $f(2) = g(2)$.

**23.** Solve

$$u_{xx} + 10u_{xy} + 9u_{yy} = 0$$
$$u(x, 0) = f(x)$$
$$u_y(x, 0) = g(x)$$

**24.** Solve

$$4u_{xx} + 5u_{xy} + u_{yy} + u_x + u_y = 2$$
$$u(x, 0) = f(g)$$
$$u_y(x, 0) = g(x)$$

**25.** Solve

$$3u_{xx} + 10u_{xy} + 3u_{yy} = 0$$
$$u(x, 0) = f(x)$$
$$u_y(x, 0) = g(x)$$

**26.** Solve

$$u_{xx} - 3u_{xy} + 2u_{yy} = 0$$
$$u(x, 0) = f(x)$$
$$u_y(x, 0) = g(x)$$

**27.** Solve

$$x^2u_{xx} - t^2u_{tt} = 0 \quad x > 0, \quad t > 0$$
$$u(x, 1) = f(x)$$
$$u_t(x, 1) = g(x)$$

**28.** Consider an initial-boundary value problem for a string of length $l$ under the action of an external force $q(x, t)$ per unit length. The displacement $u(x, t)$ satisfies the wave equation

$$\rho u_{tt} = Tu_{xx} + \rho q(x, t)$$

where $\rho$ is the line density of the string and $T$ is a constant tension of the string. The initial and boundary conditions of the problem are

$$u(x, 0) = f(x), \qquad u_t(x, 0) = g(x), \qquad 0 \leqslant x \leqslant l$$
$$u(0, t) = u(l, t) = 0, \qquad t > 0$$

Show that the energy equation is

$$\frac{dE}{dt} = [Tu_xu_t]_0^l + \int_0^l \rho q u_t \, dx$$

where $E$ represents the energy integral

$$E(t) = \frac{1}{2} \int_0^l (\rho u_t^2 + T u_x^2) \, dx$$

Explain the physical significance of the energy equation.

Hence or otherwise, derive the principle of conservation of energy, that is, that the total energy is constant for all $t \geqslant 0$ provided the string has free or fixed ends and there are no external forces.

**29.** Show that the solution of the signalling problem governed by the wave equation

$$u_{tt} = c^2 u_{xx}, \qquad x > 0, \qquad t > 0$$
$$u(x, 0) = u_t(x, 0) = 0, \qquad x > 0$$
$$u(0, t) = U(t), \qquad t > 0$$

is

$$u(x, t) = U\left(t - \frac{x}{c}\right) H\left(t - \frac{x}{c}\right)$$

where $H(t)$ is the Heaviside unit step function.

CHAPTER FIVE

# Fourier Series and Integrals

This chapter is devoted to the theory of Fourier series and integrals. Although the treatment can be extensive, the exposition of the theory here will be concise, but sufficient for its application to many problems of applied mathematics and mathematical physics.

The Fourier theory of trigonometric series is of great practical importance because certain types of discontinuous functions which cannot be expanded in power series can be expanded in Fourier series. More importantly, a wide class of problems in physics and engineering possess periodic phenomena and, as a consequence, Fourier's trigonometric series become an indispensable tool in the analyses of these problems.

We shall begin our study with the basic concepts and definitions of some properties of real-valued functions.

## 5.1 PIECEWISE CONTINUOUS FUNCTIONS AND PERIODIC FUNCTIONS

A single-valued function $f$ is said to be *piecewise continuous* in an interval $[a, b]$ if there exist finitely many points $a = x_1 < x_2 < \ldots < x_n = b$, such that $f$ is continuous in the intervals $x_j < x < x_{j+1}$ and the one-sided limits $f(x_j+)$ and $f(x_{j+1}-)$ exist for all $j = 1, 2, 3, \ldots, n-1$.

A piecewise continuous function is shown in Fig. 5.1.1. The functions such as $1/x$ and $\sin 1/x$ fail to be piecewise continuous in the closed interval $[0, 1]$ because the one-sided limit $f(0+)$ does not exist in both cases.

If $f$ is piecewise continuous in an interval $[a, b]$, then it is necessarily bounded and integrable over that interval. Also, it follows immediately

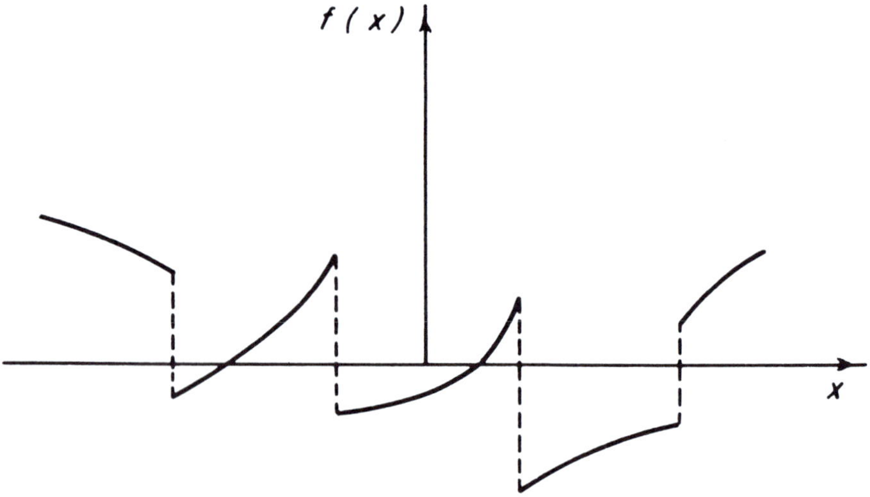


Figure 5.1.1

that the product of two piecewise continuous functions is piecewise continuous on the common interval.

If $f$ is piecewise continuous in an interval $[a, b]$ and if, in addition, the first derivative $f'$ is continuous in each of the intervals $x_j < x < x_{j+1}$, and the limits $f'(x_j+)$ and $f'(x_j-)$ exist, then $f$ is said to be *piecewise smooth*; if, in addition, the second derivative $f''$ is continuous in each of the intervals $x_j < x < x_{j+1}$, and the limits $f''(x_j+)$ and $f''(x_j-)$ exist, then $f$ is said to be *piecewise very smooth*.

A piecewise continuous function $f(x)$ in an interval $[a, b]$ is said to be *periodic* if there exists a real positive number $p$ such that

$$f(x+p)=f(x) \tag{5.1.1}$$

for all $x$. $p$ is called the *period* of $f$, and the smallest value of $p$ is termed the *fundamental period*. A sample graph of a periodic function is given in Fig. 5.1.2.

If $f$ is periodic with period $p$, then

$$\begin{aligned} f(x+p)&=f(x) \\ f(x+2p)&=f(x+p+p)=f(x+p) \\ f(x+3p)&=f(x+2p+p)=f(x+2p) \\ f(x+np)&=f(x+(n-1)p+p)=f(x+(n-1)p)=f(x) \end{aligned}$$

for any integer $n$. Hence for all integral values of $n$

$$f(x+np)=f(x) \tag{5.1.2}$$

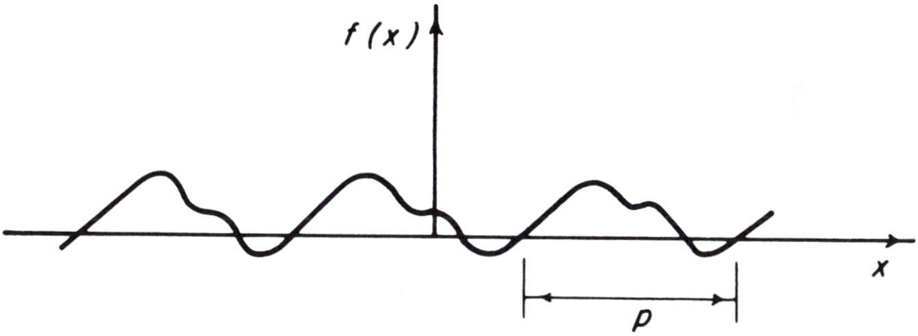


Figure 5.1.2

It can be readily shown that if $f_1, f_2, \ldots, f_k$ have the period $p$ and $c_k$ are the constants, then

$$f = c_1 f_1 + c_2 f_2 + \ldots + c_k f_k \tag{5.1.3}$$

has the period $p$.

Well known examples of periodic functions are the sine and cosine functions. As a special case, a constant function is also a periodic function with arbitrary period $p$. Thus, by the relation (5.1.3), the series

$$a_0 + a_1 \cos x + a_2 \cos 2x + \ldots + b_1 \sin x + b_2 \sin 2x + \ldots$$

if it converges, obviously has the period $2\pi$. Such types of series, which occur frequently in problems of applied mathematics and mathematical physics, will be treated later.

## 5.2 ORTHOGONALITY

The functions of a sequence $\{\phi_n(x)\}$ are said to be *orthogonal* with respect to the weight function $q(x)$ on the interval $[a, b]$ if

$$\int_a^b \phi_m(x)\phi_n(x)q(x)\,dx = 0, \qquad m \neq n \tag{5.2.1}$$

If $m = n$, then we have

$$\|\phi_n\| = \left[\int_a^b \phi_n^2 q\,dx\right]^{\frac{1}{2}} \tag{5.2.2}$$

which is called the *norm* of the orthogonal system $\{\phi_n\}$.

EXAMPLE 5.2.1. The functions $\sin mx$, $m = 1, 2, \ldots,$ form an orthogonal system on the interval $[-\pi, \pi]$, because

$$\int_{-\pi}^{\pi} \sin mx \sin nx\,dx = \begin{cases} 0, & m \neq n \\ \pi, & m = n \end{cases}$$

In this example we notice that the weight function is equal to unity, and the value of the norm is $\sqrt{\pi}$.

An orthogonal system $\phi_1, \phi_2, \ldots, \phi_n$ where $n$ may be finite or infinite, which satisfies the relations

$$\int_a^b \phi_m(x)\phi_n(x)q(x)\,dx = \begin{cases} 0, & m \neq n \\ 1, & m = n \end{cases} \tag{5.2.3}$$

is called the *orthonormal* system of functions on $[a, b]$. It is evident that an orthonormal system can be obtained from an orthogonal system by dividing each function by its norm on $[a, b]$.

EXAMPLE 5.2.2. The sequence of functions

$$1, \cos x, \sin x, \ldots, \cos nx, \sin nx$$

form an orthogonal system on $[-\pi, \pi]$ since

$$\int_{-\pi}^{\pi} \sin mx \sin nx\,dx = \begin{cases} 0, & m \neq n \\ \pi, & m = n \end{cases}$$

$$\int_{-\pi}^{\pi} \sin mx \cos nx\,dx = 0 \qquad \text{for all } m, n \tag{5.2.4}$$

$$\int_{-\pi}^{\pi} \cos mx \cos nx\,dx = \begin{cases} 0, & m \neq n \\ \pi, & m = n \end{cases}$$

for positive integers $m$ and $n$. To normalize this system, we divide the elements of the original orthogonal system by its norm. Hence

$$\frac{1}{\sqrt{2\pi}}, \frac{\cos x}{\sqrt{\pi}}, \frac{\sin x}{\sqrt{\pi}}, \ldots, \frac{\cos nx}{\sqrt{\pi}}, \frac{\sin nx}{\sqrt{\pi}}$$

form an orthonormal system.

## 5.3 FOURIER SERIES

The functions

$$1, \cos x, \sin x, \cos 2x, \sin 2x \ldots$$

are mutually orthogonal to each other in the interval $[-\pi, \pi]$ and are linearly independent. Thus, we form a formal series representing $f(x)$. We write

$$f(x) \sim \frac{a_0}{2} + \sum_{k=1}^{\infty} (a_k \cos kx + b_k \sin kx) \tag{5.3.1}$$

where the symbol $\sim$ indicates an association of $a_0$, $a_k$, and $b_k$ to $f$ in some unique manner. The series may or may not be convergent. The coefficient $a_0/2$ instead of $a_0$ is used for convenience in formulation.

Let $f(x)$ be a Riemann integrable function defined on the interval $[-\pi, \pi]$. Suppose we define the $n$th partial sum

$$s_n(x) = \frac{a_0}{2} + \sum_{k=1}^{n} (a_k \cos kx + b_k \sin kx) \tag{5.3.2}$$

that is to represent $f(x)$ on $[-\pi, \pi]$. We shall seek the coefficients $a_0$, $a_k$, and $b_k$ such that $s_n(x)$ represents the best approximation to $f(x)$ in the sense of least squares, that is, we seek to minimize the integral

$$I(a_0, a_k, b_k) = \int_{-\pi}^{\pi} [f(x) - s_n(x)]^2\, dx \tag{5.3.3}$$

This is an extremal problem. A necessary condition for $a_0$, $a_k$, $b_k$, so that $I$ be minimum, is that the first partial derivatives of $I$ with respect to these coefficients vanish. Thus, substituting Eq. (5.3.2) into (5.3.3) and differentiating with respect to $a_0$, $a_k$, and $b_k$, we obtain

$$\frac{\partial I}{\partial a_0} = -\int_{-\pi}^{\pi} \left[f(x) - \frac{a_0}{2} - \sum_{j=1}^{n} (a_j \cos jx + b_j \sin jx)\right] dx \tag{5.3.4}$$

$$\frac{\partial I}{\partial a_k} = -2\int_{-\pi}^{\pi} \left[f(x) - \frac{a_0}{2} - \sum_{j=1}^{n} (a_j \cos jx + b_j \sin jx)\right] \cos kx\, dx \tag{5.3.5}$$

$$\frac{\partial I}{\partial b_k} = -2\int_{-\pi}^{\pi} \left[f(x) - \frac{a_0}{2} - \sum_{j=1}^{n} (a_j \cos jx + b_j \sin jx)\right] \sin kx\, dx \tag{5.3.6}$$

Using the orthogonality relations of the trigonometric functions (5.2.4) and noting that

$$\int_{-\pi}^{\pi} \cos mx\, dx = \int_{-\pi}^{\pi} \sin mx\, dx = 0 \tag{5.3.7}$$

where $m$ and $n$ are positive integers, Eqs. (5.3.4), (5.3.5) and (5.3.6) become

$$\frac{\partial I}{\partial a_0} = \pi a_0 - \int_{-\pi}^{\pi} f(x)\, dx \tag{5.3.8}$$

$$\frac{\partial I}{\partial a_k} = 2\pi a_k - 2\int_{-\pi}^{\pi} f(x) \cos kx\, dx \tag{5.3.9}$$

$$\frac{\partial I}{\partial b_k} = 2\pi b_k - 2\int_{-\pi}^{\pi} f(x) \sin kx\, dx \tag{5.3.10}$$

which must vanish for $I$ to have an extremal value. Thus, we have

$$a_0 = \frac{1}{\pi}\int_{-\pi}^{\pi} f(x)\,dx \tag{5.3.11}$$

$$a_k = \frac{1}{\pi}\int_{-\pi}^{\pi} f(x)\cos kx\,dx \tag{5.3.12}$$

$$b_k = \frac{1}{\pi}\int_{-\pi}^{\pi} f(x)\sin kx\,dx \tag{5.3.13}$$

Note that $a_0$ is the special case of $a_k$ which is the reason for writing $a_0/2$ rather than $a_0$ in Eq. (5.3.1). It immediately follows from Eqs. (5.3.8), (5.3.9) and (5.3.10) that

$$\frac{\partial^2 I}{\partial a_0^2} = \pi \tag{5.3.14}$$

$$\frac{\partial^2 I}{\partial a_k^2} = \frac{\partial^2 I}{\partial b_k^2} = 2\pi \tag{5.3.15}$$

and all mixed second order and all remaining higher order derivatives vanish. Now if we expand $I$ in a Taylor series about $(a_0, a_1, \ldots, a_n, b_1, \ldots, b_n)$ we have

$$I(a_0 + \Delta a_0, \ldots, b_n + \Delta b_n) = I(a_0, \ldots, b_n) + \Delta I \tag{5.3.16}$$

where $\Delta I$ stands for the remaining terms. Since the first derivatives, all mixed second derivatives, and all remaining higher derivatives vanish, we obtain

$$\Delta I = \frac{1}{2!}\left[\frac{\partial^2 I}{\partial a_0^2}\Delta a_0^2 + \sum_{k=1}^{n}\left(\frac{\partial^2 I}{\partial a_k^2}\Delta a_k^2 + \frac{\partial^2 I}{\partial b_k^2}\Delta b_k^2\right)\right] \tag{5.3.17}$$

By virtue of Eqs. (5.3.14) and (5.3.15), $\Delta I$ is positive. Hence, for $I$ to have a minimum value, the coefficients $a_0$, $a_k$, $b_k$ must be given by Eqs. (5.3.11), (5.3.12), and (5.3.13) respectively. These coefficients are called the *Fourier coefficients* of $f(x)$ and the series in Eq. (5.3.1) is said to be the *Fourier series* corresponding to $f(x)$.

We remark that the possibility of representing the given function $f(x)$ by a Fourier series does not imply that the Fourier series converges to the function $f(x)$. As a matter of fact, there exist Fourier series which diverge. A convergent trigonometric series need not be a Fourier series. For instance, the trigonometric series

$$\sum_{n=2}^{\infty} \frac{\sin nx}{\log n}$$

which is convergent for all values of $x$, is not a Fourier series, for there exist no integrable functions corresponding to this series.

## 5.4 CONVERGENCE IN THE MEAN

Let $f(x)$ be piecewise continuous and periodic with period $2\pi$. It is obvious that

$$\int_{-\pi}^{\pi} [f(x) - s_n(x)]^2 \, dx \geqslant 0 \tag{5.4.1}$$

where

$$s_n(x) = \frac{a_0}{2} + \sum_{k=1}^{n} (a_k \cos kx + b_k \sin kx)$$

Expanding

$$\int_{-\pi}^{\pi} [f(x) - s_n(x)]^2 \, dx = \int_{-\pi}^{\pi} [f(x)]^2 \, dx - 2\int_{-\pi}^{\pi} f(x) s_n(x) \, dx + \int_{-\pi}^{\pi} [s_n(x)]^2 \, dx$$

But, by the definitions of the Fourier coefficients (5.3.11), (5.3.12), and (5.3.13) and by the orthogonal relations for the trigonometric series (5.2.4), we have

$$\int_{-\pi}^{\pi} f(x) s_n(x) \, dx = \int_{-\pi}^{\pi} f(x) \left[ \frac{a_0}{2} + \sum_{k=1}^{n} (a_k \cos kx + b_k \sin kx) \right] dx = \frac{\pi a_0^2}{2} + \pi \sum_{k=1}^{n} (a_k^2 + b_k^2) \tag{5.4.2}$$

and

$$\int_{-\pi}^{\pi} s_n^2(x) \, dx = \int_{-\pi}^{\pi} \left[ \frac{a_0}{2} + \sum_{k=1}^{n} (a_k \cos kx + b_k \sin kx) \right]^2 dx = \frac{\pi a_0^2}{2} + \pi \sum_{k=1}^{n} (a_k^2 + b_k^2) \tag{5.4.3}$$

Consequently

$$\int_{-\pi}^{\pi} [f(x) - s_n(x)]^2 \, dx = \int_{-\pi}^{\pi} f^2(x) \, dx - \left[ \frac{\pi a_0^2}{2} + \pi \sum_{k=1}^{n} (a_k^2 + b_k^2) \right] \geqslant 0 \tag{5.4.4}$$

It follows that

$$\frac{a_0^2}{2} + \sum_{k=1}^{n} (a_k^2 + b_k^2) \leqslant \frac{1}{\pi} \int_{-\pi}^{\pi} f^2(x) \, dx \tag{5.4.5}$$

for all values of $n$. Since the right hand side of Eq. (5.4.5) is independent of $n$, we obtain

$$\frac{a_0^2}{2} + \sum_{k=1}^{\infty} (a_k^2 + b_k^2) \leq \frac{1}{\pi} \int_{-\pi}^{\pi} f^2(x)\, dx \tag{5.4.6}$$

which is known as *Bessel's inequality*.

We see that the left side is nondecreasing and is bounded above, and therefore the series

$$\frac{a_0^2}{2} + \sum_{k=1}^{\infty} (a_k^2 + b_k^2) \tag{5.4.7}$$

converges. Thus, the necessary condition for the convergence of Eq. (5.4.7) is that

$$\lim_{k \to \infty} a_k = 0, \qquad \lim_{k \to \infty} b_k = 0 \tag{5.4.8}$$

The Fourier series is said to *converge in the mean* to $f(x)$ when

$$\lim_{n \to \infty} \int_{-\pi}^{\pi} \left[ f(x) - \left( \frac{a_0}{2} + \sum_{k=1}^{n} a_k \cos kx + b_k \sin kx \right) \right]^2 dx = 0 \tag{5.4.9}$$

If the Fourier series converges in the mean to $f(x)$, then

$$\frac{a_0^2}{2} + \sum_{k=1}^{\infty} (a_k^2 + b_k^2) = \frac{1}{\pi} \int_{-\pi}^{\pi} f^2(x)\, dx \tag{5.4.10}$$

which is called *Parseval's relation*. Furthermore, if the relation (5.4.9) holds true, the set of trigonometric functions 1, $\cos x$, $\sin x$, $\cos 2x$, $\sin 2x$, . . . is said to be *complete*.

## 5.5 EXAMPLES OF FOURIER SERIES

The Fourier coefficients (5.3.11), (5.3.12), and (5.3.13) of Sec. 5.3 may be obtained in a different way. Suppose the function $f(x)$ of period $2\pi$ has the expansion

$$f(x) = \frac{a_0}{2} + \sum_{k=1}^{\infty} (a_k \cos kx + b_k \sin kx) \tag{5.5.1}$$

If we assume that the infinite series is term-by-term integrable (we will see later that uniform convergence of the series is a sufficient condition for this), then

$$\begin{aligned} \int_{-\pi}^{\pi} f(x)\, dx &= \int_{-\pi}^{\pi} \left[ \frac{a_0}{2} + \sum_{k=1}^{\infty} (a_k \cos kx + b_k \sin kx) \right] dx \\ &= \pi a_0 \end{aligned}$$

Hence

$$a_0 = \frac{1}{\pi}\int_{-\pi}^{\pi} f(x)\,dx \tag{5.5.2}$$

Again, we multiply both sides of Eq. (5.5.1) by $\cos nx$ and integrate from $-\pi$ to $\pi$. We obtain

$$\int_{-\pi}^{\pi} f(x)\cos nx\,dx = \int_{-\pi}^{\pi}\left[\frac{a_0}{2} + \sum_{k=1}^{\infty}(a_k\cos kx + b_k\sin kx)\right]\cos nx\,dx$$
$$= \pi a_k$$

Thus,

$$a_k = \frac{1}{\pi}\int_{-\pi}^{\pi} f(x)\cos kx\,dx \tag{5.5.3}$$

In a similar manner, we find that

$$b_k = \frac{1}{\pi}\int_{-\pi}^{\pi} f(x)\sin kx\,dx \tag{5.5.4}$$

The coefficients $a_0$, $a_k$, $b_k$ just found are exactly the same as those obtained in Sec. 5.3.

EXAMPLE 5.5.1. Find the Fourier series expansion for the function shown in Fig. 5.5.1.

$$f(x) = x + x^2, \qquad -\pi < x < \pi$$

Here

$$a_0 = \frac{1}{\pi}\int_{-\pi}^{\pi} f(x)\,dx$$
$$= \frac{1}{\pi}\int_{-\pi}^{\pi}(x + x^2)\,dx = 2\frac{\pi^2}{3}$$

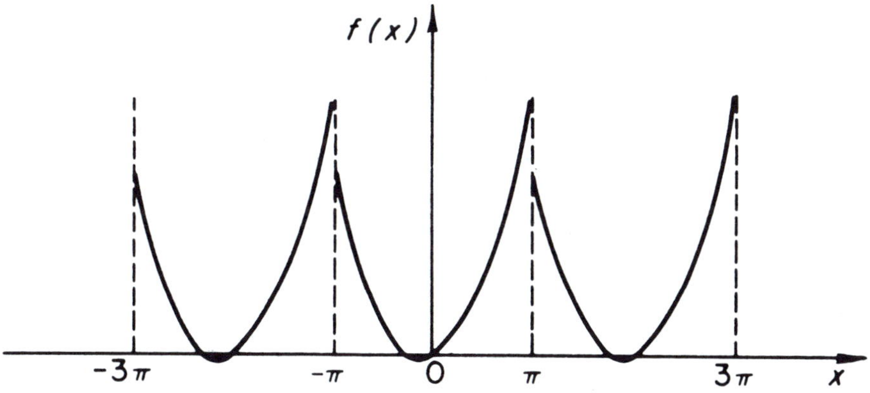


Figure 5.5.1

and

$$
\begin{aligned}
a_k &= \frac{1}{\pi}\int_{-\pi}^{\pi} f(x)\cos kx\,dx \\
&= \frac{1}{\pi}\int_{-\pi}^{\pi} (x+x^2)\cos kx\,dx \\
&= \frac{1}{\pi}\left[\frac{x\sin kx}{k}\bigg|_{-\pi}^{\pi} - \int_{-\pi}^{\pi}\frac{\sin kx}{k}\,dx\right] \\
&\quad + \frac{1}{\pi}\left[\frac{x^2\sin kx}{k}\bigg|_{-\pi}^{\pi} - \int_{-\pi}^{\pi}\frac{2x\sin kx}{k}\,dx\right] \\
&= -\frac{2}{k\pi}\left[-\frac{x\cos kx}{k}\bigg|_{-\pi}^{\pi} + \int_{-\pi}^{\pi}\frac{\cos kx}{k}\,dx\right] \\
&= \frac{4}{k^2}\cos k\pi \\
&= \frac{4}{k^2}(-1)^k \quad \text{for} \quad k = 1, 2, 3, \ldots
\end{aligned}
$$

Similarly

$$
\begin{aligned}
b_k &= \frac{1}{\pi}\int_{-\pi}^{\pi} f(x)\sin kx\,dx \\
&= \frac{1}{\pi}\int_{-\pi}^{\pi} (x+x^2)\sin kx\,dx \\
&= -\frac{2}{k}\cos k\pi \\
&= -\frac{2}{k}(-1)^k \quad \text{for} \quad k = 1, 2, 3, \ldots
\end{aligned}
$$

Therefore the Fourier series expansion for $f$ is

$$
\begin{aligned}
f(x) &= \frac{\pi^2}{3} + \sum_{k=1}^{\infty}\left[\frac{4}{k^2}(-1)^k\cos kx - \frac{2}{k}(-1)^k\sin kx\right] \\
&= \frac{\pi^2}{3} - 4\cos x + 2\sin x + \cos 2x - \sin 2x - \ldots
\end{aligned}
$$

EXAMPLE 5.5.2. Consider the periodic function shown in Fig. 5.5.2,

$$
f(x) = \begin{cases} -\pi, & -\pi < x < 0 \\ x, & 0 < x < \pi \end{cases}
$$

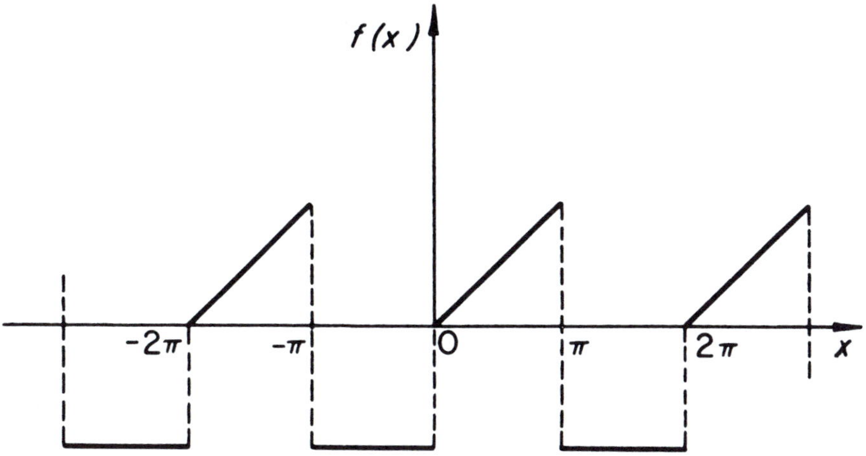


Figure 5.5.2

In this case,

$$a_0 = \frac{1}{\pi}\int_{-\pi}^{\pi} f(x)\,dx$$

$$= \frac{1}{\pi}\left[\int_{-\pi}^{0} -\pi\,dx + \int_{0}^{\pi} x\,dx\right]$$

$$= -\frac{\pi}{2}$$

and

$$a_k = \frac{1}{\pi}\int_{-\pi}^{\pi} f(x)\cos kx\,dx$$

$$= \frac{1}{\pi}\left[\int_{-\pi}^{0} -\pi\cos kx\,dx + \int_{0}^{\pi} x\cos kx\,dx\right]$$

$$= \frac{1}{k^2\pi}(\cos k\pi - 1)$$

$$= \frac{1}{k^2\pi}[(-1)^k - 1]$$

Also

$$b_k = \frac{1}{\pi}\int_{-\pi}^{\pi} f(x)\sin kx\,dx$$

$$= \frac{1}{\pi}\left[\int_{-\pi}^{0} -\pi\sin kx\,dx + \int_{0}^{\pi} x\sin kx\,dx\right]$$

$$= \frac{1}{k}(1 - 2\cos k\pi)$$

$$= \frac{1}{k}[1 - 2(-1)^k]$$

Hence the Fourier series is

$$f(x) = -\frac{\pi}{4} + \sum_{k=1}^{\infty}\left\{\frac{1}{k^2\pi}[(-1)^k - 1]\cos kx + \frac{1}{k}[1 - 2(-1)^k]\sin kx\right\}$$

EXAMPLE 5.5.3. Consider the function $f(x) = x$ in the interval $-\pi < x < \pi$. We can readily find that

$$a_k = 0, \qquad k = 0, 1, 2, \ldots$$

The coefficients $b_k$ are given by

$$b_k = \frac{1}{\pi}\int_{-\pi}^{\pi} f(x)\sin kx\,dx$$

$$= \frac{1}{\pi}\int_{-\pi}^{\pi} x\sin kx\,dx$$

$$= \frac{2}{k}(-1)^{k+1}$$

Hence

$$f(x) = 2\sum_{k=1}^{\infty}(-1)^{k+1}\frac{\sin kx}{k}$$

## 5.6 COSINE AND SINE SERIES

Let $f(x)$ be an even function defined on the interval $[-\pi, \pi]$. Since $\cos kx$ is an even function, and $\sin kx$ an odd function, the function $f(x)\cos kx$ is an even function and the function $f(x)\sin kx$ an odd function. Thus, we find that the Fourier coefficients of $f(x)$ are

$$a_k = \frac{1}{\pi}\int_{-\pi}^{\pi} f(x)\cos kx\,dx = \frac{2}{\pi}\int_0^{\pi} f(x)\cos kx\,dx, \qquad k = 0, 1, 2, \ldots$$

$$b_k = \frac{1}{\pi}\int_{-\pi}^{\pi} f(x)\sin kx\,dx = 0, \qquad k = 1, 2, 3, \ldots \tag{5.6.1}$$

Hence the Fourier series of an even function can be written as

$$f(x) \sim \frac{a_0}{2} + \sum_{k=1}^{\infty} a_k \cos kx \tag{5.6.2}$$

where the coefficients $a_k$ are given by formula (5.6.1).

In a similar manner, if $f(x)$ is an odd function, the function $f(x)\cos kx$ is an odd function and the function $f(x)\sin kx$ is an even function. As a consequence, the Fourier coefficients of $f(x)$, in this case, are

$$a_k = \frac{1}{\pi}\int_{-\pi}^{\pi} f(x)\cos kx\,dx = 0, \qquad k = 0, 1, 2, \ldots$$

$$b_k = \frac{1}{\pi}\int_{-\pi}^{\pi} f(x)\sin kx\,dx = \frac{2}{\pi}\int_0^{\pi} f(x)\sin kx\,dx, \qquad k = 1, 2, \ldots \tag{5.6.3}$$

Therefore, the Fourier series of an odd function can be written as

$$f(x) = \sum_{k=1}^{\infty} b_k \sin kx \tag{5.6.4}$$

where the coefficients $b_k$ are given by formula (5.6.3).

EXAMPLE 5.6.1. Obtain the Fourier series of the function

$$f(x) = \begin{cases} -1, & -\pi < x < 0 \\ +1, & 0 < x < \pi \end{cases}$$

In this case, $f$ is an odd function so that $a_k = 0$ for $k = 0, 1, 2, 3, \ldots$ and

$$\begin{aligned} b_k &= \frac{2}{\pi}\int_0^{\pi} f(x)\sin kx\,dx \\ &= \frac{2}{\pi}\int_0^{\pi} \sin kx\,dx \\ &= \frac{2}{k\pi}[1 - (-1)^k] \end{aligned}$$

Thus $b_{2k} = 0$ and $b_{2k-1} = [4/\pi(2k-1)]$. Therefore, the Fourier series of the function $f(x)$ is

$$f(x)=\frac{4}{\pi}\sum_{k=1}^{\infty}\frac{\sin(2k-1)x}{(2k-1)}$$

EXAMPLE 5.6.2. Expand $|\sin x|$ in Fourier series. Since $|\sin x|$ is an even function, as shown in Fig. 5.6.1, $b_k=0$ for $k=1, 2, \ldots$ and

$$\begin{aligned} a_k &= \frac{2}{\pi}\int_0^{\pi} f(x)\cos kx\,dx \\ &= \frac{2}{\pi}\int_0^{\pi} \sin x\cos kx\,dx \\ &= \frac{1}{\pi}\int_0^{\pi} [\sin(1+k)x+\sin(1-k)x]\,dx \\ &= \frac{2[1+(-1)^k]}{\pi(1-k^2)} \qquad \text{for } k=0, 2, 3, \ldots \end{aligned}$$

For $k=1$,

$$a_1=\frac{2}{\pi}\int_0^{\pi}\sin x\cos x\,dx=0.$$

Hence the Fourier series of $f(x)$ is

$$f(x)=\frac{2}{\pi}+\frac{4}{\pi}\sum_{k=1}^{\infty}\frac{\cos 2kx}{(1-4k^2)}$$

In the preceding sections, we have prescribed the function $f(x)$ in the interval $(-\pi, \pi)$ and assumed $f(x)$ to be periodic with period $2\pi$ in the entire interval $(-\infty, \infty)$. In practice, we frequently encounter problems in which a function is defined only in the interval $(-\pi, \pi)$. In such a case, we simply extend the function periodically with period $2\pi$, as in Fig.

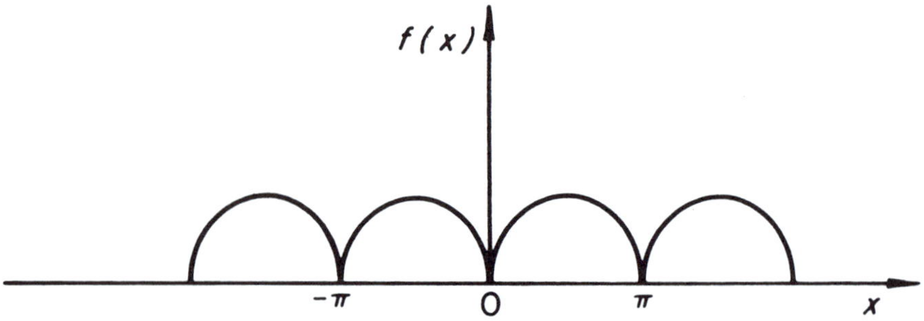


Figure 5.6.1

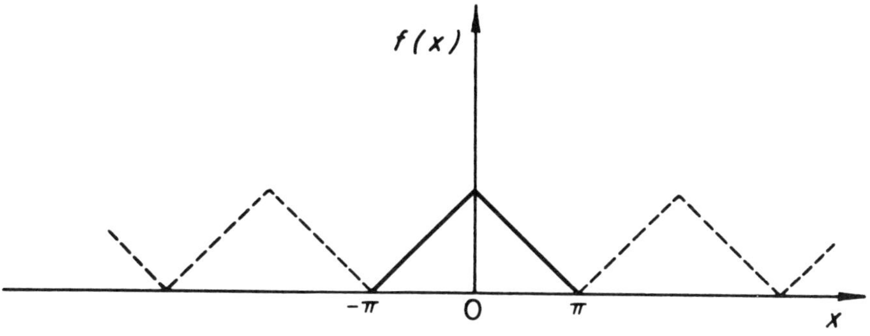


Figure 5.6.2

5.6.2. In this way, we are able to represent the function $f(x)$ by the Fourier series expansion, although we are interested only in the expansion on $(-\pi, \pi)$.

If a function $f$ is defined only in the interval $(0, \pi)$, we may extend $f$ in two ways. The first is the *even extension* of $f$, denoted and defined by (see Fig. 5.6.3)

$$F_e(x) = \begin{cases} f(x), & 0 < x < \pi \\ f(-x), & -\pi < x < 0 \end{cases}$$

while the second is the *odd extension* of $f$, denoted and defined by (see Fig. 5.6.4)

$$F_0(x) = \begin{cases} f(x), & 0 < x < \pi \\ -f(-x), & -\pi < x < 0 \end{cases}$$

Since $F_e(x)$ and $F_0(x)$ are the even and odd functions with period $2\pi$

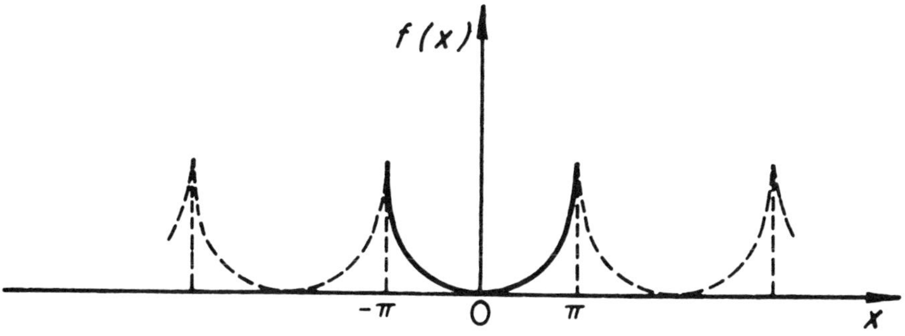


Figure 5.6.3

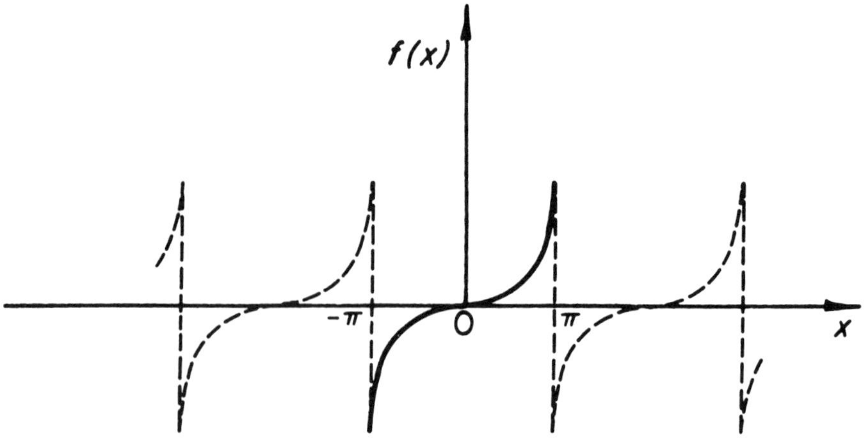


Figure 5.6.4

respectively, the Fourier series expansions of $F_e(x)$ and $F_0(x)$ are

$$F_e(x) = \frac{a_0}{2} + \sum_{k=1}^{\infty} a_k \cos kx$$

where

$$a_k = \frac{2}{\pi} \int_0^{\pi} f(x) \cos kx \, dx$$

and

$$F_0(x) = \sum_{k=1}^{\infty} b_k \sin kx$$

where

$$b_k = \frac{2}{\pi} \int_0^{\pi} f(x) \sin kx \, dx$$

## 5.7 COMPLEX FOURIER SERIES

It is sometimes convenient to represent a function by an expansion in complex form. This expansion can easily be derived from the Fourier series

$$f(x) = \frac{a_0}{2} + \sum_{k=1}^{\infty} (a_k \cos kx + b_k \sin kx)$$

Noting that

$$\sin x = \frac{e^{ix} - e^{-ix}}{2i}, \qquad \cos x = \frac{e^{ix} + e^{-ix}}{2}$$

we write

$$
\begin{aligned}
f(x) &= \frac{a_0}{2} + \sum_{k=1}^{\infty} \left[ a_k \left( \frac{e^{ikx} + e^{-ikx}}{2} \right) + b_k \left( \frac{e^{ikx} - e^{-ikx}}{2i} \right) \right] \\
&= \frac{a_0}{2} + \sum_{k=1}^{\infty} \left[ \left( \frac{a_k - ib_k}{2} \right) e^{ikx} + \left( \frac{a_k + ib_k}{2} \right) e^{-ikx} \right] \\
&= c_0 + \sum_{k=1}^{\infty} (c_k e^{ikx} + c_{-k} e^{-ikx})
\end{aligned}
$$

where

$$
\begin{aligned}
c_0 &= \frac{a_0}{2} = \frac{1}{2\pi} \int_{-\pi}^{\pi} f(x)\, dx \\
c_k &= \frac{a_k - ib_k}{2} = \frac{1}{2\pi} \int_{-\pi}^{\pi} f(x)(\cos kx - i \sin kx)\, dx \\
&= \frac{1}{2\pi} \int_{-\pi}^{\pi} f(x)\, e^{-ikx}\, dx \\
c_{-k} &= \frac{a_k + ib_k}{2} = \frac{1}{2\pi} \int_{-\pi}^{\pi} f(x)(\cos kx + i \sin kx)\, dx \\
&= \frac{1}{2\pi} \int_{-\pi}^{\pi} f(x) e^{ikx}\, dx
\end{aligned}
$$

Thus, we obtain the Fourier series expansion for $f(x)$ in complex form, namely,

$$
f(x) = \sum_{k=-\infty}^{\infty} c_k e^{ikx}, \qquad -\pi < x < \pi \tag{5.7.1}
$$

where

$$
c_k = \frac{1}{2\pi} \int_{-\pi}^{\pi} f(x) e^{-ikx}\, dx \tag{5.7.2}
$$

EXAMPLE 5.7.1. Obtain the complex Fourier series expansion for the function

$$
f(x) = e^x, \qquad -\pi < x < \pi
$$

We find

$$
\begin{aligned}
c_k &= \frac{1}{2\pi} \int_{-\pi}^{\pi} f(x) e^{-ikx}\, dx \\
&= \frac{1}{2\pi} \int_{-\pi}^{\pi} e^x e^{-ikx}\, dx \\
&= \frac{(1 + ik)(-1)^k}{\pi(1 + k^2)} \sinh \pi
\end{aligned}
$$

and hence the Fourier series is

$$f(x)=\sum_{k=-\infty}^{\infty}\frac{(1+ik)(-1)^k}{\pi(1+k^2)}\sinh\pi\, e^{ikx}$$

## 5.8 CHANGE OF INTERVAL

So far we have been concerned with functions defined on the interval $[-\pi, \pi]$. In many applications, however, this interval is restrictive, and the interval of interest may be arbitrary, say $[a, b]$.

If we introduce the new variable $t$ by the transformation

$$x=\frac{1}{2}(b+a)+\frac{(b-a)}{2\pi}t \tag{5.8.1}$$

then the interval $a\leqslant x\leqslant b$ becomes $-\pi\leqslant t\leqslant\pi$. Thus, the function $f\,[(b+a)/2+((b-a)/2\pi)t]=F(t)$ obviously has period $2\pi$. Expanding this function in a Fourier series, we obtain

$$F(t)=\frac{a_0}{2}+\sum_{k=1}^{\infty}(a_k\cos kt+b_k\sin kt) \tag{5.8.2}$$

where

$$a_k=\frac{1}{\pi}\int_{-\pi}^{\pi}F(t)\cos kt\,dt,\quad k=0,1,2,\ldots$$

$$b_k=\frac{1}{\pi}\int_{-\pi}^{\pi}F(t)\sin kt\,dt,\quad k=1,2,3,\ldots$$

On changing $t$ into $x$, we find the expansion for $f(x)$ in $[a, b]$

$$f(x)=\frac{a_0}{2}+\sum_{k=1}^{\infty}\left[a_k\cos\frac{k\pi(2x-b-a)}{(b-a)}+b_k\sin\frac{k\pi(2x-b-a)}{(b-a)}\right] \tag{5.8.3}$$

where

$$a_k=\frac{2}{b-a}\int_a^b f(x)\cos\frac{k\pi(2x-b-a)}{(b-a)}dx,\quad k=0,1,2,\ldots \tag{5.8.4}$$

$$b_k=\frac{2}{b-a}\int_a^b f(x)\sin\frac{k\pi(2x-b-a)}{(b-a)}dx,\quad k=1,2,3,\ldots \tag{5.8.5}$$

It is sometimes convenient to take the interval in which the function $f$ is defined as $[-l, l]$. It follows at once from the result just obtained that by letting $a=-l$ and $b=l$, the expansion for $f$ in $[-l, l]$ takes the form

$$f(x)=\frac{a_0}{2}+\sum_{k=1}^{\infty}\left(a_k\cos\frac{k\pi x}{l}+b_k\sin\frac{k\pi x}{l}\right) \tag{5.8.6}$$

where

$$a_k = \frac{1}{l}\int_{-l}^{l} f(x)\cos\frac{k\pi x}{l}\,dx, \quad k = 0, 1, 2, \ldots \tag{5.8.7}$$

$$b_k = \frac{1}{l}\int_{-l}^{l} f(x)\sin\frac{k\pi x}{l}\,dx, \quad k = 1, 2, 3, \ldots \tag{5.8.8}$$

If $f$ is an even function of period $2l$, then by Eq. (5.8.6), we can readily determine that

$$f(x) = \frac{a_0}{2} + \sum_{k=1}^{\infty} a_k \cos\frac{k\pi x}{l} \tag{5.8.9}$$

where

$$a_k = \frac{2}{l}\int_{0}^{l} f(x)\cos\frac{k\pi x}{l}\,dx, \quad k = 0, 1, 2, \ldots \tag{5.8.10}$$

If $f$ is an odd function of period $2l$, then by Eq. (5.8.6), the expansion for $f$ is

$$f(x) = \sum_{k=1}^{\infty} b_k \sin\frac{k\pi x}{l} \tag{5.8.11}$$

where

$$b_k = \frac{2}{l}\int_{0}^{l} f(x)\sin\frac{k\pi x}{l}\,dx \tag{5.8.12}$$

EXAMPLE 5.8.1. Consider the odd periodic function $f$

$$f(x) = x, \qquad -2 < x < 2$$

as shown in Fig. 5.8.1. Here $l = 2$. Since $f$ is odd, $a_k = 0$, and

$$\begin{aligned} b_k &= \frac{2}{l}\int_{0}^{l} f(x)\sin\frac{k\pi x}{l}\,dx \\ &= \frac{2}{2}\int_{0}^{2} x\sin\frac{k\pi x}{2}\,dx \\ &= -\frac{4}{k\pi}(-1)^k \quad \text{for} \quad k = 1, 2, 3, \ldots \end{aligned}$$

Therefore, the Fourier series of $f$ is

$$f(x) = \sum_{k=1}^{\infty} \frac{4}{k\pi}(-1)^{k+1}\sin\frac{k\pi x}{2}$$

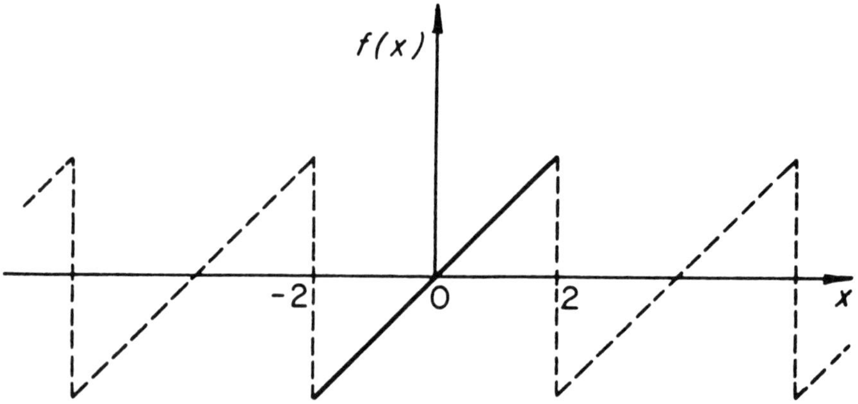


Figure 5.8.1

EXAMPLE 5.8.2. Given the function

$$f(x)=\begin{cases}1, & 0<x<\dfrac{1}{2}\\[2ex] 0, & \dfrac{1}{2}<x<1\end{cases}$$

In this case the period is $2l=2$ or $l=1$. Extend $f$ as shown in Fig. 5.8.2. Since the extension is even, we have $b_k=0$ and

$$\begin{aligned}a_0&=\frac{2}{l}\int_0^l f(x)\,dx\\ &=\frac{2}{1}\int_0^{\frac{1}{2}} dx\\ &=1\\ a_k&=\frac{2}{l}\int_0^l f(x)\cos\frac{k\pi x}{l}\,dx\\ &=\frac{2}{1}\int_0^1 \cos k\pi x\,dx\\ &=\frac{2}{k\pi}\sin\frac{k\pi}{2}\end{aligned}$$

Hence

$$f(x)=\frac{1}{2}+\sum_{k=1}^{\infty}\frac{2}{(2k-1)\pi}(-1)^{k-1}\cos(2k-1)\pi x$$

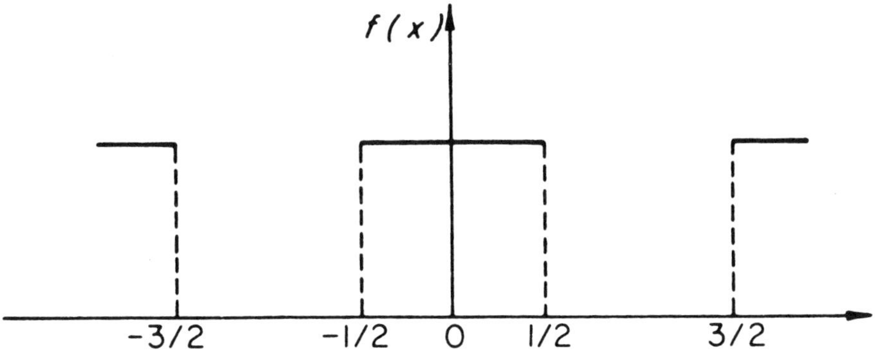


Figure 5.8.2

## 5.9 POINTWISE CONVERGENCE AND THE RIEMANN–LEBESGUE LEMMA

We have stated earlier that if $f(x)$ is piecewise continuous on the interval $[-\pi, \pi]$, then there exists a Fourier series expansion which converges in the mean to $f(x)$.

In this section, we shall state the Pointwise Convergence Theorem without proof.

***Theorem 5.9.1 (Pointwise Convergence Theorem)*** *If $f(x)$ is piecewise smooth and periodic with period $2\pi$ in $[-\pi, \pi]$, then for any $x$*

$$\frac{a_0}{2} + \sum_{k=1}^{\infty} (a_k \cos kx + b_k \sin kx) = \frac{1}{2}[f(x+) + f(x-)] \tag{5.9.1}$$

*where*

$$a_k = \frac{1}{\pi} \int_{-\pi}^{\pi} f(x) \cos kx \, dx, \qquad k = 0, 1, 2, \ldots$$

$$b_k = \frac{1}{\pi} \int_{-\pi}^{\pi} f(x) \sin kx \, dx, \qquad k = 1, 2, 3, \ldots$$

***Lemma 5.9.1 (Riemann–Lebesgue Lemma)*** *If $g(x)$ is piecewise continuous on the interval $[a, b]$, then*

$$\lim_{\lambda \to \infty} \int_a^b g(x) \sin \lambda x \, dx = 0 \tag{5.9.2}$$

PROOF. Consider the integral

$$I(\lambda) = \int_a^b g(x) \sin \lambda x \, dx \tag{5.9.3}$$

With the change of variable

$$x = t + \pi/\lambda$$

we have

$$\sin \lambda x = \sin \lambda(t + \pi/\lambda) = -\sin \lambda t$$

and

$$I(\lambda) = -\int_{a-\pi/\lambda}^{b-\pi/\lambda} g(t + \pi/\lambda) \sin \lambda t \, dt \tag{5.9.4}$$

Since $t$ is a dummy variable, we write the above integral as

$$I(\lambda) = -\int_{a-\pi/\lambda}^{b-\pi/\lambda} g(x + \pi/\lambda) \sin \lambda x \, dx \tag{5.9.5}$$

Addition of Eqs. (5.9.3) and (5.9.5) yields

$$\begin{aligned} 2I(\lambda) &= \int_a^b g(x) \sin \lambda x \, dx - \int_{a-\pi/\lambda}^{b-\pi/\lambda} g(x + \pi/\lambda) \sin \lambda x \, dx \\ &= -\int_{a-\pi/\lambda}^{a} g(x + \pi/\lambda) \sin \lambda x \, dx + \int_{b-\pi/\lambda}^{b} g(x) \sin \lambda x \, dx \\ &\quad + \int_a^{b-\pi/\lambda} [g(x) - g(x + \pi/\lambda)] \sin \lambda x \, dx \end{aligned} \tag{5.9.6}$$

First, let $g(x)$ be a continuous function in $[a, b]$. Then $g(x)$ is necessarily bounded, that is, there exists an $M$ such that $|g(x)| \leqslant M$. Hence

$$\left| \int_{a-\pi/\lambda}^{a} g(x + \pi/\lambda) \sin \lambda x \, dx \right| = \left| \int_a^{a+\pi/\lambda} g(x) \sin \lambda x \, dx \right| \leqslant \frac{\pi M}{\lambda}$$

and

$$\left| \int_{b-\pi/\lambda}^{b} g(x) \sin \lambda x \, dx \right| \leqslant \frac{\pi M}{\lambda}$$

Consequently,

$$|I(\lambda)| \leqslant \frac{\pi M}{\lambda} + \int_a^{b-\pi/\lambda} |g(x) - g(x + \pi/\lambda)| \, dx \tag{5.9.7}$$

Since $g(x)$ is a continuous function on a closed interval $[a, b]$, it is uniformly continuous on $[a, b]$ so that

$$|g(x) - g(x + \pi/\lambda)| < \varepsilon/(b - a) \tag{5.9.8}$$

for all $\lambda > \Lambda$ and all $x$ in $[a, b]$. We now choose $\lambda$ such that $\pi M/\lambda < \varepsilon/2$, whenever $\lambda > \Lambda$. Then

$$|I(\lambda)| < \frac{\varepsilon}{2} + \frac{\varepsilon}{2} = \varepsilon$$

If $g(x)$ is piecewise continuous in $[a, b]$, then the proof consists of a repeated application of the preceding argument to every subinterval of $[a, b]$ in which $g(x)$ is continuous. □

EXAMPLE 5.9.1. In Example 5.5.1, we find that the Fourier series expansion for $x + x^2$ in $[-\pi, \pi]$, as shown in Fig. 5.9.1, is

$$f(x) \sim \frac{\pi^2}{3} + \sum_{k=1}^{\infty} \frac{4}{k^2}(-1)^k \cos kx - \frac{2}{k}(-1)^k \sin kx$$

Since $f(x) = x + x^2$ is piecewise smooth, the series converges, and hence we write

$$x + x^2 = \frac{\pi^2}{3} + \sum_{k=1}^{\infty} \frac{4}{k^2}(-1)^k \cos kx - \frac{2}{k}(-1)^k \sin kx$$

at points of continuity. At points of discontinuity, such as $x = \pi$, by virtue of the Pointwise Convergence Theorem,

$$\frac{1}{2}[(\pi + \pi^2) + (-\pi + \pi^2)] = \frac{\pi^2}{3} + \sum_{k=1}^{\infty} \frac{4}{k^2}(-1)^k \cos k\pi \qquad (5.9.9)$$

since

$$f(\pi-) = \pi + \pi^2 \quad \text{and} \quad f(\pi+) = f(-\pi+) = -\pi + \pi^2$$

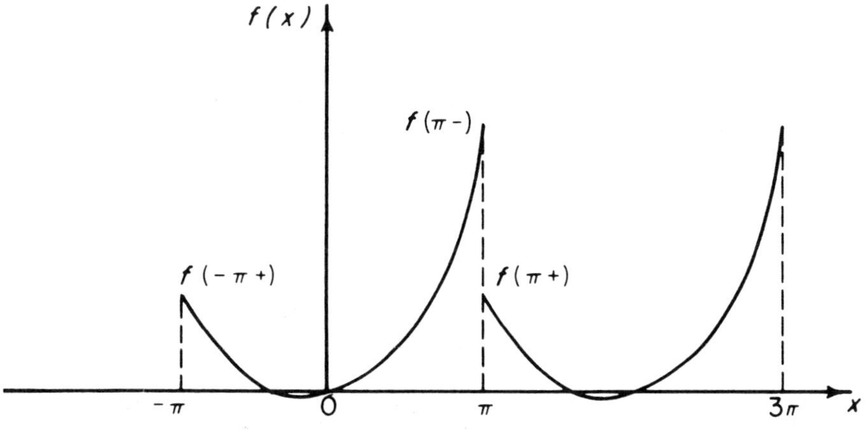


Figure 5.9.1

Simplification of Eq. (5.9.9) gives

$$\pi^2 = \frac{\pi^2}{3} + \sum_{k=1}^{\infty} \frac{4}{k^2}(-1)^{2k}$$

or

$$\frac{\pi^2}{6} = \sum_{k=1}^{\infty} 1/k^2$$

## 5.10 UNIFORM CONVERGENCE, DIFFERENTIATION, AND INTEGRATION

In the preceding section, we have proved the pointwise convergence of the Fourier series for a piecewise smooth function. Here, we shall consider several theorems without proof concerning uniform convergence, term by term differentiation, and integration of Fourier series.

***Theorem 5.10.1 (Uniform and Absolute Convergence Theorem)*** *Let $f(x)$ be a continuous function with period $2\pi$, and let $f'(x)$ be piecewise continuous in the interval $[-\pi, \pi]$. If, in addition, $f(-\pi) = f(\pi)$, then the Fourier series expansion for $f(x)$ is uniformly and absolutely convergent.*

In the preceding theorem, we have assumed that $f(x)$ is continuous and $f'(x)$ is piecewise continuous. With less stringent conditions on $f$, the following theorem can be proved.

***Theorem 5.10.2*** *Let $f(x)$ be piecewise smooth in the interval $[-\pi, \pi]$. If $f(x)$ is periodic with period $2\pi$, then the Fourier series for $f$ converges uniformly to $f$ in every closed interval containing no discontinuity.*

We note that the partial sums $s_n(x)$ of a Fourier series cannot approach the function $f(x)$ uniformly over any interval containing a point of discontinuity of $f$. The behavior of the deviation of $s_n(x)$ from $f(x)$ in such an interval is known as the *Gibbs phenomenon*. For instance, in the Example 5.6.1, the Fourier series of the function

$$f(x) = \begin{cases} -1 & -\pi < x < 0 \\ 1 & 0 < x < \pi \end{cases}$$

was given by

$$f(x) = \frac{4}{\pi} \sum_{k=1}^{\infty} \frac{\sin(2k-1)x}{(2k-1)}$$

If we plot the partial sums $s_n(x)$ against the $x$-axis, as shown in Fig. 5.10.1, we find that $s_n$ oscillate above and below the value of $f$. It can be observed that, near the discontinuous points $x = 0$ and $x = \pi$, $s_n$ deviate

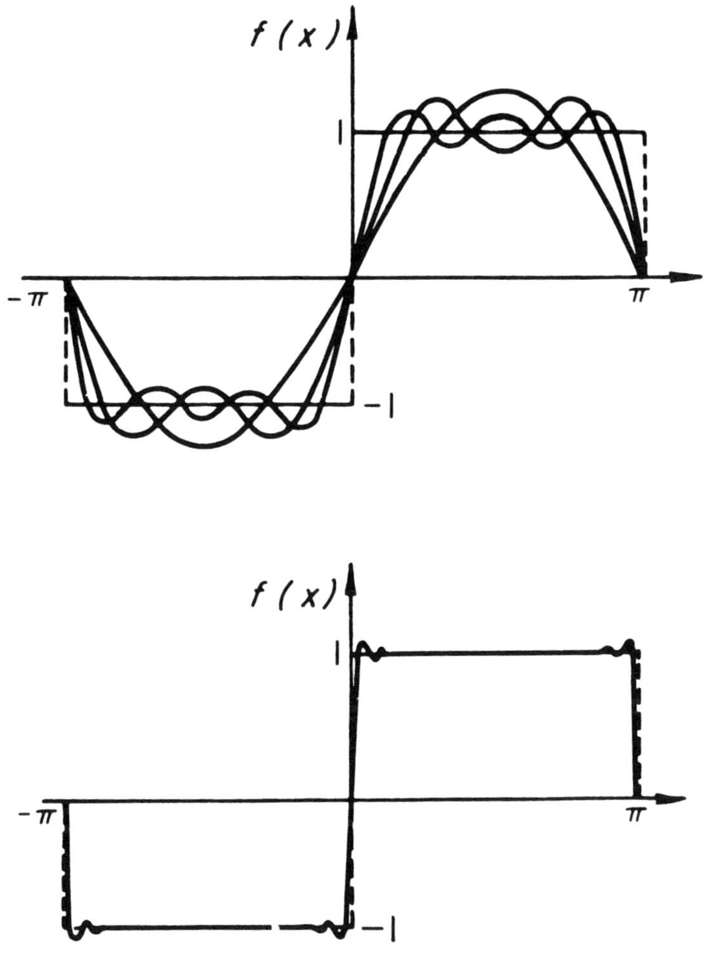


Figure 5.10.1

from the function rather significantly. Although the magnitude of oscillation decreases at all points in the interval for large $n$, very near the points of discontinuity the amplitude remains practically independent of $n$ as $n$ increases. This illustrates the fact that the Fourier series of a function $f$ does not converge uniformly on any interval which contains a discontinuity.

Termwise differentiation of Fourier series is, in general, not permissible. For example, the Fourier series for $f(x) = x$ (Example 5.5.3) is

$$x = 2\left[\sin x - \frac{\sin 2x}{2} + \frac{\sin 3x}{3} - \ldots\right]$$

which converges for all $x$, whereas the series after formal term by term differentiation,

$$2[\cos x - \cos 2x + \cos 3x - \ldots]$$

diverges for all $x$. The difficulty arises from the fact that the given function $f(x) = x$ in $[-\pi, \pi]$ when extended periodically is discontinuous at the points $\pm\pi$, $\pm 3\pi, \ldots$ We shall see below that the continuity of the periodic function is one of the conditions that must be met for the termwise differentiation of a Fourier series.

***Theorem 5.10.3 (Differentiation Theorem)*** *Let $f(x)$ be a continuous function in the interval $[-\pi, \pi]$ with $f(-\pi) = f(\pi)$, and let $f'(x)$ be piecewise smooth in that interval. Then the Fourier series for $f'$ can be obtained by termwise differentiation of the series for $f$, and the differentiated series converges pointwise to $f'$ at points of continuity and to $[f'(x) + f'(-x)]/2$ at discontinuous points.*

The termwise integration of Fourier series is possible under more general conditions than termwise differentiation. We recall that in calculus, the series of functions to be integrated must converge uniformly in order to assure the convergence of a termwise integrated series. However, in the case of Fourier series, this condition is not necessary.

***Theorem 5.10.4 (Integration Theorem)*** *Let $f(x)$ be piecewise continuous in $[-\pi, \pi]$, and periodic with period $2\pi$. Then the Fourier series of $f(x)$*

$$\frac{a_0}{2} + \sum_{k=1}^{\infty} (a_k \cos kx + b_k \sin kx)$$

*whether convergent or not, can be integrated term by term between any limits.*

EXAMPLE 5.10.1. In Example 5.6.2, we have found that $f(x) = |\sin x|$ is represented by the Fourier series

$$\sin x = \frac{2}{\pi} + \frac{4}{\pi} \sum_{k=1}^{\infty} \frac{\cos 2kx}{(1 - 4k^2)} \qquad 0 < x < \pi$$

Since $f(x) = |\sin x|$ is continuous in the interval $[-\pi, \pi]$ and $f(-\pi) = f(\pi)$, we differentiate term by term, obtaining

$$\cos x = -\frac{8}{\pi} \sum_{k=1}^{\infty} \frac{k \sin 2kx}{(1 - 4k^2)}$$

by use of Theorem 5.10.3, since $f'(x)$ is piecewise smooth in $[-\pi, \pi]$. In this way, we obtain the Fourier sine series expansion of the cosine function in $(0, \pi)$. Note that the reverse process is not permissible.

EXAMPLE 5.10.2. Consider the function $f(x) = x$ in the interval $-\pi < x < \pi$. As shown in Example 5.5.3,

$$x = 2\left[\sin x - \frac{\sin 2x}{2} + \frac{\sin 3x}{3} - \dots\right]$$

By Theorem 5.10.4, we can integrate the series term by term from $a$ to $x$ to obtain

$$\frac{1}{2}(x^2 - a^2) = 2\left[-\left(\cos x - \frac{\cos 2x}{2^2} + \frac{\cos 3x}{3^2} - \dots\right)\right.$$
$$\left. + \left(\cos a - \frac{\cos 2a}{2^2} + \frac{\cos 3a}{3^2} - \dots\right)\right]$$

To determine the sum of the series of constants, we write

$$\frac{x^2}{4} = C - \sum_{k=1}^{\infty} (-1)^{k+1} \frac{\cos kx}{k^2}$$

where $C$ is a constant. Since the series on the right is the Fourier series which converges uniformly, we can integrate the series term by term from $-\pi$ to $\pi$ to obtain

$$\int_{-\pi}^{\pi} \frac{x^2}{2}\, dx = 2\left[\int_{-\pi}^{\pi} C\, dx - \sum_{k=1}^{\infty} \frac{(-1)^{k+1}}{k^2} \int_{-\pi}^{\pi} \cos kx\, dx\right]$$

$$\frac{\pi^3}{3} = 2(2\pi C)$$

Hence

$$C = \frac{\pi^2}{12}$$

Therefore, by integrating the Fourier series of $f(x) = x$ in $(-\pi, \pi)$, we obtain the Fourier series expansion for the function $f(x) = x^2$ as

$$x^2 = 4\left[\frac{\pi^2}{12} - \sum_{k=1}^{\infty} (-1)^{k+1} \frac{\cos kx}{k^2}\right]$$

## 5.11 DOUBLE FOURIER SERIES

The theory of series expansions for functions of two variables is analogous to that of series expansions for functions of one variable. Here we shall present a short description of double Fourier series.

We have seen earlier that, if $f(x)$ is piecewise continuous and periodic with period $2\pi$, then the Fourier series

$$f(x) \sim \frac{a_0}{2} + \sum_{m=1}^{\infty} (a_m \cos mx + b_m \sin mx)$$

converges in the mean to $f(x)$. If $f$ is continuously differentiable, then its Fourier series converges uniformly.

For the sake of simplicity and convenience, let us consider the function $f(x, y)$ which is continuously differentiable (a stronger condition than necessary). Let $f(x, y)$ be periodic with period $2\pi$, that is,

$$f(x+2\pi, y) = f(x, y+2\pi) = f(x, y)$$

Then, if we hold $y$ fixed, we can expand $f(x, y)$ into a uniformly convergent Fourier series

$$f(x, y) = \frac{a_0(y)}{2} + \sum_{m=1}^{\infty} [a_m(y) \cos mx + b_m(y) \sin mx] \tag{5.11.1}$$

in which the coefficients are functions of $y$, namely,

$$a_m(y) = \frac{1}{\pi} \int_{-\pi}^{\pi} f(x, y) \cos mx \, dx$$

$$b_m(y) = \frac{1}{\pi} \int_{-\pi}^{\pi} f(x, y) \sin mx \, dx$$

These coefficients are continuously differentiable in $y$, and therefore, we can expand them in uniformly convergent series

$$\begin{aligned} a_m(y) &= \frac{a_{m0}}{2} + \sum_{n=1}^{\infty} a_{mn} \cos ny + b_{mn} \sin ny \\ b_m(y) &= \frac{c_{m0}}{2} + \sum_{n=1}^{\infty} c_{mn} \cos ny + d_{mn} \sin ny \end{aligned} \tag{5.11.2}$$

where

$$\begin{aligned} a_{mn} &= \frac{1}{\pi^2} \int_{-\pi}^{\pi} \int_{-\pi}^{\pi} f(x, y) \cos mx \cos ny \, dx \, dy \\ b_{mn} &= \frac{1}{\pi^2} \int_{-\pi}^{\pi} \int_{-\pi}^{\pi} f(x, y) \cos mx \sin ny \, dx \, dy \\ c_{mn} &= \frac{1}{\pi^2} \int_{-\pi}^{\pi} \int_{-\pi}^{\pi} f(x, y) \sin mx \cos ny \, dx \, dy \\ d_{mn} &= \frac{1}{\pi^2} \int_{-\pi}^{\pi} \int_{-\pi}^{\pi} f(x, y) \sin mx \sin ny \, dx \, dy \end{aligned} \tag{5.11.3}$$

Substitution of $a_m$ and $b_m$ into Eq. (5.11.1) yields

$$f(x, y) = \frac{a_{00}}{4} + \frac{1}{2} \sum_{n=1}^{\infty} [a_{0n} \cos ny + b_{0n} \sin ny] + \frac{1}{2} \sum_{m=1}^{\infty} [a_{m0} \cos mx + c_{m0} \sin mx] + \sum_{m=1}^{\infty} \sum_{n=1}^{\infty} [a_{mn} \cos mx \cos ny + b_{mn} \cos mx \sin ny + c_{mn} \sin mx \cos ny + d_{mn} \sin mx \sin ny] \quad (5.11.4)$$

which is called the *double Fourier series.*

a. When $f(-x, y) = f(x, y)$ and $f(x, -y) = f(x, y)$, all the coefficients vanish except $a_{mn}$, and the double Fourier series reduces to

$$f(x, y) = \sum_{m=1}^{\infty} \sum_{n=1}^{\infty} a_{mn} \cos mx \cos ny \quad (5.11.5)$$

where

$$a_{mn} = \frac{4}{\pi^2} \int_0^{\pi} \int_0^{\pi} f(x, y) \cos mx \cos ny \, dx \, dy$$

b. When $f(-x, y) = f(x, y)$ and $f(x, -y) = -f(x, y)$, we have

$$f(x, y) = \frac{1}{2} \sum_{n=1}^{\infty} b_{0n} \sin ny + \sum_{m=1}^{\infty} \sum_{n=1}^{\infty} b_{mn} \cos mx \sin ny \quad (5.11.6)$$

where

$$b_{mn} = \frac{4}{\pi^2} \int_0^{\pi} \int_0^{\pi} f(x, y) \cos mx \sin ny \, dx \, dy$$

c. When $f(-x, y) = -f(x, y)$ and $f(x, -y) = f(x, y)$, we have

$$f(x, y) = \frac{1}{2} \sum_{m=1}^{\infty} c_{m0} \sin mx + \sum_{m=1}^{\infty} \sum_{n=1}^{\infty} c_{mn} \sin mx \cos ny \quad (5.11.7)$$

where

$$c_{mn} = \frac{4}{\pi^2} \int_0^{\pi} \int_0^{\pi} f(x, y) \sin mx \cos ny \, dx \, dy$$

d. When $f(-x, y) = -f(x, y)$ and $f(x, -y) = -f(x, y)$, we have

$$f(x, y) = \sum_{m=1}^{\infty} \sum_{n=1}^{\infty} d_{mn} \sin mx \sin ny \quad (5.11.8)$$

where

$$d_{mn} = \frac{4}{\pi^2} \int_0^{\pi} \int_0^{\pi} f(x, y) \sin mx \sin ny \, dx \, dy$$

EXAMPLE 5.11.1. Expand the function $f(x, y) = xy$ into double Fourier series in the interval $-\pi < x < \pi$, $-\pi < y < \pi$.

Since $f(-x, y) = -xy = -f(x, y)$ and $f(x, -y) = -xy = -f(x, y)$, we find

$$d_{mn} = \frac{4}{\pi^2} \int_0^{\pi} \int_0^{\pi} xy \sin mx \sin ny \, dx \, dy$$

$$= (-1)^{(m+n)} \frac{4}{mn}$$

Thus, the double Fourier series for $f$ in $-\pi < x < \pi$, $-\pi < y < \pi$ is

$$f(x, y) = 4 \sum_{m=1}^{\infty} \sum_{n=1}^{\infty} (-1)^{m+n} \frac{\sin mx \sin ny}{mn}$$

## 5.12 FOURIER INTEGRALS

In earlier sections of this chapter, we have described Fourier series for functions which are periodic with period $2\pi$ in the interval $(-\infty, \infty)$. However, functions which are not periodic cannot be represented by Fourier series. In many problems of physical interest, it is desirable to develop an integral representation for such a function that is analogous to a Fourier series.

We have seen in Sec. 5.8 that the Fourier series for $f(x)$ in the interval $[-l, l]$ is

$$f(x) = \frac{a_0}{2} + \sum_{k=1}^{\infty} \left( a_k \cos \frac{k\pi x}{l} + b_k \sin \frac{k\pi x}{l} \right) \qquad (5.12.1)$$

where

$$a_k = \frac{1}{l} \int_{-l}^{l} f(t) \cos \frac{k\pi t}{l} \, dt \qquad k = 0, 1, 2, \ldots \qquad (5.12.2)$$

$$b_k = \frac{1}{l} \int_{-l}^{l} f(t) \sin \frac{k\pi t}{l} \, dt \qquad k = 1, 2, 3, \ldots \qquad (5.12.3)$$

Substituting (5.12.2) and (5.12.3) into (5.12.1), we have

$$f(x) = \frac{1}{2l}\int_{-l}^{l} f(t)\,dt + \frac{1}{l}\sum_{k=1}^{\infty}\left[\left[\int_{-l}^{l} f(t)\cos\frac{k\pi t}{l}\cdot\cos\frac{k\pi x}{l}\,dt\right]\right.$$
$$\left.+\left[\int_{-l}^{l} f(t)\sin\frac{k\pi t}{l}\cdot\sin\frac{k\pi x}{l}\,dt\right]\right]$$
$$= \frac{1}{2l}\int_{-l}^{l} f(t)\,dt + \frac{1}{l}\sum_{k=1}^{\infty}\int_{-l}^{l} f(t)\cos\left[\frac{k\pi}{l}(t-x)\right]dt \qquad (5.12.4)$$

Suppose that $f(x)$ is absolutely integrable, that is,

$$\int_{-\infty}^{\infty} |f(x)|\,dx$$

converges. Then

$$\frac{|a_0|}{2} = \frac{1}{2l}\left|\int_{-l}^{l} f(t)\,dt\right|$$
$$\leqslant \frac{1}{2l}\int_{-\infty}^{\infty} |f(t)|\,dt$$

which approaches zero as $l \to \infty$. Thus, holding $x$ fixed, as $l$ approaches infinity, Eq. (5.12.4) becomes

$$f(x) = \lim_{l\to\infty}\frac{1}{l}\sum_{k=1}^{\infty}\int_{-l}^{l} f(t)\cos\left[\frac{k\pi}{l}(t-x)\right]dt$$

Now let

$$\alpha_k = \frac{k\pi}{l} \qquad \Delta\alpha = \alpha_{k+1} - \alpha_k = \frac{\pi}{l}$$

Then $f(x)$ can be written as

$$f(x) = \lim_{l\to\infty}\sum_{k=1}^{\infty} F(\alpha_k)\Delta\alpha$$

where

$$F(\alpha) = \frac{1}{\pi}\int_{-l}^{l} f(t)\cos\left[\alpha(t-x)\right]dt$$

If we plot $F(\alpha)$ against $\alpha$, we can clearly see that the sum

$$\sum_{k=1}^{\infty} F(\alpha_k)\Delta\alpha$$

is an approximation to the area under the curve $y = F(\alpha)$ (see Fig. 5.12.1). As $l \to \infty$, $\Delta\alpha \to 0$, and the sum formally approaches the definite

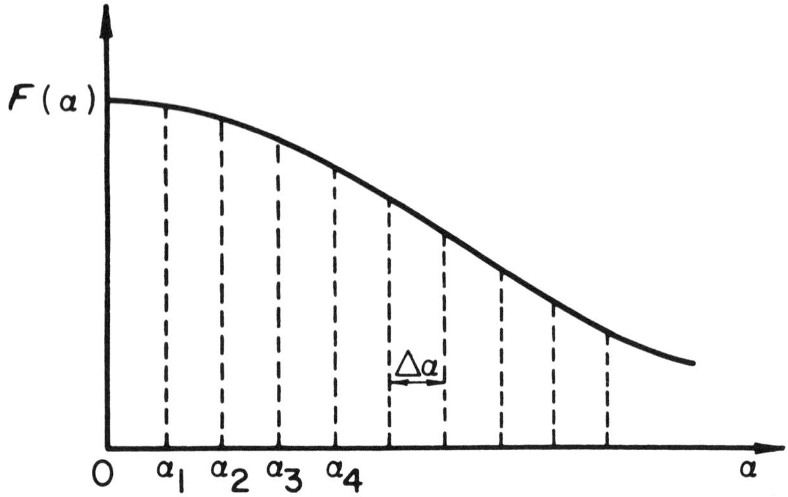


Figure 5.12.1

integral. We therefore have

$$f(x) = \int_0^\infty \left[ \frac{1}{\pi} \int_{-\infty}^\infty f(t) \cos \alpha(t-x)\, dt \right] d\alpha \tag{5.12.5}$$

which is the *Fourier integral* representation for the function $f(x)$. Its convergence to $f(x)$ is suggested but by no means established by the preceding arguments. We shall now prove that this representation is indeed valid if $f(x)$ satisfies certain conditions.

***Lemma 5.12.1*** *If $f$ is piecewise smooth in the interval $[0, b]$, then for $b > 0$,*

$$\lim_{\lambda\to\infty} \int_0^b f(x) \frac{\sin \lambda x}{x}\, dx = \frac{\pi}{2} f(0+)$$

PROOF

$$\int_0^b f(x) \frac{\sin \lambda x}{x}\, dx = \int_0^b f(0+) \frac{\sin \lambda x}{x}\, dx + \int_0^b \frac{f(x) - f(0+)}{x} \sin \lambda x\, dx$$
$$= f(0+) \int_0^{\lambda b} \frac{\sin t}{t}\, dt + \int_0^b \frac{f(x) - f(0+)}{x} \sin \lambda x\, dx$$

Since $f$ is piecewise smooth, the integrand of the last integral is bounded as $\lambda \to \infty$, and thus, by the Riemann–Lebesgue lemma, the last integral tends to zero as $\lambda \to \infty$. Hence

$$\lim_{\lambda\to\infty} \int_0^b f(x)\frac{\sin \lambda x}{x}\,dx = \frac{\pi}{2} f(0+) \tag{5.12.6}$$

since

$$\int_0^\infty \frac{\sin t}{t}\,dt = \frac{\pi}{2}$$ □

***Theorem 5.12.1 (Fourier Integral Theorem)*** *If $f$ is piecewise smooth in every finite interval, and absolutely integrable on $(-\infty, \infty)$, then*

$$\frac{1}{2}[f(x+)+f(x-)] = \frac{1}{\pi}\int_0^\infty \left[\int_{-\infty}^\infty f(t)\cos \alpha(t-x)\,dt\right] d\alpha$$

PROOF. Noting that $|\cos \alpha(t-x)| \leqslant 1$ and that by hypothesis

$$\int_{-\infty}^\infty f(t)\,dt < \infty$$

we see that the integral

$$\int_{-\infty}^\infty f(t)\cos \alpha(t-x)\,dt$$

converges independently of $\alpha$ and $x$. It therefore follows that in the double integral

$$I = \int_0^\lambda \left[\int_{-\infty}^\infty f(t)\cos \alpha(t-x)\,dt\right] d\alpha$$

the order of integration may be interchanged. We then have

$$\begin{aligned} I &= \int_{-\infty}^\infty \left[\int_0^\lambda f(t)\cos \alpha(t-x)\,d\alpha\right] dt \\ &= \int_{-\infty}^\infty f(t)\left[\frac{\sin \lambda(t-x)}{(t-x)}\right] dt \\ &= \left[\int_{-\infty}^{-M} + \int_{-M}^{x} + \int_x^M + \int_M^\infty\right] f(t)\frac{\sin \lambda(t-x)}{(t-x)}\,dt \end{aligned}$$

If we substitute $u = t - x$, we have

$$\int_x^M f(t)\frac{\sin \lambda(t-x)}{(t-x)}\,dt = \int_0^{M-x} f(u+x)\frac{\sin \lambda u}{u}\,du$$

which is equal to $\pi f(x+)/2$ in the limit $\lambda \to \infty$, by Lemma 5.12.1. Similarly, the second integral tends to $\pi f(x-)/2$ when $\lambda \to \infty$. If we make

$M$ sufficiently large, the absolute values of the first and the last integrals are each less than $\varepsilon/2$. Consequently, as $\lambda \to \infty$

$$\int_0^\infty \left[\int_{-\infty}^\infty f(t) \cos \alpha(t-x)\, dt\right] d\alpha = \frac{\pi}{2}[f(x+) + f(x-)] \qquad (5.12.7)$$

If $f$ is continuous at the point $x$, then

$$f(x+) = f(x-) = f(x)$$

so that Eq. (5.12.7) reduces to

$$f(x) = \frac{1}{\pi}\int_0^\infty \left[\int_{-\infty}^\infty f(t) \cos \alpha(t-x)\, dt\right] d\alpha \qquad (5.12.8)$$

□

We may express the Fourier integral (5.12.8) in complex form. In this case, we substitute

$$\cos \alpha(t-x) = \frac{1}{2}[e^{i\alpha(t-x)} + e^{-i\alpha(t-x)}]$$

into Eq. (5.12.8) and write it as the sum of two integrals

$$f(x) = \frac{1}{2\pi}\int_0^\infty \int_{-\infty}^\infty f(t) e^{i\alpha(t-x)}\, dt\, d\alpha + \frac{1}{2\pi}\int_0^\infty \int_{-\infty}^\infty f(t) e^{-i\alpha(t-x)}\, dt\, d\alpha$$

Changing the integration variable from $\alpha$ to $-\alpha$ in the second integral, we obtain

$$f(x) = \frac{1}{\sqrt{2\pi}}\int_{-\infty}^\infty \left[\frac{1}{\sqrt{2\pi}}\int_{-\infty}^\infty f(t) e^{i\alpha t}\, dt\right] e^{-i\alpha x}\, d\alpha \qquad (5.12.9)$$

This is the *complex form of the Fourier integral formula.* Now we assume that $f(x)$ is either an even or an odd function. Any function that is not even or odd can be expressed as a sum of two such functions. Expanding the cosine function in (5.12.8), we obtain the *Fourier cosine formula*

$$f(x) = f(-x) = \frac{2}{\pi}\int_0^\infty \cos \alpha x\, d\alpha \int_0^\infty \cos \alpha t\, f(t)\, dt \qquad (5.12.10)$$

Similarly, for an odd function, we obtain the *Fourier sine formula*

$$f(x) = -f(-x) = \frac{2}{\pi}\int_0^\infty \sin \alpha x\, d\alpha \int_0^\infty \sin \alpha t\, f(t)\, dt \qquad (5.12.11)$$

EXAMPLE 5.12.1. The rectangular pulse can be expressed as a sum of Heaviside functions

$$f(x) = H(x+1) - H(x-1)$$

Find its Fourier integral representation.

From (5.12.5) we find

$$f(x)=\frac{1}{\pi}\int_0^\infty d\alpha\left[\int_{-1}^{1}\cos\alpha(t-x)\,dt\right]d\alpha$$

$$=\frac{1}{\pi}\int_0^\infty\left[\cos\alpha x\int_{-1}^{1}\cos\alpha t\,dt+\sin\alpha x\int_{-1}^{1}\sin\alpha t\,dt\right]d\alpha$$

$$=\frac{2}{\pi}\int_0^\infty\frac{\sin\alpha}{\alpha}\cos\alpha x\,d\alpha$$

EXAMPLE 5.12.2. Find the Fourier cosine integral representation of the function

$$f(x)=\begin{cases}1, & 0<x<1\\ 0, & x\geqslant 1\end{cases}$$

We have, from (5.12.10),

$$f(x)=\frac{2}{\pi}\int_0^\infty\cos\alpha x\,d\alpha\int_0^1\cos\alpha t\,dt=\frac{2}{\pi}\int_0^\infty\frac{\sin\alpha}{\alpha}\cos\alpha x\,d\alpha$$

or

$$1=\frac{2}{\pi}\int_0^\infty\frac{\sin\alpha}{\alpha}\cos\alpha x\,d\alpha$$

## 5.13 EXERCISES

**1.** Find the Fourier series of the following functions:

a. $f(x)=\begin{cases}x & & -\pi<x<0\\ h & \text{a constant} & 0<x<\pi\end{cases}$

b. $f(x)=\begin{cases}1 & -\pi<x<0\\ x^2 & 0<x<\pi\end{cases}$

c. $f(x)=x+\sin x \qquad -\pi<x<\pi$

d. $f(x)=1+x \qquad -\pi<x<\pi$

e. $f(x)=e^x \qquad -\pi<x<\pi$

f. $f(x)=1+x+x^2 \qquad -\pi<x<\pi$

**2.** Determine the Fourier sine series of the following functions:

a. $f(x)=\pi-x \qquad 0<x<\pi$

b. $f(x)=\begin{cases}1 & 0<x<\pi/2\\ 2 & \pi/2<x<\pi\end{cases}$

c. $f(x)=x^2$ $\quad 0<x<\pi$

d. $f(x)=\cos x$ $\quad 0<x<\pi$

e. $f(x)=x^3$ $\quad 0<x<\pi$

f. $f(x)=e^x$ $\quad 0<x<\pi$

**3.** Obtain the Fourier cosine series representations for the following functions:

a. $f(x)=\pi+x$ $\quad 0<x<\pi$

b. $f(x)=x$ $\quad 0<x<\pi$

c. $f(x)=x^2$ $\quad 0<x<\pi$

d. $f(x)=\sin 3x$ $\quad 0<x<\pi$

e. $f(x)=e^x$ $\quad 0<x<\pi$

f. $f(x)=\cosh x$ $\quad 0<x<\pi$

**4.** Expand the following functions in a Fourier series:

a. $f(x)=x^2+x$ $\quad -1<x<1$

b. $f(x)=\begin{cases} 1 & 0<x<3 \\ 0 & 3<x<6 \end{cases}$

c. $f(x)=\sin \pi x/l$ $\quad 0<x<l$

d. $f(x)=x^3$ $\quad -2<x<2$

e. $f(x)=e^{-x}$ $\quad 0<x<1$

f. $f(x)=\sinh x$ $\quad -1<x<1$

**5.** Find the following functions in a complex Fourier series:

a. $f(x)=e^{2x}$ $\quad -\pi<x<\pi$

b. $f(x)=\cosh x$ $\quad -\pi<x<\pi$

c. $f(x)=\begin{cases} 1 & -\pi<x<0 \\ \cos x & 0<x<\pi \end{cases}$

d. $f(x)=x$ $\quad -1<x<1$

e. $f(x)=x^2$ $\quad -\pi<x<\pi$

f. $f(x)=\sinh \pi x/2$ $\quad -2<x<2$

**6.** a. Find the Fourier series expansion of the function

$$f(x)=\begin{cases} 0 & -\pi<x<0 \\ x/2 & 0<x<\pi \end{cases}$$

b. With the use of this series, show that

$$\frac{\pi^2}{8} = 1 + \frac{1}{3^2} + \frac{1}{5^2} + \frac{1}{7^2} + \ldots$$

**7.** a. Determine the Fourier series of the function

$$f(x) = x^2, \qquad -l < x < l$$

b. With the use of this series, show that

$$\frac{\pi^2}{12} = 1 - \frac{1}{2^2} + \frac{1}{3^2} - \frac{1}{4^2} + \ldots$$

**8.** Determine the Fourier series expansion of each of the following functions by performing the differentiation of the appropriate Fourier series:

a. $\sin^2 x \qquad 0 < x < \pi$

b. $\cos^2 x \qquad 0 < x < \pi$

c. $\sin x \cos x \qquad 0 < x < \pi$

d. $\cos x + \cos 2x \qquad 0 < x < \pi$

e. $\cos x \cos 2x \qquad 0 < x < \pi$

**9.** Find the functions represented by the new series which are obtained by termwise integration of the following series from 0 to $x$:

a. $\displaystyle\sum_{k=1}^{\infty} \frac{(-1)^{k+1}}{k} \sin kx = x/2 \qquad -\pi < x < \pi$

b. $\displaystyle\frac{3}{2} + \frac{1}{\pi} \sum_{k=1}^{\infty} \frac{1-(-1)^k}{k} \sin kx = \begin{cases} 1 & -\pi < x < 0 \\ 2 & 0 < x < \pi \end{cases}$

c. $\displaystyle\sum_{k=1}^{\infty} (-1)^{k+1} \frac{\cos kx}{k} = \ln\left(2\cos\frac{x}{2}\right) \qquad -\pi < x < \pi$

d. $\displaystyle\sum_{k=1}^{\infty} \frac{\sin(2k+1)x}{(2k+1)^3} = \frac{\pi^2 x - \pi x^2}{8} \qquad 0 < x < 2\pi$

e. $\displaystyle\frac{4}{\pi} \sum_{k=1}^{\infty} \frac{\sin(2k-1)x}{(2k-1)} = \begin{cases} -1 & -\pi < x < 0 \\ 1 & 0 < x < \pi \end{cases}$

**10.** Determine the double Fourier series of the following functions:

a. $f(x, y) = 1 \qquad 0 < x < \pi \qquad 0 < y < \pi$

b. $f(x, y) = xy^2 \qquad 0 < x < \pi \qquad 0 < y < \pi$

c. $f(x, y) = x^2 y^2 \qquad 0 < x < \pi \qquad 0 < y < \pi$

d. $f(x, y) = x^2 + y \qquad -\pi < x < \pi \qquad -\pi < y < \pi$

e. $f(x, y) = x \sin y \qquad -\pi < x < \pi \qquad -\pi < y < \pi$

f. $f(x, y) = e^{x+y} \qquad -\pi < x < \pi \qquad -\pi < y < \pi$

**11.** Deduce the general double Fourier series expansion formula for the function $f(x, y)$ in the rectangle $-a < x < a$, $-b < y < b$.

**12.** Prove the *Weierstrass Approximation Theorem*: If $f$ is a continuous function on the interval $-\pi \leqslant x \leqslant \pi$ and if $f(-\pi) = f(\pi)$, then for any $\varepsilon > 0$ there exists a trigonometric polynomial

$$T(x) = \frac{a_0}{2} + \sum_{k-1}^{n} (a_k \cos kx + b_k \sin kx)$$

such that

$$|f(x) - T(x)| < \varepsilon$$

for all $x$ in $[-\pi, \pi]$.

**13.** Use the Fourier cosine or sine integral formula to show that

(a) $e^{-\alpha x} = \dfrac{2}{\pi} \displaystyle\int_0^\infty \frac{\alpha}{\alpha^2 + \beta^2} \cos \beta x \, d\beta, \quad x \geqslant 0, \ \alpha > 0$

(b) $e^{-\alpha x} = \dfrac{2}{\pi} \displaystyle\int_0^\infty \frac{\beta}{\alpha^2 + \beta^2} \sin \beta x \, d\beta, \quad x > 0, \ \alpha > 0$

**14.** Show that the Fourier integral representation of the function

$$f(x) = \begin{Bmatrix} x^2, & 0 < x < a \\ 0, & x > a \end{Bmatrix}$$

is

$$f(x) = \frac{2}{\pi} \int_0^\infty \left[ \left( a^2 - \frac{2}{\alpha^2} \right) \sin a\alpha + \frac{2a}{\alpha} \cos a\alpha \right] \frac{\cos \alpha x}{\alpha} \, d\alpha$$

CHAPTER SIX

# Method of Separation of Variables

## 6.1 SEPARATION OF VARIABLES

We have so far been mainly concerned with initial value problems. In this section, we shall introduce one of the most common and elementary methods, called the method of separation of variables, for solving initial-boundary value problems. The class of problems for which this method is applicable contains a wide range of problems of mathematical physics, applied mathematics, and engineering sciences.

We now describe the method of separation of variables and examine the conditions of applicability of the method to problems which involve second-order partial differential equations in two independent variables.

We consider the second-order homogeneous equation

$$a^* u_{x^*x^*} + b^* u_{x^*y^*} + c^* u_{y^*y^*} + d^* u_{x^*} + e^* u_{y^*} + f^* u = 0 \qquad (6.1.1)$$

where $a^*$, $b^*$, $c^*$, $d^*$, $e^*$, and $f^*$ are functions of $x^*$ and $y^*$.

We have stated in Chapter 3 that by transformation

$$x = x(x^*, y^*)$$
$$y = y(x^*, y^*), \qquad \frac{\partial(x, y)}{\partial(x^*, y^*)} \neq 0 \qquad (6.1.2)$$

we can always transform Eq. (6.1.1) into canonical form

$$a(x, y)u_{xx} + c(x, y)u_{yy} + d(x, y)u_x + e(x, y)u_y + f(x, y)u = 0 \quad (6.1.3)$$

which when

i. $a = -c$ is hyperbolic
ii. $a = 0$ or $c = 0$ is parabolic
iii. $a = c$ is elliptic

We assume a solution in the form

$$u(x, y) = X(x)Y(y) \neq 0 \tag{6.1.4}$$

where $X$ and $Y$ are, respectively, functions of $x$ and $y$ alone, and are twice continuously differentiable. Substituting Eq. (6.1.4) into Eq. (6.1.3), we obtain

$$aX''Y + cXY'' + dX'Y + eXY + fXY = 0 \tag{6.1.5}$$

where the primes denote differentiation with the respective variables. Let there exist a function $p(x, y)$, such that, if we divide Eq. (6.1.5) by $p(x, y)$, we obtain

$$\begin{aligned} a_1(x)X''Y + b_1(y)XY'' + a_2(x)X'Y + b_2(y)XY' \\ + [a_3(x) + b_3(y)]XY = 0 \end{aligned} \tag{6.1.6}$$

Dividing Eq. (6.1.6) again by $XY$, we obtain

$$\left[a_1\frac{X''}{X} + a_2\frac{X'}{X} + a_3\right] = -\left[b_1\frac{Y''}{Y} + b_2\frac{Y'}{Y} + b_3\right] \tag{6.1.7}$$

The left side of Eq. (6.1.7) is a function of $x$ only. The right side of Eq. (6.1.7) depends only upon $y$. Thus, we differentiate Eq. (6.1.7) with respect to $x$ to obtain

$$\frac{d}{dx}\left[a_1\frac{X''}{X} + a_2\frac{X'}{X} + a_3\right] = 0 \tag{6.1.8}$$

Integration of Eq. (6.1.8) yields

$$a_1\frac{X''}{X} + a_2\frac{X'}{X} + a_3 = \lambda \tag{6.1.9}$$

where $\lambda$ is a separation constant. From Eqs. (6.1.7) and (6.1.9), we have

$$b_1\frac{Y''}{Y} + b_2\frac{Y'}{Y} + b_3 = -\lambda \tag{6.1.10}$$

We may rewrite Eqs. (6.1.9) and (6.1.10) in the form

$$a_1X'' + a_2X' + (a_3 - \lambda)X = 0 \tag{6.1.11}$$

and

$$b_1Y'' + b_2Y' + (b_3 + \lambda)Y = 0 \tag{6.1.12}$$

Thus, $u(x, y)$ is the solution of Eq. (6.1.3) if $X(x)$ and $Y(y)$ are the solutions of the ordinary differential equations (6.1.11) and (6.1.12) respectively.

If the coefficients in Eq. (6.1.1) are constant, then the reduction of Eq. (6.1.1) to canonical form is no longer necessary. To illustrate this, let us consider the second-order equation

$$Au_{xx} + Bu_{xy} + Cu_{yy} + Du_x + Eu_y + Fu = 0 \qquad (6.1.13)$$

where $A$, $B$, $C$, $D$, $E$, and $F$ are constants which are not all zero.

As before, we assume a solution in the form

$$u(x, y) = X(x)Y(y) \neq 0$$

Substituting this in Eq. (6.1.13), we obtain

$$AX''Y + BX'Y' + CXY'' + DX'Y + EXY' + FXY = 0 \quad (6.1.14)$$

Division of this equation by $AXY$ yields

$$\frac{X''}{X} + \frac{B}{A}\frac{X'}{X}\frac{Y'}{Y} + \frac{C}{A}\frac{Y''}{Y} + \frac{D}{A}\frac{X'}{X} + \frac{E}{A}\frac{Y'}{Y} + \frac{F}{A} = 0, \quad A \neq 0 \qquad (6.1.15)$$

We differentiate this equation with respect to $x$, and obtain

$$\left(\frac{X''}{X}\right)' + \frac{B}{A}\left(\frac{X'}{X}\right)'\frac{Y'}{Y} + \frac{D}{A}\left(\frac{X'}{X}\right)' = 0 \qquad (6.1.16)$$

Thus, we have

$$\frac{\left(\frac{X''}{X}\right)'}{\frac{B}{A}\left(\frac{X'}{X}\right)'} + \frac{D}{B} = -\frac{Y'}{Y} \qquad (6.1.17)$$

This equation is separated, so that both sides must be equal to a constant $\lambda$. Therefore, we obtain

$$Y' + \lambda Y = 0 \qquad (6.1.18)$$

$$\left(\frac{X''}{X}\right)' + \left(\frac{D}{B} - \lambda\right)\frac{B}{A}\left(\frac{X'}{X}\right)' = 0 \qquad (6.1.19)$$

Integrating Eq. (6.1.19) with respect to $x$, we obtain

$$\frac{X''}{X} + \left(\frac{D}{B} - \lambda\right)\frac{B}{A}\left(\frac{X'}{X}\right) = -\beta \qquad (6.1.20)$$

where $\beta$ is a constant to be determined. By substituting Eq. (6.1.18) into the original Eq. (6.1.15), we obtain

$$X'' + \left(\frac{D}{B} - \lambda\right)\frac{B}{A}X' + \left(\lambda^2 - \frac{E}{C}\lambda + \frac{F}{C}\right)\frac{C}{A}X = 0 \qquad (6.1.21)$$

Comparing Eqs. (6.1.20) and (6.1.21), we clearly find

$$\beta = \left(\lambda^2 - \frac{E}{C}\lambda + \frac{F}{C}\right)\frac{C}{A}$$

Therefore, $u(x, y)$ is a solution of Eq. (6.1.13) if $X(x)$ and $Y(y)$ satisfy the ordinary differential equations (6.1.21) and (6.1.18) respectively.

We have just described the conditions on the separability of a given partial differential equation. Now, we shall take a look at the boundary conditions involved. There are several types of boundary conditions. The ones that appear most frequently in problems of applied mathematics and mathematical physics include

i. Dirichlet condition: $u$ is prescribed on a boundary
ii. Neumann condition: $\partial u/\partial n$ is prescribed on a boundary
iii. Mixed condition: $(\partial u/\partial n) + hu$ is prescribed on a boundary where $\partial u/\partial n$ is the directional derivative of $u$ along the outward normal to the boundary, and $h$ is a given continuous function on the boundary. For details, see Chapter 8.

Besides these three boundary conditions, also known as, the first, second, and third conditions, there are other conditions, such as the Robin condition; one condition is prescribed on one portion of a boundary and the other is given on the remainder of the boundary. We shall consider a variety of conditions as we treat problems later.

To separate boundary conditions, such as the ones listed above, it is best to choose a coordinate system suitable to a boundary. For instance, we choose the Cartesian coordinate system $(x, y)$ for a rectangular region such that the boundary is described by the coordinate lines $x = \text{constant}$ and $y = \text{constant}$, and the polar coordinate system $(r, \theta)$ for a circular region so that the boundary is described by the lines $r = \text{constant}$ and $\theta = \text{constant}$.

Another condition that must be imposed on the separability of boundary conditions is that boundary conditions, say at $x = x_0$, must contain the derivatives of $u$ with respect to $x$ only, and their coefficients must depend only on $x$. For example, the boundary condition

$$[u + u_y]_{x=x_0} = 0$$

cannot be separated. It is needless to say that a mixed condition, such as $u_x + u_y$, cannot be prescribed on an axis.

## 6.2 THE VIBRATING STRING PROBLEM

As a first example, we shall consider the problem of the vibrating string stretched along the $x$-axis from 0 to $l$, fixed at its end points. We have seen in Chapter 4 that the problem is given by

$$u_{tt} - c^2 u_{xx} = 0 \qquad 0 < x < l \qquad t > 0 \tag{6.2.1}$$

$$u(x, 0) = f(x) \qquad 0 \leqslant x \leqslant l \tag{6.2.2}$$

$$u_t(x, 0) = g(x) \qquad 0 \leqslant x \leqslant l \tag{6.2.3}$$

$$u(0, t) = 0 \qquad t \geqslant 0 \tag{6.2.4}$$

$$u(l, t) = 0 \qquad t \geqslant 0 \tag{6.2.5}$$

where $f$ and $g$ are the initial displacement and initial velocity respectively.

By the method of separation of variables, we assume a solution in the form

$$u(x, t) = X(x)T(t) \neq 0 \tag{6.2.6}$$

If we substitute Eq. (6.2.6) into Eq. (6.2.1), we obtain

$$XT'' = c^2X''T$$

and hence

$$\frac{X''}{X} = \frac{1}{c^2}\frac{T''}{T} \tag{6.2.7}$$

whenever $XT \neq 0$. Since the left side of Eq. (6.2.7) is independent of $t$ and the right side is independent of $x$, we must have

$$\frac{X''}{X} = \frac{1}{c^2}\frac{T''}{T} = \lambda$$

where $\lambda$ is a separation constant. Thus,

$$X'' - \lambda X = 0 \tag{6.2.8}$$

$$T'' - \lambda c^2 T = 0 \tag{6.2.9}$$

We now separate the boundary conditions. From Eqs. (6.2.4) and (6.2.6), we obtain

$$u(0, t) = X(0)T(t) = 0$$

We know that $T(t) \neq 0$ for all values of $t$, and therefore

$$X(0) = 0 \tag{6.2.10}$$

In a similar manner, boundary condition (6.2.5) implies

$$X(l) = 0 \tag{6.2.11}$$

To determine $X(x)$ we first solve the *eigenvalue problem* (eigenvalue problems are treated in Chapter 7)

$$\begin{aligned} X'' + \lambda X &= 0 \\ X(0) &= 0 \\ X(l) &= 0 \end{aligned} \tag{6.2.12}$$

We look for values of $\lambda$ which give us nontrivial solutions. We investigate three possible cases

$$\lambda > 0, \qquad \lambda = 0, \qquad \lambda < 0$$

*Case 1.* $\lambda > 0$ The general solution in this case is of the form

$$X(x) = Ae^{-\sqrt{\lambda}x} + Be^{\sqrt{\lambda}x}$$

where $A$ and $B$ are arbitrary constants. To satisfy the boundary conditions we must have

$$\begin{aligned} A + B &= 0 \\ Ae^{-\sqrt{\lambda}l} + Be^{\sqrt{\lambda}l} &= 0 \end{aligned} \tag{6.2.13}$$

We see that the determinant of the system (6.2.13) is different from zero. Consequently, $A$ and $B$ must both be zero, and hence the general solution $X(x)$ is identically zero. The solution is trivial.

*Case 2.* $\lambda = 0$ Here, the general solution is

$$X(x) = A + Bx$$

Applying the boundary conditions, we have

$$\begin{aligned} A &= 0 \\ A + Bl &= 0 \end{aligned}$$

Hence $A = B = 0$. The solution is thus identically zero.

*Case 3.* $\lambda < 0$ In this case, the general solution assumes the form

$$X(x) = A \cos \sqrt{-\lambda}\, x + B \sin \sqrt{-\lambda}\, x$$

From the condition $X(0) = 0$, we obtain $A = 0$. The condition $X(l) = 0$ gives

$$B \sin \sqrt{-\lambda}\, l = 0$$

If $B = 0$, the solution is trivial. For nontrivial solutions,

$$\sin \sqrt{-\lambda}\, l = 0$$

This equation is satisfied when

$$\sqrt{-\lambda}\, l = n\pi \quad \text{for} \quad n = 1, 2, 3, \ldots$$

or

$$-\lambda_n = (n\pi/l)^2 \tag{6.2.14}$$

For this infinite set of discrete values of $\lambda$, the problem has a nontrivial solution. These values of $\lambda_n$ are called the *eigenvalues* of the problem, and the functions

$$\sin (n\pi/l)x, \qquad n = 1, 2, 3, \ldots$$

are the corresponding *eigenfunctions*.

We note that it is not necessary to consider negative values of $n$ since

$$\sin(-n)\pi x/l = -\sin n\pi x/l$$

No new solution is obtained in this way.

The solutions of problem (6.2.12) are, therefore,

$$X_n(x) = B_n \sin(n\pi x/l) \tag{6.2.15}$$

For $\lambda = \lambda_n$, the general solution of Eq. (6.2.9) may be written in the form

$$T_n(t) = C_n \cos\frac{n\pi c}{l}t + D_n \sin\frac{n\pi c}{l}t \tag{6.2.16}$$

where $C_n$ and $D_n$ are arbitrary constants.

Thus, the functions

$$u_n(x, t) = X_n(x)T_n(t) = \left(a_n \cos\frac{n\pi c}{l}t + b_n \sin\frac{n\pi c}{l}t\right) \sin\frac{n\pi x}{l}$$

satisfy Eq. (6.2.1) and the boundary conditions (6.2.4) and (6.2.5), where $a_n = B_n C_n$ and $b_n = B_n D_n$.

Since Eq. (6.2.1) is linear and homogeneous, by the superposition principle, the infinite series

$$u(x, t) = \sum_{n=1}^{\infty} \left(a_n \cos\frac{n\pi c}{l}t + b_n \sin\frac{n\pi c}{l}t\right) \sin\frac{n\pi x}{l} \tag{6.2.17}$$

is also a solution, provided it converges and is twice continuously differentiable with respect to $x$ and $t$. Since each term of the series satisfies the boundary conditions (6.2.4) and (6.2.5), the series satisfies these conditions. There remain two more initial conditions to be satisfied. From these conditions, we shall determine the constants $a_n$ and $b_n$.

First we differentiate the series (6.2.17) with respect to $t$. We have

$$u_t = \sum_{n=1}^{\infty} \frac{n\pi c}{l}\left(-a_n \sin\frac{n\pi c}{l}t + b_n \cos\frac{n\pi c}{l}t\right) \sin\frac{n\pi x}{l} \tag{6.2.18}$$

Then applying the initial conditions (6.2.2) and (6.2.3), we obtain

$$u(x, 0) = f(x) = \sum_{n=1}^{\infty} a_n \sin\frac{n\pi x}{l} \tag{6.2.19}$$

$$u_t(x, 0) = g(x) = \sum_{n=1}^{\infty} b_n \left(\frac{n\pi c}{l}\right) \sin\frac{n\pi x}{l} \tag{6.2.20}$$

These equations will be satisfied if $f(x)$ and $g(x)$ can be represented by Fourier sine series. The coefficients are given by

$$\begin{aligned} a_n &= \frac{2}{l}\int_0^l f(x) \sin\frac{n\pi x}{l}\,dx \\ b_n &= \frac{2}{n\pi c}\int_0^l g(x) \sin\frac{n\pi x}{l}\,dx \end{aligned} \tag{6.2.21}$$

The solution of the vibrating string problem is therefore given by the series (6.2.17) where the coefficients $a_n$ and $b_n$ are determined by the formulae (6.2.21).

EXAMPLE 6.2.1. The Plucked String

As a special case of the problem just treated, consider a stretched string fixed at both ends. Suppose the string is raised to a height $h$ at $x=a$ and then released. The string will oscillate freely. The initial conditions, as shown in Fig. 6.2.1, may be written

$$u(x,0)=f(x)=\begin{cases} hx/a, & 0\leqslant x\leqslant a \\ h(l-x)/(l-a), & a\leqslant x\leqslant l \end{cases}$$

$$u_t(x,0)=g(x)=0$$

Since $g(x)=0$ the coefficients $b_n$ are identically equal to zero. The coefficients $a_n$, according to Eq. (6.2.21), are given by

$$\begin{aligned} a_n &= \frac{2}{l}\int_0^l f(x)\sin\frac{n\pi x}{l}\,dx \\ &= \frac{2}{l}\int_0^a \frac{hx}{a}\sin\frac{n\pi x}{l}\,dx+\frac{2}{l}\int_a^l \frac{h(l-x)}{(l-a)}\sin\frac{n\pi x}{l}\,dx \end{aligned}$$

Integration and simplification yields

$$a_n=\frac{2hl^2}{\pi^2 a(l-a)}\frac{1}{n^2}\sin\frac{n\pi a}{l}$$

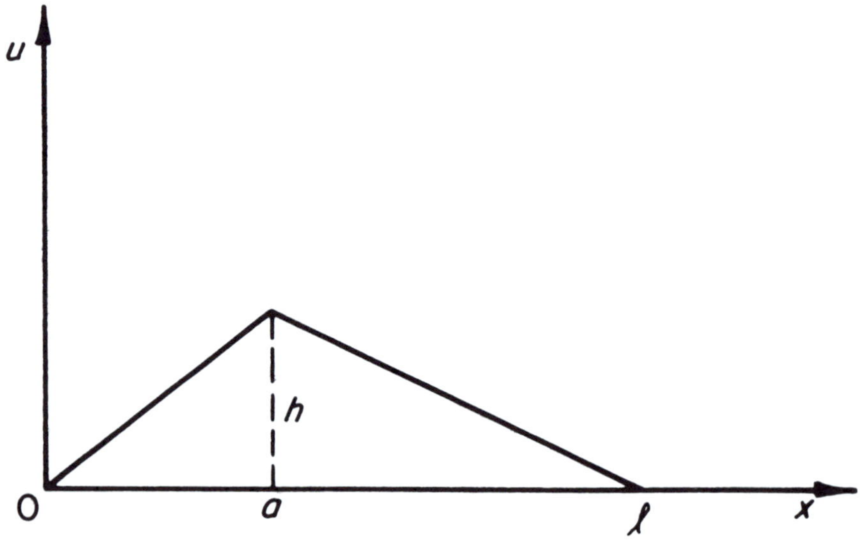


Figure 6.2.1

Thus, the displacement of the plucked string is

$$u(x, t) = \frac{2hl^2}{\pi^2 a(l-a)} \sum_{n=1}^{\infty} \frac{1}{n^2} \sin\frac{n\pi a}{l} \sin\frac{n\pi x}{l} \cos\frac{n\pi c}{l} t$$

EXAMPLE 6.2.2. The Struck String

Here, we consider the string with no initial displacement. Let the string be struck at $x = a$ so that the initial velocity is given by

$$u_t(x, 0) = \begin{cases} \dfrac{v_0}{a} x, & 0 \leqslant x \leqslant a \\ v_0(l-x)/(l-a), & a \leqslant x \leqslant l \end{cases}$$

Since $u(x, 0) = 0$, we have $a_n = 0$. By applying Eq. (6.2.21), we find that

$$\begin{aligned} b_n &= \frac{2}{n\pi c} \int_0^a \frac{v_0}{a} x \sin\frac{n\pi x}{l}\, dx + \frac{2}{n\pi c} \int_a^l v_0 \frac{(l-x)}{(l-a)} \sin\frac{n\pi x}{l}\, dx \\ &= \frac{2v_0 l^3}{\pi^3 ca(l-a)} \frac{1}{n^3} \sin\frac{n\pi a}{l} \end{aligned}$$

Hence the displacement of the struck string is

$$u(x, t) = \frac{2v_0 l^3}{\pi^3 ca(l-a)} \sum_{n=1}^{\infty} \frac{1}{n^3} \sin\frac{n\pi a}{l} \sin\frac{n\pi x}{l} \sin\frac{n\pi c}{l} t$$

## 6.3 EXISTENCE AND UNIQUENESS OF SOLUTION OF THE VIBRATING STRING PROBLEM

In the preceding section we found that the initial-boundary value problem (6.2.1)–(6.2.5) has a formal solution given by (6.2.17). We shall now show that the expression (6.2.17) is the solution of the problem under certain conditions.

First we see that

$$u_1(x, t) = \sum_{n=1}^{\infty} a_n \cos\frac{n\pi c}{l} t \sin\frac{n\pi x}{l} \tag{6.3.1}$$

is the formal solution of the problem (6.2.1)–(6.2.5) with $g(x) \equiv 0$, and

$$u_2(x, t) = \sum_{n=1}^{\infty} b_n \sin\frac{n\pi c}{l} t \sin\frac{n\pi x}{l} \tag{6.3.2}$$

is the formal solution of the above problem with $f(x) \equiv 0$. By linearity of the problem, the solution (6.2.17) may be considered as the sum of the two formal solutions (6.3.1) and (6.3.2).

We first assume that $f(x)$ and $f'(x)$ are continuous on $[0, l]$, and $f(0) = f(l) = 0$. Then by Theorem 5.10.1, the series for the function $f(x)$

given by (6.2.19) converges absolutely and uniformly on the interval $[0, l]$.

Using the trigonometric identity

$$\sin\frac{n\pi x}{l}\cos\frac{n\pi c}{l}t = \frac{1}{2}\sin\frac{n\pi}{l}(x - ct) + \frac{1}{2}\sin\frac{n\pi}{l}(x + ct), \qquad (6.3.3)$$

$u_1(x, t)$ may be written as

$$u_1(x, t) = \frac{1}{2}\sum_{n=1}^{\infty} a_n \sin\frac{n\pi}{l}(x - ct) + \frac{1}{2}\sum_{n=1}^{\infty} a_n \sin\frac{n\pi}{l}(x + ct)$$

Define

$$F(x) = \sum_{n=1}^{\infty} a_n \sin\frac{n\pi x}{l} \qquad (6.3.4)$$

and assume that $F(x)$ is the odd periodic extension of $f(x)$, that is,

$$\begin{aligned} F(x) &= f(x) \qquad 0 \leqslant x \leqslant l \\ F(-x) &= -F(x) \qquad \text{for all } x \\ F(x \pm 2l) &= F(x) \end{aligned}$$

We can now rewrite $u_1$ in the form

$$u_1(x, t) = \frac{1}{2}[F(x - ct) + F(x + ct)] \qquad (6.3.5)$$

To show that the boundary conditions are satisfied, we note that

$$\begin{aligned} u_1(0, t) &= \frac{1}{2}[F(-ct) + F(ct)] \\ &= \frac{1}{2}[-F(ct) + F(ct)] \\ &= 0 \\ u_1(l, t) &= \frac{1}{2}[F(l - ct) + F(l + ct)] \\ &= \frac{1}{2}[F(-l - ct) + F(l + ct)] \\ &= \frac{1}{2}[-F(l + ct) + F(l + ct)] \\ &= 0 \end{aligned}$$

Since

$$u_1(x, 0) = \frac{1}{2}[F(x) + F(x)]$$
$$= F(x)$$
$$= f(x) \qquad 0 \leqslant x \leqslant l$$

we see that initial condition $u_1(x, 0) = f(x)$ is satisfied. Thus Eq. (6.2.1) and conditions (6.2.2)–(6.2.3) with $g(x) \equiv 0$ are satisfied. Since $f'$ is continuous in $[0, l]$, $F'$ exists and is continuous for all $x$. Thus, if we differentiate $u_1$ with respect to $t$, we obtain

$$\frac{\partial u_1}{\partial t} = \frac{1}{2}[-cF'(x - ct) + cF'(x + ct)]$$

and

$$\frac{\partial u_1}{\partial t}(x, 0) = \frac{1}{2}[-cF'(x) + cF'(x)] = 0$$

We therefore see that initial condition (6.2.3) is also satisfied.

In order to show that $u_1(x, t)$ satisfies the differential equation (6.2.1), we impose additional restrictions on $f$. Let $f''$ be continuous on $[0, l]$ and let $f''(0) = f''(l) = 0$. Then $F''$ exists and is continuous everywhere, and therefore

$$\frac{\partial^2 u_1}{\partial t^2} = \frac{1}{2}c^2[F''(x - ct) + F''(x + ct)]$$

$$\frac{\partial^2 u_1}{\partial x^2} = \frac{1}{2}[F''(x - ct) + F''(x + ct)]$$

We find therefore that

$$\frac{\partial^2 u_1}{\partial t^2} = c^2 \frac{\partial^2 u_1}{\partial x^2}$$

Next, we shall state the assumptions which must be imposed on $g$ to make $u_2(x, t)$ the solution of problem (6.2.1)–(6.2.5) with $f(x) \equiv 0$. Let $g$ and $g'$ be continuous on $[0, l]$ and let $g(0) = g(l) = 0$. Then the series for the function $g(x)$ given by (6.2.20) converges absolutely and uniformly in the interval $[0, l]$. Introducing the new coefficient $c_n = (n\pi c/l)b_n$, we have

$$u_2(x, t) = \frac{l}{\pi c} \sum_{n=1}^{\infty} \frac{c_n}{n} \sin \frac{n\pi c}{l} t \sin \frac{n\pi x}{l} \qquad (6.3.6)$$

We shall see that term-by-term differentiation with respect to $t$ is permitted, and hence

$$\frac{\partial u_2}{\partial t} = \sum_{n=1}^{\infty} c_n \cos\frac{n\pi c}{l} t \sin\frac{n\pi x}{l} \tag{6.3.7}$$

Using the trigonometric identity (6.3.3) we obtain

$$\frac{\partial u_2}{\partial t} = \frac{1}{2}\sum_{n=1}^{\infty} c_n \sin\frac{n\pi}{l}(x - ct) + \frac{1}{2}\sum_{n=1}^{\infty} c_n \sin\frac{n\pi}{l}(x + ct) \tag{6.3.8}$$

These series are absolutely and uniformly convergent because of the assumptions on $g$, and hence the series (6.3.6) and (6.3.7) converge absolutely and uniformly on $[0, l]$. Thus the term-by-term differentiation is justified.

Let

$$G(x) = \sum_{n=1}^{\infty} c_n \sin\frac{n\pi x}{l}$$

be the odd periodic extension of the function $g(x)$. Then Eq. (6.3.8) can be written in the form

$$\frac{\partial u_2}{\partial t} = \frac{1}{2}[G(x - ct) + G(x + ct)]$$

Integration yields

$$\begin{aligned} u_2(x, t) &= \frac{1}{2}\int_0^t G(x - ct')\,dt' + \frac{1}{2}\int_0^t G(x + ct')\,dt' \\ &= \frac{1}{2c}\int_{x-ct}^{x+ct} G(\tau)\,d\tau \end{aligned} \tag{6.3.9}$$

It immediately follows that $u_2(x, 0) = 0$, and

$$\begin{aligned} \frac{\partial u_2}{\partial t}(x, 0) &= G(x) \qquad 0 \leqslant x \leqslant l \\ &= g(x) \end{aligned}$$

Moreover,

$$\begin{aligned} u_2(0, t) &= \frac{1}{2}\int_0^t G(-ct')\,dt' + \frac{1}{2}\int_0^t G(ct')\,dt' \\ &= -\frac{1}{2}\int_0^t G(ct')\,dt' + \frac{1}{2}\int_0^t G(ct')\,dt' \\ &= 0 \end{aligned}$$

and

$$\begin{aligned} u_2(l, t) &= \frac{1}{2}\int_0^t G(l - ct')\,dt' + \frac{1}{2}\int_0^t G(l + ct')\,dt' \\ &= \frac{1}{2}\int_0^t G(-l - ct')\,dt' + \frac{1}{2}\int_0^t G(l + ct')\,dt' \\ &= -\frac{1}{2}\int_0^t G(l + ct')\,dt' + \frac{1}{2}\int_0^t G(l + ct')\,dt' \\ &= 0 \end{aligned}$$

Finally, $u_2(x, t)$ must satisfy the differential equation. Since $g'$ is continuous on $[0, l]$, $G'$ exists so that

$$\frac{\partial^2 u_2}{\partial t^2} = \frac{c}{2}[-G'(x - ct) + G'(x + ct)]$$

Differentiating $u_2(x, t)$ represented by Eq. (6.3.6) with respect to $x$, we obtain

$$\begin{aligned} \frac{\partial u_2}{\partial x} &= \frac{1}{c}\sum_{n=1}^{\infty} c_n \sin\frac{n\pi c}{l}t \cos\frac{n\pi x}{l} \\ &= \frac{1}{2c}\sum_{n=1}^{\infty} c_n\left[-\sin\frac{n\pi}{l}(x - ct) + \sin\frac{n\pi}{l}(x + ct)\right] \\ &= \frac{1}{2c}[-G(x - ct) + G(x + ct)] \end{aligned}$$

Differentiating again with respect to $x$, we obtain

$$\frac{\partial^2 u_2}{\partial x^2} = \frac{1}{2c}[-G'(x - ct) + G'(x + ct)]$$

It is quite evident that

$$\frac{\partial^2 u_2}{\partial t^2} = c^2\frac{\partial^2 u_2}{\partial x^2}$$

Thus, the solution of the initial-boundary value problem (6.2.1)–(6.2.5) is established.

***Theorem 6.3.1 (Uniqueness Theorem)*** *There exists at most one solution of the wave equation*

$$u_{tt} = c^2 u_{xx}, \qquad 0 < x < l, \qquad t > 0$$

*satisfying the initial conditions*

$$\begin{aligned} u(x, 0) &= f(x), \qquad 0 \leq x \leq l \\ u_t(x, 0) &= g(x), \qquad 0 \leq x \leq l \end{aligned}$$

*and the boundary conditions*

$$u(0, t) = 0, \qquad t \geqslant 0$$
$$u(l, t) = 0, \qquad t \geqslant 0$$

*where $u(x, t)$ is a twice continuously differentiable function with respect to both $x$ and $t$.*

PROOF. Suppose that there are two solutions $u_1$ and $u_2$ and let $v = u_1 - u_2$. It can readily be seen that $v(x, t)$ is the solution of the problem

$$\begin{aligned} v_{tt} &= c^2 v_{xx}, & 0 < x < l, \quad & t > 0 \\ v(0, t) &= 0 & & t \geqslant 0 \\ v(l, t) &= 0 & & t \geqslant 0 \\ v(x, 0) &= 0 & 0 \leqslant x \leqslant l & \\ v_t(x, 0) &= 0 & 0 \leqslant x \leqslant l & \end{aligned}$$

We shall prove that the function $v(x, t)$ is identically zero. To do so, consider the function

$$I(t) = \frac{1}{2} \int_0^l (c^2 v_x^2 + v_t^2)\, dx \tag{6.3.10}$$

which physically represents the total energy of the vibrating string at time $t$.

Since the function $v(x, t)$ is twice continuously differentiable, we differentiate $I(t)$ with respect to $t$. Thus,

$$\frac{dI}{dt} = \int_0^l (c^2 v_x v_{xt} + v_t v_{tt})\, dx \tag{6.3.11}$$

Integrating by parts, we have

$$\int_0^l c^2 v_x v_{xt}\, dx = [c^2 v_x v_t]_0^l - \int_0^l c^2 v_t v_{xx}\, dx$$

But from the condition $v(0, t) = 0$ we have $v_t(0, t) = 0$, and similarly $v_t(l, t) = 0$ for $x = l$. Hence the expression in the square brackets vanishes, and Eq. (6.3.11) becomes

$$\frac{dI}{dt} = \int_0^l v_t(v_{tt} - c^2 v_{xx})\, dx \tag{6.3.12}$$

Since $v_{tt} - c^2 v_{xx} = 0$, Eq. (6.3.12) reduces to

$$\frac{dI}{dt} = 0$$

which means

$$I(t) = \text{constant} = C$$

Since $v(x, 0) = 0$ we have $v_x(x, 0) = 0$. Taking into account the condition $v_t(x, 0) = 0$, we evaluate $C$ to obtain

$$I(0) = C = \int_0^t [c^2 v_x^2 + v_t^2]_{t=0}\, dx = 0$$

This implies that $I(t) = 0$ which can happen only when $v_x = 0$ and $v_t = 0$ for $t > 0$. To satisfy both of these conditions, we must have $v(x, t) =$ constant. Employing the condition $v(x, 0) = 0$, we then find $v(x, t) = 0$. Therefore $u_1(x, t) = u_2(x, t)$ and the solution $u(x, t)$ is unique. □

## 6.4 THE HEAT CONDUCTION PROBLEM

We consider a homogeneous rod of length $l$. The rod is sufficiently thin so that the heat is distributed equally over the cross section at time $t$. The surface of the rod is insulated, and therefore there is no heat loss through the boundary. The temperature distribution of the rod is given by the solution of the initial-boundary value problem

$$\begin{aligned}
u_t &= k u_{xx}, & 0 < x < l, & \quad t > 0 \\
u(0, t) &= 0, & & \quad t \geq 0 \\
u(l, t) &= 0, & & \quad t \geq 0 \\
u(x, 0) &= f(x) & 0 \leq x \leq l &
\end{aligned} \tag{6.4.1}$$

If we assume a solution in the form

$$u(x, t) = X(x)T(t) \neq 0$$

Eq. (6.4.1) yields

$$XT' = kX''T$$

Thus, we have

$$\frac{X''}{X} = \frac{T'}{kT} = -\alpha^2$$

where $\alpha$ is a positive constant. Hence $X$ and $T$ must satisfy

$$X'' + \alpha^2 X = 0 \tag{6.4.2}$$

$$T' + \alpha^2 kT = 0 \tag{6.4.3}$$

From the boundary conditions, we have

$$u(0, t) = X(0)T(t) = 0$$
$$u(l, t) = X(l)T(t) = 0$$

Thus,

$$X(0)=0, \qquad X(l)=0$$

for arbitrary $T(t)$. Hence we must solve the eigenvalue problem

$$X''+\alpha^2 X=0$$
$$X(0)=0, \qquad X(l)=0$$

The solution of Eq. (6.4.2)

$$X(x)=A\cos\alpha x+B\sin\alpha x$$

Since $X(0)=0$, $A=0$. To satisfy the second condition, we have

$$X(l)=B\sin\alpha l=0$$

Since $B=0$ yields a trivial solution, we must have

$$\sin\alpha l=0$$

Thus,

$$\alpha=\frac{n\pi}{l} \qquad \text{for} \quad n=1,2,3,\ldots$$

Substituting these eigenvalues, we have

$$X_n(x)=B_n\sin\frac{n\pi x}{l}$$

Next, we consider Eq. (6.4.3), namely,

$$T'+\alpha^2 kT=0$$

the solution of which is

$$T(t)=Ce^{-\alpha^2 kt}$$

Substituting $\alpha=n\pi/l$, we have

$$T_n(t)=C_n e^{-(n\pi/l)^2 kt}$$

Hence the nontrivial solution of the heat equation which satisfies the two boundary conditions is

$$u_n(x,t)=X_n(x)T_n(t)=a_n e^{-(n\pi/l)^2 kt}\sin\frac{n\pi x}{l}, \qquad n=1,2,3,\ldots$$

where $a_n=B_nC_n$ is an arbitrary constant.

We formally form a series

$$u(x,t)=\sum_{n=1}^{\infty} a_n e^{-(n\pi/l)^2 kt}\sin\frac{n\pi x}{l} \tag{6.4.4}$$

which satisfies the initial condition if

$$u(x, 0) = f(x) = \sum_{n=1}^{\infty} a_n \sin \frac{n\pi x}{l}$$

This holds true if $f(x)$ can be represented by a Fourier sine series with Fourier coefficients

$$a_n = \frac{2}{l} \int_0^l f(x) \sin \frac{n\pi x}{l} \, dx \tag{6.4.5}$$

Hence

$$u(x, t) = \sum_{n=1}^{\infty} \left[ \frac{2}{l} \int_0^l f(\tau) \sin \frac{n\pi \tau}{l} \, d\tau \right] e^{-(n\pi/l)^2 kt} \sin \frac{n\pi x}{l} \tag{6.4.6}$$

is the formal solution of the heat conduction problem.

EXAMPLE 6.4.1. (a) Suppose the initial temperature distribution is $f(x) = x(l - x)$. Then from Eq. (6.4.5), we have

$$a_n = \frac{8l^2}{n^3 \pi^3}, \qquad n = 1, 3, 5, \ldots$$

Thus, the solution is

$$u(x, t) = \frac{8l^2}{\pi^3} \sum_{n=1,3,5,\ldots}^{\infty} \frac{1}{n^3} e^{-(n\pi/l)^2 kt} \sin \frac{n\pi x}{l}$$

(b) Suppose the temperature at one end of the rod is held constant, that is,

$$u(l, t) = u_0, \qquad t \geqslant 0$$

The problem here is

$$\begin{aligned} &u_t = k u_{xx}, \qquad 0 < x < l, \qquad t > 0 \\ &u(0, t) = 0 \\ &u(l, t) = u_0 \\ &u(x, 0) = f(x), \qquad 0 < x < l \end{aligned} \tag{6.4.7}$$

Let

$$u(x, t) = v(x, t) + \frac{u_0 x}{l}$$

Substitution of $u(x, t)$ in problem (6.4.7) yields

$$\begin{aligned} v_t &= kv_{xx}, \qquad 0<x<l, \qquad t>0 \\ v(0, t) &= 0 \\ v(l, t) &= 0 \\ v(x, 0) &= f(x) - \frac{u_0 x}{l}, \qquad 0<x<l \end{aligned}$$

Hence with the knowledge of solution (6.4.6), we obtain

$$u(x, t) = \sum_{n=1}^{\infty} \left[\frac{2}{l} \int_0^l \left(f - \frac{u_0 \tau}{l}\right) \sin \frac{n\pi\tau}{l}\, d\tau\right] e^{-(n\pi/l)^2 kt} \sin \frac{n\pi x}{l} + \frac{u_0 x}{l} \tag{6.4.8}$$

## 6.5 EXISTENCE AND UNIQUENESS OF SOLUTION OF THE HEAT CONDUCTION PROBLEM

In the preceding section we found that (6.4.4) is the formal solution of the heat conduction problem (6.4.1) where $a_n$ is given by (6.4.5).

We shall show that this formal solution is the solution if $f(x)$ is continuous in $[0, l]$ and $f(0) = f(l) = 0$, and $f'(x)$ is piecewise continuous in $(0, l)$. Since $f(x)$ is bounded, we have

$$|a_n| = \frac{2}{l} \left| \int_0^l f(x) \sin \frac{n\pi x}{l}\, dx \right| \leqslant \frac{2}{l} \int_0^l |f(x)|\, dx \leqslant C$$

where $C$ is a positive constant. Thus, for any $t_0 > 0$

$$\left| a_n e^{-(n\pi/l)^2 kt} \sin \frac{n\pi x}{l} \right| \leqslant C e^{-(n\pi/l)^2 kt_0} \quad \text{when} \quad t \geqslant t_0$$

According to the ratio test, the series of the constant terms $\exp[-(n\pi/l)^2 kt_0]$ converges. Hence by the Weierstrass M-test, the series (6.4.4) converges uniformly with respect to $x$ and $t$ whenever $t \geqslant t_0$ and $0 \leqslant x \leqslant l$.

Differentiating Eq. (6.4.4) termwise with respect to $t$, we obtain

$$u_t = -\sum_{n=1}^{\infty} a_n \left(\frac{n\pi}{l}\right)^2 k e^{-(n\pi/l)^2 kt} \sin \frac{n\pi x}{l} \tag{6.5.1}$$

We note that

$$\left| -a_n \left(\frac{n\pi}{l}\right)^2 k e^{-(n\pi/l)^2 kt} \sin \frac{n\pi x}{l} \right| \leqslant C \left(\frac{n\pi}{l}\right)^2 k e^{-(n\pi/l)^2 kt_0}$$

when $t \geqslant t_0$, and the series of the constant terms $C(n\pi/l)^2 k \exp[-(n\pi/l)^2 kt_0]$ converges by the ratio test. Hence Eq. (6.5.1) is uniformly convergent in the region $0 \leqslant x \leqslant l$, $t \geqslant t_0$. In a similar manner, the series

(6.4.4) can be differentiated twice with respect to $x$, and as a result

$$u_{xx} = -\sum_{n=1}^{\infty} a_n \left(\frac{n\pi}{l}\right)^2 e^{-(n\pi/l)^2 kt} \sin\frac{n\pi x}{l} \tag{6.5.2}$$

Evidently, from Eqs. (6.5.1) and (6.5.2),

$$u_t = ku_{xx}$$

Hence Eq. (6.4.4) is a solution of the one-dimensional heat equation in the region $0 \leqslant x \leqslant l$, $t > 0$.

Next, we show that the boundary conditions are satisfied. Here, we note that the series (6.4.4) representing the function $u(x, t)$ converges uniformly in the region $0 \leqslant x \leqslant l$, $t > 0$. Since the function represented by a uniformly convergent series of continuous functions is continuous, $u(x, t)$ is continuous at $x = 0$ and $x = l$. As a consequence, when $x = 0$ and $x = l$, Eq. (6.4.4) reduces to

$$u(0, t) = 0$$
$$u(l, t) = 0$$

for all $t > 0$.

It remains to show that $u(x, t)$ satisfies the initial condition

$$u(x, 0) = f(x), \qquad 0 \leqslant x \leqslant l$$

Under the assumptions stated earlier,

$$f(x) = \sum_{n=1}^{\infty} a_n \sin\frac{n\pi x}{l}$$

is uniformly and absolutely convergent. According to Abel's test [6], the series formed by the product of the terms of a uniformly convergent series

$$\sum_{n=1}^{\infty} a_n \sin\frac{n\pi x}{l}$$

and the members of a uniformly bounded and monotone sequence $\exp[-(n\pi/l)^2 kt]$ converges uniformly with respect to $t$. Hence

$$u(x, t) = \sum_{n=1}^{\infty} a_n e^{-(n\pi/l)^2 kt} \sin\frac{n\pi x}{l}$$

converges uniformly for $0 \leqslant x \leqslant l$, $t \geqslant 0$, and by the same reasoning as before, $u(x, t)$ is continuous for $0 \leqslant x \leqslant l$, $t \geqslant 0$. Thus, the initial condition

$$u(x, 0) = f(x), \qquad 0 \leqslant x \leqslant l$$

is satisfied. The solution is therefore established.

In the above discussion the condition imposed on $f(x)$ is stronger than necessary. The solution can be obtained with a less stringent condition on $f(x)$ (see Ref. 47).

***Theorem 6.5.1 (Uniqueness Theorem)*** *Let $u(x, t)$ be a continuously differentiable function. If $u(x, t)$ satisfies the differential equation*

$$u_t = ku_{xx}, \qquad 0 < x < l, \qquad t > 0$$

*the initial condition*

$$u(x, 0) = f(x), \qquad 0 \leq x \leq l$$

*and the boundary conditions*

$$u(0, t) = 0, \qquad t \geq 0$$
$$u(l, t) = 0, \qquad t \geq 0$$

*then it is unique.*

PROOF. Suppose there are two solutions $u_1(x, t)$ and $u_2(x, t)$. Let

$$v(x, t) = u_1(x, t) - u_2(x, t).$$

Then

$$\begin{aligned} v_t &= kv_{xx}, \qquad 0 < x < l, \qquad t > 0 \\ v(0, t) &= 0, \qquad t \geq 0 \\ v(l, t) &= 0, \qquad t \geq 0 \\ v(x, 0) &= 0, \qquad 0 \leq x \leq l \end{aligned} \tag{6.5.3}$$

Consider the function

$$J(t) = \frac{1}{2k} \int_0^l v^2 \, dx$$

Differentiating with respect to $t$, we have

$$J'(t) = \frac{1}{k} \int_0^l vv_t \, dx = \int_0^l vv_{xx} \, dx$$

by virtue of Eq. (6.5.3). Integrating by parts, we have

$$\int_0^l vv_{xx} \, dx = [vv_x]_0^l - \int_0^l v_x^2 \, dx$$

Since

$$v(0, t) = v(l, t) = 0$$

$$J'(t) = -\int_0^l v_x^2 \, dx \leq 0$$

From the condition $v(x, 0) = 0$, we have $J(0) = 0$. This condition and $J'(t) \leqslant 0$ implies that $J(t)$ is a nonincreasing function of $t$. Thus,

$$J(t) \leqslant 0$$

But by definition of $J(t)$,

$$J(t) \geqslant 0$$

Hence

$$J(t) = 0 \quad \text{for} \quad t \geqslant 0$$

Since $v(x, t)$ is continuous, $J(t) = 0$ implies

$$v(x, t) = 0$$

in $0 \leqslant x \leqslant l$, $t \geqslant 0$. Therefore, $u_1 = u_2$ and the solution is unique. □

## 6.6 THE LAPLACE AND BEAM EQUATIONS

EXAMPLE 6.6.1. Consider the steady state temperature distribution in a thin rectangular slab. Two sides are insulated, one side is maintained at zero temperature, and the temperature of the remaining side is prescribed to be $f(x)$. Thus, we are required to solve

$$\begin{aligned}
\nabla^2 u &= 0, \qquad 0 < x < a, \qquad 0 < y < b \\
u(x, 0) &= f(x), \qquad 0 \leqslant x \leqslant a \\
u(x, b) &= 0 \\
u_x(0, y) &= 0 \\
u_x(a, y) &= 0
\end{aligned}$$

Let $u(x, y) = X(x)Y(y)$. Substitution of this into the Laplace equation yields

$$X'' - \lambda X = 0$$
$$Y'' + \lambda Y = 0$$

Since the boundary conditions are homogeneous on $x = 0$ and $x = a$, we have $\lambda = -\alpha^2$ with $\alpha \geqslant 0$ for nontrivial solutions of the eigenvalue problem

$$X'' + \alpha^2 X = 0$$
$$X'(0) = X'(a) = 0$$

The solution is

$$X(x) = A \cos \alpha x + B \sin \alpha x$$

Application of the boundary conditions then yields $B = 0$ and $\alpha = n\pi/a$

with $n = 0, 1, 2, \ldots$ Hence

$$X_n(x) = A_n \cos \frac{n\pi x}{a}$$

The solution of the $Y$ equation is clearly

$$Y(y) = C \cosh \alpha y + D \sinh \alpha y$$

which can be written in the form

$$Y(y) = E \sinh \alpha (y + F)$$

where $E = (D^2 - C^2)^{\frac{1}{2}}$ and $F = [\tanh^{-1}(C/D)]/\alpha$.

Applying the homogeneous boundary condition $Y(b) = 0$, we obtain

$$Y(b) = E \sinh \alpha(b + F) = 0$$

which implies

$$F = -b, \qquad E \neq 0$$

for nontrivial solutions. Hence we have

$$u(x, y) = \frac{(b - y)}{b}\frac{a_0}{2} + \sum_{n=1}^{\infty} a_n \cos \frac{n\pi x}{a} \sinh \frac{n\pi}{a}(y - b)$$

Now we apply the remaining nonhomogeneous condition. We obtain

$$u(x, 0) = f(x) = \frac{a_0}{2} + \sum_{n=1}^{\infty} a_n \cos \frac{n\pi x}{a} \sinh\left(-\frac{n\pi b}{a}\right)$$

Since this is a Fourier cosine series, the coefficients are given by

$$a_0 = \frac{2}{a}\int_0^a f(x)\, dx$$

$$a_n = \frac{-2}{a \sinh \dfrac{n\pi b}{a}} \int_0^a f(x) \cos \frac{n\pi x}{a}\, dx \qquad n = 1, 2, \ldots$$

Thus the formal solution is

$$u(x, y) = \left(\frac{b - y}{b}\right)\frac{a_0}{2} + \sum_{n=1}^{\infty} a_n^* \frac{\sinh \dfrac{n\pi}{a}(b - y)}{\sinh \dfrac{n\pi b}{a}} \cos \frac{n\pi x}{a}$$

where

$$a_n^* = \frac{2}{a}\int_0^a f(x) \cos \frac{n\pi x}{a}\, dx$$

If for example $f(x) = x$ in $0 < x < \pi$, $0 < y < \pi$, then we find (note that $a = \pi$)

$$a_0 = \pi$$

$$a_n^* = \frac{2}{\pi n^2}[(-1)^n - 1], \qquad n = 1, 2, \ldots$$

and hence

$$u(x, y) = \frac{1}{2}(\pi - y) + \sum_{n=1}^{\infty} \frac{2}{\pi n^2}[(-1)^n - 1] \frac{\sinh n(\pi - y)}{\sinh n\pi} \cos nx$$

EXAMPLE 6.6.2. As another example, we consider the transverse vibration of a beam. The equation of motion is governed by

$$u_{tt} + a^2 u_{xxxx} = 0, \qquad 0 < x < l, \qquad t > 0$$

where $u(x, t)$ is the displacement and $a$ is the physical constant. Note that the equation is of the fourth order in $x$. Let the initial and boundary conditions be

$$\begin{aligned} u(x, 0) &= f(x), & 0 \leq x \leq l \\ u_t(x, 0) &= g(x), & 0 \leq x \leq l \\ u(0, t) &= u(l, t) = 0, & t > 0 \\ u_{xx}(0, t) &= u_{xx}(l, t) = 0 & \end{aligned} \tag{6.6.1}$$

The boundary conditions represent the beam being simple supported, that is, the displacements and the bending moments at the ends are zero.

Assume a nontrivial solution in the form

$$u(x, t) = X(x)T(t)$$

which transforms the equation of motion into

$$T'' + a^2\alpha^4 T = 0, \qquad \alpha > 0$$

$$X^{iv} - \alpha^4 X = 0$$

The equation for $X(x)$ has the general solution

$$X(x) = A \cosh \alpha x + B \sinh \alpha x + C \cos \alpha x + D \sin \alpha x$$

The boundary conditions require that

$$X(0) = X(l) = 0$$

$$X''(0) = X''(l) = 0$$

Differentiating $X$ twice with respect to $x$, we obtain

$$X''(x) = A\alpha^2 \cosh \alpha x + B\alpha^2 \sinh \alpha x - C\alpha^2 \cos \alpha x - D\alpha^2 \sin \alpha x$$

Now applying the conditions $X(0) = X''(0) = 0$, we obtain

$$A + C = 0$$

$$\alpha^2(A - C) = 0$$

and hence

$$A = C = 0$$

The conditions $X(l) = X''(l) = 0$ yield

$$B \sinh \alpha l + D \sin \alpha l = 0$$

$$B \sinh \alpha l - D \sin \alpha l = 0$$

These equations are satisfied if

$$B \sinh \alpha l = 0, \qquad D \sin \alpha l = 0$$

Since $\sinh \alpha l \neq 0$, $B$ must vanish. For nontrivial solutions,

$$\sin \alpha l = 0, \qquad D \neq 0$$

and hence

$$\alpha = \frac{n\pi}{l} \qquad n = 1, 2, 3, \ldots$$

We then obtain

$$X_n(x) = D_n \sin \frac{n\pi x}{l}$$

The general solution for $T(t)$ is

$$T(t) = E \cos a\alpha^2 t + F \sin a\alpha^2 t$$

Inserting the values of $\alpha^2$, we obtain

$$T_n(t) = E_n \cos a \left(\frac{n\pi}{l}\right)^2 t + F_n \sin a \left(\frac{n\pi}{l}\right)^2 t$$

Thus, the general solution of the equation for the transverse vibrations of a beam is

$$u(x, t) = \sum_{n=1}^{\infty} \left[ a_n \cos a \left(\frac{n\pi}{l}\right)^2 t + b_n \sin a \left(\frac{n\pi}{l}\right)^2 t \right] \sin \frac{n\pi x}{l} \qquad (6.6.2)$$

To satisfy the initial condition $u(x, 0) = f(x)$, we must have

$$u(x, 0) = f(x) = \sum_{n=1}^{\infty} a_n \sin \frac{n\pi x}{l}$$

from which we find

$$a_n = \frac{2}{l}\int_0^l f(x)\sin\frac{n\pi x}{l}\,dx \tag{6.6.3}$$

Now the application of the second initial condition gives

$$u_t(x, 0) = g(x) = \sum_{n=1}^{\infty} b_n a\left(\frac{n\pi}{l}\right)^2 \sin\frac{n\pi x}{l}$$

and hence

$$b_n = \frac{2}{al}\left(\frac{l}{n\pi}\right)^2 \int_0^l g(x)\sin\frac{n\pi x}{l}\,dx \tag{6.6.4}$$

Thus, the solution of the initial-boundary value problem is given by Eqs. (6.6.2)–(6.6.4).

## 6.7 NONHOMOGENEOUS PROBLEMS

The partial differential equations considered so far in this chapter are homogeneous. In practice, there is a very important class of problems involving nonhomogeneous equations. First, we shall illustrate a problem involving a time-independent nonhomogeneous equation.

EXAMPLE 6.7.1. Consider the initial-boundary value problem

$$\begin{aligned} u_{tt} &= c^2 u_{xx} + F(x), && 0<x<l, && t>0 \\ u(x, 0) &= f(x), && 0\leq x\leq l && \\ u_t(x, 0) &= g(x), && 0\leq x\leq l && \\ u(0, t) &= A, && && t>0 \\ u(l, t) &= B, && && t>0 \end{aligned} \tag{6.7.1}$$

We assume a solution in the form

$$u(x, t) = v(x, t) + U(x)$$

Substitution of $u(x, t)$ in Eq. (6.7.1) yields

$$v_{tt} = c^2(v_{xx} + U_{xx}) + F(x)$$

so that if $U(x)$ satisfies

$$c^2 U_{xx} + F(x) = 0$$

then

$$v_{tt} = c^2 v_{xx}$$

In a similar manner, if $u(x, t)$ is inserted in the initial and boundary conditions, we obtain

$$\begin{aligned} u(x, 0) &= v(x, 0) + U(x) = f(x) \\ u_t(x, 0) &= v_t(x, 0) \qquad\quad\;\; = g(x) \\ u(0, t) &= v(0, t) + U(0) \;= A \\ u(l, t) &= v(l, t) + U(l) \;\;= B \end{aligned}$$

Thus, if $U(x)$ is the solution of the problem

$$\begin{aligned} c^2 U_{xx} + F &= 0 \\ U(0) &= A \\ U(l) &= B \end{aligned}$$

then $v(x, t)$ must satisfy

$$\begin{aligned} v_{tt} &= c^2 v_{xx} \\ v(x, 0) &= f(x) - U(x) \\ v_t(x, 0) &= g(x) \\ v(0, t) &= 0 \\ v(l, t) &= 0 \end{aligned} \tag{6.7.2}$$

Now $v(x, t)$ can be solved easily as $U(x)$ is known. It can be seen that

$$\begin{aligned} U(x) = A + (B - A)\frac{x}{l} + \frac{x}{l}\int_0^l \left[\frac{1}{c^2}\int_0^\eta F(\xi)\, d\xi\right] d\eta \\ - \int_0^x \left[\frac{1}{c^2}\int_0^\eta F(\xi)\, d\xi\right] d\eta \end{aligned}$$

As a specific example, consider the problem

$$\begin{aligned} u_{tt} &= c^2 u_{xx} + h, \quad h \text{ is a constant} \\ u(x, 0) &= 0 \\ u_t(x, 0) &= 0 \\ u(0, t) &= 0 \\ u(l, t) &= 0 \end{aligned} \tag{6.7.3}$$

Then the solution of

$$\begin{aligned} c^2 U_{xx} + h &= 0 \\ U(0) &= 0 \\ U(l) &= 0 \end{aligned}$$

is

$$U(x) = \frac{h}{2c^2}(lx - x^2)$$

The function $v(x, t)$ must satisfy

$$\begin{aligned} v_{tt} &= c^2 v_{xx} \\ v(x, 0) &= -\frac{h}{2c^2}(lx - x^2) \\ v_t(x, 0) &= 0 \\ v(0, t) &= 0 \\ v(l, t) &= 0 \end{aligned}$$

The solution is given [Sec. 6.2 with $g(x) = 0$] by

$$v(x, t) = \sum_{n=1}^{\infty} a_n \cos\frac{n\pi c}{l} t \sin\frac{n\pi x}{l}$$

and the coefficient is

$$\begin{aligned} a_n &= \frac{2}{l}\int_0^l \left[-\frac{h}{2c^2}(lx - x^2)\right] \sin\frac{n\pi x}{l}\,dx \\ &= -\frac{4l^2 h}{n^3\pi^3 c^2} \qquad \text{for } n \text{ odd} \\ &= 0 \quad \text{for} \quad n \text{ even} \end{aligned}$$

The solution of the given initial-boundary value problem is, therefore, given by

$$\begin{aligned} u(x, t) &= v(x, t) + U(x) \\ &= \frac{hx}{2c^2}(l - x) + \sum_{n=1}^{\infty}\left(-\frac{4l^2 h}{c^2\pi^3}\right)\frac{\cos(2n-1)\pi ct/l}{(2n-1)^3}\sin(2n-1)\pi x/l \end{aligned} \tag{6.7.4}$$

Let us now consider the problem of a finite string with an external force acting on it. If the ends are fixed, we have

$$\begin{aligned} u_{tt} - c^2 u_{xx} &= h(x, t), && 0 < x < l, && t > 0 \\ u(x, 0) &= f(x), && 0 \leqslant x \leqslant l \\ u_t(x, 0) &= g(x), && 0 \leqslant x \leqslant l \\ u(0, t) &= 0, && && t \geqslant 0 \\ u(l, t) &= 0, && && t \geqslant 0 \end{aligned} \tag{6.7.5}$$

We assume a solution in the form[6]

$$u(x, t) = \sum_{n=1}^{\infty} u_n(t) \sin \frac{n\pi x}{l} \tag{6.7.6}$$

where the functions $u_n(t)$ are to be determined. It is evident that the boundary conditions are satisfied. Let us also assume that

$$h(x, t) = \sum_{n=1}^{\infty} h_n(t) \sin \frac{n\pi x}{l} \tag{6.7.7}$$

Thus,

$$h_n(t) = \frac{2}{l} \int_0^l h(x, t) \sin \frac{n\pi x}{l}\, dx \tag{6.7.8}$$

We assume that the series (6.7.6) is suitably convergent. We then find $u_{tt}$ and $u_{xx}$ from (6.7.6), and substitution of these values into (6.7.5) yields

$$\sum_{n=1}^{\infty} [u_n''(t) + \lambda_n^2 u_n(t)] \sin \frac{n\pi x}{l} = \sum_{n=1}^{\infty} h_n(t) \sin \frac{n\pi x}{l}$$

where $\lambda_n = n\pi c/l$. Multiplying both sides of this equation by $\sin m\pi x/l$, $m = 1,2,3, \ldots$, and integrating from $x = 0$ to $x = l$, we obtain

$$u_n''(t) + \lambda_n^2 u_n(t) = h_n(t)$$

the solution of which is

$$u_n(t) = a_n \cos \lambda_n t + b_n \sin \lambda_n t + \frac{1}{\lambda_n} \int_0^t h_n(\tau) \sin [\lambda_n (t - \tau)]\, d\tau \tag{6.7.9}$$

Hence, the formal solution is

$$u(x, t) = \sum_{n=1}^{\infty} \left\{ a_n \cos \lambda_n t + b_n \sin \lambda_n t + \frac{1}{\lambda_n} \int_0^t h_n(\tau) \sin [\lambda_n (t - \tau)]\, d\tau \right\} \cdot \sin \frac{n\pi x}{l} \tag{6.7.10}$$

Applying the initial conditions, we have

$$u(x, 0) = f(x) = \sum_{n=1}^{\infty} a_n \sin \frac{n\pi x}{l}$$

Thus,

$$a_n = \frac{2}{l} \int_0^l f(x) \sin \frac{n\pi x}{l}\, dx \tag{6.7.11}$$

[6] The functions $\sin n\pi x/l$ are the eigenfunctions of the associated eigenvalue problem. See Chapter 9, Sec 10.

Similarly,

$$u_t(x, 0) = g(x) = \sum_{n=1}^{\infty} b_n \lambda_n \sin \frac{n\pi x}{l}$$

Thus,

$$b_n = \frac{2}{l\lambda_n} \int_0^l g(x) \sin \frac{n\pi x}{l} \, dx \tag{6.7.12}$$

Hence the formal solution of the initial-boundary value problem (6.7.5) is given by (6.7.10) with $a_n$ given by (6.7.11) and $b_n$ given by (6.7.12).

EXAMPLE 6.7.2. Determine the solution of the initial-boundary value problem

$$\begin{aligned}
&u_{tt} - u_{xx} = h, && 0<x<1, && t>0, && h = \text{constant} \\
&u(x, 0) = x(1-x), && 0 \leqslant x \leqslant 1 \\
&u_t(x, 0) = 0, && 0 \leqslant x \leqslant 1 && && && (6.7.13) \\
&u(0, t) = 0, && && t \geqslant 0 \\
&u(1, t) = 0, && && t \geqslant 0
\end{aligned}$$

In this case, $b_n = 0$ and $a_n$ is given by

$$\begin{aligned}
a_n &= 2 \int_0^1 x(1-x) \sin n\pi x \, dx \\
&= \frac{4}{(n\pi)^3} [1 - (-1)^n]
\end{aligned}$$

We also have

$$\begin{aligned}
h_n &= 2 \int_0^1 h \sin \frac{n\pi x}{l} \, dx \\
&= \frac{2h}{n\pi} [1 - (-1)^n]
\end{aligned}$$

Hence

$$\phi_n(t) = \frac{1}{\lambda_n} \int_0^t h_n \sin [\lambda_n(t-\tau)] \, d\tau = \frac{2h}{n\pi\lambda_n^2} [1 - (-1)^n](1 - \cos \lambda_n t)$$

The solution is thus given by

$$\begin{aligned}
u(x, t) = \sum_{n=1}^{\infty} \Big\{ &\frac{4}{n^3\pi^3} [1 - (-1)^n] \cos n\pi t \\
&+ \frac{2h}{n^3\pi^3} [1 - (-1)^n](1 - \cos n\pi t) \Big\} \cdot \sin n\pi x
\end{aligned} \tag{6.7.14}$$

We have treated the initial-boundary problem with the fixed end conditions. Problems with other boundary conditions can also be solved in a similar manner.

We will now consider the initial-boundary value problem with the time-dependent boundary conditions, namely,

$$\begin{aligned} u_{tt} - c^2 u_{xx} &= h(x, t), \qquad 0 < x < l, \qquad t > 0 \\ u(x, 0) &= f(x), \qquad 0 \leqslant x \leqslant l \\ u_t(x, 0) &= g(x), \qquad 0 \leqslant x \leqslant l \\ u(0, t) &= p(t), \qquad t \geqslant 0 \\ u(l, t) &= q(t), \qquad t \geqslant 0 \end{aligned} \tag{6.7.15}$$

We assume a solution in the form

$$u(x, t) = v(x, t) + U(x, t) \tag{6.7.16}$$

Substituting this into Eq. (6.7.15), we obtain

$$v_{tt} - c^2 v_{xx} = h - U_{tt} + c^2 U_{xx}$$

From the initial and boundary conditions, we have

$$\begin{aligned} v(x, 0) &= f(x) - U(x, 0) \\ v_t(x, 0) &= g(x) - U_t(x, 0) \\ v(0, t) &= p(t) - U(0, t) \\ v(l, t) &= q(t) - U(l, t) \end{aligned}$$

In order to make the boundary conditions homogeneous, we set

$$\begin{aligned} U(0, t) &= p(t) \\ U(l, t) &= q(t) \end{aligned}$$

Thus, $U(x, t)$ must take the form

$$U(x, t) = p(t) + \frac{x}{l}[q(t) - p(t)] \tag{6.7.17}$$

The problem now is to find the function $v(x, t)$ which satisfies

$$\begin{aligned} v_{tt} - c^2 v_{xx} &= h - U_{tt} = H(x, t) \\ v(x, 0) &= f(x) - U(x, 0) = F(x) \\ v_t(x, 0) &= g(x) - U_t(x, 0) = G(x) \\ v(0, t) &= 0 \\ v(l, t) &= 0 \end{aligned} \tag{6.7.18}$$

This is the same type of problem as the homogeneous boundary condition that has previously been treated.

EXAMPLE 6.7.3. Find the solution of the problem

$$\begin{aligned}
&u_{tt} - u_{xx} = h, && 0<x<1, \quad t>0, \quad h = \text{constant}\\
&u(x, 0) = x(1-x), && 0 \leqslant x \leqslant 1\\
&u_t(x, 0) = 0, && 0 \leqslant x \leqslant 1\\
&u(0, t) = t, && t \geqslant 0\\
&u(1, t) = \sin t, && t \geqslant 0
\end{aligned} \tag{6.7.19}$$

In this case,

$$U(x, t) = t + x\,[\sin t - t] \tag{6.7.20}$$

Then $v$ must satisfy

$$\begin{aligned}
v_{tt} - v_{xx} &= h + x \sin t\\
v(x, 0) &= x(1-x)\\
v_t(x, 0) &= 1\\
v(0, t) &= 0\\
v(1, t) &= 0
\end{aligned} \tag{6.7.21}$$

From (6.7.8)

$$\begin{aligned}
h_n(t) &= 2\int_0^1 (h + x \sin t) \sin n\pi x \, dx\\
&= \frac{2h}{n\pi}[1 - (-1)^n] + \frac{2(-1)^{n+1}}{n\pi} \sin t\\
&= a + b \sin t \text{ (say)}
\end{aligned} \tag{6.7.22}$$

We also find

$$\begin{aligned}
a_n &= 2\int_0^1 x(1-x) \sin n\pi x \, dx\\
&= \frac{4}{(n\pi)^3}[1 - (-1)^n]
\end{aligned}$$

and

$$\begin{aligned}
b_n &= \frac{2}{n\pi}\int_0^1 \sin n\pi x \, dx\\
&= \frac{2}{(n\pi)^2}[1 - (-1)^n]
\end{aligned}$$

Then we determine

$$\phi_n(t) = \frac{1}{n\pi}\int_0^t (a + b\sin\tau)\sin[n\pi(t-\tau)]\,d\tau$$

$$= \frac{1}{n\pi}\left\{\frac{a}{n\pi}(1-\cos n\pi t) + \frac{b}{4}[(\sin 2t - 2t)\cos n\pi t\right.$$

$$\left. - (\cos 2t - 1)\sin n\pi t]\right\} \qquad (6.7.23)$$

Hence the solution of the problem (6.7.21) is

$$v(x, t) = \sum_{n=1}^{\infty} [a_n \cos n\pi t + b_n \sin n\pi t + \phi_n(t)] \sin n\pi x \qquad (6.7.24)$$

Thus, the solution of problem (6.7.19) is given by

$$u(x, t) = v(x, t) + U(x, t)$$

where $v(x, t)$ is given by (6.7.24) and $U(x, t)$ is given by (6.7.20).

## 6.8 EXERCISES

**1.** Solve the following initial-boundary value problems:

a. $u_{tt} = c^2 u_{xx}, \quad 0 < x < 1 \quad t > 0$

$u(x, 0) = x(1-x), \quad 0 \leq x \leq 1$

$u_t(x, 0) = 0$

$u(0, t) = u(1, t) = 0$

b. $u_{tt} = c^2 u_{xx}, \quad 0 < x < \pi, \quad t > 0$

$u(x, 0) = 3\sin x, \quad 0 \leq x \leq \pi$

$u_t(x, 0) = 0$

$u(0, t) = u(\pi, t) = 0$

**2.** Determine the solutions of the following initial-boundary value problems:

a. $u_{tt} - c^2 u_{xx} = 0, \quad 0 < x < \pi \quad t > 0$

$u(x, 0) = 0$

$u_t(x, 0) = 8\sin^2 x, \quad 0 \leq x \leq \pi$

$u(0, t) = u(\pi, t) = 0$

b. $u_{tt} - c^2 u_{xx} = 0, \quad 0 < x < 1 \quad t > 0$

$u(x, 0) = 0$

$u_t(x, 0) = x\sin x, \quad 0 \leq x \leq 1$

$u(0, t) = u(1, t) = 0$

**3.** Find the solution of each of the following problems:

a. $u_{tt} - c^2 u_{xx} = 0, \qquad 0 < x < 1 \qquad t > 0$

$u(x, 0) = x(1 - x), \qquad 0 \leqslant x \leqslant 1$

$u_t(x, 0) = x - \tan \dfrac{\pi x}{4}, \qquad 0 \leqslant x \leqslant 1$

$u(0, t) = u(1, t) = 0$

b. $u_{tt} - c^2 u_{xx} = 0, \qquad 0 < x < \pi \qquad t > 0$

$u(x, 0) = \sin x, \qquad 0 \leqslant x \leqslant \pi$

$u_t(x, 0) = x^2 - \pi x, \qquad 0 \leqslant x \leqslant \pi$

$u(0, t) = u(\pi, t) = 0$

**4.** Solve the following problems:

a. $u_{tt} - c^2 u_{xx} = 0, \qquad 0 < x < \pi \qquad t > 0$

$u(x, 0) = x + \sin x, \qquad 0 \leqslant x \leqslant \pi$

$u_t(x, 0) = 0$

$u(0, t) = 0$

$u_x(\pi, t) = 0$

b. $u_{tt} - c^2 u_{xx} = 0, \qquad 0 < x < \pi \qquad t > 0$

$u(x, 0) = \cos x, \qquad 0 \leqslant x \leqslant \pi$

$u_t(x, 0) = 0$

$u_x(0, t) = 0$

$u_x(\pi, t) = 0$

**5.** By the method of separation of variables, solve the telegraph problem:

$$u_{tt} + a u_t + b u = c^2 u_{xx}, \qquad 0 < x < l \qquad t > 0$$
$$u(x, 0) = f(x)$$
$$u_t(x, 0) = 0$$
$$u(0, t) = u(l, t) = 0$$

**6.** Obtain the solution of the damped wave motion problem:

$$u_{tt} + a u_t = c^2 u_{xx}, \qquad 0 < x < l \qquad t > 0$$
$$u(x, 0) = 0$$
$$u_t(x, 0) = g(x)$$
$$u(0, t) = u(l, t) = 0$$

**7.** The torsional oscillation of a shaft of circular cross section is governed by the partial differential equation

$$\theta_{tt} = a^2\theta_{xx}$$

where $\theta(x, t)$ is the angular displacement of the cross section and $a$ is a physical constant. The ends of the shaft are fixed elastically, that is,

$$\theta_x(0, t) - h\theta(0, t) = 0, \qquad \theta_x(l, t) + h\theta(l, t) = 0$$

Determine the angular displacement if the initial angular displacement is $f(x)$.

**8.** Solve the initial-boundary value problem of the longitudinal vibration of a truncated cone of length $l$ and base radius $a$. The equation of motion is given by

$$\left(1 - \frac{x}{h}\right)^2 \frac{\partial^2 u}{\partial t^2} = c^2 \frac{\partial}{\partial x}\left[\left(1 - \frac{x}{h}\right)^2 \frac{\partial u}{\partial x}\right], \qquad 0 < x < l, \qquad t > 0$$

where $c^2 = E/\rho$, $E$ is the elastic modulus, $\rho$ is the density of the material and $h = la/(a - r)$. The two ends are rigidly fixed. If the initial displacement is $f(x)$, find $u(x, t)$.

**9.** Establish the validity of the formal solution of the initial-boundary value problem:

$$\begin{aligned} u_{tt} &= c^2 u_{xx}, && 0 < x < \pi && t > 0 \\ u(x, 0) &= f(x), && 0 \leq x \leq \pi && \\ u_t(x, 0) &= g(x), && 0 \leq x \leq \pi && \\ u_x(0, t) &= 0, && && t > 0 \\ u_x(\pi, t) &= 0, && && t > 0 \end{aligned}$$

**10.** Prove the uniqueness of the solution of the initial-boundary value problem:

$$\begin{aligned} u_{tt} &= c^2 u_{xx}, && 0 < x < \pi && t > 0 \\ u(x, 0) &= f(x), && 0 \leq x \leq \pi && \\ u_t(x, 0) &= g(x), && 0 \leq x \leq \pi && \\ u_x(0, t) &= 0, && && t > 0 \\ u_x(\pi, t) &= 0, && && t > 0 \end{aligned}$$

**11.** Determine the solution of

$$\begin{aligned} u_{tt} &= c^2 u_{xx} + A \sinh x, && 0 < x < l && t > 0 \\ u(x, 0) &= 0, && 0 \leq x \leq l && \\ u_t(x, 0) &= 0, && 0 \leq x \leq l && \\ u(0, t) &= h, && && t > 0 \\ u(l, t) &= k, && && t > 0 \end{aligned}$$

where $h$, $k$, and $A$ are constants.

**12.** Solve the problem:

$$
\begin{array}{llll}
u_{tt} = c^2 u_{xx} + Ax, & 0 < x < 1, & t > 0, & A = \text{constant} \\
u(x, 0) = 0, & 0 \leq x \leq 1 & & \\
u_t(x, 0) = 0, & & & \\
u(0, t) = 0, & & t > 0 & \\
u(1, t) = 0, & & &
\end{array}
$$

**13.** Solve the problem:

$$
\begin{array}{lll}
u_{tt} = c^2 u_{xx} + x^2, & 0 < x < 1 & t > 0 \\
u(x, 0) = x, & 0 \leq x \leq 1 & \\
u_t(x, 0) = 0, & & \\
u(0, t) = 0, & & t \geq 0 \\
u(1, t) = 1, & &
\end{array}
$$

**14.** Find the solution of the problem:

$$
\begin{array}{llll}
u_t = k u_{xx} + h, & 0 < x < 1, & t > 0, & h = \text{constant} \\
u(x, 0) = u_0(1 - \cos \pi x), & 0 \leq x \leq 1, & & u_0 = \text{constant} \\
u(0, t) = 0, & & t \geq 0 & \\
u(1, t) = 2u_0, & & &
\end{array}
$$

**15.** Obtain the solution of each of the following initial-boundary value problems:

a.
$$
\begin{array}{lll}
u_t = 4u_{xx}, & 0 < x < 1 & t > 0 \\
u(x, 0) = x^2(1 - x), & 0 \leq x \leq 1 & \\
u(0, t) = 0, & & t \geq 0 \\
u(1, t) = 0, & &
\end{array}
$$

b.
$$
\begin{array}{lll}
u_t = k u_{xx}, & 0 < x < \pi & t > 0 \\
u(x, 0) = \sin^2 x, & 0 \leq x \leq \pi & \\
u(0, t) = 0, & & t \geq 0 \\
u(\pi, t) = 0, & &
\end{array}
$$

c.
$$
\begin{array}{lll}
u_t = u_{xx}, & 0 < x < 2 & t > 0 \\
u(x, 0) = x, & 0 \leq x \leq 2 & \\
u(0, t) = 0, & & t \geq 0 \\
u_x(2, t) = 1, & &
\end{array}
$$

d.
$$
\begin{array}{lll}
u_t = k u_{xx}, & 0 < x < l & t > 0 \\
u(x, 0) = \sin \pi x / 2l, & 0 \leq x \leq l & \\
u(0, t) = 0, & & t \geq 0 \\
u(l, t) = 1, & &
\end{array}
$$

**16.** Find the temperature distribution in a rod of length $l$. The faces are insulated, and the initial temperature distribution is given by $x(l-x)$.

**17.** Find the temperature distribution in a rod of length $\pi$, one end of which is kept at zero temperature and the other end of which loses heat at a rate proportional to the temperature at that end $x=\pi$. The initial temperature distribution is given by $f(x)=x$.

**18.** The voltage distribution in an electric transmission line is given by

$$v_t = kv_{xx}, \qquad 0<x<l, \qquad t>0$$

The voltage equal to zero is maintained at $x=l$, while at the end $x=0$, the voltage varies according to the law

$$v(0, t) = Ct, \qquad t>0$$

where $C$ is a constant. Find $v(x, t)$ if the initial voltage distribution is zero.

**19.** Establish the validity of the formal solution of the initial–boundary value problem:

$$\begin{aligned} u_t &= ku_{xx}, && 0<x<l, && t>0 \\ u(x, 0) &= f(x) && 0\leq x\leq l && \\ u(0, t) &= 0, && && t\geq 0 \\ u_x(l, t) &= 0, && && \end{aligned}$$

**20.** Prove the uniqueness of the solution of the problem:

$$\begin{aligned} u_t &= ku_{xx}, && 0<x<l, && t>0 \\ u(x, 0) &= f(x) && 0\leq x\leq l && \\ u_x(0, t) &= 0, && && t\geq 0 \\ u_x(l, t) &= 0, && && \end{aligned}$$

**21.** Solve the radioactive decay problem:

$$\begin{aligned} u_t - ku_{xx} &= Ae^{-ax}, && 0<x<\pi, && t>0 \\ u(x, 0) &= \sin x, && 0\leq x\leq \pi && \\ u(0, t) &= 0, && && t\geq 0 \\ u(\pi, t) &= 0, && && \end{aligned}$$

**22.** Determine the solution of the initial-boundary value problem:

$$\begin{aligned} u_t - ku_{xx} &= h(x, t), && 0<x<l, && t>0, && k = \text{constant} \\ u(x, 0) &= f(x), && 0\leq x\leq l && && \\ u(0, t) &= p(t), && && t\geq 0 && \\ u(l, t) &= q(t), && && t\geq 0 && \end{aligned}$$

**23.** Determine the solution of the initial-boundary value problem:

$$\begin{aligned} &u_t - ku_{xx} = h(x, t), && 0<x<l, && t>0 \\ &u(x, 0) = f(x), && 0\leqslant x\leqslant l && \\ &u(0, t) = p(t), && && t\geqslant 0 \\ &u_x(l, t) = q(t), && && t\geqslant 0 \end{aligned}$$

**24.** Solve the problem:

$$\begin{aligned} &u_t - ku_{xx} = 0, && 0<x<1, && t>0 \\ &u(x, 0) = x(1-x), && 0\leqslant x\leqslant 1 && \\ &u(0, t) = t, && && t\geqslant 0 \\ &u(1, t) = \sin t, && && t\geqslant 0 \end{aligned}$$

**25.** Solve the problem:

$$\begin{aligned} &u_t - 4u_{xx} = xt, && 0<x<1, && t\geqslant 0 \\ &u(x, 0) = \sin \pi x, && 0\leqslant x\leqslant 1 && \\ &u(0, t) = t, && && t\geqslant 0 \\ &u(1, t) = t^2, && && t\geqslant 0 \end{aligned}$$

**26.** Solve the problem:

$$\begin{aligned} &u_t - ku_{xx} = x\cos t, && 0<x<\pi, && t>0 \\ &u(x, 0) = \sin x, && 0\leqslant x\leqslant \pi && \\ &u(0, t) = t^2, && && t\geqslant 0 \\ &u(\pi, t) = 2t, && && t\geqslant 0 \end{aligned}$$

**27.** Solve the problem:

$$\begin{aligned} &u_t - u_{xx} = 2x^2 t, && 0<x<1, && t>0 \\ &u(x, 0) = \cos 3\pi x/2, && 0\leqslant x\leqslant 1 && \\ &u(0, t) = 1, && && t\geqslant 0 \\ &u_x(1, t) = \frac{3\pi}{2}, && && t\geqslant 0 \end{aligned}$$

**28.** Solve the problem:

$$\begin{aligned} &u_t - 2u_{xx} = h, && 0<x<1, && t>0, && h = \text{constant} \\ &u(x, 0) = x, && 0\leqslant x\leqslant 1 && \\ &u(0, t) = \sin t, && && t\geqslant 0 \\ &u_x(1, t) + u(1, t) = 2, && && t\geqslant 0 \end{aligned}$$

**29.** Determine the solution of the initial-boundary value problem:

$$\begin{aligned} &u_{tt} - c^2 u_{xx} = h(x, t), && 0 < x < l, && t > 0 \\ &u(x, 0) = f(x), && 0 \leq x \leq l && \\ &u_t(x, 0) = g(x), && 0 \leq x \leq l && \\ &u(0, t) = p(t), && && t \geq 0 \\ &u_x(1, t) = q(t), && && t \geq 0 \end{aligned}$$

**30.** Determine the solution of the initial-boundary value problem:

$$\begin{aligned} &u_{tt} - c^2 u_{xx} = h(x, t). && 0 < x < l, && t > 0 \\ &u(x, 0) = f(x), && 0 \leq x \leq l && \\ &u_t(x, 0) = g(x), && 0 \leq x \leq l && \\ &u_x(0, t) = p(t), && && t \geq 0 \\ &u_x(l, t) = q(t), && && t \geq 0 \end{aligned}$$

**31.** Solve

$$\begin{aligned} &u_{tt} - u_{xx} = 0, && 0 < x < 1, && t > 0 \\ &u(x, 0) = x, && 0 \leq x \leq 1 && \\ &u_t(x, 0) = 0, && 0 \leq x \leq 1 && \\ &u(0, t) = t^2, && && t \geq 0 \\ &u(1, t) = \cos t, && && t \geq 0 \end{aligned}$$

**32.** Solve

$$\begin{aligned} &u_{tt} - 4u_{xx} = xt, && 0 < x < 1, && t > 0 \\ &u(x, 0) = x, && 0 \leq x \leq 1 && \\ &u_t(x, 0) = 0, && 0 \leq x \leq 1 && \\ &u(0, t) = 0, && && t \geq 0 \\ &u_x(l, t) = 1 + t, && && t \geq 0 \end{aligned}$$

**33.** Solve

$$\begin{aligned} &u_{tt} - 9u_{xx} = 0, && 0 < x < 1, && t > 0 \\ &u(x, 0) = \sin \frac{\pi x}{2}, && 0 \leq x \leq 1 && \\ &u_t(x, 0) = 1 + x, && 0 \leq x \leq 1 && \\ &u_x(0, t) = \pi/2, && && t \geq 0 \\ &u_x(1, t) = 0, && && t \geq 0 \end{aligned}$$

**34.** Find the solution of the problem:

$$
\begin{aligned}
u_{tt} + 2ku_t - c^2 u_{xx} &= 0, & 0 < x < l, \quad & t > 0 \\
u(x, 0) &= 0, & 0 \leqslant x \leqslant l & \\
u_t(x, 0) &= 0, & 0 \leqslant x \leqslant l & \\
u_x(0, t) &= 0, & & t \geqslant 0 \\
u(l, t) &= h, & & t \geqslant 0, \quad h = \text{constant}
\end{aligned}
$$

**35.** Solve

$$
\begin{aligned}
u_t - c^2 u_{xx} + hu &= hu_0, & -\pi < x < \pi, \quad & t > 0 \\
u(x, 0) &= f(x), & -\pi \leqslant x \leqslant \pi & \\
u(-\pi, t) &= u(\pi, t), & & t \geqslant 0 \\
u_x(-\pi, t) &= u_x(\pi, t), & & t \geqslant 0
\end{aligned}
$$

where $u_0$ is a constant.

CHAPTER SEVEN

# Eigenvalue Problems and Special Functions

## 7.1 STURM–LIOUVILLE SYSTEMS

In the preceding chapter, we determined the solutions of partial differential equations by the method of separation of variables. Under separable conditions we transformed the second order homogeneous partial differential equation into two ordinary differential equations (6.1.11) and (6.1.12) which are of the form

$$a_1(x)\frac{d^2y}{dx^2}+a_2(x)\frac{dy}{dx}+[a_3(x)+\lambda]y=0 \tag{7.1.1}$$

If we introduce

$$p(x)=\exp\left[\int^x \frac{a_2(t)}{a_1(t)}\,dt\right], \qquad q(x)=\frac{a_3(x)}{a_1(x)}p(x), \qquad s(x)=\frac{p(x)}{a_1(x)} \tag{7.1.2}$$

into Eq. (7.1.1), we obtain

$$\frac{d}{dx}\left(p\frac{dy}{dx}\right)+(q+\lambda s)y=0 \tag{7.1.3}$$

which is known as the *Sturm–Liouville equation*. In terms of the operator

$$L\equiv\frac{d}{dx}\left(p\frac{d}{dx}\right)+q$$

Eq. (7.1.3) can be written as

$$L[y]+\lambda s(x)y=0 \tag{7.1.4}$$

where $\lambda$ is a parameter independent of $x$, and $p$, $q$, and $s$ are real-valued functions of $x$. To ensure the existence of solutions, we let $q$ and $s$ be continuous and $p$ be continuously differentiable in a closed finite interval $[a, b]$.

The Sturm–Liouville equation is called *regular* in the interval $[a, b]$ if the functions $p(x)$ and $s(x)$ are positive in the interval $[a, b]$. Thus, for a given $\lambda$, there exist two linearly independent solutions of a regular Sturm–Liouville equation in the interval $[a, b]$.

The Sturm–Liouville equation

$$L[y] + \lambda s(x)y = 0, \qquad a \leqslant x \leqslant b$$

together with the separated end conditions

$$\begin{aligned} a_1 y(a) + a_2 y'(a) &= 0 \\ b_1 y(b) + b_2 y'(b) &= 0 \end{aligned} \tag{7.1.5}$$

where the constants $a_1$ and $a_2$, and likewise $b_1$ and $b_2$, are not both zero and are given real numbers, is called a *regular Sturm–Liouville system.*

The values of $\lambda$ for which the Sturm–Liouville system has a nontrivial solution are called the *eigenvalues,* and the corresponding solutions are called the *eigenfunctions.*

EXAMPLE 7.1.1. Consider the Sturm–Liouville system

$$\begin{aligned} y'' + \lambda y = 0, &\qquad 0 \leqslant x \leqslant \pi \\ y(0) = 0 & \\ y'(\pi) = 0 & \end{aligned}$$

When $\lambda \leqslant 0$, it can be readily shown that $\lambda$ is not an eigenvalue. However, when $\lambda > 0$, the solution of the Sturm–Liouville equation is

$$y(x) = A \cos \sqrt{\lambda}\, x + B \sin \sqrt{\lambda}\, x$$

Applying the condition $y(0) = 0$, we obtain $A = 0$. The condition $y'(\pi) = 0$ yields

$$B\sqrt{\lambda} \cos \sqrt{\lambda}\, \pi = 0$$

Since $\lambda \neq 0$ and $B = 0$ yields a trivial solution, we must have

$$\cos \sqrt{\lambda}\, \pi = 0, \qquad B \neq 0$$

This equation is satisfied if

$$\sqrt{\lambda} = \frac{2n-1}{2}, \qquad n = 1, 2, 3, \ldots$$

and hence the eigenvalues are $\lambda_n = (2n-1)^2/4$, and the corresponding eigenfunctions are

$$\sin\left(\frac{2n-1}{2}\right)x, \qquad n = 1, 2, 3, \ldots$$

EXAMPLE 7.1.2. Consider the Euler equation

$$x^2y'' + xy' + \lambda y = 0, \qquad 1 \leqslant x \leqslant e$$

with the end conditions

$$y(1) = 0, \qquad y(e) = 0$$

By using the transformation (7.1.2), the Euler equation can be put into Sturm–Liouville form:

$$\frac{d}{dx}\left(x\frac{dy}{dx}\right) + \frac{1}{x}\lambda y = 0$$

The solution of the Euler equation is

$$y(x) = c_1x^{i\sqrt{\lambda}} + c_2x^{-i\sqrt{\lambda}}$$

Noting that $x^{ia} = e^{ia\,ln\,x} = \cos(a\,ln\,x) + i\sin(a\,ln\,x)$, $y(x)$ becomes

$$y(x) = A\cos(\sqrt{\lambda}\,ln\,x) + B\sin(\sqrt{\lambda}\,ln\,x)$$

where $A$ and $B$ are constants related to $c_1$ and $c_2$. The end condition $y(1) = 0$ gives $A = 0$, and the end condition $y(e) = 0$ gives

$$\sin\sqrt{\lambda} = 0, \qquad B \neq 0$$

which in turn yields the eigenvalues

$$\lambda_n = n^2\pi^2, \qquad n = 1, 2, 3, \ldots$$

and the corresponding eigenfunctions

$$\sin(n\pi\,ln\,x), \qquad n = 1, 2, 3, \ldots$$

Another type of problem that often occurs in practice is the periodic Sturm–Liouville system.

The Sturm–Liouville equation

$$\frac{d}{dx}\left(p(x)\frac{dy}{dx}\right) + [q(x) + \lambda s(x)]y = 0, \qquad a \leqslant x \leqslant b$$

in which $p(a) = p(b)$, together with the periodic end conditions

$$y(a) = y(b)$$
$$y'(a) = y'(b)$$

is called a *periodic Sturm–Liouville system.*

EXAMPLE 7.1.3. Consider the periodic Sturm–Liouville system

$$y'' + \lambda y = 0, \qquad -\pi \leqslant x \leqslant \pi$$
$$y(-\pi) = y(\pi)$$
$$y'(-\pi) = y'(\pi)$$

Here we note that $p(x) = 1$ and hence $p(-\pi) = p(\pi)$. When $\lambda > 0$, we see that the solution of the Sturm–Liouville equation is

$$y(x) = A \cos \sqrt{\lambda}\, x + B \sin \sqrt{\lambda}\, x$$

Application of the periodic end conditions yields

$$(2 \sin \sqrt{\lambda}\, \pi) B = 0$$
$$(2\sqrt{\lambda} \sin \sqrt{\lambda}\, \pi) A = 0$$

Thus to obtain a nontrivial solution, we must have

$$\sin \sqrt{\lambda}\, \pi = 0, \qquad A \neq 0, \qquad B \neq 0$$

Consequently,

$$\lambda_n = n^2, \qquad n = 1, 2, 3, \ldots$$

Since $\sin \sqrt{\lambda}\, \pi = 0$ is satisfied for arbitrary $A$ and $B$, we obtain two linearly independent eigenfunctions $\cos nx$, $\sin nx$ corresponding to the same eigenvalue $n^2$.

It can be readily shown that if $\lambda < 0$, the solution of the Sturm–Liouville equation does not satisfy the periodic end conditions. However, when $\lambda = 0$ the corresponding eigenfunction is 1. Thus, the eigenvalues of the periodic Sturm–Liouville system are 0, $\{n^2\}$, and the corresponding eigenfunctions are 1, $\{\cos nx\}$, $\{\sin nx\}$, where $n$ is a positive integer.

## 7.2 EIGENVALUES AND EIGENFUNCTIONS

In Examples 7.1.1 and 7.1.2 of the regular Sturm–Liouville systems in the preceding section, we see that there exists only one linearly independent eigenfunction corresponding to the eigenvalue $\lambda$, which is called an eigenvalue of multiplicity one (or a simple eigenvalue). An eigenvalue is said to be of multiplicity $k$ if there exist $k$ linearly independent eigenfunctions corresponding to the same eigenvalue. In Example 7.1.3 of the periodic Sturm–Liouville system, the eigenfunctions $\cos nx$, $\sin nx$ correspond to the same eigenvalue $n^2$. Thus, this eigenvalue is of multiplicity two.

In the preceding examples, we see that the eigenfunctions are $\cos nx$ and $\sin nx$ for $n = 1, 2, 3, \ldots$. It can be easily shown by using trigonometric identities that

$$\int_{-\pi}^{\pi} \cos mx \cos nx \, dx = 0, \qquad m \neq n$$

$$\int_{-\pi}^{\pi} \cos mx \sin nx \, dx = 0, \qquad \text{for all integers } m, n$$

$$\int_{-\pi}^{\pi} \sin mx \sin nx \, dx = 0, \qquad m \neq n$$

We say that these functions are orthogonal to each other in the interval $[-\pi, \pi]$. The orthogonality relation holds in general for the eigenfunctions of Sturm–Liouville systems.

Let $\phi(x)$ and $\psi(x)$ be any real-valued integrable functions on an interval $I$. Then $\phi$ and $\psi$ are said to be *orthogonal* on $I$ with respect to a weight function $\rho(x) > 0$, if and only if,

$$(\phi, \psi) = \int_I \phi(x)\psi(x)\rho(x) \, dx = 0 \tag{7.2.1}$$

The interval $I$ may be of infinite extent, or it may be either open or closed at one or both ends of the finite interval.

When $\phi = \psi$ in (7.2.1) we have the *norm* of $\phi$

$$\|\phi\| = \left[\int_I \phi^2(x)\rho(x) \, dx\right]^{\frac{1}{2}} \tag{7.2.2}$$

***Theorem 7.2.1*** *Let the coefficients $p$, $q$, and $s$ in the Sturm–Liouville system be continuous in $[a, b]$. Let the eigenfunctions $\phi_j$ and $\phi_k$, corresponding to $\lambda_j$ and $\lambda_k$, be continuously differentiable. Then $\phi_j$ and $\phi_k$ are orthogonal with respect to the weight function $s(x)$ in $[a, b]$.*

PROOF. Since $\phi_j$ corresponding to $\lambda_j$ satisfies the Sturm–Liouville equation, we have

$$\frac{d}{dx}(p\phi_j') + (q + \lambda_j s)\phi_j = 0 \tag{7.2.3}$$

and for the same reason

$$\frac{d}{dx}(p\phi_k') + (q + \lambda_k s)\phi_k = 0 \tag{7.2.4}$$

Multiplying Eq. (7.2.3) by $\phi_k$ and Eq. (7.2.4) by $\phi_j$, and subtracting, we

obtain

$$(\lambda_j - \lambda_k)s\phi_j\phi_k = \phi_j \frac{d}{dx}(p\phi_k') - \phi_k \frac{d}{dx}(p\phi_j')$$

$$= \frac{d}{dx}[(p\phi_k')\phi_j - (p\phi_j')\phi_k]$$

and integration yields

$$\begin{aligned}(\lambda_j - \lambda_k)\int_a^b s\phi_j\phi_k\,dx &= [p(\phi_j\phi_k' - \phi_j'\phi_k)]_a^b \\ &= p(b)[\phi_j(b)\phi_k'(b) - \phi_j'(b)\phi_k(b)] \\ &\quad - p(a)[\phi_j(a)\phi_k'(a) - \phi_j'(a)\phi_k(a)] \qquad (7.2.5)\end{aligned}$$

the right side of which is called the boundary term of the Sturm–Liouville system. The end conditions for the eigenfunctions $\phi_j$ and $\phi_k$ are

$$b_1\phi_j(b) + b_2\phi_j'(b) = 0$$
$$b_1\phi_k(b) + b_2\phi_k'(b) = 0$$

If $b_2 \neq 0$, we multiply the first condition by $\phi_k(b)$ and the second condition by $\phi_j(b)$, and subtract to obtain

$$[\phi_j(b)\phi_k'(b) - \phi_j'(b)\phi_k(b)] = 0 \qquad (7.2.6)$$

In a similar manner, if $a_2 \neq 0$, we obtain

$$[\phi_j(a)\phi_k'(a) - \phi_j'(a)\phi_k(a)] = 0 \qquad (7.2.7)$$

We see by virtue of (7.2.6) and (7.2.7) that

$$(\lambda_j - \lambda_k)\int_a^b s\phi_j\phi_k\,dx = 0 \qquad (7.2.8)$$

If $\lambda_j$ and $\lambda_k$ are distinct eigenvalues, then

$$\int_a^b s\phi_j\phi_k\,dx = 0 \qquad (7.2.9)$$

□

***Theorem 7.2.2*** *The eigenfunctions of a periodic Sturm–Liouville system in* $[a, b]$ *are orthogonal with respect to the weight function* $s(x)$ *in* $[a, b]$.

PROOF. The periodic conditions for the eigenfunctions $\phi_j$ and $\phi_k$ are

$$\phi_j(a) = \phi_j(b), \qquad \phi_j'(a) = \phi_j'(b)$$
$$\phi_k(a) = \phi_k(b), \qquad \phi_k'(a) = \phi_k'(b)$$

Substitution of these into Eq. (7.2.5) yields

$$(\lambda_j - \lambda_k)\int_a^b s\phi_j\phi_k\,dx = [p(b) - p(a)][\phi_j(a)\phi_k'(a) - \phi_j'(a)\phi_k(a)]$$

Since $p(a) = p(b)$, we have

$$(\lambda_j - \lambda_k)\int_a^b s\phi_j\phi_k\,dx = 0 \tag{7.2.10}$$

For distinct eigenvalues $\lambda_j \neq \lambda_k$, and thus

$$\int_a^b s\phi_j\phi_k\,dx = 0 \tag{7.2.11}$$

□

***Theorem 7.2.3*** *All the eigenvalues of a regular Sturm–Liouville system with $s(x) > 0$ are real.*

PROOF. Suppose that there is a complex eigenvalue $\lambda_j = \alpha + i\beta$ with eigenfunction $\phi_j = u + iv$. Then, because the coefficients of the equation are real, the complex conjugate of the eigenvalue is also an eigenvalue. Thus, there exists an eigenfunction $\phi_k = u - iv$ corresponding to the eigenvalue $\lambda_k = \alpha - i\beta$.

By using the relation (7.2.8), we have

$$2i\beta\int_a^b s(u^2 + v^2)\,dx = 0$$

This implies that $\beta$ must vanish for $s > 0$, and hence the eigenvalues are real. □

***Theorem 7.2.4*** *If $\phi_1(x)$ and $\phi_2(x)$ are any two solutions of $L[y] + \lambda sy = 0$ on $[a, b]$, then $p(x)W(x; \phi_1, \phi_2) = \text{constant}$, where $W$ is the Wronskian.*

PROOF. Since $\phi_1$ and $\phi_2$ are solutions of $L[y] + \lambda sy = 0$, we have

$$\frac{d}{dx}\left(p\frac{d\phi_1}{dx}\right) + (q + \lambda s)\phi_1 = 0$$

$$\frac{d}{dx}\left(p\frac{d\phi_2}{dx}\right) + (q + \lambda s)\phi_2 = 0$$

Multiplying the first equation by $\phi_2$ and the second by $\phi_1$, and subtracting, we obtain

$$\phi_1\frac{d}{dx}\left(p\frac{d\phi_2}{dx}\right) - \phi_2\frac{d}{dx}\left(p\frac{d\phi_1}{dx}\right) = 0$$

Integrating this equation from $a$ to $x$, we obtain

$$p(x)[\phi_1(x)\phi_2'(x) - \phi_1'(x)\phi_2(x)] = p(a)[\phi_1(a)\phi_2'(a) - \phi_1'(a)\phi_2(a)]$$
$$= \text{constant} \qquad (7.2.12)$$

which is called *Abel's formula.* □

***Theorem 7.2.5*** *An eigenfunction of a regular Sturm–Liouville system is unique except for a constant factor.*

PROOF. Let $\phi_1(x)$ and $\phi_2(x)$ be eigenfunctions corresponding to an eigenvalue $\lambda$. Then according to Abel's formula (7.2.12), we have

$$p(x)W(x; \phi_1, \phi_2) = \text{constant}, \qquad p(x) > 0$$

where $W$ is the Wronskian. Thus, if $W$ vanishes at a point in $[a, b]$, it must vanish for all $x \in [a, b]$.

Since $\phi_1$ and $\phi_2$ satisfy the end condition at $x = a$, we have

$$a_1\phi_1(a) + a_2\phi_1'(a) = 0$$
$$a_1\phi_2(a) + a_2\phi_2'(a) = 0$$

Since $a_1$ and $a_2$ are not both zero, we have

$$\begin{vmatrix} \phi_1(a) & \phi_1'(a) \\ \phi_2(a) & \phi_2'(a) \end{vmatrix} = +W(a; \phi_1, \phi_2) = 0$$

Therefore $W(x; \phi_1, \phi_2) = 0$ for all $x \in [a, b]$, which is a sufficient condition for the linear dependence of two functions $\phi_1$ and $\phi_2$. Hence $\phi_1(x)$ differs from $\phi_2(x)$ only by a constant factor. □

Theorem 7.2.3 states that all eigenvalues of a regular Sturm–Liouville system are real, but it does not guarantee that any eigenvalue exists. However, it can be proved that a self-adjoint regular Sturm–Liouville system has a denumerably infinite number of eigenvalues. To illustrate this, let us consider the following example:

EXAMPLE 7.2.1. Consider the Sturm–Liouville system

$$y'' + \lambda y = 0, \qquad 0 \leq x \leq 1$$
$$y(0) = 0,$$
$$y(1) + hy'(1) = 0, \qquad h > 0 \text{ a constant}$$

Here $p = 1$, $q = 0$, $s = 1$. The solution of the Sturm–Liouville equation is

$$y(x) = A \cos \sqrt{\lambda}\, x + B \sin \sqrt{\lambda}\, x$$

Since $y(0) = 0$ gives $A = 0$, we have

$$y(x) = B \sin \sqrt{\lambda}\, x$$

Applying the second end condition, we have

$$\sin\sqrt{\lambda} + h\sqrt{\lambda}\cos\sqrt{\lambda} = 0 \quad \text{for} \quad B \neq 0$$

which can be rewritten as

$$\tan\sqrt{\lambda} = -h\sqrt{\lambda}$$

If $\alpha = \sqrt{\lambda}$ is introduced in this equation, we have

$$\tan\alpha = -h\alpha$$

This equation does not possess an explicit solution. Thus, we determine the solution graphically by plotting the functions $\xi = \tan\alpha$ and $\xi = -h\alpha$ against $\alpha$, as shown in Fig. 7.2.1. The roots are given by the intersection of two curves, and as is evident from the graph, there are infinitely many roots $\alpha_n$ for $n = 1, 2, 3, \ldots$. To each root $\alpha_n$, there corresponds an eigenvalue

$$\lambda_n = \alpha_n^2, \qquad n = 1, 2, 3, \ldots.$$

Thus, there exists a sequence of eigenvalues

$$\lambda_1 < \lambda_2 < \lambda_3 < \ldots$$

with

$$\lim_{n\to\infty} \lambda_n = \infty$$

The corresponding eigenfunctions are $\sin\sqrt{\lambda_n}\,x$.

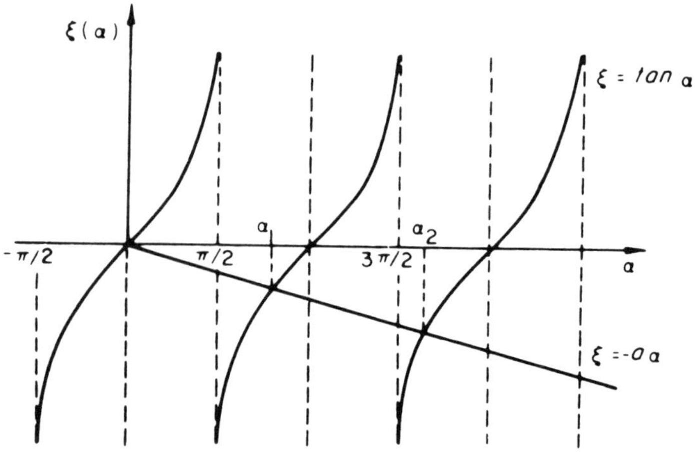


**Figure** 7.2.1

***Theorem 7.2.6*** *A self-adjoint regular Sturm–Liouville system has an infinite sequence of real eigenvalues*

$$\lambda_0 < \lambda_1 < \lambda_2 < \lambda_3 < \ldots$$

*with*

$$\lim_{n\to\infty} \lambda_n = \infty$$

*For each n the corresponding eigenfunction $\phi_n(x)$, uniquely determined up to a constant factor, has exactly n zeros in the interval $(a, b)$.*

Proof of this theorem can be found in [67].

## 7.3 EIGENFUNCTION EXPANSIONS

A real-valued function $\phi(x)$ is said to be *square-integrable* with respect to a weight function $\rho(x) > 0$, if on an interval $I$,

$$\int_I \phi^2(x)\rho(x)\,dx < +\infty \tag{7.3.1}$$

An immediate consequence of this definition is the *Schwarz inequality*

$$\left|\int_I \phi(x)\psi(x)\rho(x)\,dx\right|^2 \leq \int_I \phi^2(x)\rho(x)\,dx \int_I \psi^2(x)\rho(x)\,dx \tag{7.3.2}$$

for square-integrable functions $\phi(x)$ and $\psi(x)$.

Let $\{\phi_n(x)\}$, for a positive integer $n$, be an orthogonal set of square-integrable functions with a positive weight function $\rho(x)$ on an interval $I$. Let $f(x)$ be a given function that can be represented by a uniformly convergent series of the form

$$f(x) = \sum_{n=1}^{\infty} c_n\phi_n(x) \tag{7.3.3}$$

where the coefficients $c_n$ are constants. Now multiplying both sides of (7.3.3) by $\phi_m(x)\rho(x)$ and integrating term by term over the interval $I$ (uniform convergence of the series is a sufficient condition for this), we obtain

$$\int_I f(x)\phi_m(x)\rho(x)\,dx = \sum_{n=1}^{\infty} \int_I c_n\phi_n(x)\phi_m(x)\rho(x)\,dx$$

and hence

$$\int_I f(x)\phi_n(x)\rho(x)\,dx = c_n \int_I \phi_n^2(x)\rho(x)\,dx$$

Thus

$$c_n = \frac{\int_I f\phi_n\rho\, dx}{\int_I \phi_n^2\rho\, dx} \tag{7.3.4}$$

Hence we have the following theorem:

***Theorem 7.3.1*** *If $f$ is represented by a uniformly convergent series*

$$f(x) = \sum_{n=1}^{\infty} c_n\phi_n(x)$$

*on an interval $I$, where $\phi_n$ are square-integrable functions orthogonal with respect to a positive weight function $\rho(x)$, then $c_n$ are determined by*

$$c_n = \frac{\int_I f\phi_n\rho\, dx}{\int_I \phi_n^2\rho\, dx}$$

EXAMPLE 7.3.1. The Legendre polynomials $P_n(x)$ are orthogonal with respect to the weight function $\rho(x) = 1$ on $(-1, 1)$. If we assume that $f(x)$ can be represented by the *Fourier-Legendre series*

$$f(x) = \sum_{n=1}^{\infty} c_n P_n(x)$$

then $c_n$ are given by

$$c_n = \frac{\int_{-1}^{1} f(x)P_n(x)\, dx}{\int_{-1}^{1} P_n^2(x)\, dx}$$

$$= \frac{2n+1}{2}\int_{-1}^{1} f(x)\, P_n(x)\, dx$$

In the above discussion, we assumed that the given function $f(x)$ is represented by a uniformly convergent series. This is rather restrictive, and we will show in the following section that $f(x)$ can be represented by a mean-square convergent series.

## 7.4 CONVERGENCE IN THE MEAN

Let $\{\phi_n\}$ be the set of square-integrable functions orthogonal with respect to the weight function $\rho(x)$ on $[a, b]$. Let

$$s_n(x) = \sum_{k=1}^{n} c_k\phi_k(x)$$

be the $n$th partial sum of the series $\sum_{k=1}^{\infty} c_k\phi_k(x)$.

Let $f$ be a square-integrable function. The sequence $\{s_n\}$ is said to *converge in the mean* to $f(x)$ on the interval $I$ with respect to the weight function $\rho(x)$ if

$$\lim_{n\to+\infty}\int_I [f(x) - s_n(x)]^2\rho(x)\,dx = 0 \qquad (7.4.1)$$

We shall now seek the coefficients $c_k$ such that $s_n(x)$ represents the best approximation to $f(x)$ in the sense of least squares, that is, we seek to minimize the integral

$$\begin{aligned} E(c_k) &= \int_I [f(x) - s_n(x)]^2\rho(x)\,dx \\ &= \int_I f^2\rho\,dx - 2\sum_{k=1}^{n} c_k\int_I f\phi_k\rho\,dx + \sum_{k=1}^{n} c_k^2\int_I \phi_k^2\rho\,dx \end{aligned} \qquad (7.4.2)$$

This is an extremal problem. A necessary condition on the $c_k$ for $E$ to be a minimum is that the first partial derivatives of $E$ with respect to these coefficients vanish. Thus, differentiating (7.4.2) with respect to $c_k$, we obtain

$$\frac{\partial E}{\partial c_k} = -2\int_I f\phi_k\rho\,dx + 2c_k\int_I \phi_k^2\rho\,dx = 0 \qquad (7.4.3)$$

and hence

$$c_k = \frac{\displaystyle\int_I f\phi_k\rho\,dx}{\displaystyle\int_I \phi_k^2\rho\,dx} \qquad (7.4.4)$$

Now if we complete the square, the right side of (7.4.2) becomes

$$E = \int_I f^2\rho\,dx + \sum_{k=1}^{n}\int_I \phi_k^2\rho\,dx\left[c_k - \frac{\displaystyle\int_I f\phi_k\rho\,dx}{\displaystyle\int_I \phi_k^2\rho\,dx}\right]^2 - \sum_{k=1}^{n}\frac{\left(\displaystyle\int_I f\phi_k\rho\,dx\right)^2}{\displaystyle\int_I \phi_k^2\rho\,dx}$$

The right side shows that $E$ is a minimum if and only if $c_k$ is given by

(7.4.4). Therefore, this choice of $c_k$ yields the best approximation to $f(x)$ in the sense of least squares.

For series convergent in the mean to $f(x)$, we conventionally write

$$f(x) \sim \sum_{k=1}^{\infty} c_k \phi_k(x)$$

where the coefficients $c_k$ are the *Fourier coefficients* and the series is the *Fourier series*. This series may or may not be pointwise or uniformly convergent.

## 7.5 COMPLETENESS AND PARSEVAL'S EQUALITY

Substituting the Fourier coefficients (7.4.4) into (7.4.2), we obtain

$$\int_I \left(f(x) - \sum_{k=1}^{n} c_k \phi_k(x)\right)^2 \rho(x)\, dx = \int_I f^2 \rho\, dx - \sum_{k=1}^{n} c_k^2 \int_I \phi_k^2 \rho\, dx$$

Since the left side is nonnegative, we have

$$\sum_{k=1}^{n} c_k^2 \int_I \phi_k^2 \rho\, dx \leqslant \int_I f^2 \rho\, dx \tag{7.5.1}$$

The integral on the right side is finite, and hence the series on the left side is bounded above for any $n$. Thus, as $n \to \infty$, the inequality (7.5.1) may be written as

$$\sum_{k=1}^{\infty} c_k^2 \int_I \phi_k^2 \rho\, dx \leqslant \int_I f^2 \rho\, dx \tag{7.5.2}$$

This is called *Bessel's inequality*.

If the series converges in the mean to $f(x)$, that is,

$$\lim_{n \to \infty} \int_I \left(f(x) - \sum_{k=1}^{n} c_k \phi_k(x)\right)^2 \rho(x)\, dx = 0$$

then it follows from the above derivation that

$$\sum_{k=1}^{\infty} c_k^2 \int_I \phi_k^2 \rho\, dx = \int_I f^2 \rho\, dx$$

which is called *Parseval's equality*. Sometimes it is known as the completeness relation. Thus, when every continuous square-integrable function $f(x)$ can be expanded into an infinite series

$$f(x) = \sum_{k=1}^{\infty} c_k \phi_k(x),$$

the sequence of continuous square-integrable functions $\{\phi_k\}$ orthogonal with respect to the weight function $\rho$ is said to be *complete*.

Next we state the following theorem:

***Theorem 7.5.1*** *The eigenfunctions of any regular Sturm–Liouville system are complete in the space of functions that are piecewise continuous on the interval $[a, b]$ with respect to the weight function $s(x)$. Moreover, any piecewise smooth function on $[a, b]$ that satisfies the end conditions of the regular Sturm–Liouville system can be expanded in an absolutely and uniformly convergent series*

$$f(x) = \sum_{k=1}^{\infty} c_k \phi_k(x)$$

*where $c_k$ are given by*

$$c_k = \int_a^b f\phi_k \, s(x)\, dx \Big/ \int_a^b \phi_k^2 \, s(x)\, dx$$

Proof of a more general theorem can be found in Coddington and Levinson [59].

EXAMPLE 7.5.1. Consider a cylindrical wire of length $l$ whose surface is perfectly insulated against the flow of heat. The end $l = 0$ is maintained at the zero degree temperature while the other end radiates freely into the surrounding medium of zero degree temperature. Let the initial temperature distribution in the wire be $f(x)$. Find the temperature distribution $u(x, t)$.

The initial-boundary value problem is

$$u_t = ku_{xx}, \qquad 0 < x < l, \qquad t > 0 \tag{7.5.3}$$

$$u(x, 0) = f(x), \qquad 0 < x \leq l \tag{7.5.4}$$

$$u(0, t) = 0, \qquad t > 0 \tag{7.5.5}$$

$$hu(l, t) + u'(l, t) = 0, \qquad t > 0, \qquad h > 0 \tag{7.5.6}$$

By the method of separation of variables, we assume a nontrivial solution in the form

$$u(x, t) = X(x)T(t)$$

and substituting it in the heat equation, we obtain

$$X'' + \lambda X = 0$$

$$T' + k\lambda T = 0$$

where $\lambda > 0$ is a separation constant. The solution of the latter equation is

$$T(t) = Ce^{-k\lambda t}$$

where $C$ is an arbitrary constant. The former equation has to be solved subject to the boundary conditions

$$X(0)=0$$
$$hX(l)+X'(l)=0$$

This is a Sturm–Liouville system which gives the solution with $X(0)=0$

$$X(x)=B\sin\sqrt{\lambda}\,x$$

where $B$ is a constant to be determined.
Application of the second end condition (7.5.6) yields

$$h\sin\sqrt{\lambda}\,l+\sqrt{\lambda}\cos\sqrt{\lambda}\,l=0 \quad \text{for} \quad B\neq 0$$

which can be rewritten as

$$\tan\sqrt{\lambda}\,l=-\sqrt{\lambda}\,/h$$

If $\alpha=\sqrt{\lambda}\,l$ is introduced in the preceding equation, we have

$$\tan\alpha=-a\alpha$$

where $a=1/hl$. As in Example 7.2.1, there exists a sequence of eigenvalues

$$\lambda_1<\lambda_2<\lambda_3<\ldots$$

with $\lim_{n\to\infty}\lambda_n=\infty$. The corresponding eigenfunctions are $\sin\sqrt{\lambda_n}\,x$, and hence

$$X_n(x)=B_n\sin\sqrt{\lambda_n}\,x$$

Therefore, the solution takes the form

$$u_n(x,t)=a_n e^{-k\lambda_n t}\sin\sqrt{\lambda_n}\,x, \qquad a_n=B_nC_n$$

which satisfies the heat equation and the boundary conditions. Since the heat equation is linear and homogeneous, we form a series of solutions

$$u(x,t)=\sum_{n=1}^{\infty}a_n e^{-k\lambda_n t}\sin\sqrt{\lambda_n}\,x \tag{7.5.7}$$

which is also a solution, provided it converges and is twice differentiable with respect to $x$ and once differentiable with respect to $t$. According to Theorem 7.2.1, the eigenfunctions $\sin\sqrt{\lambda_n}\,x$ form an orthogonal system over the interval $(0,l)$. Application of the initial condition yields

$$u(x,0)=f(x)\sim\sum_{n=1}^{\infty}a_n\sin\sqrt{\lambda_n}\,x$$

If we assume that $f$ is a piecewise smooth function on $[a,b]$, then by

Theorem 7.5.1 we can expand $f(x)$ in terms of the eigenfunctions, and formally write

$$f(x) = \sum_{n=1}^{\infty} a_n \sin \sqrt{\lambda_n}\, x$$

where the coefficient $a_n$ is given by

$$a_n = \int_0^l f(x) \sin \sqrt{\lambda_n}\, x \, dx \Big/ \int_0^l \sin^2 \sqrt{\lambda_n}\, x \, dx$$

With this value of $a_n$, the temperature distribution is given by (7.5.7).

## 7.6 BESSEL'S EQUATION AND BESSEL'S FUNCTION

Bessel's equation frequently occurs in problems of applied mathematics and mathematical physics involving cylindrical symmetry.

The standard form of Bessel's equation is given by

$$x^2 y'' + xy' + (x^2 - \nu^2)y = 0 \tag{7.6.1}$$

where $\nu$ is a nonnegative real number. We shall first restrict our attention to $x > 0$. Since $x = 0$ is the regular singular point, a solution is taken in accordance with the Frobenius method to be

$$y(x) = \sum_{n=0}^{\infty} a_n x^{s+n} \tag{7.6.2}$$

where the index $s$ is to be determined. Substitution of this series into Eq. (7.6.1) then yields

$$[s^2 - \nu^2]a_0 x^s + [(s+1)^2 - \nu^2]a_1 x^{s+1} + \sum_{n=2}^{\infty} \{[(s+n)^2 - \nu^2]a_n + a_{n-2}\} x^{s+n} = 0 \tag{7.6.3}$$

The requirement that the coefficient of $x^s$ vanish leads to the *indicial equation*

$$(s^2 - \nu^2)a_0 = 0 \tag{7.6.4}$$

from which it follows that $s = \pm\nu$ for arbitrary $a_0 \neq 0$. Since the leading term in the series (7.6.2) is $a_0 x^s$, it is clear that for $\nu > 0$ the solution of Bessel's equation corresponding to the choice $s = \nu$ vanishes at the origin, whereas the solution corresponding to $s = -\nu$ is infinite at that point.

We consider first the regular solution of Bessel's equation, that is, the solution corresponding to the choice $s = \nu$. The vanishing of the coefficient of $x^{s+1}$ in Eq. (7.6.3) requires that

$$(2\nu + 1)a_1 = 0 \tag{7.6.5}$$

which in turn implies that $a_1 = 0$ (since $\nu \geq 0$). From the requirement that the coefficient of $x^{s+n}$ in Eq. (7.6.3) be zero, we obtain the two-term recurrence relation

$$a_n = -\frac{a_{n-2}}{n(2\nu + n)} \tag{7.6.6}$$

Since $a_1 = 0$, it is obvious that $a_n = 0$ for $n = 3, 5, 7, \ldots$. The remaining coefficients are given by

$$a_{2k} = \frac{(-1)^k a_0}{2^{2k} k!\,(\nu + k)(\nu + k - 1) \ldots (\nu + 1)} \tag{7.6.7}$$

for $k = 1, 2, 3, \ldots$. This relation may also be written as

$$a_{2k} = \frac{(-1)^k 2^\nu \Gamma(\nu + 1) a_0}{2^{2k+\nu} k!\,\Gamma(\nu + k + 1)}, \qquad k = 1, 2, \ldots \tag{7.6.8}$$

where $\Gamma(\alpha)$ is the gamma function, whose properties are described in Appendix I.

Hence, the regular solution of Bessel's equation takes the form

$$y(x) = a_0 \sum_{k=0}^{\infty} \frac{(-1)^k 2^\nu \Gamma(\nu + 1)}{2^{2k+\nu} k!\,\Gamma(\nu + k + 1)} x^{2k+\nu} \tag{7.6.9}$$

It is customary to choose

$$a_0 = \frac{1}{2^\nu \Gamma(\nu + 1)} \tag{7.6.10}$$

and to denote the corresponding solution by $J_\nu(x)$. This solution, called the *Bessel function of the first kind of order* $\nu$, is therefore given by

$$J_\nu(x) = \sum_{k=0}^{\infty} \frac{(-1)^k x^{2k+\nu}}{2^{2k+\nu} k!\,\Gamma(\nu + k + 1)} \tag{7.6.11}$$

To determine the irregular solution of the Bessel equation for $s = -\nu$, we proceed as above. In this way, we obtain as the analogue of Eq. (7.6.5) the relation

$$(-2\nu + 1) a_1 = 0$$

from which it follows, without loss of generality, that $a_1 = 0$. Using the recurrence relation

$$a_n = -\frac{a_{n-2}}{n(n - 2\nu)}, \qquad n \geq 2 \tag{7.6.12}$$

we obtain as the irregular solution the Bessel function of the first kind of order $-\nu$:

$$J_{-\nu}(x) = \sum_{k=0}^{\infty} \frac{(-1)^k x^{2k-\nu}}{2^{2k-\nu} k!\, \Gamma(-\nu + k + 1)} \tag{7.6.13}$$

It can be easily proved that, if $\nu$ is not an integer, $J_\nu$ and $J_{-\nu}$ converge for all values of $x$, and are linearly independent. Thus, the general solution of the Bessel equation for nonintegral $\nu$ is

$$y(x) = c_1 J_\nu(x) + c_2 J_{-\nu}(x) \tag{7.6.14}$$

If $\nu$ is an integer, say $\nu = n$, then from Eq. (7.6.13), noting that, when gamma functions in the coefficients of the first $n$ terms become infinite, the coefficients become zero, we have

$$\begin{aligned} J_{-n}(x) &= \sum_{k=n}^{\infty} \frac{(-1)^k x^{2k-n}}{2^{2k-\nu} k!\, \Gamma(-n + k + 1)} \\ &= (-1)^n \sum_{k=0}^{\infty} \frac{(-1)^k x^{2k+n}}{2^{2k+n} k!\, \Gamma(n + k + 1)} \\ &= (-1)^n J_n(x) \end{aligned} \tag{7.6.15}$$

It shows that $J_{-n}$ is not independent of $J_n$, and therefore a second linearly independent solution is required.

A number of distinct irregular solutions are discussed in the literature, but the one most commonly used, as defined by Weber [85], is

$$Y_\nu(x) = \frac{(\cos \nu\pi) J_\nu(x) - J_{-\nu}(x)}{\sin \nu\pi} \tag{7.6.16}$$

For nonintegral $\nu$, it is obvious that $Y_\nu(x)$, being a linear combination of $J_\nu(x)$ and $J_{-\nu}(x)$, is linearly independent of $J_\nu(x)$. When $\nu$ is a nonnegative integer $n$, $Y_\nu(x)$ is indeterminate. But

$$Y_n(x) = \lim_{\nu \to \infty} Y_\nu(x)$$

exists and is a solution of the Bessel equation. Moreover, it is linearly independent of $J_n(x)$ (For an extended treatment, see Watson [85]). The function $Y_\nu(x)$ is called *the Bessel function of the second kind of order* $\nu$. Thus, the general solution of the Bessel equation is

$$y(x) = c_1 J_\nu(x) + c_2 Y_\nu(x), \quad \text{for} \quad \nu \geq 0 \tag{7.6.17}$$

Like elementary functions, Bessel's functions are tabulated (see Jahnke and Emde [114]). For illustration, the functions $J_0$, $J_1$, $Y_0$, and $Y_1$ are plotted for small values of $x$, as shown in Fig. 7.6.1.

It should be noted that $J_\nu(x)$ for $\nu \geq 0$ and $J_{-\nu}(x)$ for a positive integer

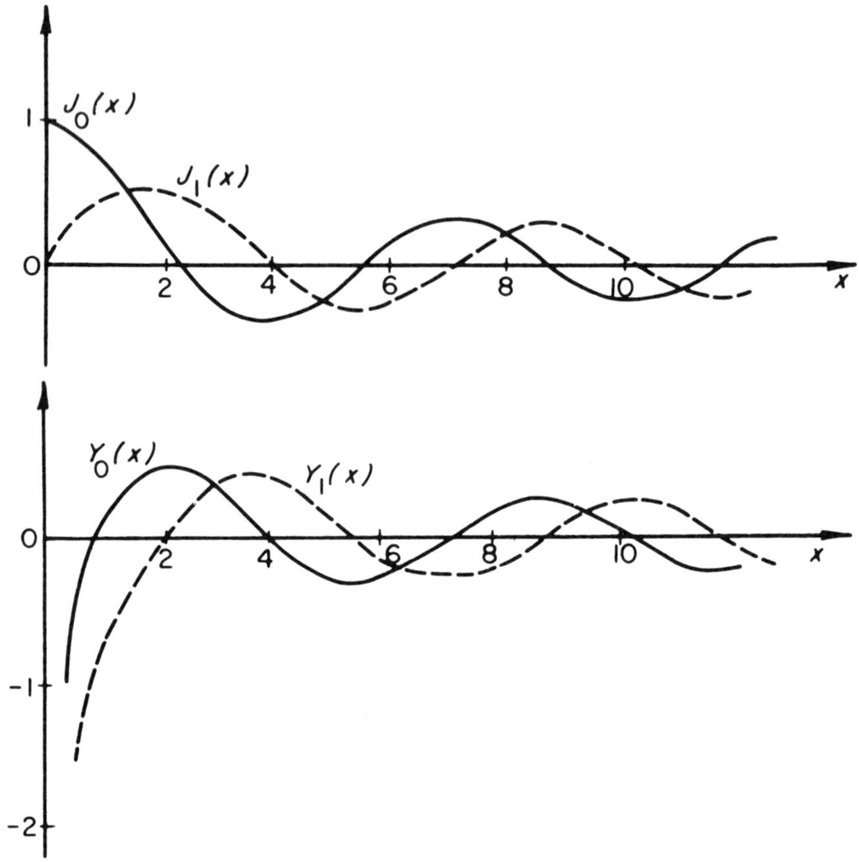


Figure 7.6.1

$\nu$ are finite at the origin, but $J_{-\nu}(x)$ for nonintegral $\nu$ and $Y_\nu(x)$ for $\nu \geqslant 0$ approach infinity as $x$ tends to zero.

Some of the useful recurrence relations are

$$J_{\nu-1}(x) + J_{\nu+1}(x) = \frac{2\nu}{x} J_\nu(x) \tag{7.6.18}$$

$$\nu J_\nu(x) + xJ'_\nu(x) = xJ_{\nu-1}(x) \tag{7.6.19}$$

$$J_{\nu-1}(x) - J_{\nu+1}(x) = 2J'_\nu(x) \tag{7.6.20}$$

$$\nu J_\nu(x) - xJ'_\nu(x) = xJ_{\nu+1}(x) \tag{7.6.21}$$

$$\frac{d}{dx}[x^\nu J_\nu(x)] = x^\nu J_{\nu-1}(x) \tag{7.6.22}$$

$$\frac{d}{dx}[x^{-\nu} J_\nu(x)] = -x^{-\nu} J_{\nu+1}(x) \tag{7.6.23}$$

All of these relations also hold true for $Y_\nu(x)$.

For $|x| \gg 1$ and $|x| \gg \nu$, the asymptotic expansion of $J_\nu(x)$ is

$$J_\nu(x) \sim \sqrt{\frac{2}{\pi x}}\left[\left\{1 - \frac{(4\nu^2-1^2)(4\nu^2-3^2)}{2!\,(8x)^2} + \frac{(4\nu^2-1^2)(4\nu^2-3^2)(4\nu^2-5^2)(4\nu^2-7^2)}{4!\,(8x)^4} - \dots\right\}\cos\phi - \left\{\frac{(4\nu^2-1^2)}{8x} - \frac{(4\nu^2-1)(4\nu^2-3^2)(4\nu^2-5^2)}{3!\,(8x)^3} + \dots\right\}\sin\phi\right] \tag{7.6.24}$$

where

$$\phi = x - \left(\nu + \frac{1}{2}\right)\frac{\pi}{2}.$$

For $|x| \gg 1$ and $x \gg \nu_1$ the asymptotic expansion of $Y_\nu(x)$ is

$$Y_\nu(x) \sim \sqrt{\frac{2}{\pi x}}\left[\left\{1 - \frac{(4\nu^2-1^2)(4\nu^2-3^2)}{2!\,(8x)^2} + \frac{(4\nu^2-1^2)(4\nu^2-3^2)(4\nu^2-5^2)(4\nu^2-7^2)}{4!\,(8x)^4} - \dots\right\}\sin\phi + \left\{\frac{(4\nu^2-1^2)}{8x} - \frac{(4\nu^2-1^2)(4\nu^2-3^2)(4\nu^2-5^2)}{3!\,(8x)^3} + \dots\right\}\cos\phi\right] \tag{7.6.25}$$

When $\nu = \pm 1/2$, Bessel's functions may be expressed in the form

$$J_{\frac{1}{2}}(x) = \sqrt{\frac{2}{\pi x}}\sin x \tag{7.6.26}$$

$$J_{-\frac{1}{2}}(x) = \sqrt{\frac{2}{\pi x}}\cos x \tag{7.6.27}$$

Bessel functions which satisfy the condition

$$J_\nu(ak_m) + hJ'_\nu(ak_m) = 0, \qquad h,\ a = \text{constant} \tag{7.6.28}$$

are orthogonal to each other with respect to the weight function $x$, that is, for a nonnegative integer $\nu$, the orthogonal relation is

$$\int_0^a xJ_\nu(xk_n)J_\nu(xk_m)\,dx = 0, \qquad n \neq m \tag{7.6.29}$$

When $n = m$, we have the norm

$$\|J_\nu(xk_m)\|^2 = \int_0^a x[J_\nu(xk_m)]^2\,dx = \frac{1}{2k_m^2}\{a^2k_m^2[J'_\nu(ak_m)]^2 + (a^2k_m^2 - \nu^2)[J_\nu(ak_m)]^2\} \tag{7.6.30}$$

where $k_m$ are the roots of (7.6.28).

We now give a particular example of the eigenfunction expansion theorem discussed in § 7.4–7.5. Assume a formal expansion of the function $f(x)$ defined in $0 \leq x \leq a$ in the form

$$f(x) = \sum_{m=1}^{\infty} a_m J_\nu(xk_m) \tag{7.6.31}$$

where the summation is taken over all the positive roots $k_1, k_2, k_3, \ldots$ of Eq. (7.6.28). Multiplying (7.6.31) by $xJ_\nu(xk_n)$, integrating, and utilizing (7.6.30), we obtain

$$\begin{aligned}\int_0^a xf(x)J_\nu(xk_m)\,dx &= a_m \int_0^a x[J_\nu(xk_m)]^2\,dx \\ &= \frac{a_m}{2k_m^2}\{a^2k_m^2[J'_\nu(ak_m)]^2 \\ &\quad + (a^2k_m^2 - \nu^2)[J_\nu(ak_m)]^2\}\end{aligned} \tag{7.6.32}$$

Thus we have the following theorem:

***Theorem 7.6.1*** *If*

$$b_m = \int_0^a xf(x)J_\nu(xk_m)\,dx \tag{7.6.33}$$

*then the expansion* (7.6.31) *of* $f(x)$ *takes the form*

$$f(x) = \sum_{m=1}^{\infty} \frac{2k_m^2 b_m J_\nu(xk_m)}{a^2k_m^2[J'_\nu(ak_m)]^2 + (a^2k_m^2 - \nu^2)[J_\nu(ak_m)]^2} \tag{7.6.34}$$

*In particular, when* $h = 0$ *in* (7.6.28), *that is,* $k_m$ *are the positive roots of* $J_\nu(ak_m) = 0$, *then* (7.6.34) *becomes*

$$f(x) = \frac{2}{a^2}\sum_{m=1}^{\infty} \frac{b_m J_\nu(xk_m)}{[J'_\nu(ak_m)]^2} = \frac{2}{a^2}\sum_{m=1}^{\infty} \frac{b_m J_\nu(xk_m)}{[J_{\nu+1}(ak_m)]^2} \tag{7.6.35}$$

*These expansions are known as the Bessel–Fourier series for* $f(x)$. *They are generated by Sturm–Liouville problems involving the Bessel equation, and arise from problems associated with partial differential equations.*

Closely related to Bessel's functions are *Hankel's functions of the first and second kind*, defined by

$$\begin{aligned} H_\nu^{(1)}(x) &= J_\nu(x) + iY_\nu(x) \\ H_\nu^{(2)}(x) &= J_\nu(x) - iY_\nu(x) \end{aligned} \tag{7.6.36ab}$$

respectively, where $i = \sqrt{-1}$.

Other closely related functions are the modified Bessel functions. Consider Bessel's equation containing a parameter $\lambda$, namely,

$$x^2y'' + xy' + (\lambda^2x^2 - \nu^2)y = 0 \tag{7.6.37}$$

The general solution of this equation is

$$y(x) = c_1 J_\nu(\lambda x) + c_2 Y_\nu(\lambda x)$$

If $\lambda = i$, then

$$y(x) = c_1 J_\nu(ix) + c_2 Y_\nu(ix)$$

We write

$$\begin{aligned} J_\nu(ix) &= \sum_{k=0}^{\infty} \frac{(-1)^k (ix)^{2k+\nu}}{2^{2k+\nu} k!\, \Gamma(\nu+k+1)} \\ &= i^\nu I_\nu(x) \end{aligned}$$

where

$$I_\nu(x) = \sum_{k=0}^{\infty} \frac{x^{2k+\nu}}{2^{2k+\nu} k!\, \Gamma(\nu+k+1)} \tag{7.6.38}$$

$I_\nu(x)$ is called *the modified Bessel function of the first kind of order $\nu$*. As in the case of $J_\nu$ and $J_{-\nu}$, $I_\nu$ and $I_{-\nu}$ (which is defined in a similar manner) are linearly independent solutions except when $\nu$ is an integer. Consequently, we define *the modified Bessel function of the second kind of order $\nu$* as

$$K_\nu(x) = \frac{\pi}{2}\left(\frac{I_{-\nu}(x) - I_\nu(x)}{\sin \nu\pi}\right) \tag{7.6.39}$$

Thus, we obtain the general solution of the modified Bessel equation

$$x^2 y'' + xy' - (x^2 + \nu^2) y = 0$$

as

$$y(x) = c_1 I_\nu(x) + c_2 K_\nu(x) \tag{7.6.40}$$

We should note that

$$I_\nu(0) = \begin{cases} 1, & \nu = 0 \\ 0, & \nu > 0 \end{cases} \tag{7.6.41}$$

and that $K_\nu$ approaches infinity as $x \to 0$.

For a detailed treatment of Bessel and related functions, refer to Watson's Theory of Bessel Functions [85].

The eigenvalue problems which involve Bessel's functions will be described in Sec. 7.8 on singular Sturm–Liouville systems.

## 7.7 ADJOINT FORMS: LAGRANGE IDENTITY

Self-adjoint equations play a very important role in many areas of applied mathematics and mathematical physics. Here we will give a brief account of self-adjoint operators and the Lagrange identity.

Let us consider the equation

$$L[y] = a_0(x)y'' + a_1(x)y' + a_2(x)y = 0$$

defined on an interval $I$. Integrating $z(x)L[y]$ by parts from $a$ to $x$, we have

$$\int_a^x zL[y]\,dx = [(za_0)y' - (za_0)'y + (za_1)y]_a^x + \int_a^x [(za_0)'' - (za_1)' + (za_2)]y\,dx \tag{7.7.1}$$

Now, if we define the second-order operator $L^*$ by

$$L^*[z] = (za_0)'' - (za_1)' + (za_2) = a_0z'' + (2a_0' - a_1)z' + (a_0'' + a_1' + a_2)z$$

the relation (7.7.1) takes the form

$$\int_a^x (zL[y] - yL^*[z])\,dx = [a_0(y'z - yz') + (a_1 - a_0')yz]_a^x \tag{7.7.2}$$

The operator $L^*$ is called the *adjoint operator* corresponding to the operator $L$. It can be readily verified that the adjoint of $L^*$ is $L$ itself. If $L$ and $L^*$ are the same, $L$ is said to be *self-adjoint*. The necessary and sufficient condition for this is that

$$a_1 = 2a_0' - a_1$$
$$a_2 = a_0'' - a_1' + a_2$$

which is satisfied if

$$a_1 = a_0'$$

Thus, if $L$ is self-adjoint, we have

$$L(y) = a_0y'' + a_0'y' + a_2y = (a_0y')' + a_2y \tag{7.7.3}$$

In general $L[y]$ is not self-adjoint. But if we let

$$h(x) = \frac{1}{a_0}\exp\left\{\int^x \frac{a_1(t)}{a_0(t)}\,dt\right\} \tag{7.7.4}$$

then $h(x)L[y]$ is self-adjoint. Thus any second-order linear differential equation

$$a_0(x)y'' + a_1(x)y' + a_2(x)y = 0 \tag{7.7.5}$$

can be made self-adjoint. Multiplying by $h(x)$, given by Eq. (7.7.4), Eq. (7.7.5) is transformed into the self-adjoint form

$$\frac{d}{dx}\left[p(x)\frac{dy}{dx}\right] + q(x)y = 0 \tag{7.7.6}$$

where

$$p(x) = \exp\left\{\int^x \frac{a_1(t)}{a_0(t)}\,dt\right\}$$
$$q(x) = \frac{a_2}{a_0}\exp\left\{\int^x \frac{a_1(t)}{a_0(t)}\,dt\right\} \tag{7.7.7}$$

For example, the self-adjoint form of the Legendre equation

$$(1-x^2)y'' - 2xy' + n(n+1)y = 0$$

is

$$\frac{d}{dx}\left[(1-x^2)\frac{dy}{dx}\right] + n(n+1)y = 0 \tag{7.7.8}$$

and the self-adjoint form of the Bessel equation

$$x^2y'' + xy' + (x^2 - \nu^2)y = 0$$

is

$$\frac{d}{dx}\left(x\frac{dy}{dx}\right) + \left(x - \frac{\nu^2}{x}\right)y = 0 \tag{7.7.9}$$

Now, if we differentiate both sides of Eq. (7.7.2), we obtain

$$zL[y] - yL^*[z] = \frac{d}{dx}[a_0(y'z - yz') + (a_1 - a_0')yz] \tag{7.7.10}$$

which is known as the *Lagrange identity* for the operator $L$.

If we consider the integral from $a$ to $b$ of Eq. (7.7.2), we obtain *Green's identity*

$$\int_a^b (zL[y] - yL^*[z])\,dx = [a_0(y'z - yz') + (a_1 - a_0')yz]_a^b \tag{7.7.11}$$

When $L$ is self-adjoint, this relation becomes

$$\int_a^b (zL[y] - yL[z])\,dx = [a_0(y'z - yz')]_a^b \tag{7.7.12}$$

## 7.8 SINGULAR STURM–LIOUVILLE SYSTEMS

A Sturm–Liouville equation is called *singular* when it is given on a semi-infinite or infinite interval, or when the coefficient $p(x)$ or $s(x)$ vanishes, or when one of the coefficients becomes infinite at one end or both ends of a finite interval. A singular Sturm–Liouville equation together with appropriate linear homogeneous end conditions is called a *singular Sturm–Liouville system.* The conditions imposed in this case are not like the separated end conditions in the regular Sturm–Liouville system. The condition that is often necessary to prescribe is the boundedness of the function $y(x)$ at the singular end point. To exhibit this, let us consider a problem with a singularity at the end point $x = a$. By the relation (7.7.12), for any twice continuously differentiable functions $y(x)$ and $z(x)$, we have on $(a, b)$

$$\int_{a+\varepsilon}^{b} \{zL[y] - yL[z]\}\, dx = p(b)[y'(b)z(b) - y(b)z'(b)]$$
$$-p(a+\varepsilon)[y'(a+\varepsilon)z(a+\varepsilon) - y(a+\varepsilon)z'(a+\varepsilon)]$$

where $\varepsilon$ is a small positive number. If the conditions

$$\lim_{x\to a+} p(x)[y'(x)z(x) - y(x)z'(x)] = 0 \tag{7.8.1}$$

$$p(b)[y'(b)z(b) - y(b)z'(b)] = 0 \tag{7.8.2}$$

are imposed on $y$ and $z$, it follows that

$$\int_{a}^{b} \{zL[y] - yL[z]\}\, dx = 0 \tag{7.8.3}$$

For example, when $p(a) = 0$, the relations (7.8.1) and (7.8.2) are replaced by the conditions

1. $y(x)$ and $y'(x)$ are finite as $x \to a$
2. $b_1 y(b) + b_2 y'(b) = 0$

Thus, we say that the singular Sturm–Liouville system is *self-adjoint,* if any functions $y(x)$ and $z(x)$ that satisfy the end conditions satisfy

$$\int_{a}^{b} \{zL[y] - yL[z]\}\, dx = 0$$

EXAMPLE 7.8.1. Consider the singular Sturm–Liouville system involving Legendre's equation

$$\frac{d}{dx}\left[(1-x^2)\frac{dy}{dx}\right] + \lambda y = 0, \qquad -1 < x < 1$$

with the conditions that $y$ and $y'$ are finite as $x \to \pm 1$.

In this case $p(x) = 1 - x^2$ and $s(x) = 1$, and $p(x)$ vanishes at $x = \pm 1$. The Legendre functions of the first kind, $P_n(x)$, $n = 0, 1, 2, \ldots,$ are the eigenfunctions which are finite as $x \to \pm 1$. The corresponding eigenvalues are $\lambda_n = n(n+1)$ for $n = 0, 1, 2, \ldots.$ We observe here that the singular Sturm–Liouville system has infinitely many real eigenvalues, and the eigenfunctions $P_n(x)$ are orthogonal to each other.

EXAMPLE 7.8.2. Another example of a singular Sturm–Liouville system is the Bessel equation for fixed $\nu$,

$$\frac{d}{dx}\left(x\frac{dy}{dx}\right)+\left(\lambda x-\frac{\nu^2}{x}\right)y=0, \qquad 0<x<a$$

with the end conditions that $y(a)=0$ and $y$ and $y'$ are finite as $x\to 0+$.

Here $p(x)=x$, $q(x)=-\nu^2/x$, $s(x)=x$. Now $p(0)=0$, $q(x)$ becomes infinite as $x\to 0+$, and $s(0)=0$; therefore the system is singular. If $\lambda=k^2$, the eigenfunctions of the system are Bessel functions of the first kind of order $\nu$, namely $J_\nu(k_n x)$, $n=1, 2, 3, \ldots$, where $k_n a$ is the $n$th zero of $J_\nu$. The Bessel function $J_\nu$ and its derivative are both finite as $x\to 0+$. The eigenvalues are $\lambda_n=k_n^2$. Thus, the system has infinitely many eigenvalues, and the eigenfunctions are orthogonal with respect to the weight function $x$.

In the preceding examples, we see that the eigenfunctions are orthogonal with respect to the weight function $s(x)$. In general, the eigenfunctions of a singular Sturm–Liouville system are orthogonal if they are square-integrable with respect to the weight function $s(x)$.

***Theorem 7.8.1*** *The square-integrable eigenfunctions corresponding to distinct eigenvalues of a singular Sturm–Liouville system are orthogonal with respect to a weight function $s(x)$.*

PROOF. Proceeding as in Theorem 7.2.1, we arrive at

$$(\lambda_j-\lambda_k)\int_a^b s\phi_j\phi_k\,dx=p(b)[\phi_j(b)\phi_k'(b)-\phi_j'(b)\phi_k(b)]$$
$$-p(a)[\phi_j(a)\phi_k'(a)-\phi_j'(a)\phi_k(a)]$$

Suppose the boundary term vanishes, as in the case mentioned earlier, where $p(a)=0$, $y$ and $y'$ are finite as $x\to a$, and at the other end $b_1y(b)+b_2y'(b)=0$. Then we have

$$(\lambda_j-\lambda_k)\int_a^b s\phi_j\phi_k\,dx=0$$

This integral exists by virtue of (7.3.2). Thus, for distinct eigenvalues $\lambda_j\neq\lambda_k$, the square-integrable functions $\phi_j$ and $\phi_k$ are orthogonal with respect to the weight function $s(x)$. □

EXAMPLE 7.8.3. Consider the singular Sturm–Liouville system involving the *Hermite equation*

$$u''-2xu'+\lambda u=0, \qquad -\infty<x<\infty \tag{7.8.4}$$

which is not self-adjoint.

If we let $y(x) = e^{-x^2/2}u(x)$, the Hermite equation takes the self-adjoint form

$$y'' + [(1 - x^2) + \lambda]y = 0, \qquad -\infty < x < \infty$$

Here $p = 1$, $q(x) = 1 - x^2$, $s = 1$. The eigenvalues are $\lambda_n = 2n$ for a nonnegative integer $n$, and the corresponding eigenfunctions are $\phi_n(x) = e^{-x^2/2}H_n(x)$, where $H_n(x)$ are the Hermite polynomials of Eq. (7.8.4). (See Magnus and Oberhettinger [115].)

Now, we impose the end condition that $y$ tends to zero as $x \to \pm\infty$. This is satisfied because $H_n(x)$ are polynomials in $x$ and in fact $x^n e^{-x^2/2} \to 0$ as $x \to \pm\infty$. Since $\phi_n(x)$ are square-integrable, we have

$$\int_{-\infty}^{\infty} H_m(x) H_n(x) e^{-x^2}\, dx = 0, \qquad m \neq n$$

EXAMPLE 7.8.4. Consider the problem of the transverse vibration of a thin elastic circular membrane

$$\begin{aligned}
u_{tt} &= c^2\left(u_{rr} + \frac{1}{r}u_r\right), && r < 1, && t > 0\\
u(r, 0) &= f(r), && 0 \leqslant r \leqslant 1 &&\\
u_t(r, 0) &= 0 && 0 \leqslant r \leqslant 1 && \qquad (7.8.5)\\
u(1, t) &= 0, && && t \geqslant 0\\
\lim_{r\to 0} u(r, t) &< \infty
\end{aligned}$$

By the method of separation of variables, we seek a nontrivial solution in the form

$$u(r, t) = R(r)T(t)$$

Substitution of this in the wave equation yields

$$\frac{R'' + (1/r)R'}{R} = \frac{1}{c^2}\frac{T''}{T} = -\alpha^2$$

where $\alpha$ is a positive constant. The negative sign in front of $\alpha^2$ is chosen to obtain the solution periodic in time. Thus, we have

$$rR'' + R' + \alpha^2 rR = 0$$

$$T'' + \alpha^2 c^2 T = 0$$

The solution $T(t)$ is therefore given by

$$T(t) = A\cos \alpha ct + B \sin \alpha ct$$

Next, it is required to determine the solution $R(r)$ of the following singular Sturm–Liouville system

$$\frac{d}{dr}\left[r\frac{dR}{dr}\right]+\alpha^2 rR=0 \tag{7.8.6}$$

$$R(1)=0 \tag{7.8.7}$$

$$\lim_{r\to 0} R(r)<\infty \tag{7.8.8}$$

We note that in this case, $p=r$ which vanishes at $r=0$. The condition on the boundedness of the function $R(r)$ is obtained from the fact that

$$\lim_{r\to 0} u(r, t)=\lim_{r\to 0} R(r)T(t)<\infty$$

which implies that

$$\lim_{r\to 0} R(r)<\infty \tag{7.8.9}$$

for arbitrary $T(t)$. Equation (7.8.6) is Bessel's equation of order zero, the solution of which is given by

$$R(r)=CJ_0(\alpha r)+DY_0(\alpha r) \tag{7.8.10}$$

where $J_0$ and $Y_0$ are Bessel functions of the first and second kinds respectively of order zero. The condition (7.8.8) requires that $D=0$ since $Y_0(\alpha r)\to -\infty$ as $r\to 0$. Hence

$$R(r)=CJ_0(\alpha r)$$

The remaining condition $R(1)=0$ yields

$$J_0(\alpha)=0$$

This transcendental equation has infinitely many positive zeros

$$\alpha_1<\alpha_2<\alpha_3<\ldots$$

Thus, the solution of problem (7.8.5) is given by

$$u_n(r, t)=J_0(\alpha_n r)(A_n\cos\alpha_n ct+B_n\sin\alpha_n ct), \qquad n=1, 2, 3, \ldots$$

Since the Bessel equation is linear and homogeneous,

$$u(r, t)=\sum_{n=1}^{\infty} J_0(\alpha_n r)(A_n\cos\alpha_n ct+B_n\sin\alpha_n ct) \tag{7.8.11}$$

is also a solution, provided the series converges and is sufficiently differentiable with respect to $r$ and $t$. Differentiating (7.8.11) formally with respect to $t$, we obtain

$$u_t(r, t) = \sum_{n=1}^{\infty} J_0(\alpha_n r)(-A_n\alpha_n c \sin \alpha_n ct + B_n\alpha_n c \cos \alpha_n ct)$$

Application of the initial condition $u_t(r, 0) = 0$ yields $B_n = 0$. Consequently, we have

$$u(r, t) = \sum_{n=1}^{\infty} A_n J_0(\alpha_n r) \cos \alpha_n ct \tag{7.8.12}$$

It now remains to show that $u(r, t)$ satisfies the initial condition $u(r, 0) = f(r)$. For this, we have

$$u(r, 0) = f(r) \sim \sum_{n=1}^{\infty} A_n J_0(\alpha_n r)$$

If $f(r)$ is piecewise smooth on $[0, 1]$, then the eigenfunctions $J_0(\alpha_n r)$ form a complete orthogonal system with respect to the weight function $r$ over the interval $(0, 1)$. Hence we can formally expand $f(r)$ in terms of the eigenfunctions. Thus

$$f(r) = \sum_{n=1}^{\infty} A_n J_0(\alpha_n r)$$

where the coefficient $A_n$ is represented by

$$A_n = \int_0^1 r f(r) J_0(\alpha_n r)\, dr \bigg/ \int_0^1 r[J_0(\alpha_n r)]^2\, dr \tag{7.8.13}$$

The solution of the problem (7.8.5) is therefore given by (7.8.12) with the coefficient $A_n$ given by (7.8.13).

## 7.9 LEGENDRE'S EQUATION AND LEGENDRE'S FUNCTION

Legendre's equation is

$$(1 - x^2)y'' - 2xy' + \nu(\nu + 1)y = 0 \tag{7.9.1}$$

where $\nu$ is a real number. This equation arises in problems with spherical symmetry in mathematical physics. Its coefficients are analytic at $x = 0$. Thus, if we expand near the point $x = 0$, the coefficients are

$$p(x) = -\frac{2x}{1 - x^2} = -2x \sum_{m=0}^{\infty} x^{2m} = \sum_{m=0}^{\infty} (-2)x^{2m+1}$$

and

$$q(x) = \frac{\nu(\nu + 1)}{1 - x^2} = \nu(\nu + 1) \sum_{m=0}^{\infty} x^{2m} = \sum_{m=0}^{\infty} \nu(\nu + 1)x^{2m}$$

We see that these series converge for $|x|<1$. Thus the Legendre equation on $|x|<1$ has convergent power series solution at $x=0$.

Now to find the solution near the ordinary point $x=0$, we assume

$$y(x)=\sum_{m=0}^{\infty} a_m x^m$$

Substituting $y$, $y'$, and $y''$ in the Legendre equation, we obtain

$$(1-x^2)\sum_{m=0}^{\infty} m(m-1)a_m x^{m-2} - 2x\sum_{m=0}^{\infty} m a_m x^{m-1} + \nu(\nu+1)\sum_{m=0}^{\infty} a_m x^m = 0$$

Simplification gives

$$\sum_{m=0}^{\infty} [(m+1)(m+2)a_{m+2} + (\nu-m)(\nu+m+1)a_m]x^m = 0$$

The coefficients in the power series must therefore satisfy the recurrence relation

$$a_{m+2} = -\frac{(\nu-m)(\nu+m+1)}{(m+1)(m+2)}a_m, \qquad m \geqslant 0 \tag{7.9.2}$$

This relation determines $a_2, a_4, a_6, \ldots$ in terms of $a_0$, and $a_3, a_5, a_7, \ldots$ in terms of $a_1$. It can easily be verified that $a_{2k}$ and $a_{2k+1}$ can be expressed in terms of $a_0$ and $a_1$ respectively as

$$a_{2k} = \frac{(-1)^k \nu(\nu-2)\ldots(\nu-2k+2)(\nu+1)(\nu+3)\ldots(\nu+2k-1)}{(2k)!}a_0$$

and

$$a_{2k+1} = \frac{(-1)^k(\nu-1)(\nu-3)\ldots(\nu-2k+1)(\nu+2)(\nu+4)\ldots(\nu+2k)}{(2k+1)!}a_1$$

Hence the solution of the Legendre equation is

$$\begin{aligned} y(x) &= a_0\Bigg[1 \\ &+ \sum_{k=1}^{\infty} \frac{(-1)^k \nu(\nu-2)\ldots(\nu-2k+2)(\nu+1)(\nu+3)\ldots(\nu+2k-1)x^{2k}}{(2k)!}\Bigg] \\ &+ a_1\Bigg[x \\ &+ \sum_{k=1}^{\infty} \frac{(-1)^k(\nu-1)(\nu-3)\ldots(\nu-2k+1)(\nu+2)(\nu+4)\ldots(\nu+2k)x^{2k+1}}{(2k+1)!}\Bigg] \\ &= a_0\phi_\nu(x) + a_1\psi_\nu(x) \end{aligned} \tag{7.9.3}$$

It can easily be proved that the functions $\phi_\nu(x)$ and $\psi_\nu(x)$ converge for $|x|<1$ and are linearly independent.

Now consider the case in which $\nu = n$, with $n$ a nonnegative integer. It is then evident from the recurrence relation (7.9.2) that when $m = n$.

$$a_{n+2} = a_{n+4} = \dots = 0$$

Consequently, when $n$ is even, the series $\phi_n(x)$ terminates with $x^n$, whereas the series for $\psi_n(x)$ does not terminate. When $n$ is odd, it is the series for $\psi_n(x)$ which terminates with $x^n$, while that for $\phi_n(x)$ does not terminate. In the first case ($n$ even), $\phi_n(x)$ is a polynomial of degree $n$; the same is true for $\psi_n(x)$ in the second case ($n$ odd).

Thus, for any nonnegative integer $n$, either $\phi_n(x)$ or $\psi_n(x)$, but not both, is a polynomial of degree $n$. Consequently, the general solution of the Legendre equation contains a polynomial solution $P_n(x)$ and an infinite series solution $Q_n(x)$ for a nonnegative integer $n$. To find the polynomial solution $P_n(x)$ it is convenient to choose $a_n$ so that $P_n(1) = 1$. Let this $a_n$ be

$$a_n = \frac{(2n)!}{2^n (n!)^2} \tag{7.9.4}$$

Rewriting the recurrence relation (7.9.2), we have

$$a_{n-2} = -\frac{(n-1)n}{2(2n-1)} a_n$$

Substituting $a_n$ from (7.9.4) into this relation, we obtain

$$a_{n-2} = -\frac{(2n-2)!}{2^n (n-1)!\,(n-2)!}$$

and

$$a_{n-4} = \frac{(2n-4)!}{2^n 2!\,(n-2)!\,(n-4)!}$$

It follows by induction that

$$a_{n-2k} = \frac{(-1)^k (2n-2k)!}{2^n k!\,(n-k)!\,(n-2k)!}$$

Hence we may write $P_n(x)$ in the form

$$P_n(x) = \sum_{k=0}^{N} \frac{(-1)^k (2n-2k)!}{2^n k!\,(n-k)!\,(n-2k)!} x^{n-2k} \tag{7.9.5}$$

where $N = n/2$ when $n$ is even and $N = (n-1)/2$ when $n$ is odd. This polynomial $P_n(x)$ is called the *Legendre function of the first kind of order* $n$. It is also known as the *Legendre polynomial of degree* $n$.

The first few Legendre polynomials are

$$P_0(x) = 1$$
$$P_1(x) = x$$
$$P_2(x) = \frac{1}{2}(3x^2 - 1)$$
$$P_3(x) = \frac{1}{2}(5x^3 - 3x)$$
$$P_4(x) = \frac{1}{8}(35x^4 - 30x^2 + 3)$$

These polynomials are plotted in Fig. 7.9.1 for small values of $x$.

Recall that for a given nonnegative integer $n$, only one of the two solutions $\phi_n(x)$ and $\psi_n(x)$ of Legendre's equation is a polynomial, while the other is an infinite series. This infinite series, when appropriately normalized, is called the *Legendre function of the second kind*. It is defined for $|x| < 1$ by

$$Q_n(x) = \begin{cases} \phi_n(1)\psi_n(x) & \text{for } n \text{ even} \\ -\psi_n(1)\phi_n(x) & \text{for } n \text{ odd} \end{cases} \tag{7.9.6}$$

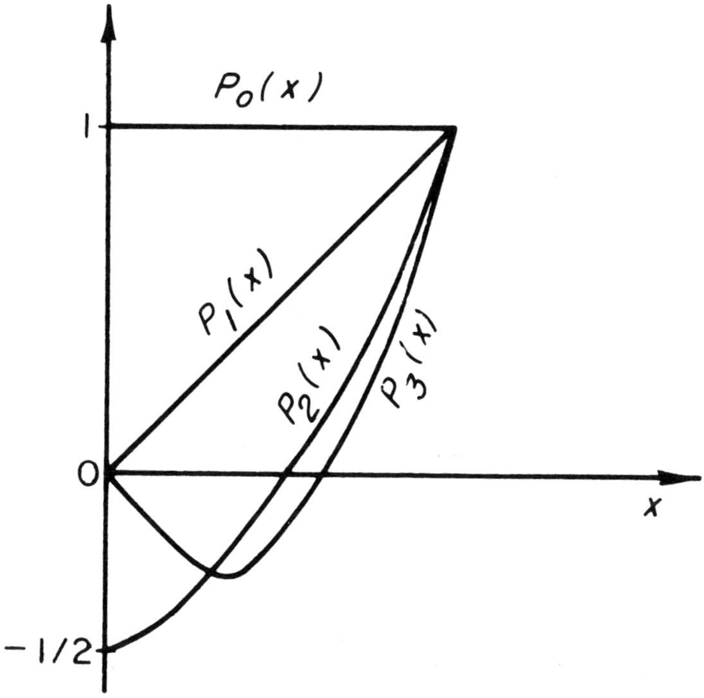


Figure 7.9.1

Thus, when $n$ is a nonnegative integer, the general solution of the Legendre equation is given by

$$y(x) = c_1 P_n(x) + c_2 Q_n(x) \tag{7.9.7}$$

The Legendre polynomial may also be expressed in the form

$$P_n(x) = \frac{1}{2^n n!} \frac{d^n}{dx^n} (x^2 - 1)^n \tag{7.9.8}$$

This expression is known as the *Rodriguez formula*.

Like Bessel's functions, Legendre polynomials satisfy certain recurrence relations. Some of the important relations are

$$(n+1)P_{n+1}(x) - (2n+1)xP_n(x) + nP_{n-1}(x) = 0, \qquad n \geqslant 1 \tag{7.9.9}$$

$$(x^2-1)P_n'(x) = nxP_n(x) - nP_{n-1}(x), \qquad n \geqslant 1 \tag{7.9.10}$$

$$nP_n(x) + P_{n-1}'(x) - xP_n'(x) = 0, \qquad n \geqslant 1 \tag{7.9.11}$$

$$P_{n+1}'(x) = xP_n'(x) + (n+1)P_n(x), \qquad n \geqslant 0 \tag{7.9.12}$$

In addition,

$$P_{2n}(-x) = P_{2n}(x) \tag{7.9.13}$$

$$P_{2n+1}(-x) = -P_{2n+1}(x) \tag{7.9.14}$$

These indicate that $P_n(x)$ is an even function for $n$ even, and an odd function for $n$ odd.

It can easily be shown that the Legendre polynomials form a sequence of orthogonal functions on the interval $[1, -1]$. Thus, we have

$$\int_{-1}^{1} P_n(x) P_m(x)\, dx = 0 \quad \text{for} \quad n \neq m \tag{7.9.15}$$

The norm of the function $P_n(x)$ is given by

$$\|P_n(x)\|^2 = \int_{-1}^{1} P_n^2(x)\, dx = \frac{2}{2n+1} \tag{7.9.16}$$

Another important equation in mathematical physics, one which is closely related to the Legendre equation (7.9.1), is *Legendre's associated equation.*

$$(1-x^2)y'' - 2xy' + \left[n(n+1) - \frac{m^2}{1-x^2}\right]y = 0 \tag{7.9.17}$$

where $m$ is an integer. Although this equation is independent of the algebraic sign of the integer $m$, it is often convenient to have the solutions for negative $m$ differ somewhat from those for positive $m$.

We consider first the case for a nonnegative integer $m$. Introducing the change of variable

$$y = (1-x^2)^{m/2}u, \qquad |x| < 1$$

Legendre's associated equation becomes

$$(1-x^2)u'' - 2(m+1)xu' + (n-m)(n+m+1)u = 0$$

But this is the same as the equation obtained by differentiating the Legendre equation (7.9.1) $m$ times. Thus, the general solution of (7.9.17) is given by

$$y(x) = (1-x^2)^{m/2}\frac{d^mY(x)}{dx^m} \tag{7.9.18}$$

where

$$Y(x) = c_1P_n(x) + c_2Q_n(x) \tag{7.9.19}$$

is the general solution of (7.9.1). Hence we have the linearly independent solutions of (7.9.17), known as the *associated Legendre functions of the first and second kind,* respectively given by

$$P_n^m(x) = (1-x^2)^{m/2}\frac{d^mP_n(x)}{dx^m} \tag{7.9.20}$$

and

$$Q_n^m(x) = (1-x^2)^{m/2}\frac{d^mQ_n(x)}{dx^m} \tag{7.9.21}$$

We observe that

$$P_n^0(x) = P_n(x)$$
$$Q_n^0(x) = Q_n(x)$$

and that $P_n^m(x)$ vanishes for $m > n$.

$P_n^{-m}(x)$ and $Q_n^{-m}(x)$ are defined by

$$P_n^{-m}(x) = (-1)^m\frac{(n-m)!}{(n+m)!}P_n^m(x), \qquad m = 0, 1, 2, \ldots, n \tag{7.9.22}$$

$$Q_n^{-m}(x) = (-1)^m\frac{(n-m)!}{(n+m)!}Q_n^m(x), \qquad m = 0, 1, 2, \ldots, n \tag{7.9.23}$$

The first few associated Legendre functions are

$$P_1^1(x) = (1-x^2)^{\frac{1}{2}}$$
$$P_2^1(x) = 3x(1-x^2)^{\frac{1}{2}}$$
$$P_2^2(x) = 3(1-x^2)$$

The associated Legendre functions of the first kind also form a sequence of orthogonal functions in the interval $[-1, 1]$. Their orthogonality, as well as their norm, is expressed by the equation

$$\int_{-1}^{1} P_n^m(x)P_k^m(x)\,dx = \frac{2(n+m)!}{(2n+1)(n-m)!}\,\delta_{nk} \tag{7.9.24}$$

Note that (7.9.15) and (7.9.16) are special cases of (7.9.24), corresponding to the choice $m = 0$.

We finally observe that, whereas $P_n^m(x)$ is bounded everywhere in the interval $[-1, 1]$, $Q_n^m(x)$ is unbounded at the end points $x = \pm 1$.

Problems in which Legendre's polynomials arise will be treated in Chapter 9.

## 7.10 BOUNDARY-VALUE PROBLEMS INVOLVING ORDINARY DIFFERENTIAL EQUATIONS

A *boundary-value problem* consists in finding an unknown solution which satisfies an ordinary differential equation and appropriate boundary conditions at two or more points. This is in contrast to an initial-value problem for which a unique solution exists for an equation satisfying prescribed initial conditions at one point.

The linear two-point boundary-value problem, in general, may be written in the form

$$\begin{aligned} L[y] &= f(x), \quad && a < x < b \\ U_i[y] &= \alpha_i, && 1 \leq i \leq n \end{aligned} \tag{7.10.1}$$

where $L$ is the linear operator of order $n$ and $U_i$ is the boundary operator defined by

$$U_i[y] = \sum_{j=1}^{n} a_{ij} y^{(j-1)}(a) + \sum_{j=1}^{n} b_{ij} y^{(j-1)}(b) \tag{7.10.2}$$

Here $a_{ij}$, $b_{ij}$, and $\alpha_i$ are constants. The treatment of this problem can be found in Coddington and Levinson [59]. More complicated boundary conditions occur in practice. Treating a general differential system is rather complex and difficult.

A large class of boundary-value problems that occur often in the physical sciences consists of the second-order equations of the type

$$y'' = f(x, y, y'), \qquad a < x < b$$

with the boundary conditions

$$U_1[y] = a_1 y(a) + a_2 y'(a) = \alpha$$

$$U_2[y] = b_1 y(b) + b_2 y'(b) = \beta$$

where $a_1$, $a_2$, $b_1$, $b_2$, $\alpha$, and $\beta$ are constants. The existence and uniqueness of solutions to this problem are treated by Keller [64]. Here we are interested in considering a special case where the linear boundary-value problem consists of the differential equation

$$L[y] = y'' + p(x)y' + q(x)y = f(x) \tag{7.10.3}$$

and the boundary conditions

$$\begin{aligned} U_1[y] &= a_1 y(a) + a_2 y'(a) = \alpha \\ U_2[y] &= b_1 y(b) + b_2 y'(b) = \beta \end{aligned} \tag{7.10.4}$$

where the constants $a_1$ and $a_2$, and likewise $b_1$ and $b_2$, are not both zero, and $\alpha$ and $\beta$ are constants.

In general a boundary-value problem may not possess a solution, and if it does, the solution may not be unique. We will illustrate this with a simple problem.

EXAMPLE 7.10.1. We first consider the boundary-value problem

$$y'' + y = 1$$

$$y(0) = 0, \qquad y\left(\frac{\pi}{2}\right) = 0$$

By the method of variation of parameters, we find a unique solution

$$y(x) = 1 - \cos x - \sin x$$

We observe that the solution of the associated homogeneous boundary-value problem

$$y'' + y = 0$$

$$y(0) = 0, \qquad y\left(\frac{\pi}{2}\right) = 0$$

is trivial.

Next we consider the boundary-value problem

$$y'' + y = 1$$

$$y(0) = 0, \qquad y(\pi) = 0$$

The general solution is

$$y(x) = c_1 \cos x + c_2 \sin x + 1$$

Applying the boundary conditions, we see that

$$y(0) = c_1 + 1 = 0$$

$$y(\pi) = -c_1 + 1 = 0$$

This is not possible, and hence the boundary-value problem has no solution. However, if we consider its associated homogeneous boundary-value problem

$$y'' + y = 0$$
$$y(0) = 0, \qquad y(\pi) = 0$$

we can easily determine that solutions exist and are given by

$$y(x) = c_2 \sin x$$

where $c_2$ is an arbitrary constant.

We are thus led to the following alternative theorem:

***Theorem 7.10.1*** *Let $p(x)$, $q(x)$, and $f(x)$ be continuous on $[a, b]$. Then either the boundary-value problem*

$$\begin{aligned} L[y] &= f \\ U_1[y] = \alpha, &\quad U_2[y] = \beta \end{aligned} \tag{7.10.5}$$

*has a unique solution for any given constants $\alpha$ and $\beta$, or else the associated homogeneous boundary-value problem*

$$\begin{aligned} L[y] &= 0 \\ U_1[y] = 0, &\quad U_2[y] = 0 \end{aligned} \tag{7.10.6}$$

*has a nontrivial solution.*

Proof of this theorem can be found in [67].

## 7.11 GREEN'S FUNCTIONS

In the present section we will introduce Green's functions. Let us consider the linear nonhomogeneous ordinary differential equation of second order:

$$L[y] = -f(x) \tag{7.11.1}$$

in $(a, b)$, where

$$L \equiv \frac{d}{dx}\left[p(x)\frac{d}{dx}\right] + q(x),$$

with the homogeneous boundary conditions

$$a_1 y(a) + a_2 y'(a) = 0 \tag{7.11.2}$$

$$b_1 y(b) + b_2 y'(b) = 0 \tag{7.11.3}$$

where the constants $a_1$ and $a_2$, and likewise $b_1$ and $b_2$, are not both zero. We shall assume that $f$ and $q$ are continuous and that $p$ is continuously differentiable and does not vanish in the interval $[a, b]$.

According to the theory of ordinary differential equations, the general solution of (7.11.1) is given by

$$y(x) = Ay_1(x) + By_2(x) + y_p(x) \tag{7.11.4}$$

where $A$ and $B$ are arbitrary constants, $y_1 \equiv y_1(x)$ and $y_2 \equiv y_2(x)$ are two linearly independent solutions of the corresponding homogeneous equation $L[y] = 0$, and $y_p(x)$ is any particular solution of (7.11.1).

Using the method of variation of parameters, the particular solution will be sought by replacing the constants $A$ and $B$ by arbitrary functions of $x$, $u_1(x)$ and $u_2(x)$, to get

$$y_p(x) = u_1(x)y_1(x) + u_2(x)y_2(x) \tag{7.11.5}$$

These arbitrary functions are to be determined so that (7.11.5) satisfies Eq. (7.11.1). The substitution of the above trial form for $y_p(x)$ into (7.11.1) imposes one condition that (7.11.5) must satisfy. However, there are two arbitrary functions, and hence two conditions are needed to determine them. It follows that another condition is available in solving the problem.

The first task is to substitute the trial solution into Eq. (7.11.1). Differentiating $y_p(x)$, we obtain

$$y'_p = u_1y'_1 + u_2y'_2 + u'_1y_1 + u'_2y_2$$

It is convenient to set the second condition noted above to require that

$$u'_1y_1 + u'_2y_2 = 0 \tag{7.11.6}$$

leaving

$$y'_p = u_1y'_1 + u_2y'_2 \tag{7.11.7}$$

A second differentiation of $y_p$ gives

$$y''_p = u_1y''_1 + u_2y''_2 + u'_1y'_1 + u'_2y'_2$$

Putting these results into Eq. (7.11.1) and grouping yields

$$u_1[p(x)y''_1 + p'y'_1 + qy_1] + u_2[p(x)y''_2 + p'y'_2 + qy_2] + p(x)\{u'_1y'_1 + u'_2y'_2\} = -f(x)$$

Since $y_1(x)$ and $y_2(x)$ are solutions of the homogeneous equations, both the square brackets in the above expression vanish. The result is that

$$u'_1y'_1 + u'_2y'_2 = -\frac{f(x)}{p(x)} \tag{7.11.8}$$

Thus two equations ((7.11.6) and (7.11.8)) determine $u_1(x)$ and $u_2(x)$. Solving these equations algebraically for $u'_1$ and $u'_2$ produces

$$u'_1(x) = \frac{f(x)y_2(x)}{p(x)W(x)}, \qquad u'_2(x) = -\frac{f(x)y_1(x)}{p(x)W(x)}$$

where $W(x) = y_1 y_2' - y_2 y_1'$ is the non-zero Wronskian of the solutions $y_1$ and $y_2$. Integration of these results yields

$$u_1(x) = +\int \frac{f(x) y_2(x)\, dx}{p(x) W(x)}, \qquad u_2(x) = -\int \frac{f(x) y_1(x)\, dx}{p(x) W(x)}$$

The substitution of these results into (7.11.5) gives the solution in $(a, b)$

$$\begin{aligned} y_p(x) &= y_1(x) \int_b^x \frac{f(\xi) y_2(\xi)}{p(\xi) W(\xi)}\, d\xi - y_2(x) \int_a^x \frac{f(\xi) y_1(\xi)}{p(\xi) W(\xi)}\, d\xi \\ &= -\int_a^x \frac{y_2(x) y_1(\xi)}{p(\xi) W(\xi)} f(\xi)\, d\xi - \int_x^b \frac{y_1(x) y_2(\xi)}{p(\xi) W(\xi)} f(\xi)\, d\xi \end{aligned} \tag{7.11.9}$$

This form suggests the definition

$$G(x, \xi) = \begin{bmatrix} -\dfrac{y_2(x) y_1(\xi)}{p(\xi) W(\xi)}, & a \leqslant \xi < x \\ -\dfrac{y_1(x) y_2(\xi)}{p(\xi) W(\xi)}, & x < \xi \leqslant b \end{bmatrix} \tag{7.11.10ab}$$

This is called a *Green's function*. Thus the solution becomes

$$y_p(x) = \int_a^b G(x, \xi) f(\xi)\, d\xi \tag{7.11.11}$$

provided $G(x, \xi)$ is continuous and $f(\xi)$ is at least piecewise continuous in $(a, b)$. The existence of a Green's function is evident from Eqs. (7.11.10ab) provided $W(\xi) \neq 0$.

In order to obtain a deeper insight into the role of the Green's function, certain properties are important to note. These are

(i) $G(x, \xi)$ is continuous at $x = \xi$ and, consequently, throughout the interval $(a, b)$. This follows from the fact that $G(x, \xi)$ is constructed from the solutions of the homogeneous equation which are continuous in the intervals $a \leqslant \xi < x$ and $x < \xi \leqslant b$, $W(\xi) \neq 0$.

(ii) Its first and second derivatives are continuous for all $x \neq \xi$ in $a \leqslant x$, $\xi \leqslant b$.

(iii) $G(x, \xi)$ is symmetric in $x$ and $\xi$, that is, if $x$ and $\xi$ are interchanged in (7.11.10ab), the definition is not changed.

(iv) The first derivative of $G(x, \xi)$ has a finite discontinuity at $x = \xi$

$$\lim_{x \to \xi +} \frac{\partial G}{\partial x} - \lim_{x \to \xi -} \frac{\partial G}{\partial x} = -\frac{1}{p(\xi)} \tag{7.11.12}$$

(v) $G(x, \xi)$ is a solution of the homogeneous equation throughout the interval except at $x = \xi$. This point can be pressed further by substitution of (7.11.11) into (7.11.1), note that $y_p(x)$ is a solution. This leads to

$$\int_a^b \left[\frac{d}{dx}\left[p(x)\frac{dG(x,\xi)}{dx}\right] + q(x)G(x,\xi)\right] f(\xi)\,d\xi = -f(x)$$

where the quantity in the square bracket is zero except at $x = \xi$. It follows that, in order for this result to hold,

$$\frac{d}{dx}\left[p(x)\frac{dG(x,\xi)}{dx}\right] + q(x)G(x,\xi) = -\delta(\xi - x) \quad (7.11.13)$$

where $\delta(\xi - x)$ is the Dirac function which has the following properties:

$$\delta(\xi - x) = 0, \quad \text{except at} \quad x = \xi$$

$$\int_a^b \delta(\xi - x)\,d\xi = 1, \qquad a < x < b$$

for any continuous function $f(\xi)$ which is continuous in $(a, b)$ and for $a \leqslant x \leqslant b$,

$$\int_a^b f(\xi)\delta(\xi - x)\,d\xi = f(x)$$

(vi) For fixed $\xi$, $G(x, \xi)$ satisfies the prescribed boundary conditions (7.11.2)–(7.11.3).

Differentiating (7.11.11) with respect to $x$ by Leibnitz's rule, we obtain

$$\begin{aligned} y_p'(x) &= \int_a^x G'(x,\xi)f(\xi)\,d\xi + G(x, x-)f(x) \\ &\quad + \int_x^b G'(x,\xi)f(\xi)\,d\xi - G(x, x+)f(x) \\ &= \int_a^b G'(x,\xi)f(\xi)\,d\xi, \end{aligned}$$

since $G(x, \xi)$ is continuous in $\xi$, that is, $G(x, x-) = G(x, x+)$.

Because $G(x, \xi)$ satisfies the boundary conditions, hence

$$a_1 y_p(a) + a_2 y_p'(a) = \int_a^b [a_1 G(a,\xi) + a_2 G'(a,\xi)] f(\xi)\,d\xi = 0$$

Similarly

$$b_1 y_p(b) + b_2 y_p'(b) = 0$$

(vii) $G(x, \xi)$ is symmetric in $x$ and $\xi$, that is, if $x$ and $\xi$ are interchanged in (7.11.10ab), the definition remains unchanged.

The result embodied in Eq (7.11.13) permits a meaningful physical interpretation of the role of the Green's function. If we assume that the differential equation (7.11.1) represents a vibrating mechanical system driven by a uniformly distributed force $-f(x)$ in the interval $(a, b)$ where the system is defined, then $G(x, \xi)$ governed by (7.11.13) represents the displacement of the system at a point $x$ resulting from a unit impulse of force at $x = \xi$. The displacement at the point $x$ due to uniformly distributed force $f(\xi)$ per unit length over an elementary interval $(\xi, \xi + d\xi)$ is given by $f(\xi)\, G(x, \xi)\, d\xi$. Finally, the total displacement of the system at the point $x$ results from superposition (addition) of these contributions so that the total solution over the entire interval $(a, b)$ is given by (7.11.11), that is,

$$y_p(x) = \int_a^b G(x, \xi) f(\xi)\, d\xi$$

Combining all results, we state the fundamental theorem for the Green's function:

***Theorem 7.11.1*** *If $f(x)$ is continuous in the interval $[a, b]$, then the function*

$$y(x) = \int_a^b G(x, \xi) f(\xi)\, d\xi$$

*is a solution of the boundary-value problem*

$$L[y] = -f(x)$$
$$a_1 y(a) + a_2 y'(a) = 0$$
$$b_1 y(b) + b_2 y'(b) = 0$$

EXAMPLE 7.11.1. Consider the problem

$$y'' = -x$$
$$y(0) = 0 \qquad (7.11.14)$$
$$y(1) = 0$$

For a fixed value of $\xi$, the Green's function $G(x, \xi)$ satisfies the associated homogeneous equation

$$G'' = 0$$

in $0 < x < \xi$, $\xi < x < 1$, and the boundary conditions

$$G(0, \xi) = 0$$
$$G(1, \xi) = 0$$

In addition, it satisfies

$$\frac{dG}{dx}(x, \xi)\Big|_{x=\xi-}^{x=\xi+} = -\frac{1}{p(\xi)}$$

Now if we choose $G(x, \xi)$ such that

$$G(x, \xi) = \begin{cases} G_1(x, \xi) = (1-\xi)x & \text{for } 0 \leqslant x \leqslant \xi \\ G_2(x, \xi) = (1-x)\xi & \text{for } \xi \leqslant x \leqslant 1 \end{cases}$$

It can be seen that $G'' = 0$ over the intervals $0 < x < \xi$, $\xi < x < 1$. Also

$$G_1(0, \xi) = 0$$
$$G_2(1, \xi) = 0$$

Moreover

$$G_2'(x, \xi) - G_1'(x, \xi) = -\xi - (1-\xi) = -1$$

which is the value of the jump $-1/p(\xi)$, because in this case $p = 1$. Hence from Theorem 7.11.1, keeping in mind that $\xi$ is the variable in $G(x, \xi)$, the solution of (7.11.14) is

$$\begin{aligned} y(x) &= \int_0^x G(x, \xi) f(\xi)\, d\xi + \int_x^1 G(x, \xi) f(\xi)\, d\xi \\ &= \int_0^x (1-x)\xi^2\, d\xi + \int_x^1 x(1-\xi)\xi\, d\xi \\ &= \frac{x}{6}(1-x^2) \end{aligned}$$

## 7.12 CONSTRUCTION OF GREEN'S FUNCTION

In the above example, we see that the solution was obtained immediately as soon as the Green's function was selected properly. Thus, the real problem is not that of finding the solution but that of determining the Green's function for the problem. We will now show by construction that there exists a Green's function for $L[y]$ satisfying the prescribed boundary conditions.

We first assume that the associated homogeneous equation satisfying the conditions (7.11.2) and (7.11.3) has the trivial solution only, as in Example 7.11.1. We construct the solution $y_1(x)$ of

$$L[y] = 0$$

satisfying $a_1 y(a) + a_2 y'(a) = 0$. We see that $c_1 y_1(x)$ is the most general such solution, where $c_1$ is an arbitrary constant.

In a similar manner, we let $c_2 y_2(x)$, with $c_2$ as an arbitrary constant, be the most general solution of

$$L[y]=0$$

satisfying $b_1y(b)+b_2y'(b)=0$. Thus, $y_1$ and $y_2$ exist in the interval $(a, b)$ and are linearly independent. For, if they were linearly dependent, then $y_1=cy_2$, which shows that $y_1$ would satisfy both the boundary conditions at $x=a$ and $x=b$. This contradicts our assumption about the trivial solution. Consequently, the Green's functon can take the form

$$G(x, \xi)=\begin{cases} c_1(\xi)y_1(x) & \text{for } x<\xi \\ c_2(\xi)y_2(x) & \text{for } x>\xi \end{cases} \tag{7.12.1}$$

Since $G(x, \xi)$ is continuous at $x=\xi$, we have

$$c_2(\xi)y_2(\xi)-c_1(\xi)y_1(\xi)=0 \tag{7.12.2}$$

The discontinuity in the derivative of $G$ at that point requires that

$$\frac{dG}{dx}(x, \xi)\bigg|_{x=\xi-}^{x=\xi+}=c_2(\xi)y_2'(\xi)-c_1(\xi)y_1'(\xi)=-\frac{1}{p(\xi)} \tag{7.12.3}$$

Solving Eqs. (7.12.2) and (7.12.3) for $c_1$ and $c_2$, we find

$$\begin{aligned} c_1(\xi)&=\frac{-y_2(\xi)}{p(\xi)W(y_1, y_2; \xi)} \\ c_2(\xi)&=\frac{-y_1(\xi)}{p(\xi)W(y_1, y_2; \xi)} \end{aligned} \tag{7.12.4}$$

where $W(y_1, y_2; \xi)$ is the Wronskian given by $W(y_1, y_2; \xi)=y_1(\xi)y_2'(\xi)-y_2(\xi)y_1'(\xi)$. Since the two solutions are linearly independent, the Wronskian differs from zero.

From Theorem 7.2.4, with $\lambda=0$, we have

$$pW=\text{constant}=C \tag{7.12.5}$$

Hence the Green's function is given by

$$G(x, \xi)=\begin{cases} -y_1(x)y_2(\xi)/C & \text{for } x\leqslant\xi \\ -y_2(x)y_1(\xi)/C & \text{for } x\geqslant\xi \end{cases} \tag{7.12.6}$$

Thus, we state the following theorem:

***Theorem 7.12.1*** *If the associated homogeneous boundary-value problem of* (7.11.1)–(7.11.3) *has the trivial solution only, then the Green's function exists and is unique.*

The proof for uniqueness of Green's function is left as an exercise for the reader.

EXAMPLE 7.12.1. Consider the problem

$$y'' + y = -1$$
$$y(0) = 0 \tag{7.12.7}$$
$$y\left(\frac{\pi}{2}\right) = 0$$

The solution of $L[y] = dy'/dx + y = 0$ satisfying $y(0) = 0$ is

$$y_1(x) = \sin x, \qquad 0 \leqslant x < \xi$$

and the solution of $L[y] = 0$ satisfying $y(\pi/2) = 0$ is

$$y_2(x) = \cos x, \qquad \xi < x \leqslant \frac{\pi}{2}$$

The Wronskian of $y_1$ and $y_2$ is then given by

$$W(\xi) = y_1(\xi)y_2'(\xi) - y_2(\xi)y_1'(\xi) = -1$$

Since in this case $p = 1$, (7.12.6) becomes

$$G(x, \xi) = \begin{cases} \sin x \cos \xi & \text{for } x \leqslant \xi \\ \cos x \sin \xi & \text{for } x \geqslant \xi \end{cases}$$

Therefore the solution of (7.12.7) is

$$\begin{aligned} y(x) &= \int_0^x G(x, \xi)f(\xi)\,d\xi + \int_x^{\pi/2} G(x, \xi)f(\xi)\,d\xi \\ &= \int_0^x \cos x \sin \xi\,d\xi + \int_x^{\pi/2} \sin x \cos \xi\,d\xi \\ &= -1 + \sin x + \cos x \end{aligned}$$

It can be seen in the formula (7.12.6) that the Green's function is symmetric in $x$ and $\xi$.

EXAMPLE 7.12.2. Construct the Green's function for the two-point boundary–value problem

$$y''(x) + \omega^2 y = f(x), \qquad y(a) = y(b) = 0$$

This describes the forced oscillation of an elastic string with fixed ends at $x = a$ and $x = b$.

It is easy to check that $\sin \omega x$ and $\cos \omega x$ are two functions which satisfy the homogeneous equation $y'' + \omega^2 y = 0$. These are to be used to construct two functions $y_1(x)$ and $y_2(x)$ satisfying the boundary conditions $y_1(a) = y_2(b) = 0$. Accordingly, $y_1(x) = A \sin \omega x + B \cos \omega x$ and $y_2(x) = C \sin \omega x + D \cos \omega x$, and the resulting functions are

$$y_1(x) = \sin \omega(x - a), \qquad y_2(x) = \sin \omega(x - b)$$

The corresponding Wronskian is $W = -\omega \sin \omega(a-b)$. Substituting these results into (7.11.10) yields

$$G(x, \xi) = \begin{cases} \dfrac{\sin \omega(\xi - a) \sin \omega(x - b)}{-\omega \sin \omega(a-b)}, & a \leqslant \xi < x \\ \dfrac{\sin \omega(x - a) \sin \omega(\xi - b)}{-\omega \sin \omega(a-b)}, & x \leqslant \xi \leqslant b \end{cases}$$

provided $\sin \omega(a-b) \neq 0$.

## 7.13 THE SCHRÖDINGER EQUATION FOR THE LINEAR HARMONIC OSCILLATOR

The quantum mechanical motion of the harmonic oscillator is described by the one-dimensional Schrödinger equation

$$H\psi(x) = E\psi(x) \tag{7.13.1}$$

where the Hamiltonian $H$ is given by

$$H = -\frac{\hbar^2}{2M}\frac{d^2}{dx^2} + V(x), \qquad V(x) = \frac{1}{2} M\omega^2 x^2 \tag{7.13.2ab}$$

and $V(x)$ is the potential, $h = 2\pi\hbar$ is the Planck constant, $M$ is the mass of the particle, and $\omega$ is the classical frequency of the oscillator.

We solve Eq. (7.13.1) subject to the requirement that the solution be bounded as $|x| \to \infty$. The solution of (7.13.1) is facilitated by first solving the equation for large $x$. In terms of the constants

$$\beta = \frac{2ME}{\hbar^2}, \qquad \alpha = \frac{M\omega}{\hbar} > 0,$$

the equation (7.13.1) takes the form

$$\frac{d^2\psi}{dx^2} + (\beta - \alpha^2 x^2)\psi = 0 \tag{7.13.3}$$

For small $\beta$ and large $x$, $\beta - \alpha^2 x^2 \sim -\alpha^2 x^2$ so that the equation becomes

$$\frac{d^2\psi}{dx^2} - \alpha^2 x^2 \psi = 0$$

As $|x| \to \infty$, $\psi(x) = x^n \exp\left(\pm \dfrac{\alpha x^2}{2}\right)$ satisfies (7.13.3) for a finite $n$ so far as leading terms ($\sim \alpha^2 x^2$) are concerned. The positive exponential factor is unacceptable because of the boundary conditions, so the asymptotic solution $\psi(x) = x^n \exp\left(-\dfrac{\alpha x^2}{2}\right)$ suggests the possibility of the exact solution in the form $\psi(x) = v(x) \exp\left(-\frac{\alpha x^2}{2}\right)$ where $v(x)$ is to be determined. Substituting this result into (7.13.3), we obtain

$$\frac{d^2v}{dx^2} - 2\alpha x \frac{dv}{dx} + (\beta - \alpha)v = 0 \tag{7.13.4}$$

In terms of a new independent variable $\zeta = x\sqrt{\alpha}$, this equation reduces to the form

$$\frac{d^2v}{d\zeta^2} - 2\zeta \frac{dv}{d\zeta} + \left(\frac{\beta}{\alpha} - 1\right)v = 0 \tag{7.13.5}$$

We seek a power series solution

$$v(\zeta) = \sum_{n=0}^{\infty} a_n \zeta^n \tag{7.13.6}$$

Substituting this series into Eq. (7.13.5) and equating the coefficient of $\zeta^n$ to zero, we obtain the recurrence relation

$$a_{n+2} = \frac{(2n + 1 - \beta/\alpha)}{(n+1)(n+2)} a_n \tag{7.13.7}$$

which gives

$$\frac{a_{n+2}}{a_n} \sim \frac{2}{n} \quad \text{as} \quad n \to \infty \tag{7.13.8}$$

This ratio is the same as that of the series for $\zeta^n e^{\zeta^2} (\sim x^n e^{\alpha x^2})$ with finite $n$. This leads to the fact that $\psi(x) = v(x) e^{-\alpha x^2/2} \sim x^n e^{\alpha x^2/2}$ which does not satisfy the basic requirement for $|x| \to \infty$. This unacceptable result can only be avoided if $n$ is an integer and the series terminates so that it becomes a polynomial of degree $n$. This means that $a_{n+2} = 0$ but $a_n \neq 0$ so that

$$2n + 1 - \frac{\beta}{\alpha} = 0 \tag{7.13.9}$$

or

$$\frac{\beta}{\alpha} = (2n + 1)$$

Substituting the values for $\alpha$ and $\beta$, it turns out that

$$E \equiv E_n = \left(n + \frac{1}{2}\right)\omega\hbar, \qquad n = 0, 1, 2, \ldots \tag{7.13.10}$$

This represents a discrete set of energies. Thus, in quantum mechanics, a stationary state of the harmonic oscillator can assume only one of the values from the set $E_n$. The energy is thus quantized, and forms a discrete spectrum. According to the classical theory, the energy forms a continuous spectrum, that is, all non-negative numbers are allowed for the energy of a harmonic oscillator. This shows a remarkable contrast between the results of the classical and quantum theory.

The number $n$ which characterizes the energy eigenvalues and eigenfunctions is called the *quantum number*. The value of $n = 0$ corresponds to the minimum value of the quantum number with the energy

$$E_0 = \frac{1}{2}\omega\hbar \tag{7.13.11}$$

This is called the *lowest* (or *ground*) *state energy* which never vanishes, as the lowest possible classical energy would. $E_0$ is proportional to $\hbar$, representing a quantum phenomenon. The discrete energy spectrum is in perfect agreement with the quantization rules of the quantum theory.

To determine the eigenfunctions for the harmonic oscillator associated with the eigenvalues $E_n$, we obtain the solution of Eq. (7.13.5) which has the form

$$\frac{d^2v}{d\zeta^2} - 2\zeta\frac{dv}{d\zeta} + 2nv(\zeta) = 0 \tag{7.13.12}$$

This is a well-known differential equation for the *Hermite polynomials* $H_n(\zeta)$ of degree $n$. Thus the complete eigenfunctions can be expressed in terms of $H_n(\zeta)$ as

$$\psi_n(x) = A_n H_n(x\sqrt{\alpha})\exp\left(-\frac{\alpha x^2}{2}\right) \tag{7.13.13}$$

where $A_n$ are arbitrary constants.

The Hermite polynomials $H_n(x)$ are usually defined by

$$H_n(x) = (-1)^n e^{x^2} D^n(e^{-x^2}), \qquad D \equiv \frac{d}{dx} \tag{7.13.14}$$

They form an orthogonal system in $(-\infty, \infty)$ with the weight function $e^{-x^2}$.

The orthogonal relation for these polynomials is

$$\int_{-\infty}^{\infty} e^{-x^2} H_m(x) H_n(x)\, dx = \begin{Bmatrix} 0, & n \neq m \\ 2^n n!\sqrt{\pi}, & n = m \end{Bmatrix}$$

The Hermite polynomials $H_n(x)$ for $n = 0, 1, 2, 3, 4$ are

$$H_0(x) = 1$$
$$H_1(x) = 2x$$
$$H_2(x) = -2 + 4x^2$$
$$H_3(x) = -12x + 8x^3$$
$$H_4(x) = 12 - 48x^2 + 16x^4$$

Finally, the eigenfunctions $\psi_n$ of the linear harmonic oscillator for the quantum number $n = 0, 1, 2, 3$ are given in Figure 7.13.1.

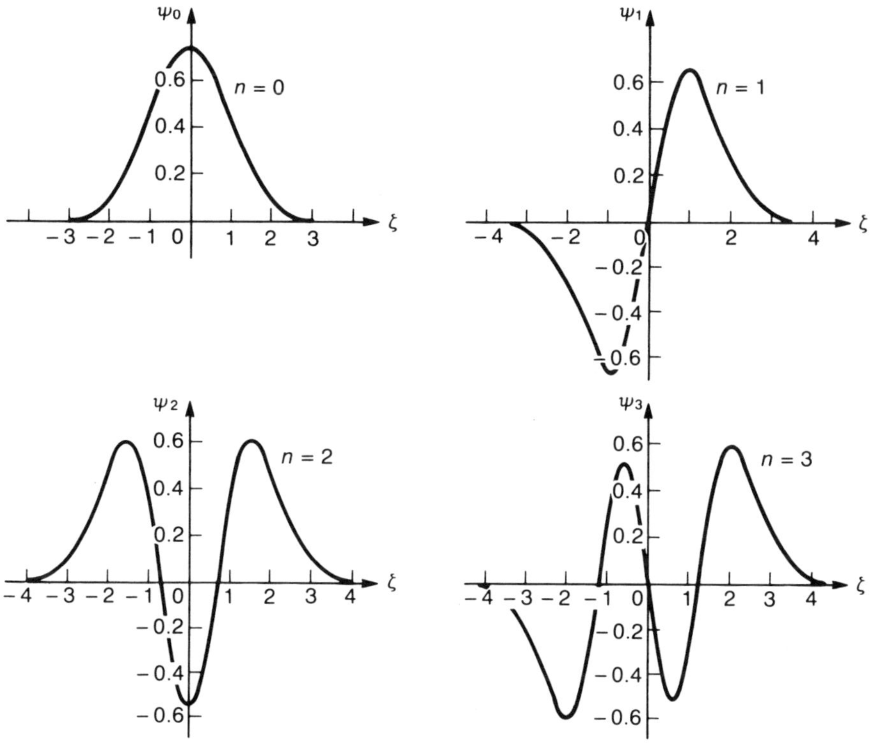


Figure 7.13.1

## 7.14 EXERCISES

**1.** Determine the eigenvalues and eigenfunctions of the following regular Sturm–Liouville systems:

a. $y'' + \lambda y = 0$
$y(0) = 0,\ y(\pi) = 0$

b. $y'' + \lambda y = 0$
$y(0) = 0,\ y'(1) = 0$

c. $y'' + \lambda y = 0$
$y'(0) = 0,\ y'(\pi) = 0$

d. $y'' + \lambda y = 0$
$y(1) = 0,\ y(0) + y'(0) = 0$

**2.** Find the eigenvalues and eigenfunctions of the following periodic Sturm–Liouville systems:

a. $y'' + \lambda y = 0$
$y(-1) = y(1),\ y'(-1) = y'(1)$

b. $y'' + \lambda y = 0$
$y(0) = y(2\pi),\ y'(0) = y'(2\pi)$

c. $y'' + \lambda y = 0$
$y(0) = y(\pi),\ y'(0) = y'(\pi)$

**3.** Obtain the eigenvalues and eigenfunctions of the following Sturm–Liouville systems:

a. $y'' + y' + (1 + \lambda)y = 0$
$y(0) = 0,\ y(1) = 0$

b. $y'' + 2y' + (1 - \lambda)y = 0$
$y(0) = 0,\ y'(1) = 0$

c. $y'' - 3y' + 3(1 + \lambda)y = 0$
$y'(0) = 0,\ y'(\pi) = 0$

**4.** Find the eigenvalues and eigenfunctions of the following regular Sturm–Liouville systems:

a. $x^2y'' + 3xy' + \lambda y = 0, \quad 1 \leqslant x \leqslant e$
$y(1) = 0, \quad y(e) = 0$

b. $\dfrac{d}{dx}[(2 + x)^2 y'] + \lambda y = 0, \quad -1 \leqslant x \leqslant 1$
$y(-1) = 0, \quad y(1) = 0$

c. $(1 + x)^2 y'' + 2(1 + x)y' + 3\lambda y = 0, \quad 0 \leqslant x \leqslant 1$
$y(0) = 0, \quad y(1) = 0$

**5.** Determine all eigenvalues and eigenfunctions of the singular Sturm–Liouville systems

a. $x^2 y'' + xy' + \lambda y = 0$
$y(1) = 0$, $y$, $y'$ are bounded at $x = 0$

b. $y'' + \lambda y = 0$
$y(0) = 0$, $y$, $y'$ are bounded at infinity

**6.** Expand the function

$$f(x) = \sin x, \quad 0 \leqslant x \leqslant \pi$$

in terms of the eigenfunctions of the Sturm–Liouville problems

$$y'' + \lambda y = 0$$
$$y(0) = 0, \quad y(\pi) + y'(\pi) = 0$$

**7.** Find the expansion of

$$f(x) = x, \quad 0 \leqslant x \leqslant \pi$$

in a series of eigenfunctions of the Sturm–Liouville system

$$y'' + \lambda y = 0$$
$$y'(0) = 0, \quad y'(\pi) = 0$$

**8.** Transform each of the following equations into the equivalent self-adjoint form:

a. The Laguerre equation

$$xy'' + (1 - x)y' + ny = 0, \quad n = 0, 1, 2, \ldots$$

b. The Hermite equation

$$y'' - 2xy' + 2ny = 0, \quad n = 0, 1, 2, \ldots$$

c. The Tchebycheff equation

$$(1 - x^2)y'' - xy' + n^2y = 0, \quad n = 0, 1, 2, \ldots$$

**9.** If $q(x)$ and $s(x)$ are continuous and $p(x)$ is twice continuously differentiable in $[a, b]$, then the solutions of the fourth-order Sturm–Liouville system

$$[p(x)y'']'' + [q(x) + \lambda s(x)]y = 0$$
$$[a_1 y + a_2(py'')']_{x=a} = 0, \quad [b_1 y + b_2(py'')']_{x=b} = 0$$
$$[c_1 y' + c_2(py'')]_{x=a} = 0, \quad [d_1 y' + d_2(py'')]_{x=b} = 0$$

with $a_1^2 + a_2^2 \neq 0$, $b_1^2 + b_2^2 \neq 0$, $c_1^2 + c_2^2 \neq 0$, $d_1^2 + d_2^2 \neq 0$ are orthogonal with respect to $s(x)$ in $[a, b]$.

**10.** If the eigenfunctions of the problem

$$\frac{1}{r}\frac{d}{dr}(ry') + \lambda y = 0, \quad 0 < r < a$$

$$c_1 y(a) + c_2 y'(a) = 0$$

$$\lim_{r \to 0+} y(r) < \infty$$

satisfy

$$\lim_{r \to 0+} ry'(r) = 0$$

show that all the eigenvalues are real for real $c_1$ and $c_2$.

**11.** Find the Green's functions for the following problems:

a. $L[y] = y'' = 0$
$y(0) = 0, \quad y'(1) = 0$

b. $L[y] = (1 - x^2)y'' - 2xy' = 0$
$y(0) = 0, \quad y'(1) = 0$

c. $L[y] = y'' + a^2 y = 0, \quad a = \text{constant}$
$y(0) = 0, \quad y(1) = 0$

**12.** Determine the solution of each of the following boundary-value problems:

a. $y'' + y = 1$
$y(0) = 0, \quad y(1) = 0$

b. $y'' + 4y = e^x$
$y(0) = 0, \quad y'(1) = 0$

c. $y'' = \sin x$
$y(0) = 0, \quad y(1) + 2y'(1) = 0$

d. $y'' + 4y = -2$
$y(0) = 0, \quad y\left(\frac{\pi}{4}\right) = 0$

e. $y'' = -x$
$y(0) = 2, \quad y(1) + y'(1) = 4$

f. $y'' = -x^2$
$y(0) + y'(0) = 4, \quad y'(1) = 2$

g. $y'' = -x$
$y(0) = 1, \quad y'(1) = 2$

**13.** Determine the solution of the following boundary-value problems:

a. $y'' = -f(x), \quad y(0) = 0, \quad y'(1) = 0$

b. $y'' = -f(x), \quad y(-1) = 0, \quad y(1) = 0$

**14.** Find the solution of the following boundary-value problems:

a. $y'' - y = -f(x), \quad y(0) = y(1) = 0$

b. $y'' - y = -f(x), \quad y'(0) = y'(1) = 0$

**15.** Show that Green's function $G(t, \xi)$ for the forced harmonic oscillator described by the initial–value problem

$$\ddot{x} + \omega^2 x = \frac{F}{m} \sin \Omega t$$

$$x(0) = a, \qquad \dot{x}(0) = 0$$

is

$$G(t, \xi) = \frac{1}{\omega} \sin \omega(t - \xi)$$

Hence the particular solution is

$$x_p(t) = \frac{F}{m\omega} \int_0^t \sin \omega(t - \xi) \sin \Omega\xi \, d\xi$$

**16.** Determine the Green's function for the boundary value problem

$$xy'' + y' = -f(x)$$
$$y(1) = 0$$
$$\lim_{x \to 0} |y(x)| < \infty$$

**17.** Determine the Green's function for the boundary-value problem

$$xy'' + y' - \frac{n^2}{x} y = -f(x)$$
$$y(1) = 0$$
$$\lim_{x \to 0} |y(x)| < \infty$$

**18.** Determine the Green's function for the boundary-value problem

$$[(1 - x^2)y']' - \frac{h^2}{1 - x^2} y = -f(x), \quad h = 1, 2, 3, \ldots$$
$$\lim_{r \to \pm 1} |y(x)| < \infty$$

**19.** Prove the uniqueness of the Green's function for the boundary value problem

$$L[y] = -f(x)$$
$$a_1 y(a) + a_2 y'(a) = 0$$
$$b_1 y(b) + b_2 y'(b) = 0$$

**20.** Find the Green's function for the boundary-value problem

$$L[y] = y^{\text{iv}} = -f(x)$$
$$y(0) = y(1) = y'(0) = y'(1) = 0$$

Prove that the problem has a trivial solution only, and prove that the nonhomogeneous problem has a unique solution.

**21.** Determine the Green's function for the boundary-value problem

$$y'' = -f(x)$$
$$y(-1) = y(1)$$
$$y'(-1) = y'(1)$$

**22.** Consider the nonself-adjoint boundary-value problem

$$L[y] = y'' + 3y' + 2y = -f(x)$$
$$2y(0) - y(1) = 0$$
$$y'(1) = 2$$

By direct integration of $GL[y]$ from 0 to 1, show that

$$y(x) = -2G(1, x) - \int_0^1 G(x, \xi)f(\xi)\, d\xi$$

is the solution of the boundary-value problem if $G$ satisfies

$$G_{\xi\xi} - 3G_\xi + 2G = 0, \quad \xi \neq x$$
$$G(0, x) = 0$$
$$6G(1, x) - 2G_\xi(1, x) + G_\xi(0, x) = 0$$

Find $G(x, \xi)$.

**23.** Show that

$$\frac{dG(x, \xi)}{dx}\bigg|_{\xi=x-}^{\xi=x+} = \frac{1}{p(x)}$$

is equivalent to

$$\frac{dG(x, \xi)}{dx}\bigg|_{x=\xi-}^{x=\xi+} = -\frac{1}{p(\xi)}$$

# CHAPTER EIGHT

# Boundary-Value Problems

## 8.1 BOUNDARY-VALUE PROBLEMS

In the preceding chapters, we have treated the initial-value and initial-boundary-value problems. In this chapter, we shall be concerned with boundary-value problems. Mathematically, a boundary-value problem is finding a function which satisfies a given partial differential equation and particular boundary conditions. Physically speaking, the problem is independent of time, involving only space coordinates. Just as initial-value problems are associated with hyperbolic partial differential equations, boundary-value problems are associated with partial differential equations of elliptic type. In marked contrast to initial-value problems, boundary-value problems are considerably more difficult to solve. This is due to the physical requirement that solutions must attain in the large unlike the case of initial-value problems, where solutions in the small, say over a short interval of time, may still be of physical interest.

The second-order partial differential equation of the elliptic type in $n$ independent variables $x_1, x_2, \ldots, x_n$ is of the form[9]

$$\nabla^2 u = F(x_1, x_2, \ldots, x_n, u_{x_1}, u_{x_2}, \ldots, u_{x_n}) \tag{8.1.1}$$

Some well-known elliptic equations include

A. Laplace equation

$$\nabla^2 u = 0 \tag{8.1.2}$$

B. Poisson equation

$$\nabla^2 u = g(x) \tag{8.1.3}$$

---

[9] $\nabla^2 u = \sum_{i=1}^{n} u_{x_i x_i}$

where

$$g(x) = g(x_1, x_2, \ldots, x_n)$$

C. Helmholtz equation

$$\nabla^2 u + \lambda u = 0 \tag{8.1.4}$$

where $\lambda$ is a positive constant.

D. Schrödinger equation (time independent)

$$\nabla^2 u + [\lambda - q(x)]u = 0 \tag{8.1.5}$$

We shall not attempt to treat general elliptic partial differential equations. Instead, we shall begin by presenting the simplest boundary-value problems for the Laplace equation in two dimensions.

Let us first define a harmonic function. A function is said to be *harmonic* in a domain $D$ if it satisfies the Laplace equation and if it and its first two derivatives are continuous in $D$.

We may note here that, since the Laplace equation is linear and homogeneous, a linear combination of harmonic functions is harmonic.

## 1. The First Boundary-Value Problem

(The Dirichlet Problem): Find a function $u(x, y)$, harmonic in $D$, which satisfies

$$u = f(s) \qquad \text{on } B \tag{8.1.6}$$

where $f(s)$ is a prescribed continuous function on the boundary $B$ of the domain $D$. $D$ is the interior of a simple closed piecewise smooth curve $B$.

We may physically interpret the solution $u$ of the Dirichlet problem as the steady-state temperature distribution in a body containing no sources or sinks of heat, with the temperature prescribed at all points on the boundary.

## 2. The Second Boundary-Value Problem

(The Neumann Problem): Find a function $u(x, y)$, harmonic in $D$, which satisfies

$$\frac{\partial u}{\partial n} = f(s) \qquad \text{on } B \tag{8.1.7}$$

with

$$\int_B f(s)\, ds = 0 \tag{8.1.8}$$

The symbol $\partial u / \partial n$ denotes the directional derivative of $u$ along the outward normal to the boundary $B$. The last condition (8.1.8) is known as *the compatibility condition,* since it is a consequence of (8.1.7) and the equation $\nabla^2 u = 0$. Here the solution $u$ may be interpreted as the steady-state temperature distribution in a body containing no heat sources or heat sinks when the heat flux across the boundary is prescribed.

The compatibility condition, in this case, may be interpreted physically as the heat requirement that the net heat flux across the boundary be zero.

### 3. The Third Boundary-Value Problem

Find a function $u(x, y)$ harmonic in $D$ which satisfies

$$\frac{\partial u}{\partial n} + h(s)u = f(s) \qquad \text{on } B \tag{8.1.9}$$

where $h$ and $f$ are given continuous functions. In this problem, the solution $u$ may be interpreted as the steady-state temperature distribution in a body, from the boundary of which the heat radiates freely into the surrounding medium of prescribed temperature.

### 4. The Fourth Boundary-Value Problem

(The Robin Problem): Find a function $u(x, y)$, harmonic in $D$, which satisfies boundary conditions of different types on different portions of the boundary $B$. An example involving such boundary conditions is

$$\begin{aligned} u &= f_1(s) \qquad \text{on } B_1 \\ \frac{\partial u}{\partial n} &= f_2(s) \qquad \text{on } B_2 \end{aligned} \tag{8.1.10}$$

where $B = B_1 + B_2$.

Problems 1 through 4 are called *interior boundary-value problems.* These differ from *exterior boundary-value problems* in two respects:

i. For problems of the latter variety, part of the boundary is at infinity.
ii. Solutions of exterior problems must satisfy an additional requirement, namely, that of boundedness at infinity.

## 8.2 MAXIMUM AND MINIMUM PRINCIPLES

Before we prove the uniqueness and continuity theorems for the interior Dirichlet problem for the two-dimensional Laplace equation, we first prove the maximum and minimum principles.

***Theorem 8.2.1 (The Maximum Principle)*** *Suppose that $u(x, y)$ is harmonic in a bounded domain $D$ and continuous in $\mathbf{D} = D + B$. Then $u$ attains its maximum on the boundary $B$ of $D$.*

*Physically, we may interpret this as meaning that the temperature of a body which has neither a source nor a sink of heat acquires its largest (and smallest) values on the surface of the body, and the electrostatic potential in a region which does not contain any free charge attains its maximum (and minimum) values on the boundary of the region.*

PROOF. Let the maximum of $u$ on $B$ be $M$. Let us now suppose that the maximum of $u$ in $\mathbf{D}$ is not attained at any point of $B$. Then it must be attained at some point $P_0(x_0, y_0)$ in $D$. If $M_0 = u(x_0, y_0)$ denotes the maximum of $u$ in $D$, then $M_0$ must also be the maximum of $u$ in $\mathbf{D}$.

Consider the function

$$v(x, y) = u(x, y) + \frac{M_0 - M}{4R^2}[(x - x_0)^2 + (y - y_0)^2] \tag{8.2.1}$$

where the point $P(x, y)$ is in $D$ and where $R$ is the radius of a circle containing $D$. Note that

$$v(x_0, y_0) = u(x_0, y_0) = M_0$$

We have $v(x, y) \leqslant M + (M_0 - M)/2 = \frac{1}{2}(M + M_0) < M_0$ on $B$. Thus, $v(x, y)$ like $u(x, y)$ must attain its maximum at a point in $D$. It follows from the definition of $v$ that

$$v_{xx} + v_{yy} = u_{xx} + u_{yy} + \frac{(M_0 - M)}{R^2} = \frac{(M_0 - M)}{R^2} > 0 \tag{8.2.2}$$

But for $v$ to be a maximum in $D$,

$$v_{xx} \leqslant 0, \qquad v_{yy} \leqslant 0$$

Thus,

$$v_{xx} + v_{yy} \leqslant 0$$

which contradicts Eq. (8.2.2). Hence the maximum of $u$ must be attained on $B$. □

***Theorem 8.2.2 (The Minimum Principle)*** *If $u(x, y)$ is harmonic in a bounded domain D and continuous in $\mathbf{D} = D + B$, then $u$ attains its minimum on the boundary $B$ of $D$.*

PROOF. The proof follows directly by applying the preceding theorem to the harmonic function $-u(x, y)$. □

As a result of the above theorems, we see that $u =$ constant which is evidently harmonic attains the same value in the domain $D$ as on the boundary $B$.

## 8.3 UNIQUENESS AND CONTINUITY THEOREMS

***Theorem 8.3.1* (*Uniqueness Theorem*)** *The solution of the Dirichlet problem, if it exists, is unique.*

PROOF. Let $u_1(x, y)$ and $u_2(x, y)$ be two solutions of the Dirichlet problem. Then $u_1$ and $u_2$ satisfy

$$\nabla^2 u_1 = 0, \qquad \nabla^2 u_2 = 0 \qquad \text{in } D$$
$$u_1 = f, \qquad u_2 = f \qquad \text{on } B$$

Since $u_1$ and $u_2$ are harmonic in $D$, $u_1 - u_2$ is also harmonic in $D$. But

$$u_1 - u_2 = 0 \qquad \text{on } B$$

By the maximum–minimum principles

$$u_1 - u_2 = 0$$

at all interior points of $D$. Thus, we have

$$u_1 = u_2$$

Therefore, the solution is unique. □

***Theorem 8.3.2* (*Continuity Theorem*)** *The solution of the Dirichlet problem depends continuously on the boundary data.*

PROOF. Let $u_1$ and $u_2$ be the solutions of

$$\nabla^2 u_1 = 0 \qquad \text{in } D$$
$$u_1 = f_1 \qquad \text{on } B$$

and

$$\nabla^2 u_2 = 0 \qquad \text{in } D$$
$$u_2 = f_2 \qquad \text{on } B$$

If $v = u_1 - u_2$, then $v$ satisfies

$$\nabla^2 v = 0 \qquad \text{in } D$$
$$v = f_1 - f_2 \qquad \text{on } B$$

By the maximum and minimum principles, $f_1 - f_2$ attains the maximum and minimum of $v$ on $B$. Thus, if $|f_1 - f_2| < \varepsilon$, then

$$-\varepsilon < v_{\min} \leqslant v_{\max} < \varepsilon \qquad \text{on } B$$

Thus, at any interior points in $D$, we have

$$-\varepsilon < v_{\min} \leqslant v \leqslant v_{\max} < \varepsilon$$

Therefore $|v| < \varepsilon$ in $D$. Hence

$$|u_1 - u_2| < \varepsilon \qquad \square$$

***Theorem 8.3.3*** *Let $\{u_n\}$ be a sequence of functions harmonic in $D$ and continuous in $\mathbf{D}$. Let $f_i$ be the values of $u_i$ on $B$. If $\{u_n\}$ converges uniformly on $B$, then it converges uniformly in $\mathbf{D}$.*

PROOF. By hypothesis $\{f_n\}$ converges uniformly on $B$. Thus, for $\varepsilon > 0$, there exists an integer $N$ such that everywhere on $B$

$$|f_n - f_m| < \varepsilon \quad \text{for} \quad n, m > N$$

It follows from the continuity theorem that for all $n, m > N$

$$|u_n - u_m| < \varepsilon$$

in $D$, and hence the theorem is proved. $\square$

## 8.4 DIRICHLET PROBLEM FOR A CIRCLE

### 1. Interior Problem

We shall now establish the existence of the solution of the Dirichlet problem for a circle.

The Dirichlet problem is

$$\nabla^2 u = u_{rr} + \frac{1}{r} u_r + \frac{1}{r^2} u_{\theta\theta} = 0, \qquad 0 \leqslant r < a \tag{8.4.1}$$

$$u(a, \theta) = f(\theta) \tag{8.4.2}$$

By the method of separation of variables, we seek a solution in the form

$$u(r, \theta) = R(r)\Theta(\theta)$$

Substitution of this in Eq. (8.4.1) yields

$$r^2 \frac{R''}{R} + r\frac{R'}{R} = -\frac{\Theta''}{\Theta} = \lambda$$

Hence

$$r^2 R'' + rR' - \lambda R = 0 \tag{8.4.3}$$

$$\Theta'' + \lambda\Theta = 0 \tag{8.4.4}$$

Because of the periodicity conditions $\Theta(0) = \Theta(2\pi)$ and $\Theta'(0) = \Theta'(2\pi)$ which ensure that the function $\Theta$ is single-valued, the case $\lambda < 0$

does not yield an acceptable solution. When $\lambda = 0$, we have

$$u(r, \theta) = (A + B \log r)(C\theta + D)$$

Since $\log r \rightarrow -\infty$ as $r \rightarrow 0+$ (note that $r = 0$ is a singular point of Eq. (8.4.1)), $B$ must vanish in order for $u$ to be finite at $r = 0$. $C$ must also vanish in order for $u$ to be periodic with period $2\pi$. Hence the solution for $\lambda = 0$ is $u =$ constant. When $\lambda > 0$, the solution of Eq. (8.4.4) is

$$\Theta(\theta) = A \cos \sqrt{\lambda}\, \theta + B \sin \sqrt{\lambda}\, \theta$$

The periodicity conditions imply

$$\sqrt{\lambda} = n \quad \text{for} \quad n = 1, 2, 3, \ldots .$$

Equation (8.4.3) is the Euler equation and therefore the general solution is

$$R(r) = Cr^{\sqrt{\lambda}} + Dr^{-\sqrt{\lambda}}$$

Since $r^{-\sqrt{\lambda}} \rightarrow \infty$ as $r \rightarrow 0$, $D$ must vanish for $u$ to be continuous at $r = 0$. Thus, the solution is

$$u(r, \theta) = Cr^{\sqrt{\lambda}}(A \cos \sqrt{\lambda}\, \theta + B \sin \sqrt{\lambda}\, \theta) \quad \text{for} \quad \sqrt{\lambda} = 1, 2, \ldots .$$

Hence the general solution of Eq. (8.4.1) may be written in the form

$$u(r, \theta) = \frac{a_0}{2} + \sum_{n=1}^{\infty} \left(\frac{r}{a}\right)^n (a_n \cos n\theta + b_n \sin n\theta) \tag{8.4.5}$$

where the constant term $a_0/2$ represents the solution for $\lambda = 0$, and where $a_n$ and $b_n$ are constants. Letting $\rho = r/a$, we have

$$u(\rho, \theta) = \frac{a_0}{2} + \sum_{n=1}^{\infty} \rho^n (a_n \cos n\theta + b_n \sin n\theta) \tag{8.4.6}$$

Our next task is to show that $u(r, \theta)$ is harmonic in $0 \leqslant r < a$ and continuous in $0 \leqslant r \leqslant a$. We must also show that $u$ satisfies the boundary condition (8.4.2).

We first assume that $a_n$ and $b_n$ are the Fourier coefficients of $f(\theta)$, that is,

$$\begin{aligned} a_n &= \frac{1}{\pi} \int_0^{2\pi} f(\theta) \cos n\theta \, d\theta \qquad n = 0, 1, 2, 3, \ldots \\ b_n &= \frac{1}{\pi} \int_0^{2\pi} f(\theta) \sin n\theta \, d\theta \qquad n = 1, 2, 3, \ldots \end{aligned} \tag{8.4.7}$$

Thus, from their very definitions, $a_n$ and $b_n$ are bounded, that is, there exists some number $M > 0$ such that

$$|a_0| < M, \qquad |a_n| < M, \qquad |b_n| < M, \qquad n = 1, 2, 3, \ldots$$

Thus, if we consider the sequence of functions $\{u_n\}$ defined by

$$u_n(\rho, \theta) = \rho^n(a_n \cos n\theta + b_n \sin n\theta) \tag{8.4.8}$$

we see that

$$|u_n| < 2\rho_0^n M, \qquad 0 \leqslant \rho \leqslant \rho_0 < 1$$

Hence in any closed circular region, series (8.4.6) converges uniformly.

Next, differentiate $u_n$ with respect to $r$. Then for $0 \leqslant \rho \leqslant \rho_0 < 1$

$$\left|\frac{\partial u_n}{\partial r}\right| = \left|\frac{n}{a}\rho^{n-1}(a_n \cos n\theta + b_n \sin n\theta)\right| < 2\frac{n}{a}\rho_0^{n-1}M$$

Thus, the series obtained by differentiating series (8.4.6) term by term with respect to $r$ converges uniformly. In a similar manner, we can prove that the series obtained by twice differentiating series (8.4.6) term by term with respect to $r$ and $\theta$ converge uniformly. Consequently,

$$\begin{aligned}\nabla^2 u &= u_{rr} + \frac{1}{r}u_r + \frac{1}{r^2}u_{\theta\theta}\\ &= \sum_{n=1}^{\infty} \frac{\rho^{n-2}}{a^2}(a_n \cos n\theta + b_n \sin n\theta)[n(n-1) + n - n^2]\\ &= 0, \quad 0 \leqslant \rho \leqslant \rho_0 < 1\end{aligned}$$

Since each term of series (8.4.6) is a harmonic function, and since the series converges uniformly, $u(r, \theta)$ is harmonic at any interior point of the region $0 \leqslant \rho < 1$. It now remains to show that $u$ satisfies the boundary data $f(\theta)$.

Substitution of the Fourier coefficients $a_n$ and $b_n$ into Eq. (8.4.6) yields

$$\begin{aligned}u(\rho, \theta) &= \frac{1}{2\pi}\int_0^{2\pi} f(\theta)\, d\theta + \frac{1}{\pi}\sum_{n=1}^{\infty} \rho^n \int_0^{2\pi} f(\tau)\\ &\qquad \times [\cos n\tau \cos n\theta + \sin n\tau \sin n\theta]\, d\tau\\ &= \frac{1}{2\pi}\int_0^{2\pi} \left[1 + 2\sum_{n=1}^{\infty} \rho^n \cos n(\theta - \tau)\right] f(\tau)\, d\tau\end{aligned} \tag{8.4.9}$$

The interchange of summation and integration is permitted due to the uniform convergence of the series. For $0 \leqslant \rho \leqslant 1$

$$\begin{aligned}1 + 2\sum_{n=1}^{\infty} [\rho^n \cos n(\theta - \tau)] &= 1 + \sum_{n=1}^{\infty} [\rho^n e^{in(\theta-\tau)} + \rho^n e^{-in(\theta-\tau)}]\\ &= 1 + \frac{\rho e^{i(\theta-\tau)}}{1 - \rho e^{i(\theta-\tau)}} + \frac{\rho e^{-i(\theta-\tau)}}{1 - \rho e^{-i(\theta-\tau)}}\\ &= \frac{1 - \rho^2}{1 - \rho e^{i(\theta-\tau)} - \rho e^{-i(\theta-\tau)} + \rho^2}\\ &= \frac{1 - \rho^2}{1 - 2\rho \cos(\theta - \tau) + \rho^2}\end{aligned}$$

Hence

$$u(\rho, \theta) = \frac{1}{2\pi} \int_0^{2\pi} \frac{1-\rho^2}{1-2\rho\cos(\theta-\tau)+\rho^2} f(\tau)\, d\tau \qquad (8.4.10)$$

The integral on the right side of (8.4.10) is called the *Poisson integral formula for a circle.*

Now if $f(\theta)=1$, then according to series (8.4.9), $u(r, \theta)=1$ for $0 \leqslant \rho \leqslant 1$. Thus, Eq. (8.4.10) gives

$$1 = \frac{1}{2\pi} \int_0^{2\pi} \frac{1-\rho^2}{1-2\rho\cos(\theta-\tau)+\rho^2}\, d\tau$$

Hence,

$$f(\theta) = \frac{1}{2\pi} \int_0^{2\pi} \frac{1-\rho^2}{1-2\rho\cos(\theta-\tau)+\rho^2} f(\theta)\, d\tau, \qquad 0 \leqslant \rho < 1$$

Therefore

$$u(\rho, \theta) - f(\theta) = \frac{1}{2\pi} \int_0^{2\pi} \frac{(1-\rho^2)[f(\tau)-f(\theta)]}{1-2\rho\cos(\theta-\tau)+\rho^2}\, d\tau \qquad (8.4.11)$$

Since $f(\theta)$ is uniformly continuous on $[0, 2\pi]$, for given $\varepsilon > 0$, there exists a positive number $\delta(\varepsilon)$ such that $|\theta-\tau| < \delta$ implies $|f(\theta) - f(\tau)| < \varepsilon$. If $|\theta-\tau| \geqslant \delta$ so that $\theta - \tau \neq 2n\pi$ for $n = 0, 1, 2, \ldots,$ then

$$\lim_{\rho \to 1-} \frac{1-\rho^2}{1-2\rho\cos(\theta-\tau)+\rho^2} = 0$$

In other words, there exists $\rho_0$ such that if $|\theta-\tau| \geqslant \delta$, then

$$\frac{1-\rho^2}{1-2\rho\cos(\theta-\tau)+\rho^2} < \varepsilon$$

for $0 \leqslant \rho \leqslant \rho_0 < 1$. Hence Eq. (8.4.10) yields

$$\begin{aligned} |u(r, \theta) - f(\theta)| &\leqslant \frac{1}{2\pi} \int_{|\theta-\tau|\geqslant\delta}^{2\pi} \frac{(1-\rho^2)\,|f(\theta)-f(\tau)|}{1-2\rho\cos(\theta-\tau)+\rho^2}\, d\tau \\ &\quad + \frac{1}{2\pi} \int_{|\theta-\tau|<\delta}^{2\pi} \frac{(1-\rho^2)\,|f(\theta)-f(\tau)|}{1-2\rho\cos(\theta-\tau)+\rho^2}\, d\tau \\ &\leqslant \frac{1}{2\pi} 2\pi\varepsilon \left[2 \max_{0\leqslant\theta\leqslant 2\pi} |f(\theta)|\right] + \frac{\varepsilon}{2\pi} \cdot 2\pi \\ &\leqslant \varepsilon\left[1 + 2\left(\max_{0\leqslant\theta\leqslant 2\pi} |f(\theta)|\right)\right] \end{aligned}$$

which implies that

$$\lim_{\rho\to 1-} u(r,\theta)=f(\theta)$$

uniformly in $\theta$. Therefore we state the following theorem.

***Theorem 8.4.1*** *There exists one and only one harmonic function $u(r,\theta)$ which satisfies the continuous boundary data $f(\theta)$. This function is either given by*

$$u(r,\theta)=\frac{1}{2\pi}\int_0^{2\pi}\frac{a^2-r^2}{a^2-2ar\cos(\theta-\tau)+r^2}f(\tau)\,d\tau \qquad (8.4.12)$$

*or*

$$u(r,\theta)=\frac{a_0}{2}+\sum_{n=1}^{\infty}\frac{r^n}{a^n}(a_n\cos n\theta+b_n\sin n\theta) \qquad (8.4.13)$$

*where $a_n$ and $b_n$ are the Fourier coefficients of $f(\theta)$.*

For $\rho=0$, the Poisson integral formula (8.4.10) becomes

$$u(0,\theta)=u(0)=\frac{1}{2\pi}\int_0^{2\pi}f(\tau)\,d\tau \qquad (8.4.14)$$

This result may be stated as follows:

***Theorem 8.4.2 (Mean Value Theorem)*** *If $u$ is harmonic in a circle, then the value of $u$ at the center is equal to the mean value of $u$ on the boundary of the circle.*

Several comments are in order. First, the Continuity Theorem 8.3.2 for the Dirichlet problem for the Laplace equation is a special example of the general result that the Dirichlet problems for all elliptic equations are well-posed. Second, the formula (8.4.12) represents a unique continuous solution of the Laplace equation in $0\leqslant r<a$ even when $f(\theta)$ is discontinuous. This means that for Laplace's equation, discontinuities in boundary conditions are smoothed out in the interior of the domain. This is a remarkable contrast to the linear hyperbolic equations where any discontinuity in the data propagates along the characteristics. Third, the integral solution (8.4.12) can be written as

$$u(r,\theta)=\int_{-\pi}^{\pi}P(r,\tau-\theta)f(\tau)\,d\tau$$

where $P(r,\tau-\theta)$ is called the *Poisson kernel* given by

$$P(r,\tau-\theta)=\frac{1}{2\pi}\frac{(a^2-r^2)}{[a^2-2ar\cos(\tau-\theta)+r^2]}$$

Clearly, $P(a, \tau - \theta) = 0$ except at $\tau = \theta$. Also

$$f(\theta) = \lim_{r \to a^-} u(r, \theta) = \int_{-\pi}^{\pi} \lim_{r \to a^-} P(r, \tau - \theta) f(\tau)\, d\tau$$

This implies that

$$\lim_{r \to a^-} P(r, \tau - \theta) = \delta(\tau - \theta)$$

where $\delta(x)$ is the Dirac delta function.

As in the preceding section, the *exterior Dirichlet problem* for a circle can readily be solved. For the exterior problem $u$ must be bounded as $r \to \infty$. The general solution, therefore, is

$$u(r, \theta) = \frac{a_0}{2} + \sum_{n=1}^{\infty} \left(\frac{r^{-n}}{a^{-n}}\right)(a_n \cos n\theta + b_n \sin n\theta) \tag{8.4.15}$$

Applying the boundary condition $u(a, \theta) = f(\theta)$, we obtain

$$f(\theta) = \frac{a_0}{2} + \sum_{n=1}^{\infty} (a_n \cos n\theta + b_n \sin n\theta)$$

Hence, we find

$$a_n = \frac{1}{\pi} \int_0^{2\pi} f(\tau) \cos n\tau\, d\tau, \qquad n = 0, 1, 2, \ldots \tag{8.4.16}$$

$$b_n = \frac{1}{\pi} \int_0^{2\pi} f(\tau) \sin n\tau\, d\tau, \qquad n = 1, 2, 3, \ldots \tag{8.4.17}$$

Substitution of $a_n$ and $b_n$ into Eq. (8.4.15) yields

$$u(r, \theta) = \frac{1}{2\pi} \int_0^{2\pi} \left[1 + 2 \sum_{n=1}^{\infty} \left(\frac{a}{r}\right)^n \cos n(\theta - \tau)\right] f(\tau)\, d\tau$$

Comparing with Eq. (8.4.9), we see that the only difference between the exterior and interior problem is that $\rho^n$ is replaced by $\rho^{-n}$. Therefore, the final result takes the form

$$u(\rho, \theta) = \frac{1}{2\pi} \int_0^{2\pi} \frac{\rho^2 - 1}{1 - 2\rho \cos(\theta - \tau) + \rho^2} f(\tau)\, d\tau \tag{8.4.18}$$

for $\rho > 1$.

## 8.5 DIRICHLET PROBLEM FOR A CIRCULAR ANNULUS

The natural extension of the Dirichlet problem for a circle is the Dirichlet problem for a circular annulus, that is

$$\nabla^2 u = 0, \qquad r_2 < r < r_1 \tag{8.5.1}$$

$$u(r_1, \theta) = f(\theta) \tag{8.5.2}$$

$$u(r_2, \theta) = g(\theta) \tag{8.5.3}$$

In addition $u(r, \theta)$ must satisfy the periodicity condition. Accordingly, $f(\theta)$ and $g(\theta)$ must also be periodic with period $2\pi$.

Proceeding as in the case of the Dirichlet problem for a circle, we obtain for $\lambda = 0$

$$u(r, \theta) = (A + B \log r)(C\theta + D)$$

The periodicity condition on $u$ requires that $C = 0$. Then $u(r, \theta)$ becomes

$$u(r, \theta) = \frac{a_0}{2} + \frac{b_0}{2} \log r$$

where $a_0 = 2AD$ and $b_0 = 2BD$.

The solution for the case $\lambda > 0$ is

$$u(r, \theta) = (Cr^{\sqrt{\lambda}} + Dr^{-\sqrt{\lambda}})(A \cos \sqrt{\lambda}\, \theta + B \sin \sqrt{\lambda}\, \theta)$$

for $\sqrt{\lambda} = n = 1, 2, 3, \ldots$. Thus, the general solution is

$$u(r, \theta) = \frac{1}{2}(a_0 + b_0 \log r) + \sum_{n=1}^{\infty} [(a_n r^n + b_n r^{-n}) \cos n\theta + (c_n r^n + d_n r^{-n}) \sin n\theta] \qquad (8.5.4)$$

where $a_n$, $b_n$, $c_n$, and $d_n$ are constants.

Applying the boundary conditions (8.5.2) and (8.5.3), we find that the coefficients are given by

$$a_0 + b_0 \log r_1 = \frac{1}{\pi} \int_0^{2\pi} f(\tau)\, d\tau$$

$$a_n r_1^n + b_n r_1^{-n} = \frac{1}{\pi} \int_0^{2\pi} f(\tau) \cos n\tau\, d\tau$$

$$c_n r_1^n + d_n r_1^{-n} = \frac{1}{\pi} \int_0^{2\pi} f(\tau) \sin n\tau\, d\tau$$

and

$$a_0 + b_0 \log r_2 = \frac{1}{\pi} \int_0^{2\pi} g(\tau)\, d\tau$$

$$a_n r_2^n + b_n r_2^{-n} = \frac{1}{\pi} \int_0^{2\pi} g(\tau) \cos n\tau\, d\tau$$

$$c_n r_2^n + d_n r_2^{-n} = \frac{1}{\pi} \int_0^{2\pi} g(\tau) \sin n\tau\, d\tau$$

The constants $a_0$, $b_0$, $a_n$, $b_n$, $c_n$, $d_n$ for $n = 1, 2, 3, \ldots$ can then be determined. Hence the solution of the Dirichlet problem for an annulus is given by (8.5.4).

## 8.6 NEUMANN PROBLEM FOR A CIRCLE

Let $u$ be a solution of the Neumann problem

$$\nabla^2 u = 0 \qquad \text{in } D$$

$$\frac{\partial u}{\partial n} = f \qquad \text{on } B$$

It is evident that $u$ + constant is also a solution. Thus, we see that the solution of the Neumann problem is not unique, and it differs from another by a constant.

Consider the interior Neumann problem

$$\nabla^2 u = 0, \qquad r < R \tag{8.6.1}$$

$$\frac{\partial u}{\partial n} = \frac{\partial u}{\partial r} = f(\theta), \qquad r = R \tag{8.6.2}$$

Before we determine a solution of the Neumann problem, a necessary condition for the existence of a solution will be established.

In Green's second formula

$$\iint_D (v\nabla^2 u - u\nabla^2 v)\, dS = \int_B \left(v\frac{\partial u}{\partial n} - u\frac{\partial v}{\partial n}\right) ds \tag{8.6.3}$$

we put $v = 1$, so that $\nabla^2 v = 0$ in $D$ and $\partial v/\partial n = 0$ on $B$. Thus, the result is

$$\iint_D \nabla^2 u\, dS = \int_B \frac{\partial u}{\partial n}\, ds \tag{8.6.4}$$

Substitution of Eqs. (8.6.1) and (8.6.2) into Eq. (8.6.4) yields

$$\int_B f\, ds = 0 \tag{8.6.5}$$

which may also be written in the form

$$R\int_0^{2\pi} f(\theta)\, d\theta = 0 \tag{8.6.6}$$

As in the case of the interior Dirichlet problem for a circle, the solution of the Laplace equation is

$$u(r, \theta) = \frac{a_0}{2} + \sum_{k=1}^{\infty} r^k (a_k \cos k\theta + b_k \sin k\theta) \tag{8.6.7}$$

Differentiating this with respect to $r$ and applying the boundary condition

(8.6.2), we obtain

$$\frac{\partial u}{\partial r}(R, \theta)=\sum_{k=1}^{\infty} kR^{k-1}(a_k \cos k\theta + b_k \sin k\theta)=f(\theta) \tag{8.6.8}$$

Hence the coefficients are given by

$$\begin{aligned} a_k &= \frac{1}{k\pi R^{k-1}}\int_0^{2\pi} f(\tau)\cos k\tau\, d\tau, \qquad k=1, 2, 3, \ldots \\ b_k &= \frac{1}{k\pi R^{k-1}}\int_0^{2\pi} f(\tau)\sin k\tau\, d\tau, \qquad k=1, 2, 3, \ldots \end{aligned} \tag{8.6.9}$$

Note that the expansion of $f(\theta)$ in a series of the form (8.6.8) is possible only by virtue of the compatibility condition (8.6.6) since

$$a_0=\frac{1}{\pi}\int_0^{2\pi} f(\tau)\, d\tau = 0$$

Inserting $a_k$ and $b_k$ in Eq. (8.6.7), we obtain

$$u(r, \theta)=\frac{a_0}{2}+\frac{R}{\pi}\int_0^{2\pi}\left[\sum_{k=1}^{\infty}\left(\frac{r}{R}\right)^k \cos k(\theta-\tau)\right]f(\tau)\, d\tau$$

Using the identity

$$-\frac{1}{2}\log\left[1+\rho^2-2\rho\cos(\theta-\tau)\right]=\sum_{k=1}^{\infty}\frac{1}{k}\rho^k\cos k(\theta-\tau)$$

with $\rho = r/R$, we find that

$$u(r, \theta)=\frac{a_0}{2}-\frac{R}{2\pi}\int_0^{2\pi}\log\left[R^2-2rR\cos(\theta-\tau)+r^2\right]f(\tau)\, d\tau \tag{8.6.10}$$

in which a constant factor $R^2$ in the argument of the logarithm was eliminated by virtue of Eq. (8.6.6).

In a similar manner, for the exterior Neumann problem, we can readily find that

$$u(r, \theta)=\frac{a_0}{2}+\frac{R}{2\pi}\int_0^{2\pi}\log\left[R^2-2rR\cos(\theta-\tau)+r^2\right]f(\tau)\, d\tau \tag{8.6.11}$$

## 8.7 DIRICHLET PROBLEM FOR A RECTANGLE

Let us first consider the problem

$$\nabla^2 u = u_{xx}+u_{yy}=0, \qquad 0<x<a, \quad 0<y<b \tag{8.7.1}$$

$$u(x, 0)=f(x), \qquad 0\leqslant x\leqslant a \tag{8.7.2}$$

$$u(x, b)=0 \tag{8.7.3}$$

$$u(0, y)=0 \tag{8.7.4}$$

$$u(a, y)=0 \tag{8.7.5}$$

We seek a solution in the form

$$u(x, y) = X(x)Y(y)$$

Substituting $u(x, y)$ in the Laplace equation, we obtain

$$X'' - \lambda X = 0 \tag{8.7.6}$$

$$Y'' + \lambda Y = 0 \tag{8.7.7}$$

where $\lambda$ is a separation constant. Since the boundary conditions are homogeneous on $x = 0$ and $x = a$, we choose $\lambda = -\alpha^2$ with $\alpha > 0$ in order to obtain nontrivial solutions of the eigenvalue problem

$$X'' + \alpha^2 X = 0$$

$$X(0) = X(a) = 0$$

It is easily found that the eigenvalues are

$$\alpha = \frac{n\pi}{a}, \qquad n = 1, 2, 3, \ldots$$

and the corresponding eigenfunctions are $\sin n\pi x/a$. Hence

$$X_n(x) = B_n \sin \frac{n\pi x}{a}$$

The solution of Eq. (8.7.7) is $Y(y) = C \cosh \alpha y + D \sinh \alpha y$, which may also be written in the form

$$Y(y) = E \sinh \alpha(y + F)$$

where $E = (D^2 - C^2)^{\frac{1}{2}}$ and $F = 1/\alpha \tanh^{-1}(C/D)$. Applying the remaining homogeneous boundary condition

$$u(x, b) = X(x)Y(b) = 0$$

we obtain

$$Y(b) = E \sinh \alpha(b + F) = 0$$

and hence

$$F = -b, \qquad E \neq 0$$

for a nontrivial solution $u(x, y)$. Thus, we have

$$Y_n(y) = E_n \sinh \frac{n\pi}{a}(y - b)$$

Because of linearity, the solution is

$$u(x, y) = \sum_{n=1}^{\infty} a_n \sin \frac{n\pi x}{a} \sinh \frac{n\pi}{a}(y - b)$$

where $a_n = B_n E_n$. Now, we apply the nonhomogeneous boundary condition to obtain

$$u(x, 0) = f(x) = \sum_{n=1}^{\infty} a_n \sinh\left(\frac{-n\pi b}{a}\right) \sin\frac{n\pi x}{a}$$

This is a Fourier sine series and hence

$$a_n = \frac{-2}{a \sinh\left(\dfrac{n\pi b}{a}\right)} \int_0^a f(x) \sin\frac{n\pi x}{a}\, dx$$

Thus, the formal solution is given by

$$u(x, y) = \sum_{n=1}^{\infty} a_n^* \frac{\sinh\dfrac{n\pi}{a}(b-y)}{\sinh\dfrac{n\pi b}{a}} \sin\frac{n\pi x}{a} \tag{8.7.8}$$

where

$$a_n^* = \frac{2}{a}\int_0^a f(x)\sin\frac{n\pi x}{a}\, dx$$

To prove the existence of solution (8.7.8), we first note that

$$\frac{\sinh\dfrac{n\pi}{a}(b-y)}{\sinh\dfrac{n\pi b}{a}} = e^{-n\pi y/a}\left[\frac{1-e^{-(2n\pi/a)(b-y)}}{1-e^{-2n\pi b/a}}\right]$$

$$\leqslant C_1 e^{-n\pi y/a}$$

where $C_1$ is a constant. Since $f(x)$ is bounded, we have

$$|a_n^*| \leqslant \frac{2}{a}\int_0^a |f(x)|\, dx = C_2$$

Thus, the series for $u(x, y)$ is dominated by the series

$$\sum_{n=1}^{\infty} M e^{-n\pi y_0/a} \quad \text{for} \quad y \geqslant y_0 > 0, \quad M = \text{constant}$$

and hence $u(x, y)$ converges uniformly in $x$ and $y$ whenever $0 \leqslant x \leqslant a$, $y \geqslant y_0 > 0$. Consequently, $u(x, y)$ is continuous in this region and satisfies the boundary values $u(0, y) = u(a, y) = u(x, b) = 0$.

Now differentiating $u$ twice with respect to $x$, we obtain

$$u_{xx}(x, y) = \sum_{n=1}^{\infty} -a_n^*\left(\frac{n\pi}{a}\right)^2 \frac{\sinh\dfrac{n\pi}{a}(b-y)}{\sinh\dfrac{n\pi b}{a}} \sin\frac{n\pi x}{a}$$

and differentiating $u$ twice with respect to $y$, we obtain

$$u_{yy}(x, y) = \sum_{n=1}^{\infty} a_n^* \left(\frac{n\pi}{a}\right)^2 \frac{\sinh \frac{n\pi}{a}(b-y)}{\sinh \frac{n\pi b}{a}} \sin \frac{n\pi x}{a}$$

It is evident that the series for $u_{xx}$ and $u_{yy}$ are both dominated by

$$\sum_{n=1}^{\infty} M^* n^2 e^{-n\pi y_0/a}$$

and hence converge uniformly for any $0 < y_0 < b$. It follows that $u_{xx}$ and $u_{yy}$ exist, and $u$ satisfies the Laplace equation.

It now remains to be shown that $u(x, 0) = f(x)$. Let $f(x)$ be a continuous function and let $f'(x)$ be piecewise continuous on $[0, a]$. If, in addition, $f(0) = f(a) = 0$, then the Fourier series for $f(x)$ converges uniformly. Putting $y = 0$ in the series for $u(x, y)$, we obtain

$$u(x, 0) = \sum_{n=1}^{\infty} a_n^* \sin \frac{n\pi x}{a}$$

Since $u(x, 0)$ converges uniformly to $f(x)$, we write for $\varepsilon > 0$

$$|s_m(x, 0) - s_n(x, 0)| < \varepsilon \quad \text{for} \quad m, n > N_\varepsilon$$

where

$$s_m(x, y) = \sum_{n=1}^{m} a_n^* \sin \frac{n\pi x}{a}$$

We also know that $s_m(x, y) - s_n(x, y)$ satisfies the Laplace equation and the boundary conditions on $x = 0$, $x = a$ and $y = b$. Then by the maximum principle,

$$|s_m(x, y) - s_n(x, y)| < \varepsilon \quad \text{for} \quad m, n > N_\varepsilon$$

in the region $0 \leqslant x \leqslant a$, $0 \leqslant y \leqslant b$. Thus, the series for $u(x, y)$ converges uniformly, and as a consequence, $u(x, y)$ is continuous in the region $0 \leqslant x \leqslant a$, $0 \leqslant y \leqslant b$. Hence, we obtain

$$u(x, 0) = \sum_{n=1}^{\infty} a_n^* \sin \frac{n\pi x}{a} = f(x)$$

Thus the solution (8.7.8) is established.

The general Dirichlet problem

$$\nabla^2 u = 0, \qquad 0 < x < a, \qquad 0 < y < b$$

$$u(x, 0) = f_1(x)$$

$$u(x, a) = f_2(x)$$

$$u(0, y) = f_3(y)$$

$$u(b, y) = f_4(y)$$

can be solved by separating it into four problems, each of which has one nonhomogeneous boundary condition and the rest zero. Thus, determining each solution as in the preceding problem and then adding the four solutions, the solution of the Dirichlet problem for a rectangle is obtained.

## 8.8 DIRICHLET PROBLEM INVOLVING POISSON EQUATION

The solution of the Dirichlet problem involving the Poisson equation can be obtained for simple regions when the solution of the corresponding Dirichlet problem for the Laplace equation is known.

Consider the Poisson equation

$$\nabla^2 u = u_{xx} + u_{yy} = f(x, y) \quad \text{in } D$$

with the condition

$$u = g(x, y) \quad \text{on } B$$

Assume that the solution can be written in the form

$$u = v + w$$

where $v$ is a particular solution of the Poisson equation and $w$ is the solution of the associated homogeneous equation, that is,

$$\nabla^2 v = f$$
$$\nabla^2 w = 0$$

As soon as $v$ is ascertained, the solution of the Dirichlet problem

$$\nabla^2 w = 0 \qquad \text{in } D$$
$$w = -v + g(x, y) \quad \text{on } B$$

can be determined. The usual method of finding a particular solution for the case in which $f(x, y)$ is a polynomial of degree $n$ is to seek a solution in the form of a polynomial of degree $(n+2)$ with undetermined coefficients.

As an example, consider the torsion problem

$$\nabla^2 u = -2, \qquad 0 < x < a, \qquad 0 < y < b$$
$$u(0, y) = 0$$
$$u(a, y) = 0$$
$$u(x, 0) = 0$$
$$u(x, b) = 0$$

We let $u = v + w$. Now assume $v$ to be of the form

$$v(x, y) = A + Bx + Cy + Dx^2 + Exy + Fy^2$$

Substituting this in the Poisson equation, we obtain

$$2D + 2F = -2$$

The simplest way of satisfying this equation is to choose

$$D = -1 \quad \text{and} \quad F = 0$$

The remaining coefficients are arbitrary. Thus we take

$$v(x, y) = ax - x^2$$

so that $v$ reduces to zero on the sides $x = 0$ and $x = a$. Next, we find $w$ from

$$\nabla^2 w = 0, \qquad 0 < x < a, \qquad 0 < y < b$$

$$w(0, y) = -v(0, y) = 0$$

$$w(a, y) = -v(a, y) = 0$$

$$w(x, 0) = -v(x, 0) = -(ax - x^2)$$

$$w(x, b) = -v(x, b) = -(ax - x^2)$$

As in the Dirichlet problem (Sec. 8.7), the solution is found to be

$$w(x, y) = \sum_{n=1}^{\infty} \left(a_n \cosh \frac{n\pi y}{a} + b_n \sinh \frac{n\pi y}{a}\right) \sin \frac{n\pi x}{a}$$

Application of the nonhomogeneous boundary conditions yield

$$w(x, 0) = -(ax - x^2) = \sum_{n=1}^{\infty} a_n \sin \frac{n\pi x}{a}$$

$$w(x, b) = -(ax - x^2) = \sum_{n=1}^{\infty} \left(a_n \cosh \frac{n\pi b}{a} + b_n \sinh \frac{n\pi b}{a}\right) \sin \frac{n\pi x}{a}$$

from which we find

$$a_n = \frac{2}{a} \int_0^a (x^2 - ax) \sin \frac{n\pi x}{a}\, dx$$

$$= \begin{cases} 0 & \text{if } n \text{ is even} \\ \dfrac{-8a^2}{\pi^3 n^3} & \text{if } n \text{ is odd} \end{cases}$$

and

$$\left(a_n \cosh \frac{n\pi b}{a} + b_n \sinh \frac{n\pi b}{a}\right) = \frac{2}{a} \int_0^a (x^2 - ax) \sin \frac{n\pi x}{a}\, dx$$

Thus, we have

$$b_n = \frac{\left(1 - \cosh \dfrac{n\pi b}{a}\right) a_n}{\sinh \dfrac{n\pi b}{a}}$$

Hence the solution of the Dirichlet problem for the Poisson equation is given by

$$u(x, y) = (a - x)x$$

$$- \frac{8a^2}{\pi^3} \sum_{n=1}^{\infty} \frac{\left[\sinh (2n-1) \dfrac{\pi(b-y)}{a} + \sinh (2n-1) \dfrac{\pi y}{a}\right]}{\sinh (2n-1) \dfrac{\pi b}{a}} \frac{\sin (2n-1) \dfrac{\pi x}{a}}{(2n-1)^3}$$

## 8.9 THE NEUMANN PROBLEM FOR A RECTANGLE

Consider the Neumann problem

$$\nabla^2 u = 0, \qquad 0 < x < a, \qquad 0 < y < b \tag{8.9.1}$$

$$u_x(0, y) = f_1(y) \tag{8.9.2}$$

$$u_x(a, y) = f_2(y) \tag{8.9.3}$$

$$u_y(x, 0) = g_1(x) \tag{8.9.4}$$

$$u_y(x, b) = g_2(x) \tag{8.9.5}$$

The compatibility condition that must be fulfilled in this case is

$$\int_0^a [g_1(x) - g_2(x)]\, dx + \int_0^b [f_1(y) - f_2(y)]\, dy = 0 \tag{8.9.6}$$

We assume a solution in the form

$$u(x, y) = u_1(x, y) + u_2(x, y) \tag{8.9.7}$$

where $u_1(x, y)$ is a solution of

$$\begin{aligned} &\nabla^2 u_1 = 0 \\ &\frac{\partial u_1}{\partial x}(0, y) = 0 \\ &\frac{\partial u_1}{\partial x}(a, y) = 0 \\ &\frac{\partial u_1}{\partial y}(x, 0) = g_1(x) \\ &\frac{\partial u_1}{\partial y}(x, b) = g_2(x) \end{aligned} \tag{8.9.8}$$

and where $g_1$ and $g_2$ satisfy the compatibility condition

$$\int_0^a [g_1(x) - g_2(x)]\, dx = 0 \tag{8.9.9}$$

The function $u_2(x, y)$ is a solution of

$$\begin{aligned}
&\nabla^2 u_2 = 0 \\
&\frac{\partial u_2}{\partial x}(0, y) = f_1(y) \\
&\frac{\partial u_2}{\partial x}(a, y) = f_2(y) \\
&\frac{\partial u_2}{\partial y}(x, 0) = 0 \\
&\frac{\partial u_2}{\partial y}(x, b) = 0
\end{aligned} \tag{8.9.10}$$

where $f_1$ and $f_2$ satisfy the compatibility condition

$$\int_0^b [f_1(y) - f_2(y)]\, dy = 0 \tag{8.9.11}$$

$u_1(x, y)$ and $u_2(x, y)$ can be determined. Conditions (8.9.9) and (8.9.11) ensure that condition (8.9.6) is fulfilled. Thus the problem is solved.

However, the solution obtained in this manner is rather restrictive. In general, condition (8.9.6) does not imply conditions (8.9.9) and (8.9.11). Thus, generally speaking, it is not possible to obtain a solution of the Neumann problem for a rectangle by the method described above.

To obtain a general solution, Grunberg [17] proposed the following method. Suppose we assume a solution in the form

$$u(x, y) = \frac{Y_0}{2}(y) + \sum_{n=1}^{\infty} X_n(x) Y_n(y) \tag{8.9.12}$$

where $X_n(x) = \cos n\pi x/a$ is an eigenfunction of the eigenvalue problem

$$\begin{aligned}
&X'' + \lambda X = 0 \\
&X'(0) = X'(a) = 0
\end{aligned}$$

corresponding to the eigenvalue $\lambda_n = (n\pi/a)^2$. Then from Eq. (8.9.12), we see that

$$\begin{aligned}
Y_n(y) &= \frac{2}{a}\int_0^a u(x, y) X_n(x)\, dx \\
&= \frac{2}{a}\int_0^a u(x, y) \cos\frac{n\pi x}{a}\, dx
\end{aligned} \tag{8.9.13}$$

Multiplying both sides of Eq. (8.9.1) by $2\cos(n\pi x/a)$ and integrating with respect to $x$ from 0 to $a$, we obtain

$$\frac{2}{a}\int_0^a (u_{xx}+u_{yy})\cos\frac{n\pi x}{a}\,dx = 0$$

or

$$Y_n'' + \frac{2}{a}\int_0^a u_{xx}\cos\frac{n\pi x}{a}\,dx = 0$$

Integrating the second term by parts and applying the boundary conditions (8.9.2) and (8.9.3), we obtain

$$Y_n'' - \left(\frac{n\pi}{a}\right)^2 Y_n = F_n(y) \tag{8.9.14}$$

where $F_n(y) = 2[f_1(y) - (-1)^n f_2(y)]/a$. This is an ordinary differential equation whose solution may be written in the form

$$Y_n(y) = A_n\cosh\frac{n\pi y}{a} + B_n\sinh\frac{n\pi y}{a} + \frac{2}{\pi n}\int_0^y F_n(\tau)\sinh\frac{n\pi}{a}(y-\tau)\,d\tau \tag{8.9.15}$$

The coefficients $A_n$ and $B_n$ are determined from the boundary conditions

$$\begin{aligned} Y_n'(0) &= \frac{2}{a}\int_0^a u_y(x,0)\cos\frac{n\pi x}{a}\,dx \\ &= \frac{2}{a}\int_0^a g_1(x)\cos\frac{n\pi x}{a}\,dx \end{aligned} \tag{8.9.16}$$

and

$$Y_n'(b) = \frac{2}{a}\int_0^a g_2(x)\cos\frac{n\pi x}{a}\,dx \tag{8.9.17}$$

For $n=0$, Eq. (8.9.14) takes the form

$$Y_0'' = \frac{2}{a}[f_1(y) - f_2(y)]$$

and hence

$$Y_0' = \frac{2}{a}\int_0^y [f_1(\tau) - f_2(\tau)]\,d\tau + C$$

where $C$ is an integration constant. Employing the condition (8.9.16) for $n=0$, we find

$$C = \frac{2}{a}\int_0^a g_1(x)\,dx$$

Thus, we have

$$Y_0'(y) = \frac{2}{a}\left\{\int_0^y [f_1(\tau) - f_2(\tau)]\,d\tau + \int_0^a g_1(x)\,dx\right\}$$

Consequently,

$$Y_0'(b) = \frac{2}{a}\left\{\int_0^b [f_1(\tau) - f_2(\tau)]\,d\tau + \int_0^a g_1(x)\,dx\right\}$$

Also from Eq. (8.9.16), we have

$$Y_0'(b) = \frac{2}{a}\int_0^a g_2(x)\,dx$$

It follows from these two expressions for $Y_0'(b)$ that

$$\int_0^b [f_1(y) - f_2(y)]\,dy + \int_0^a [g_1(x) - g_2(x)]\,dx = 0$$

which is the necessary condition for the existence of a solution to the Neumann problem for a rectangle.

## 8.10 EXERCISES

**1.** Reduce the Neumann problem to the Dirichlet problem in the two-dimensional case.

**2.** Reduce the wave equation

$$u_{tt} = c^2(u_{xx} + u_{yy} + u_{zz})$$

to the Laplace equation

$$u_{xx} + u_{yy} + u_{zz} + u_{\tau\tau} = 0$$

by letting $\tau = ict$ where $i = \sqrt{-1}$. Obtain the solution of the wave equation in cylindrical coordinates via the solution of the Laplace equation. Assume that $u(r, \theta, z, \tau)$ is independent of $z$.

**3.** Prove that, if $u(x, t)$ satisfies

$$u_t = ku_{xx}$$

for $0 \leq x \leq l$, $0 \leq t \leq t_0$, then the maximum value of $u$ is attained either at $t = 0$ or at the end points $x = 0$ or $x = l$ for $0 \leq t \leq t_0$. This is called the maximum principle for the heat equation.

**4.** Prove that a function which is harmonic everywhere on a plane and is bounded either above or below is a constant. This is called the Liouville theorem.

**5.** Show that the compatibility condition for the Neumann problem

$$\nabla^2 u = f \quad \text{in } D$$

$$\frac{\partial u}{\partial n} = g \quad \text{on } B$$

is

$$\int_D f\, dS + \int_B g\, ds = 0$$

where $B$ is the boundary of domain $D$.

**6.** Show that the second degree polynomial

$$P = Ax^2 + Bxy + Cy^2 + Dyz + Fz^2 + Fxz$$

is harmonic if

$$E = -(A + C)$$

and obtain

$$P = A(x^2 - z^2) + Bxy + C(y^2 - z^2) + Dyz + Fxz$$

**7.** Prove that a solution of the Neumann problem

$$\nabla^2 u = f \quad \text{in } D$$

$$u = g \quad \text{on } B$$

differs from another solution by at most a constant.

**8.** Determine the solution of each of the following problems:

a. $\nabla^2 u = 0, \qquad 1 < r < 2, \qquad 0 < \theta < \pi$

$u(1, \theta) = \sin \theta, \qquad 0 \leqslant \theta \leqslant \pi$

$u(2, \theta) = 0$

$u(r, 0) = 0$

$u(r, \pi) = 0$

b. $\nabla^2 u = 0, \qquad 1 < r < 2, \qquad 0 < \theta < \pi$

$u(1, \theta) = 0$

$u(2, \theta) = \theta(\theta - \pi), \qquad 0 \leqslant \theta \leqslant \pi$

$u(r, 0) = 0$

$u(r, \pi) = 0$

c. $\nabla^2 u = 0, \qquad 1 < r < 3, \qquad 0 < \theta < \pi/2$

$u(1, \theta) = 0$

$u(3, \theta) = 0$

$u(r, 0) = (r - 1)(r - 3), \qquad 1 \leqslant r \leqslant 3$

$u\left(r, \frac{\pi}{2}\right) = 0$

d. $$\nabla^2 u = 0, \qquad 1 < r < 3, \qquad 0 < \theta < \pi/2$$
$$u(1, \theta) = 0$$
$$u(3, \theta) = 0$$
$$u(r, 0) = 0$$
$$u\left(r, \frac{\pi}{2}\right) = f(r), \qquad 1 \leq r \leq 3$$

**9.** Solve

$$\nabla^2 u = 0, \qquad a < r < b, \qquad 0 < \theta < \alpha$$
$$u(a, \theta) = f(\theta), \qquad 0 \leq \theta \leq \alpha$$
$$u(b, \theta) = 0$$
$$u(r, 0) = f(r), \qquad a \leq r \leq b$$
$$u(r, \alpha) = 0$$

**10.** Verify directly that the Poisson integral is a solution of the Laplace equation.

**11.** Solve

$$\nabla^2 u = 0, \qquad 0 < r < a, \qquad 0 < \theta < \pi$$
$$u(r, 0) = 0$$
$$u(r, \pi) = 0$$
$$u(a, \theta) = \theta(\pi - \theta), \qquad 0 \leq \theta \leq \pi$$
$u(0, \theta)$ is bounded

**12.** Solve

$$\nabla^2 u + u = 0, \qquad 0 < r < a, \qquad 0 < \theta < \alpha$$
$$u(r, 0) = 0$$
$$u(r, \alpha) = 0$$
$$u(a, \theta) = f(\theta), \qquad 0 \leq \theta \leq \alpha$$
$u(0, \theta)$ is bounded

**13.** Find the solution of the Dirichlet problem

$$\nabla^2 u = -2, \qquad r < a, \qquad 0 < \theta < 2\pi$$
$$u(a, 0) = 0$$

**14.** Solve the following problems:

a. $$\nabla^2 u = 0, \qquad 1 < r < 2, \qquad 0 < \theta < 2\pi$$
$$u_r(1, \theta) = \sin\theta, \qquad 0 \leq \theta \leq 2\pi$$
$$u_r(2, \theta) = 0$$

b. $$\nabla^2 u = 0, \quad 1<r<2, \quad 0<\theta<2\pi$$
$$u_r(1, \theta) = 0$$
$$u_r(2, \theta) = \theta - \pi, \quad 0 \leqslant \theta \leqslant 2\pi$$

**15.** Solve

$$\nabla^2 u = 0, \quad a<r<b, \quad 0<\theta<2\pi$$
$$u_r(a, \theta) = f(\theta), \quad 0 \leqslant \theta \leqslant 2\pi$$
$$u_r(b, \theta) = g(\theta), \quad 0 \leqslant \theta \leqslant 2\pi$$

where

$$\int_{r=a} f\, ds + \int_{r=b} g\, ds = 0$$

**16.** Solve the Robin problem for a semicircular disk

$$\nabla^2 u = 0, \quad r<R, \quad 0<\theta<\pi$$
$$u_r(R, \theta) = \sin\theta, \quad 0 \leqslant \theta \leqslant \pi$$
$$u(r, 0) = 0$$
$$u(r, \pi) = 0$$

**17.** Solve

$$\nabla^2 u = 0, \quad a<r<b, \quad 0<\theta<\alpha$$
$$u_r(a, \theta) = 0$$
$$u_r(b, \theta) = f(\theta), \quad 0 \leqslant \theta \leqslant \alpha$$
$$u(r, 0) = 0$$
$$u(r, \alpha) = 0$$

**18.** Determine the solution of the mixed boundary-value problem

$$\nabla^2 u = 0, \quad r<R, \quad 0<\theta<2\pi$$
$$u_r(R, \theta) + hu(R, \theta) = f(\theta), \quad h = \text{constant}$$

**19.** Solve

$$\nabla^2 u = 0, \quad a<r<b, \quad 0<\theta<2\pi$$
$$u_r(a, \theta) + hu(a, \theta) = f(\theta)$$
$$u_r(b, \theta) + hu(b, \theta) = g(\theta)$$

**20.** Find a solution of the Neumann problem

$$\nabla^2 u = -r^2 \sin 2\theta, \quad r_1<r<r_2, \quad 0<\theta<2\pi$$
$$u_r(r_1, \theta) = 0$$
$$u_r(r_2, \theta) = 0$$

**21.** Solve the Robin problem

$$\nabla^2 u = -r^2 \sin 2\theta$$
$$u(r_1, \theta) = 0$$
$$u_r(r_2, \theta) = 0$$

**22.** Solve the following Dirichlet problems:

a. $\nabla^2 u = 0, \quad 0 < x < 1, \quad 0 < y < 1$

$u(x, 0) = x(x-1), \quad 0 \leq x \leq 1$

$u(x, 1) = 0$

$u(0, y) = 0$

$u(1, y) = 0$

b. $\nabla^2 u = 0, \quad 0 < x < 1, \quad 0 < y < 1$

$u(x, 0) = 0$

$u(x, 1) = \sin \pi x, \quad 0 \leq x \leq 1$

$u(0, y) = 0$

$u(1, y) = 0$

c. $\nabla^2 u = 0, \quad 0 < x < 1, \quad 0 < y < 1$

$u(x, 0) = 0$

$u(x, 1) = 0$

$u(0, y) = \left(\cos \frac{\pi y}{2} - 1\right) \cos \frac{\pi y}{2}, \quad 0 \leq y \leq 1$

$u(1, y) = 0$

d. $\nabla^2 u = 0, \quad 0 < x < 1, \quad 0 < y < 1$

$u(x, 0) = 0$

$u(x, 1) = 0$

$u(0, y) = 0$

$u(1, y) = \sin \pi y \cos \pi y, \quad 0 \leq y \leq 1$

**23.** Solve the following Neumann problems:

a. $\nabla^2 u = 0, \quad 0 < x < \pi, \quad 0 < y < \pi$

$u_x(0, y) = \left(y - \frac{\pi}{2}\right), \quad 0 \leq y \leq \pi$

$u_x(\pi, y) = 0$

$u_y(x, 0) = x$

$u_y(x, \pi) = x$

b. $\nabla^2 u = 0, \qquad 0 < x < \pi, \qquad 0 < y < \pi$

$$u_x(0, y) = 0$$
$$u_x(\pi, y) = 2\cos y, \qquad 0 \leq y \leq \pi$$
$$u_y(x, 0) = 0$$
$$u_y(x, \pi) = 0$$

c. $\nabla^2 u = 0, \qquad 0 < x < \pi, \qquad 0 < y < \pi$

$$u_x(0, y) = 0$$
$$u_x(\pi, y) = 0$$
$$u_y(x, 0) = \cos x, \qquad 0 \leq x \leq \pi$$
$$u_y(x, \pi) = 0$$

d. $\nabla^2 u = 0, \qquad 0 < x < \pi, \qquad 0 < y < \pi$

$$u_x(0, y) = y$$
$$u_x(\pi, y) = y$$
$$u_y(x, 0) = x$$
$$u_y(x, \pi) = x$$

**24.** The steady-state temperature distribution in a rectangular plate of length $a$ and width $b$ is described by

$$\nabla^2 u = 0, \qquad 0 < x < a, \qquad 0 < y < b$$

At $x = 0$, the temperature is kept at zero degrees, while at $x = a$, the plate is insulated. The temperature is prescribed at $y = 0$. At $y = b$, heat is allowed to radiate freely into the surrounding medium of zero degree temperature. That is,

$$u(0, y) = 0$$
$$u_x(a, y) = 0$$
$$u(x, 0) = f(x), \qquad 0 \leq x \leq a$$
$$u_y(x, b) + hu(x, b) = 0$$

Determine the temperature distribution.

**25.** Solve the Dirichlet problem

$$\nabla^2 u = -2y, \qquad 0 < x < 1, \qquad 0 < y < 1$$
$$u(0, y) = 0$$
$$u(1, y) = 0$$
$$u(x, 0) = 0$$
$$u(x, 1) = 0$$

**26.** Find the harmonic function which vanishes on the hypotenuse and has prescribed values on the other two sides of an isosceles right-angled triangle formed by $x = 0$, $y = 0$, and $y = a - x$ where $a =$ constant.

**27.** Find a solution of the Neumann problem

$$\nabla^2 u = x^2 - y^2, \qquad 0 < x < a, \qquad 0 < y < a$$
$$u_x(0, y) = 0$$
$$u_x(a, y) = 0$$
$$u_y(x, 0) = 0$$
$$u_y(x, a) = 0$$

**28.** Solve the third boundary value problem

$$\nabla^2 u = 0, \qquad 0 < x < 1, \qquad 0 < y < 1$$
$$u_x(0, y) + hu(0, y) = 0, \qquad h = \text{constant}$$
$$u_x(1, y) + hu(1, y) = 0$$
$$u_y(x, 0) + hu(x, 0) = 0$$
$$u_y(x, 1) + hu(x, 1) = f(x), \qquad 0 \leqslant x \leqslant 1$$

**29.** Determine the solution of

$$\nabla^2 u = 1, \qquad 0 < x < \pi, \qquad 0 < y < \pi$$
$$u(0, y) = 0$$
$$u_x(\pi, y) = 0$$
$$u_y(x, 0) = 0$$
$$u_y(x, \pi) + hu(x, \pi) = f(x), \qquad 0 \leqslant x \leqslant \pi$$

**30.** Obtain the integral representation of the Neumann problem

$$\nabla^2 u = f \quad \text{in } D$$
$$\frac{\partial u}{\partial n} = g \quad \text{on } B$$

**31.** Find the solution in terms of Green's function of

$$\nabla^2 u = f \quad \text{in } D$$
$$\frac{\partial u}{\partial n} + hu = g \quad \text{on } B$$

CHAPTER NINE

# Higher Dimensional Boundary-Value Problems

The treatment of problems in more than two space variables is much more involved than problems in two space variables. Here a number of multi-dimensional problems with various boundary conditions will be presented.

## 9.1 DIRICHLET PROBLEM FOR A CUBE

The steady-state temperature distribution in a cube is described by the Laplace equation

$$\nabla^2 u = u_{xx} + u_{yy} + u_{zz} = 0 \tag{9.1.1}$$

for $0 < x < \pi$, $0 < y < \pi$, $0 < z < \pi$. The faces are kept at zero degree temperature except for the face $z = 0$, that is,

$$\begin{aligned} &u(0, y, z) = u(\pi, y, z) = 0 \\ &u(x, 0, z) = u(x, \pi, z) = 0 \\ &u(x, y, \pi) = 0 \\ &u(x, y, 0) = f(x, y), \qquad 0 \leq x \leq \pi, \qquad 0 \leq y \leq \pi \end{aligned}$$

By the method of separation of variables, we assume a nontrivial solution in the form

$$u(x, y, z) = X(x)Y(y)Z(z)$$

Substituting this in the Laplace equation, we obtain

$$X''YZ + XY''Z + XYZ'' = 0$$

Division by $XYZ$ yields

$$\frac{X''}{X}+\frac{Y''}{Y}=-\frac{Z''}{Z}$$

Since the right side depends only on $z$ and the left side is independent of $z$, both terms must be equal to a constant. Thus we have

$$\frac{X''}{X}+\frac{Y''}{Y}=-\frac{Z''}{Z}=\lambda$$

By the same reasoning, we have

$$\frac{X''}{X}=\lambda-\frac{Y''}{Y}=\mu$$

Hence we obtain the equations

$$X''-\mu X=0$$
$$Y''-(\lambda-\mu)Y=0$$
$$Z''+\lambda Z=0$$

With use of the boundary conditions, the eigenvalue problem for $X$

$$X''-\mu X=0$$
$$X(0)=X(\pi)=0$$

yields the eigenvalues $\mu=-m^2$ for $m=1, 2, 3, \ldots$ and the corresponding eigenfunctions $\sin mx$.

Similarly, the eigenvalue problem for $Y$

$$Y''-(\lambda-\mu)Y=0$$
$$Y(0)=Y(\pi)=0$$

gives the eigenvalues $\lambda-\mu=-n^2$ where $n=1, 2, 3, \ldots$ and the corresponding eigenfunctions $\sin ny$.

Since $\lambda$ is given by $-(m^2+n^2)$, it follows that the solution of $Z''+\lambda Z=0$ satisfying the condition $Z(\pi)=0$ is

$$Z(z)=C\sinh\left[\sqrt{m^2+n^2}\,(\pi-z)\right]$$

Thus, the solution of the Laplace equation satisfying the homogeneous boundary conditions takes the form

$$u(x, y, z)=\sum_{m=1}^{\infty}\sum_{n=1}^{\infty} a_{mn}\sinh\sqrt{m^2+n^2}\,(\pi-z)\sin mx\sin ny$$

Applying the nonhomogeneous boundary condition, we formally obtain

$$f(x, y)=\sum_{m=1}^{\infty}\sum_{n=1}^{\infty} a_{mn}\sinh\left(\sqrt{m^2+n^2}\,\pi\right)\sin mx\sin ny$$

The coefficient of the double Fourier series is thus given by

$$a_{mn} \sinh \sqrt{m^2+n^2}\,\pi = \frac{4}{\pi^2}\int_0^{\pi}\int_0^{\pi} f(x, y) \sin mx \sin ny\, dx\, dy$$

Therefore, the formal solution to the Dirichlet problem for a cube may be written in the form

$$u(x, y, z) = \sum_{m=1}^{\infty}\sum_{n=1}^{\infty} b_{mn} \frac{\sinh \sqrt{m^2+n^2}\,(\pi - z)}{\sinh \sqrt{m^2+n^2}\,\pi} \sin mx \sin ny \tag{9.1.2}$$

where

$$b_{mn} = a_{mn} \sinh \sqrt{m^2+n^2}\,\pi$$

## 9.2 DIRICHLET PROBLEM FOR A CYLINDER

EXAMPLE 9.2.1. We consider the problem of determining the electric potential $u$ inside a charge-free cylinder. The potential $u$ satisfies the Laplace equation

$$\nabla^2 u = u_{rr} + \frac{1}{r} u_r + \frac{1}{r^2} u_{\theta\theta} + u_{zz} = 0 \tag{9.2.1}$$

for $0 \leqslant r < a$, $0 < z < l$. Let the lateral surface $r = a$ and the top $z = l$ be grounded, that is, at zero potential. Let the potential on the base $z = 0$ be given by

$$u(r, \theta, 0) = f(r, \theta) \tag{9.2.2}$$

where $f(a, \theta) = 0$.

We assume a solution in the form

$$u(r, \theta, z) = R(r)\Theta(\theta)Z(z)$$

Substitution of this in the Laplace equation yields

$$\frac{R'' + \frac{1}{r} R'}{R} + \frac{1}{r^2}\frac{\Theta''}{\Theta} = -\frac{Z''}{Z} = \lambda$$

It follows that

$$\frac{r^2 R'' + rR'}{R} - r^2\lambda = -\frac{\Theta''}{\Theta} = \mu$$

Thus, we obtain the equations

$$r^2 R'' + rR' - (\lambda r^2 + \mu)R = 0 \tag{9.2.3}$$

$$\Theta'' + \mu\Theta = 0 \tag{9.2.4}$$

$$Z'' + \lambda Z = 0 \tag{9.2.5}$$

Using the periodicity conditions, the eigenvalue problem for $\Theta(\theta)$

$$\Theta'' + \mu\Theta = 0$$
$$\Theta(0) = \Theta(2\pi)$$
$$\Theta'(0) = \Theta'(2\pi)$$

yields the eigenvalues $\mu = n^2$ for $n = 0, 1, 2, \ldots$ with the corresponding eigenfunctions $\sin n\theta$, $\cos n\theta$. Thus

$$\Theta(\theta) = A \cos n\theta + B \sin n\theta \tag{9.2.6}$$

Suppose $\lambda$ is real and negative and let $\lambda = -\beta^2$ where $\beta > 0$. If the condition $Z(l) = 0$ is imposed, then the solution of Eq. (9.2.5) can be written in the form

$$Z(z) = C \sinh \beta(l - z) \tag{9.2.7}$$

Next we introduce the new independent variable $\xi = \beta r$. Equation (9.2.3) transforms into

$$\xi^2 \frac{d^2R}{d\xi^2} + \xi \frac{dR}{d\xi} + (\xi^2 - n^2)R = 0$$

which is the Bessel equation of order $n$. The general solution is

$$R_n(\xi) = DJ_n(\xi) + EY_n(\xi)$$

where $J_n$ and $Y_n$ are the Bessel functions of the first and second kind respectively. In terms of the original variable, we have

$$R_n(r) = DJ_n(\beta r) + EY_n(\beta r)$$

Since $Y_n(\beta r)$ is unbounded at $r = 0$, we choose $E = 0$. The condition $R(a) = 0$ requires that

$$J_n(\beta a) = 0$$

For each $n \geqslant 0$, there exist positive zeros. Arranging these in an infinite increasing sequence, we have

$$0 < \alpha_{n1} < \alpha_{n2} < \ldots < \alpha_{nm} < \ldots$$

Thus, we obtain

$$\beta_{nm} = \alpha_{nm}/a$$

Consequently

$$R_n(r) = DJ_n(\alpha_{nm} r/a)$$

The solution $u$ then finally takes the form

$$u(r, \theta, z) = \sum_{n=0}^{\infty} \sum_{m=1}^{\infty} J_n\left(\alpha_{nm} \frac{r}{a}\right)(a_{nm} \cos n\theta + b_{nm} \sin n\theta) \sinh \alpha_{nm} \frac{(l - z)}{a}$$

To satisfy the nonhomogeneous boundary condition, it is required that

$$f(r, \theta) = \sum_{n=0}^{\infty} \sum_{m=1}^{\infty} J_n\left(\alpha_{nm}\frac{r}{a}\right)(a_{nm}\cos n\theta + b_{nm}\sin n\theta)\sinh \alpha_{nm}\frac{l}{a}$$

The coefficients $a_{nm}$ and $b_{nm}$ are given by

$$a_{0m} = \frac{1}{\pi a^2 \sinh\left(\alpha_{0m}\frac{1}{a}\right)[J_1(\alpha_{0m})]^2} \int_0^a \int_0^{2\pi} f(r, \theta)\, J_0\left(\alpha_{0m}\frac{r}{a}\right) r\, dr\, d\theta$$

$$a_{nm} = \frac{2}{\pi a^2 \sinh\left(\alpha_{nm}\frac{1}{a}\right)[J_{n+1}(\alpha_{nm})]^2} \int_0^a \int_0^{2\pi} f(r, \theta) J_n\left(\alpha_{nm}\frac{r}{a}\right) \cos n\theta\, r\, dr\, d\theta$$

$$b_{nm} = \frac{2}{\pi a^2 \sinh\left(\alpha_{nm}\frac{1}{a}\right)[J_{n+1}(\alpha_{nm})]^2} \int_0^a \int_0^{2\pi} f(r, \theta) J_n\left(\alpha_{nm}\frac{r}{a}\right) \sin n\theta\, r\, dr\, d\theta$$

EXAMPLE 9.2.2. We shall illustrate the same problem with different boundary conditions. Consider the problem

$$\nabla^2 u = 0, \qquad 0 \le r < a, \qquad 0 < z < \pi$$
$$u(r, \theta, 0) = 0$$
$$u(r, \theta, \pi) = 0$$
$$u(a, \theta, z) = f(\theta, z)$$

As before, by the separation of variables we obtain

$$r^2 R'' + rR' - (\lambda r^2 + \mu)R = 0$$
$$\Theta'' + \mu\Theta = 0$$
$$Z'' + \lambda Z = 0$$

By the periodicity conditions, again as in the previous example, the $\Theta$ equation yields the eigenvalues $\mu = n^2$ with $n = 0, 1, 2, \ldots$; the corresponding eigenfunctions are $\sin n\theta$, $\cos n\theta$. Thus, we have

$$\Theta(\theta) = A_n \cos n\theta + B_n \sin n\theta$$

Now let $\lambda = \beta^2$ with $\beta > 0$. Then the problem

$$Z'' + \beta^2 Z = 0$$
$$Z(0) = 0, \qquad Z(\pi) = 0$$

has the solution

$$Z(z) = C_m \sin mz, \qquad m = 1, 2, 3, \ldots$$

Finally, we have

$$r^2R'' + rR' - (m^2r^2 + n^2)R = 0$$

or

$$R'' + \frac{1}{r}R' - \left(m^2 + \frac{n^2}{r^2}\right)R = 0$$

the general solution of which is

$$R(r) = DI_n(mr) + EK_n(mr)$$

where $I_n$ and $K_n$ are the modified Bessel functions of the first and second kind, respectively.

Since $R$ must remain finite at $r = 0$, we set $E = 0$. Then $R$ appears in the form

$$R(r) = DI_n(mr)$$

Applying the nonhomogeneous condition, we find the solution to be

$$u(r, \theta, z) = \sum_{m=1}^{\infty} \frac{a_{m0}}{2} \frac{I_0(mr)}{I_0(ma)} \sin mz$$

$$+ \sum_{m=1}^{\infty} \sum_{n=1}^{\infty} (a_{mn} \cos n\theta + b_{mn} \sin n\theta) \frac{I_n(mr)}{I_n(ma)} \sin mz$$

where

$$a_{mn} = \frac{2}{\pi^2} \int_0^{\pi} \int_0^{2\pi} f(\theta, z) \sin mz \cos n\theta \, d\theta \, dz$$

$$b_{mn} = \frac{2}{\pi^2} \int_0^{\pi} \int_0^{2\pi} f(\theta, z) \sin mz \sin n\theta \, d\theta \, dz$$

## 9.3 DIRICHLET PROBLEM FOR A SPHERE

EXAMPLE 9.3.1. To determine the potential in a sphere, we transform the Laplace equation in spherical coordinates. It has the form

$$\nabla^2 u = u_{rr} + \frac{2}{r} u_r + \frac{1}{r^2} u_{\theta\theta} + \frac{\cot \theta}{r^2} u_\theta + \frac{1}{r^2 \sin^2 \theta} u_{\varphi\varphi} = 0 \qquad (9.3.1)$$

$0 \leqslant r < a$, $0 < \theta < \pi$, $0 < \varphi < 2\pi$.

Let the prescribed potential on the sphere be

$$u(a, \theta, \varphi) = f(\theta, \varphi) \qquad (9.3.2)$$

By the method of separation of variables, we assume a solution in the form

$$u(r, \theta, \varphi) = R(r)\Theta(\theta)\Phi(\varphi)$$

Substitution of $u$ in the Laplace equation yields

$$r^2R'' + 2rR' - \lambda R = 0 \tag{9.3.3}$$

$$\sin^2 \theta\, \Theta'' + \sin \theta \cos \theta\, \Theta' + (\lambda \sin^2 \theta - \mu)\Theta = 0 \tag{9.3.4}$$

$$\Phi'' + \mu\Phi = 0 \tag{9.3.5}$$

The general solution of Eq. (9.3.5) is

$$\Phi(\varphi) = A \cos \sqrt{\mu}\, \varphi + B \sin \sqrt{\mu}\, \varphi \tag{9.3.6}$$

The periodicity condition requires that

$$\sqrt{\mu} = m, \qquad m = 0, 1, 2, \ldots$$

Since Eq. (9.3.3) is of the Euler type, the solution is of the form

$$R(r) = r^\beta$$

Inserting this in Eq. (9.3.3), we obtain

$$\beta^2 + \beta - \lambda = 0$$

The roots are $\beta = (-1 + \sqrt{1 + 4\lambda})/2$ and $-(1 + \beta)$. Hence the general solution of Eq. (9.3.3) is

$$R(r) = Cr^\beta + Dr^{-(1+\beta)} \tag{9.3.7}$$

The variable $\xi = \cos \theta$ transforms Eq. (9.3.4) into

$$(1 - \xi^2)\Theta'' - 2\xi\Theta' + \left[\beta(\beta + 1) - \frac{m^2}{1 - \xi^2}\right]\Theta = 0 \tag{9.3.8}$$

which is the Legendre's associated equation. The general solution with $\beta = n$ for $n = 0, 1, 2, \ldots$ is

$$\Theta(\theta) = EP_n^m(\cos \theta) + FQ_n^m(\cos \theta)$$

Continuity of $\Theta(\theta)$ at $\theta = 0, \pi$ corresponds to continuity of $\Theta(\xi)$ at $\xi = \pm 1$. Since $Q_n^m(\xi)$ has a logarithmic singularity at $\xi = 1$, we choose $F = 0$. Thus, the solution of Eq. (9.3.8) becomes

$$\Theta(\theta) = EP_n^m(\cos \theta)$$

Consequently, the solution of the Laplace equation in spherical coordinates is

$$u(r, \theta, \varphi) = \sum_{n=0}^{\infty} \sum_{m=0}^{n} r^n P_n^m(\cos \theta)[a_{nm} \cos m\varphi + b_{nm} \sin m\varphi]$$

In order for $u$ to satisfy the prescribed function on the boundary, it is necessary that

$$f(\theta, \varphi) = \sum_{n=0}^{\infty} \sum_{m=0}^{n} a^n P_n^m(\cos \theta)[a_{nm} \cos m\varphi + b_{nm} \sin m\varphi]$$

for $0 \leqslant \theta \leqslant \pi$, $0 \leqslant \varphi \leqslant 2\pi$. By the orthogonal properties of the functions $P_n^m(\cos\theta)\cos m\varphi$ and $P_n^m(\cos\theta)\sin m\varphi$, the coefficients are given by

$$a_{nm} = \frac{(2n+1)}{2\pi a^n}\frac{(n-m)!}{(n+m)!}\int_0^{2\pi}\int_0^{\pi} f(\theta, \varphi)P_n^m(\cos\theta)\cos m\varphi \sin\theta \, d\theta \, d\varphi$$

$$b_{nm} = \frac{(2n+1)}{2\pi a^n}\frac{(n-m)!}{(n+m)!}\int_0^{2\pi}\int_0^{\pi} f(\theta, \varphi)P_n^m(\cos\theta)\sin m\varphi \sin\theta \, d\theta \, d\varphi$$

for $m = 1, 2, \ldots$ and $n = 1, 2, \ldots$, and

$$a_{n0} = \frac{2n+1}{4\pi a^n}\int_0^{2\pi}\int_0^{\pi} f(\theta, \varphi)P_n(\cos\theta)\sin\theta \, d\theta \, d\varphi$$

for $n = 0, 1, 2, \ldots$.

EXAMPLE 9.3.2. To determine the potential of a grounded conducting sphere in a uniform field, we are required to solve

$$\nabla^2 u = 0, \qquad 0 \leqslant r < a, 0 < \theta < \pi, 0 < \phi < 2\pi$$

$$u(a, \theta) = 0$$

$$u \to -E_0 r \cos\theta \qquad \text{as } r \to \infty$$

Let the field be in the $z$ direction so that the potential $u$ will be independent of $\phi$. Then the Laplace equation takes the form

$$u_{rr} + \frac{2}{r}u_r + \frac{1}{r^2}u_{\theta\theta} + \frac{\cot\theta}{r^2}u_\theta = 0$$

We assume a solution in the form

$$u(r, \theta) = R(r)\Theta(\theta)$$

Substitution of this in the Laplace equation yields

$$r^2R'' + 2rR' - \lambda R = 0$$

$$\sin^2\theta\,\Theta'' + \sin\theta\cos\theta\,\Theta' + \lambda\sin^2\theta\,\Theta = 0$$

If we set $\lambda = n(n+1)$, with $n = 0, 1, 2, \ldots$, then the second equation is the Legendre equation. The general solution of this equation is

$$\Theta(\theta) = A_n P_n(\cos\theta) + B_n Q_n(\cos\theta)$$

where $P_n$ and $Q_n$ are the Legendre functions of the first and second kind respectively. In order for the solution not to be singular at $\theta = 0$ and $\theta = \pi$, we set $B_n = 0$. Thus, $\Theta(\theta)$ becomes

$$\Theta(\theta) = A_n P_n(\cos\theta)$$

The solution of the $R$-equation is obtained in the form

$$R(r) = C_n r^n + D_n r^{-(n+1)}$$

Thus, the potential function is

$$u(r, \theta) = \sum_{n=0}^{\infty} (a_n r^n + b_n r^{-(n+1)}) P_n(\cos \theta)$$

To satisfy the condition at infinity, we must have

$$a_1 = -E_0 \qquad \text{and } a_n = 0 \qquad \text{for } n \geqslant 2$$

and hence

$$u(r, \theta) = -E_0 r \cos \theta + \sum_{n=1}^{\infty} \frac{b_n}{r^{n+1}} P_n(\cos \theta)$$

The condition $u(a, \theta) = 0$ yields

$$0 = -E_0 a \cos \theta + \sum_{n=1}^{\infty} \frac{b_n}{a^{n+1}} P_n(\cos \theta)$$

Using the orthogonality of the Legendre functions, we find that $b_n$ are given by

$$\begin{aligned} b_n &= \frac{2n+1}{2} E_0 a^{n+2} \int_{-\pi}^{\pi} \cos \theta P_n(\cos \theta) d(\cos \theta) \\ &= E_0 a^3 \delta_{n1} \end{aligned}$$

since the integral vanishes for all $n$ except $n = 1$. Hence the potential function is given by

$$u(r, \theta) = -E_0 r \cos \theta + E_0 \frac{a^3}{r^2} \cos \theta$$

EXAMPLE 9.3.3. A dielectric sphere of radius $a$ is placed in a uniform electric field $E_0$. Determine the potentials inside and outside the sphere.

The problem is

$$\nabla^2 u_1 = \nabla^2 u_2 = 0$$

$$K \frac{\partial u_1}{\partial r} = \frac{\partial u_2}{\partial r} \qquad \text{on } r = a$$

$$u_1 = u_2 \qquad \text{on } r = a$$

$$u_2 \to -E_0 r \cos \theta \text{ as } r \to \infty$$

where $u_1$ and $u_2$ are the potentials inside and outside the sphere respectively, and $K$ is the dielectric constant.

As in the preceding example, the potential function is

$$u(r, \theta) = \sum_{n=0}^{\infty} (a_n r^n + b_n r^{-(n+1)}) P_n(\cos \theta) \tag{9.3.9}$$

Since $u_1$ must be finite at the origin, we take

$$u_1(r, \theta) = \sum_{n=0}^{\infty} a_n r^n P_n(\cos \theta) \qquad \text{for} \quad r \leqslant a \tag{9.3.10}$$

For $u_2$, which must approach infinity in the prescribed manner, we choose

$$u_2(r, \theta) = -E_0 r \cos \theta + \sum_{n=0}^{\infty} b_n r^{-(n+1)} P_n(\cos \theta) \tag{9.3.11}$$

From the two continuity conditions at $r = a$, we obtain

$$a_1 = -E_0 + \frac{b_1}{a^3}$$

$$Ka_1 = -E_0 - \frac{2b_1}{a^3}$$

$$a_n = b_n = 0, \qquad n \geqslant 2$$

The coefficients $a_1$ and $b_1$ are then found to be

$$a_1 = -\frac{3E_0}{K+2}, \qquad b_1 = E_0 a^3 \frac{(K-1)}{(K+2)}$$

Hence the potential for $r \leqslant a$ is given by

$$u_1(r, \theta) = -\frac{3E_0}{K+2} r \cos \theta$$

and the potential for $r \geqslant a$ is given by

$$u_2(r, \theta) = -E_0 r \cos \theta + E_0 a^3 \frac{(K-1)}{(K+2)} r^{-2} \cos \theta$$

EXAMPLE 9.3.4. Determine the potential between concentric spheres held at different constant potentials.

Here we need to solve

$$\nabla^2 u = 0, \qquad a < r < b$$

$$u = A \quad \text{on} \quad r = a$$

$$u = B \quad \text{on} \quad r = b$$

In this case, the potential depends only on the radial distance. Hence we have

$$\frac{1}{r^2} \frac{\partial}{\partial r} \left( r^2 \frac{\partial u}{\partial r} \right) = 0$$

By elementary integration, we obtain

$$u(r) = c_1 + \frac{c_2}{r}$$

Applying the boundary conditions, we obtain

$$c_1 = \frac{Bb - Aa}{b - a}$$

$$c_2 = (A - B)\frac{ab}{b - a}$$

Thus, the solution is

$$u(r) = \frac{Bb - Aa}{(b - a)} + \frac{(A - B)ab}{(b - a)r}$$

$$= \frac{Bb}{r}\frac{r - a}{b - a} + \frac{Aa}{r}\frac{b - r}{b - a}$$

## 9.4 WAVE AND HEAT EQUATIONS

The wave equation in three space variables may be written

$$u_{tt} = c^2\nabla^2 u \tag{9.4.1}$$

where $\nabla^2$ is the Laplace operator.

By the method of separation of variables, we assume a solution in the form

$$u(x, y, z, t) = U(x, y, z)T(t)$$

Substituting this in Eq. (9.4.1) we obtain

$$T'' + \lambda c^2 T = 0 \tag{9.4.2}$$

$$\nabla^2 U + \lambda U = 0 \tag{9.4.3}$$

where $-\lambda$ is a separation constant. The variables are separated and the solutions of Eqs. (9.4.2) and (9.4.3) are to be determined.

Next we consider the heat equation

$$u_t = k\nabla^2 u \tag{9.4.4}$$

As before, we seek a solution in the form

$$u(x, y, z, t) = U(x, y, z)T(t)$$

Substitution of this in Eq. (9.4.4) yields

$$T' + \lambda kT = 0$$

$$\nabla^2 U + \lambda U = 0$$

Thus, we see that the problem here, as in the previous case, is essentially that of solving the Helmholtz equation

$$\nabla^2 U + \lambda U = 0$$

## 9.5 VIBRATING MEMBRANE

As a specific example of the higher dimensional wave equation, let us determine the solution of the problem of the vibrating membrane of length $a$ and width $b$. The initial-boundary value problem for the displacement function $u(x, y, t)$ is

$$u_{tt} = c^2(u_{xx} + u_{yy}), \quad 0 < x < a, \quad 0 < y < b, \quad t > 0 \tag{9.5.1}$$

$$u(x, y, 0) = f(x, y), \quad 0 \leqslant x \leqslant a, \quad 0 \leqslant y \leqslant b \tag{9.5.2}$$

$$u_t(x, y, 0) = g(x, y), \quad 0 \leqslant x \leqslant a, \quad 0 \leqslant y \leqslant b \tag{9.5.3}$$

$$u(0, y, t) = 0 \tag{9.5.4}$$

$$u(a, y, t) = 0 \tag{9.5.5}$$

$$u(x, 0, t) = 0 \tag{9.5.6}$$

$$u(x, b, t) = 0 \tag{9.5.7}$$

We have just shown that the separated equations for the wave equation are

$$T'' + \lambda c^2 T = 0 \tag{9.5.8}$$

$$\nabla^2 U + \lambda U = 0 \tag{9.5.9}$$

where, in this case, $\nabla^2 U = U_{xx} + U_{yy}$. Let $\lambda = \alpha^2$. Then the solution of Eq. (9.5.8) is

$$T(t) = A \cos \alpha ct + B \sin \alpha ct$$

Now we look for a solution of Eq. (9.5.9) in the form

$$U(x, y) = X(x)Y(y)$$

Substitution of this into Eq. (9.5.9) yields

$$X'' - \mu X = 0$$

$$Y'' + (\lambda + \mu)Y = 0$$

If we let $\mu = -\beta^2$, then the solutions of these equations take the form

$$X(x) = C \cos \beta x + D \sin \beta x$$

$$Y(y) = E \cos \gamma y + F \sin \gamma y$$

where

$$\gamma^2 = (\lambda + \mu) = \alpha^2 - \beta^2$$

The homogeneous boundary conditions in $x$ require that $C = 0$ and

$$D \sin \beta a = 0$$

which implies that $\beta = m\pi/a$ with $D \neq 0$. Similarly, the homogeneous boundary conditions in $y$ require that $E = 0$ and

$$F \sin \gamma b = 0$$

which implies that $\gamma = n\pi/b$ with $F \neq 0$. Noting that $m$ and $n$ are independent integers, we obtain the displacement function in the form

$$u(x, y, t) = \sum_{m=1}^{\infty} \sum_{n=1}^{\infty} (a_{mn} \cos \alpha_{mn} ct + b_{mn} \sin \alpha_{mn} ct) \sin \frac{m\pi x}{a} \sin \frac{n\pi y}{b} \tag{9.5.10}$$

where $\alpha_{mn} = m^2\pi^2/a^2 + n^2\pi^2/b^2$, $a_{mn}$ and $b_{mn}$ are constants.

Now applying the nonhomogeneous initial conditions, we have

$$u(x, y, 0) = f(x, y) = \sum_{m=1}^{\infty} \sum_{n=1}^{\infty} a_{mn} \sin \frac{m\pi x}{a} \sin \frac{n\pi y}{b}$$

and thus

$$a_{mn} = \frac{4}{ab} \int_0^a \int_0^b f(x, y) \sin \frac{m\pi x}{a} \sin \frac{n\pi y}{b} \, dx \, dy$$

In a similar manner, the initial condition on $u_t$ implies

$$u_t(x, y, 0) = g(x, y) = \sum_{m=1}^{\infty} \sum_{n=1}^{\infty} b_{mn} \alpha_{mn} c \sin \frac{m\pi x}{a} \sin \frac{n\pi y}{b}$$

from which it follows that

$$b_{mn} = \frac{4}{\alpha_{mn} abc} \int_0^a \int_0^b g(x, y) \sin \frac{m\pi x}{a} \sin \frac{n\pi y}{b} \, dx \, dy$$

The solution of the rectangular membrane problem is, therefore, given by Eq. (9.5.10).

## 9.6 HEAT FLOW IN A RECTANGULAR PLATE

Another example of a two-dimensional problem is the conduction of heat in a thin rectangular plate. Let the plate of length $a$ and width $b$ be perfectly insulated at the faces $x = 0$ and $x = a$. Let the two other sides be maintained at zero temperature. Let the initial temperature distribution be $f(x, y)$. Then we seek the solution of the initial-boundary value problem

$$u_t = k\nabla^2 u, \qquad 0<x<a, \quad 0<y<b, \quad t>0 \tag{9.6.1}$$
$$u(x, y, 0) = f(x, y), \quad 0\leqslant x\leqslant a, \quad 0\leqslant y\leqslant b \tag{9.6.2}$$
$$u_x(0, y, t) = 0 \tag{9.6.3}$$
$$u_x(a, y, t) = 0 \tag{9.6.4}$$
$$u(x, 0, t) = 0 \tag{9.6.5}$$
$$u(x, b, t) = 0 \tag{9.6.6}$$

As shown earlier, the separated equations for this problem are found to be

$$T' + \lambda k T = 0 \tag{9.6.7}$$
$$\nabla^2 U + \lambda U = 0 \tag{9.6.8}$$

We assume a solution in the form

$$U(x, y) = X(x)Y(y)$$

Inserting this in Eq. (9.6.8), we obtain

$$X'' - \mu X = 0 \tag{9.6.9}$$
$$Y'' + (\lambda + \mu)Y = 0 \tag{9.6.10}$$

Because the conditions in $x$ are homogeneous, we choose $\mu = -\alpha^2$ so that

$$X(x) = A \cos \alpha x + B \sin \alpha x$$

Since $X'(0) = 0$, $B = 0$ and since $X'(a) = 0$

$$\sin \alpha a = 0, \qquad A \neq 0$$

which gives

$$\alpha = m\pi/a, \qquad m = 1, 2, 3, \ldots$$

We note that $\mu = 0$ is also an eigenvalue. Consequently,

$$X_m(x) = A_m \cos(m\pi x/a), \qquad m = 0, 1, 2, \ldots$$

Similarly, for nontrivial solution $Y$, we select $\beta^2 = \lambda + \mu = \lambda - \alpha^2$ so that the solution of Eq. (9.6.10) is

$$Y(y) = C \cos \beta y + D \sin \beta y$$

Applying the homogeneous conditions, we find $C = 0$ and

$$\sin \beta b = 0, \qquad D \neq 0$$

Thus, we obtain

$$\beta = n\pi/b, \qquad n = 1, 2, 3, \ldots$$

and

$$Y_n(y) = D_n \sin (n\pi y/b)$$

Recalling that $\lambda = \alpha^2 + \beta^2$, the solution of Eq. (9.6.7) may be written in the form

$$T_{mn}(t) = E_{mn} e^{-(m^2/a^2+n^2/b^2)\pi^2 kt}$$

Thus, the solution of the heat equation satisfying the prescribed boundary conditions may be written as

$$u(x, y, t) = \sum_{m=0}^{\infty} \sum_{n=1}^{\infty} a_{mn} e^{-(m^2/a^2+n^2/b^2)\pi^2 kt} \cos \frac{m\pi x}{a} \sin \frac{n\pi y}{b} \qquad (9.6.11)$$

where $a_{mn} = A_m D_m E_{mn}$ is an arbitrary constant.

Applying the initial condition, we obtain

$$u(x, y, 0) = f(x, y) = \sum_{m=0}^{\infty} \sum_{n=1}^{\infty} a_{mn} \cos \frac{m\pi x}{a} \sin \frac{n\pi y}{b}$$

This is a double Fourier series and the coefficients given by

$$a_{0n} = \frac{2}{ab} \int_0^a \int_0^b f(x, y) \sin \frac{n\pi y}{b}\, dx\, dy$$

and for $m \geqslant 1$

$$a_{mn} = \frac{4}{ab} \int_0^a \int_0^b f(x, y) \cos \frac{m\pi x}{a} \sin \frac{n\pi y}{b}\, dx\, dy$$

The solution of the heat equation is thus given by Eq. (9.6.11).

## 9.7 WAVES IN THREE DIMENSIONS

The propagation of waves due to an initial disturbance in a rectangular volume is best described by the solution of the initial-boundary value problem

$$u_{tt} = c^2 \nabla^2 u, \qquad 0 < x < a,\ 0 < y < b,\ 0 < z < d,\ t > 0 \qquad (9.7.1)$$

$$u(x, y, z, 0) = f(x, y, z), \qquad 0 \leqslant x \leqslant a,\ 0 \leqslant y \leqslant b,\ 0 \leqslant z \leqslant d \qquad (9.7.2)$$

$$u_t(x, y, z, 0) = g(x, y, z), \qquad 0 \leqslant x \leqslant a,\ 0 \leqslant y \leqslant b,\ 0 \leqslant z \leqslant d \qquad (9.7.3)$$

$$u(0, y, z, t) = 0 \qquad (9.7.4)$$

$$u(a, y, z, t) = 0 \qquad (9.7.5)$$

$$u(x, 0, z, t) = 0 \qquad (9.7.6)$$

$$u(x, b, z, t) = 0 \qquad (9.7.7)$$

$$u(x, y, 0, t) = 0 \qquad (9.7.8)$$

$$u(x, y, d, t) = 0 \qquad (9.7.9)$$

We assume a separable solution to be in the form

$$u(x, y, z, t) = U(x, y, z)T(t)$$

The separated equations are

$$T'' + \lambda c^2 T = 0 \tag{9.7.10}$$

$$\nabla^2 U + \lambda U = 0 \tag{9.7.11}$$

Assume $U$ of the form

$$U(x, y, z) = X(x)Y(y)Z(z)$$

Substitution of this in Eq. (9.7.11) yields

$$X'' - \mu X = 0 \tag{9.7.12}$$

$$Y'' - \nu Y = 0 \tag{9.7.13}$$

$$Z'' + (\lambda + \mu + \nu)Z = 0 \tag{9.7.14}$$

Because of the homogeneous conditions in $x$, we let $\mu = -\alpha^2$ so that

$$X(x) = A \cos \alpha x + B \sin \alpha x$$

As in the preceding examples, we obtain

$$X_l(x) = B_l \sin \frac{l\pi x}{a}, \qquad l = 1, 2, 3, \ldots$$

In a similar manner, we let $\nu = -\beta^2$ to obtain

$$Y(y) = C \cos \beta y + D \sin \beta y$$

and accordingly

$$Y_m(y) = D_m \sin \frac{m\pi y}{b}, \qquad m = 1, 2, 3, \ldots$$

We again choose $\gamma^2 = \lambda + \mu + \nu = \lambda - \alpha^2 - \beta^2$ so that

$$Z(z) = E \cos \gamma z + F \sin \gamma z$$

Applying the homogeneous conditions in $z$, we obtain

$$Z_n(z) = F_n \sin \frac{n\pi z}{d}$$

Since the solution of Eq. (9.7.10) is

$$T(t) = G \cos \sqrt{\lambda}\, ct + H \sin \sqrt{\lambda}\, ct$$

the solution of the wave equation is

$$u(x, y, z, t) = \sum_{l=1}^{\infty} \sum_{m=1}^{\infty} \sum_{n=1}^{\infty} [a_{lmn} \cos \sqrt{\lambda}\, ct + b_{lmn} \sin \sqrt{\lambda}\, ct] \sin \frac{l\pi x}{a} \sin \frac{m\pi y}{b} \sin \frac{n\pi z}{d}$$

where $a_{lmn}$ and $b_{lmn}$ are arbitrary constants. The coefficient $a_{lmn}$ is determined from the initial condition $u(x, y, z, 0) = f(x, y, z)$ and found to be

$$a_{lmn} = \frac{8}{abd} \int_0^a \int_0^b \int_0^d f(x, y, z) \sin \frac{l\pi x}{a} \sin \frac{m\pi y}{b} \sin \frac{n\pi z}{d}\, dx\, dy\, dz$$

whereas $b_{lmn}$ is determined from the initial condition $u_t(x, y, z, 0) = g(x, y, z)$ and found to be

$$b_{lmn} = \frac{8}{\sqrt{\lambda}\, cabd} \int_0^a \int_0^b \int_0^d g(x, y, z) \sin \frac{l\pi x}{a} \sin \frac{m\pi y}{b} \sin \frac{n\pi z}{d}\, dx\, dy\, dz$$

where

$$\lambda = \left(\frac{l^2}{a^2} + \frac{m^2}{b^2} + \frac{n^2}{d^2}\right)\pi^2$$

## 9.8 HEAT CONDUCTION IN A RECTANGULAR VOLUME

As in the case of the wave equation, the solution of the heat equation in three space variables can be determined. Consider the problem of heat distribution in a rectangular volume. The faces are maintained at zero degree temperature. The solid is initially heated so that the problem may be written as

$$\begin{aligned}
u_t &= k\nabla^2 u, && 0 < x < a, \quad 0 < y < b, \quad 0 < z < d, \quad t > 0 \\
u(x, y, z, 0) &= f(x, y, z), && 0 \leq x \leq a, \quad 0 \leq y \leq b, \quad 0 \leq z \leq d \\
u(0, y, z, t) &= 0 \\
u(a, y, z, t) &= 0 \\
u(x, 0, z, t) &= 0 \\
u(x, b, z, t) &= 0 \\
u(x, y, 0, t) &= 0 \\
u(x, y, d, t) &= 0
\end{aligned}$$

As before, the separated equations are

$$T' + \lambda k T = 0 \tag{9.8.1}$$

$$\nabla^2 U + \lambda U = 0 \tag{9.8.2}$$

If we assume $U$ of the form

$$U(x, y, z) = X(x)Y(y)Z(z)$$

then the solution of the Helmholtz equation is

$$U_{lmn}(x, y, z) = B_l D_m F_n \sin\frac{l\pi x}{a} \sin\frac{m\pi y}{b} \sin\frac{n\pi z}{d}$$

Since the solution of Eq. (9.8.1) is

$$T(t) = Ge^{-\lambda kt}$$

the solution of the heat equation takes the form

$$u(x, y, z, t) = \sum_{l=1}^{\infty} \sum_{m=1}^{\infty} \sum_{n=1}^{\infty} a_{lmn} e^{-\lambda kt} \sin\frac{l\pi x}{a} \sin\frac{m\pi y}{b} \sin\frac{n\pi z}{d}$$

where $\lambda = [(l^2/a^2) + (m^2/b^2) + (n^2/d^2)]\pi^2$ and $a_{lmn}$ is a constant. Application of the initial condition yields

$$a_{lmn} = \frac{8}{abd} \int_0^a \int_0^b \int_0^d f(x, y, z) \sin\frac{l\pi x}{a} \sin\frac{m\pi y}{b} \sin\frac{n\pi z}{d}\, dx\, dy\, dz$$

## 9.9 THE THREE-DIMENSIONAL SCHRÖDINGER EQUATION IN A CENTRAL FIELD OF FORCE AND THE HYDROGEN ATOM

In quantum mechanics the Hamiltonian (or energy operator) is usually denoted by $H$ and is defined by

$$H = \frac{\mathbf{p}^2}{2M} + V(\mathbf{r}) \tag{9.9.1}$$

where $\mathbf{p} = (\hbar/i)\nabla = -i\hbar\nabla$ is the momentum of a particle of mass $M$, $h = 2\pi\hbar$ is the Planck constant, and $V(\mathbf{r})$ is the potential energy.

The physical state of a particle at time $t$ is described as fully as possible by the wave function $\psi(\mathbf{r}, t)$. The probability of finding the particle at position $\mathbf{r} = (x, y, z)$ within a finite volume $dV = dx\, dy\, dz$ is

$$\iiint |\psi|^2\, dx\, dy\, dz$$

The particle must always be somewhere in the space, so the probability of finding the particle within the whole space is one, that is,

$$\iiint_{-\infty}^{\infty} |\psi|^2\, dx\, dy\, dz = 1$$

The time dependent Schrödinger equation for the wave function $\psi(\mathbf{r}, t)$ is

$$i\hbar\psi_t = H\psi \tag{9.9.2}$$

where $H$ is explicitly given by

$$H = -\frac{\hbar^2}{2M}\nabla^2 + V(\mathbf{r}) \tag{9.9.3}$$

Given the potential $V(\mathbf{r})$, the fundamental problem of quantum mechanics is to obtain a solution of (9.9.2) which agrees with a given initial state $\psi(\mathbf{r}, 0)$.

For the stationary state solutions, we seek a solution of the form

$$\psi(\mathbf{r}, t) = f(t)\psi(\mathbf{r})$$

Substitution of this into (9.9.2) gives

$$\frac{df}{dt} + \frac{iE}{\hbar} f = 0 \tag{9.9.4}$$

$$H\psi(\mathbf{r}) = E\psi(\mathbf{r}) \tag{9.9.5}$$

where $E$ is a separation constant and has the dimension of energy. Integration of (9.9.4) gives

$$f(t) = A \exp\left(-\frac{iEt}{\hbar}\right) \tag{9.9.6}$$

where $A$ is an arbitrary constant.

Equation (9.9.5) is called the time independent Schrödinger equation. The great importance of this equation follows from the fact that the separation of variables gives not just some particular solution of (9.9.5), but generally yields all solutions of physical interest. If $\psi_E(\mathbf{r})$ represents one particular solution of (9.9.5), then most general solutions of (9.9.2) can be obtained by the principle of superposition of such particular solutions. In fact, the general solution is

$$\psi(\mathbf{r}, t) = \sum_E A_E \exp\left(-\frac{iEt}{\hbar}\right)\psi_E(\mathbf{r}) \tag{9.9.7}$$

where the summation is taken over all admissible values of $E$, and $A_E$ is an arbitrary constant to be determined from the initial conditions.

We now solve the eigenvalue problem for the Schrödinger equation for the spherically symmetric potential so that $V(\mathbf{r}) = V(r)$. The equation for $\psi(r)$ is

$$\nabla^2\psi + \frac{2M}{\hbar^2}[E - V(r)]\psi = 0 \tag{9.9.8}$$

where $\nabla^2$ is the three-dimensional Laplacian.

To determine the wave function $\psi$, it is convenient to introduce spherical polar coordinates $(r, \theta, \phi)$ so that Eq. (9.9.8) takes the form

$$\frac{1}{r^2}\frac{\partial}{\partial r}\left(r^2\frac{\partial\psi}{\partial r}\right)+\frac{1}{r^2\sin\theta}\frac{\partial}{\partial\theta}\left(\sin\theta\frac{\partial\psi}{\partial\theta}\right)+\frac{1}{r^2\sin^2\theta}\frac{\partial^2\psi}{\partial\phi^2}+K[E-V(r)]\psi=0 \qquad (9.9.9)$$

where $K=2M/\hbar^2$, $\psi\equiv\psi(r, \theta, \phi)$, $0\leqslant r<\infty$, $0\leqslant\theta\leqslant\pi$, and $0\leqslant\phi\leqslant 2\pi$.

We seek a separable solution of the form

$$\psi=R(r)Y(\theta, \phi)$$

and then substitute into (9.9.9) to obtain the following equations

$$\frac{d}{dr}\left(r^2\frac{dR}{dr}\right)+[K(E-V)r^2-\lambda]R=0 \qquad (9.9.10)$$

$$\left[\frac{1}{\sin\theta}\frac{\partial}{\partial\theta}\left(\sin\theta\frac{\partial}{\partial\theta}\right)+\frac{1}{\sin^2\theta}\frac{\partial^2}{\partial\phi^2}\right]Y+\lambda Y=0 \qquad (9.9.11)$$

where $\lambda$ is a separation constant.

We first solve (9.9.11) by separation of variables through $Y=\Theta(\theta)\Phi(\phi)$ so that the equation becomes

$$\sin\theta\frac{d}{d\theta}\left(\sin\theta\frac{d\Theta}{d\theta}\right)+(\lambda\sin^2\theta-m^2)\Theta=0 \qquad (9.9.12)$$

$$\frac{d^2\Phi}{d\phi^2}+m^2\Phi=0 \qquad (9.9.13)$$

where $m^2$ is a separation constant.

The general solution of (9.9.13) is

$$\Phi=Ae^{im\phi}+Be^{-im\phi}$$

where $A$ and $B$ are arbitrary constants to be determined by the boundary conditions on $\psi(r, \theta, \phi)=R(r)\Theta(\theta)\Phi(\phi)$ which will now be formulated.

According to the fundamental postulate of quantum mechanics, the wave function for a particle without spin must have a definite value at every point in space. Hence we assume that $\psi$ is a single-valued function of position. In particular, $\psi$ must have the same value whether the azimuthal coordinate $\phi$ is given by $\phi$ or $\phi+2\pi$, that is, $\Phi(\phi)=\Phi(\phi+2\pi)$. Consequently, the solution for $\Phi$ has the form

$$\Phi=Ce^{im\phi}, \qquad m=0, \pm1, \pm2, \ldots \qquad (9.9.14)$$

where $C$ is an arbitrary constant.

In order to solve (9.9.12), it is convenient to change the variable $x=\cos\theta$, $\Theta(\theta)=u(x)$, $-1\leqslant x\leqslant 1$ so that this equation becomes

$$\frac{d}{dx}\left[(1-x^2)\frac{du}{dx}\right]+\left(\lambda-\frac{m^2}{1-x^2}\right)u=0 \qquad (9.9.15)$$

For the particular case $m = 0$, this equation becomes

$$\frac{d}{dx}\left[(1-x^2)\frac{du}{dx}\right] + \lambda u = 0 \tag{9.9.16}$$

This is known as the Legendre equation, which gives the Legendre polynomials $P_l(x)$ of degree $l$ as solutions provided $\lambda = l(l+1)$ where $l$ is a positive integer or zero.

When $m \neq 0$, Eq. (9.9.15) with $\lambda = l(l+1)$ admits solutions which are well known as associated Legendre functions $P_l^m(x)$ of degree $l$ and order $m$ defined by

$$P_l^m(x) = (1-x^2)^{m/2}\frac{d^m}{dx^m}P_l^m(x), \qquad x = \cos\theta$$

Clearly, $P_l^m(x)$ vanishes when $m > l$. As for the negative integral values of $m$, it can be readily shown that

$$P_l^{-m}(x) = (-1)^m\frac{(l-m)!}{(l+m)!}P_l^m(x)$$

Hence the functions $P_l^{-m}(x)$ differ from $P_l^m(x)$ by a constant factor, and as a consequence, $m$ is restricted to a positive integer or zero. Thus the associated Legendre functions $P_l^m(x)$ with $|m| \leq l$ are the only non-singular and physically acceptable solutions of (9.9.15). Since $|m| \leq l$, when $l = 0$, $m = 0$; when $l = 1$, $m = -1, 0, +1$; when $l = 2$, $m = -2, -1, 0, 1, 2$ etc. This means that, given $l$, there are exactly $(2l+1)$ different values of $m = -l, \ldots -1, 0, 1, \ldots l$. The numbers $l$ and $m$ are called the *orbital quantum member* and the *magnetic quantum number* respectively.

It is convenient to write down the solutions of (9.9.11) as functions which are normalized with respect to an integration over the whole solid angle. They are called *spherical harmonics* and are given by, for $m \geq 0$,

$$Y_l^m(\theta, \phi) = \left[\frac{2l+1}{4\pi}\frac{(l-m)!}{(l+m)!}\right]^{\frac{1}{2}}(-1)^m e^{im\phi}P_l^m(\cos\theta) \tag{9.9.17}$$

Spherical harmonics with negative $m$ with $|m| \leq l$ are defined by

$$Y_l^m(\theta, \phi) = (-1)^m\overline{Y_l^{-m}(\theta, \phi)} \tag{9.9.18}$$

We now return to a general discussion of the radial equation (9.9.10) which becomes, under the transformation $R(r) = P(r)/r$,

$$\frac{d^2P}{dr^2} + \left[K(E-V) - \frac{\lambda}{r^2}\right]P(r) = 0 \tag{9.9.19}$$

Most cases of physical interest require $V(r)$ to be finite everywhere except at the origin $r = 0$. Also, $V(r) \to 0$ as $r \to \infty$. The Coulomb and square well potentials are typical examples of this kind. In the neighborhood of $r=0$, $V(r)$ can be neglected compared to the centrifugal term ($\sim 1/r^2$) so that Eq. (9.9.19) takes the form

$$\frac{d^2P}{dr^2} - \frac{l(l+1)}{r^2}P(r) = 0 \tag{9.9.20}$$

for all states with $l \neq 0$. The general solution of this equation is

$$P(r) = Ar^{l+1} + Br^{-l} \tag{9.9.21}$$

where $A$ and $B$ are arbitrary constants. With the boundary condition $P(0)=0$, $B=0$ so that the solution is proportional to $r^{l+1}$.

On the other hand, in view of the assumption that $V(r)\to 0$ as $r\to\infty$, the radial equation (9.9.19) reduces to

$$\frac{d^2P}{dr^2} + KE\,P(r) = 0 \tag{9.9.22}$$

The general solution of this equation is

$$P(r) = Ce^{ir\sqrt{KE}} + De^{-ir\sqrt{KE}} \tag{9.9.23}$$

This solution is oscillatory for $E>0$, and exponential in nature for $E<0$. The oscillatory solutions are not physically acceptable because the wave function does not tend to zero as $r\to\infty$. When $E<0$, the second term in (9.9.23) tends to infinity as $r\to\infty$. Consequently, only physically acceptable solutions for $E<0$ have the asymptotic form

$$P(r) = Ce^{-\alpha r/2} \tag{9.9.24}$$

where $KE = -\alpha^2/4$.

Thus the general solution of (9.9.19) can be written as

$$P(\mathbf{r}) = f(r)e^{-(\alpha/2)r}$$

so that $f(r)$ satisfies the equation

$$\frac{d^2f}{dr^2} - \alpha\frac{df}{dr} - \left[KV + \frac{l(l+1)}{r^2}\right]f = 0 \tag{9.9.25}$$

Note that this general solution is physically acceptable because the wave function tends to zero as $r\to 0$ and $r\to\infty$.

We now specify the form of the potential $V(r)$. One of the most common potentials is the Coulomb potential $V(r) = -Ze^2/r$ representing the attraction between an atomic nucleus of charge $+Ze$ and a moving electron of charge $-e$. For the hydrogen atom $Z=1$, and for the singly charged helium ion $Z=2$, where $Z$ represents the number of unit charges on the nucleus. Consequently, Eq. (9.9.25) reduces to

$$\frac{d^2f}{dr^2} - \alpha\frac{df}{dr} + \left[\frac{KZe^2}{r} - \frac{l(l+1)}{r^2}\right]f(r) = 0 \tag{9.9.26}$$

We seek a power series solution of this equation in the form

$$f(r) = r^k \sum_{s=1}^{\infty} a_s r^s, \qquad k \neq 0 \tag{9.9.27}$$

Substituting this series into (9.9.26), we obtain

$$\sum_{s=1}^{\infty} [(s+k)(s+k-1) - l(l+1)]a_s r^{s+k-2} + \sum_{s=1}^{\infty} [Zke^2 - \alpha(s+k)]a_s r^{s+k-1} = 0$$

Clearly, the lowest power of $r$ is $k-1$, so that

$$[k(k+1) - l(l+1)]a_1 = 0$$

This implies that $k = l$ or $-(l+1)$ provided $a_1 \neq 0$. The negative root of $k$ is not acceptable because it leads to an unbounded solution. Equating the coefficient of $r^{s+k-1}$, we get the recurrence relation for the coefficients as

$$a_{s+1} = \frac{\alpha(s+l) - ZKe^2}{s(s+2l+1)} a_s, \qquad s = 1, 2, 3, \ldots. \tag{9.9.28}$$

The asymptotic nature of this result is

$$\frac{a_{s+1}}{a_s} \sim \frac{\alpha}{s} \quad \text{as} \quad s \to \infty$$

This ratio is the same as that of the series for $e^{\alpha r}$. This means that $R(r)$ is unbounded as $r \to \infty$, which is physically unacceptable. Hence the series for $f(r)$ must terminate and $f(r)$ must be a polynomial so that $a_{s+1} = 0$, but $a_s \neq 0$. Hence

$$\alpha(s+l) - ZKe^2 = 0, \qquad s = 1, 2, 3, \ldots.$$

or

$$\frac{\alpha^2}{4} = \frac{Z^2K^2e^4}{4(s+l)^2} = -KE \tag{9.9.29}$$

Putting $K = 2M/\hbar^2$, the energy levels are given by

$$E = E_n = -\frac{Z^2K^2e^4}{4n^2K} = -\frac{MZ^2e^4}{2\hbar^2n^2} \tag{9.9.30}$$

where $n = (s+l)$ is called the *principle quantum number* and $n = 1, 2, 3, \ldots$.

For the hydrogen atom, $Z = 1$, the discrete energy spectrum is

$$E_n = -\frac{Me^4}{2\hbar^2n^2} = -\frac{e^2}{2an^2} \tag{9.9.31}$$

where $a = \hbar^2/e^2M$ is called the Bohr radius of the hydrogen atom of mass $M$ and charge of the electron $-e$. This discrete energy spectrum depends only on the principal quantum number $n$ (but not on $m$) and has an excellent agreement with experiments predicting discrete spectral lines.

For a given $n$, there are $n$ sets of $l$ and $s$

$$n=1,\ \{l=0,\ s=1\}; \qquad n=2,\ \begin{Bmatrix} l=0,\ s=2 \\ l=1,\ s=1 \end{Bmatrix};$$

$$n=3,\ \begin{Bmatrix} l=0,\ s=3 \\ l=1,\ s=2 \\ l=2,\ s=1 \end{Bmatrix};\ \text{etc.}$$

Given $n$, there are exactly $n$ values of $l(l=0, 1, 2, \ldots n-1)$ and the highest value of $l$ is $n-1$.

Thus the three numbers $n$, $l$, $m$ determine a unique eigenfunction $\psi_{n,l,m}(r, \theta, \phi) = R_{n,l}(r)Y_l^m(\theta, \phi)$. Since the energy levels depend only on the principal quantum number $n$, there are, in general, several linearly independent eigenfunctions of the Schrödinger equation for the hydrogen atom corresponding to each energy level, so the energy levels are said to be degenerate. There are $(2l+1)$ different eigenfunctions of the same energy obtained by varying the magnetic quantum number $m$ from $-l$ to $l$. The total degeneracy of the energy level $E_n$ is then

$$\sum_{l=0}^{n-1} (2l+1) = 2\frac{n(n-1)}{2} + n = n^2$$

## 9.10 METHOD OF EIGENFUNCTIONS

Consider the nonhomogeneous initial-boundary value problem

$$L[u] = \rho u_{tt} - G \qquad \text{in } D \tag{9.10.1}$$

with prescribed homogeneous boundary conditions on the boundary $B$ of $D$, and the initial conditions

$$u(x_1, x_2, \ldots, x_n, 0) = f(x_1, x_2, \ldots, x_n) \tag{9.10.2}$$

$$u_t(x_1, x_2, \ldots, x_n, 0) = g(x_1, x_2, \ldots, x_n) \tag{9.10.3}$$

Here $\rho \equiv \rho(x_1, x_2, \ldots, x_n)$ is a real-valued positive continuous function and $G \equiv G(x_1, x_2, \ldots, x_n)$ is a real-valued continuous function.

We assume the only solution of the associated homogeneous problem

$$L[u] = \rho u_{tt} \tag{9.10.4}$$

with the prescribed boundary conditions is the trivial solution. Then, if there exists a solution of the given problem (9.10.1)–(9.10.3), it can be represented by a series of eigenfunctions of the associated eigenvalue problem

$$L[\varphi] + \lambda\rho\varphi = 0 \tag{9.10.5}$$

with $\varphi$ satisfying the boundary conditions given for $u$. For problems with one space variable, see Sec. 6.7.

## 9.11 FORCED VIBRATION OF MEMBRANE

As a specific example, we shall determine the solution of the problem of forced vibration of a rectangular membrane of length $a$ and width $b$. The problem is

$$u_{tt} - c^2\nabla^2 u = F(x, y, t) \qquad \text{in } D \tag{9.11.1}$$

$$u(x, y, 0) = f(x, y), \qquad 0 \leqslant x \leqslant a, \quad 0 \leqslant y \leqslant b \tag{9.11.2}$$

$$u_t(x, y, 0) = g(x, y), \qquad 0 \leqslant x \leqslant a, \quad 0 \leqslant y \leqslant b \tag{9.11.3}$$

$$u(0, y, t) = 0 \tag{9.11.4}$$

$$u(a, y, t) = 0 \tag{9.11.5}$$

$$u(x, 0, t) = 0 \tag{9.11.6}$$

$$u(x, b, t) = 0 \tag{9.11.7}$$

The associated eigenvalue problem is

$$\nabla^2\varphi + \lambda\varphi = 0 \qquad \text{in } D$$

$$\varphi = 0 \qquad \text{on } B$$

The eigenvalues for this problem according to Sec. 9.5 are

$$\alpha_{mn} = \frac{m^2\pi^2}{a^2} + \frac{n^2\pi^2}{b^2}, \qquad m, n = 1, 2, 3, \ldots$$

and the corresponding eigenfunctions are

$$\varphi_{mn}(x, y) = \sin\frac{m\pi x}{a}\sin\frac{n\pi y}{b}$$

Thus, we assume the solution

$$u(x, y, t) = \sum_{m=1}^{\infty}\sum_{n=1}^{\infty} u_{mn}(t)\sin\frac{m\pi x}{a}\sin\frac{n\pi y}{b}$$

and the forcing function

$$F(x, y, t) = \sum_{m=1}^{\infty}\sum_{n=1}^{\infty} F_{mn}(t)\sin\frac{m\pi x}{a}\sin\frac{n\pi y}{b}$$

Here $F_{mn}(t)$ are given by

$$F_{mn}(t) = \frac{4}{ab}\int_0^a \int_0^b F(x, y, t) \sin\frac{m\pi x}{a} \sin\frac{n\pi y}{b}\, dx\, dy$$

Note that $u$ automatically satisfies the homogeneous boundary conditions. Now inserting $u(x, y, t)$ and $F(x, y, t)$ in Eq. (9.11.1), we obtain

$$u''_{mn} + c^2\alpha_{mn}^2 u_{mn} = F_{mn}$$

where $\alpha_{mn}^2 = (m\pi/a)^2 + (n\pi/b)^2$. We have assumed that $u$ is twice continuously differentiable with respect to $t$. Thus the solution of the preceding ordinary differential equation takes the form

$$u_{mn}(t) = A_{mn} \cos \alpha_{mn}ct + B_{mn} \sin \alpha_{mn}ct$$
$$+ \frac{1}{\alpha_{mn}c}\int_0^t F_{mn}(\tau) \sin \alpha_{mn}c(t - \tau)\, d\tau$$

The first initial condition gives

$$u(x, y, 0) = f(x, y) = \sum_{m=1}^{\infty}\sum_{n=1}^{\infty} A_{mn} \sin\frac{m\pi x}{a} \sin\frac{n\pi y}{b}$$

Assuming that $f(x, y)$ is continuous in $x$ and $y$, the coefficient $A_{mn}$ of the double Fourier series is given by

$$A_{mn} = \frac{4}{ab}\int_0^a \int_0^b f(x, y) \sin\frac{m\pi x}{a} \sin\frac{n\pi y}{b}\, dx\, dy$$

Similarly, from the remaining initial condition, we have

$$u_t(x, y, 0) = g(x, y) = \sum_{m=1}^{\infty}\sum_{n=1}^{\infty} B_{mn}\alpha_{mn}c \sin\frac{m\pi x}{a} \sin\frac{n\pi y}{b}$$

and hence for continuous $g(x, y)$

$$B_{mn} = \frac{4}{ab\alpha_{mn}c}\int_0^a \int_0^b g(x, y) \sin\frac{m\pi x}{a} \sin\frac{n\pi y}{b}\, dx\, dy$$

The solution of the given initial-boundary value problem is therefore given by

$$u(x, y, t) = \sum_{m=1}^{\infty}\sum_{n=1}^{\infty} u_{mn}(t) \sin\frac{m\pi x}{a} \sin\frac{n\pi y}{b}$$

provided the series for $u$ and its first and second derivatives converge uniformly.

If $F(x, y, t) = e^{x+y} \cos \omega t$, then

$$F_{mn}(t) = \frac{4mn\pi^2}{(m^2\pi^2 + a^2)(n^2\pi^2 + b^2)}[1 + (-1)^{m+1}e^a][1 + (-1)^{n+1}e^b] \cos \omega t$$
$$= C_{mn} \cos \omega t$$

Hence we have

$$u_{mn}(t) = \frac{1}{\alpha_{mn}c}\int_0^t C_{mn}\cos\omega\tau\sin\alpha_{mn}c(t-\tau)\,d\tau$$

$$= \frac{C_{mn}}{(\alpha_{mn}^2c^2-\omega^2)}(\cos\omega t - \cos\alpha_{mn}ct)$$

provided $\omega \neq \alpha_{mn}c$. Thus, the solution may be written in the form

$$u(x, y, t) = \sum_{m=1}^{\infty}\sum_{n=1}^{\infty}\frac{C_{mn}}{(\alpha_{mn}^2c^2-\omega^2)}(\cos\omega t - \cos\alpha_{mn}ct)\sin\frac{m\pi x}{a}\sin\frac{n\pi y}{b}$$

## 9.12 TIME-DEPENDENT BOUNDARY CONDITIONS

The preceding chapters have been devoted to problems with homogeneous boundary conditions. Due to occurrence of problems with time dependent boundary conditions in practice, we consider the forced vibration of a rectangular membrane with moving boundaries. The problem here is to determine the displacement function $u$ of

$$u_{tt} - c^2\nabla^2u = F(x, y, t), \qquad 0<x<a, \qquad 0<y<b \tag{9.12.1}$$

$$u(x, y, 0) = f(x, y), \qquad 0\leq x\leq a, \qquad 0\leq y\leq b \tag{9.12.2}$$

$$u_t(x, y, 0) = g(x, y), \qquad 0\leq x\leq a, \qquad 0\leq y\leq b \tag{9.12.3}$$

$$u(0, y, t) = p_1(y, t), \qquad 0\leq y\leq b, \qquad t\geq 0 \tag{9.12.4}$$

$$u(a, y, t) = p_2(y, t), \qquad 0\leq y\leq b, \qquad t\geq 0 \tag{9.12.5}$$

$$u(x, 0, t) = q_1(x, t), \qquad 0\leq x\leq a, \qquad t\geq 0 \tag{9.12.6}$$

$$u(x, b, t) = q_2(x, t), \qquad 0\leq x\leq a, \qquad t\geq 0 \tag{9.12.7}$$

For such problems, we seek a solution in the form

$$u(x, y, t) = U(x, y, t) + v(x, y, t) \tag{9.12.8}$$

where $v$ is the new dependent variable to be determined. Before finding $v$, we must first determine $U$. If we substitute Eq. (9.12.8) in Eqs. (9.12.1)–(9.12.7), we respectively obtain

$$v_{tt} - c^2(v_{xx} + v_{yy}) = F - U_{tt} + c^2(U_{xx} + U_{yy})$$
$$= \tilde{F}(x, y, t)$$

and

$$v(x, y, 0) = f(x, y) - U(x, y, 0) = \tilde{f}(x, y)$$
$$v_t(x, y, 0) = g(x, y) - U_t(x, y, 0) = \tilde{g}(x, y)$$
$$v(0, y, t) = p_1(y, t) - U(0, y, t) = \tilde{p}_1(y, t)$$
$$v(a, y, t) = p_2(y, t) - U(a, y, t) = \tilde{p}_2(y, t)$$
$$v(x, 0, t) = q_1(x, t) - U(x, 0, t) = \tilde{q}_1(x, t)$$
$$v(x, b, t) = q_2(x, t) - U(x, b, t) = \tilde{q}_2(x, t)$$

In order to make the conditions on $v$ homogeneous, we set

$$\tilde{p}_1 = \tilde{p}_2 = \tilde{q}_1 = \tilde{q}_2 = 0$$

so that

$$\begin{aligned} U(0, y, t) &= p_1(y, t) \\ U(a, y, t) &= p_2(y, t) \\ U(x, 0, t) &= q_1(x, t) \\ U(x, b, t) &= q_2(x, t) \end{aligned} \tag{9.12.9}$$

In order that the boundary conditions be compatible, we assume the prescribed functions take the forms

$$\begin{aligned} p_1(y, t) &= \varphi(y)p_1^*(y, t) \\ p_2(y, t) &= \varphi(y)p_2^*(y, t) \\ q_1(x, t) &= \psi(x)q_1^*(x, t) \\ q_2(x, t) &= \psi(x)q_2^*(x, t) \end{aligned}$$

where the function $\varphi$ must vanish at the end points $y = 0$, $y = b$ and the function $\psi$ must vanish at $x = 0$, $x = a$. Thus, $U(x, y, t)$ which satisfies Eq. (9.12.9) takes the form

$$U(x, y, t) = \varphi(y)\left[p_1^* + \frac{x}{a}(p_2^* - p_1^*)\right] + \psi(x)\left[q_1^* + \frac{y}{b}(q_2^* - q_1^*)\right]$$

The problem then is to find the function $v(x, y, t)$ of

$$\begin{aligned} v_{tt} - c^2(v_{xx} + v_{yy}) &= \tilde{F}(x, y, t) \\ v(x, y, 0) &= \tilde{f}(x, y) \\ v_t(x, y, 0) &= \tilde{g}(x, y) \end{aligned}$$

$$\begin{aligned} v(0, y, t) &= 0 \\ v(a, y, t) &= 0 \\ v(x, 0, t) &= 0 \\ v(x, b, t) &= 0 \end{aligned}$$

This is an initial-boundary value problem with homogeneous boundary conditions, which has already been solved.

As a particular case, consider the following problem

$$\begin{aligned} u_{tt} - c^2(u_{xx} + u_{yy}) &= 0 \\ u(x, y, 0) &= 0 \\ u_t(x, y, 0) &= \frac{y}{b}\sin\frac{\pi x}{a} \\ u(0, y, t) &= 0 \\ u(a, y, t) &= 0 \\ u(x, 0, t) &= 0 \\ u(x, b, t) &= \sin\frac{\pi x}{a}\sin t \end{aligned}$$

We assume a solution in the form

$$u(x, y, t) = v(x, y, t) + U(x, y, t)$$

The function which satisfies

$$U(0, y, t) = 0$$
$$U(a, y, t) = 0$$
$$U(x, 0, t) = 0$$
$$U(x, b, t) = \sin \frac{\pi x}{a} \sin t$$

is

$$U(x, y, t) = \sin \frac{\pi x}{a} \left(\frac{y}{b} \sin t\right)$$

Thus the new problem to be solved is

$$v_{tt} - c^2(v_{xx} + v_{yy}) = \left(1 - \frac{c^2\pi^2}{a^2}\right) \frac{y}{b} \sin \frac{\pi x}{a} \sin t$$
$$v(x, y, 0) = 0$$
$$v_t(x, y, 0) = 0$$
$$v(0, y, t) = 0$$
$$v(a, y, t) = 0$$
$$v(x, 0, t) = 0$$
$$v(x, b, t) = 0$$

Then we find $F_{mn}$ from

$$F_{mn}(t) = \frac{4}{ab} \int_0^a \int_0^b F(x, y, t) \sin \frac{m\pi x}{a} \sin \frac{n\pi y}{b}\, dx\, dy$$

where

$$F(x, y, t) = \left(1 - \frac{c^2\pi^2}{a^2}\right) \frac{y}{b} \sin \frac{\pi x}{a} \sin t$$

and obtain

$$F_{mn}(t) = \frac{2(-1)^{n+1}}{an} \left(1 - \frac{c^2\pi^2}{a^2}\right) \sin t$$

Now we determine $v_{mn}(t)$ which is given by

$$v_{mn}(t) = A_{mn} \cos \alpha_{mn}ct + B_{mn} \sin \alpha_{mn}ct + \frac{1}{\alpha_{mn}c} \int_0^t F_{mn}(\tau) \sin \alpha_{mn}c(t - \tau)\, d\tau$$

Since $v(x, y, 0) = 0$, $A_{mn} = 0$, but

$$B_{mn} = \frac{4}{ab\alpha_{mn}c} \int_0^a \int_0^b \left(-\frac{y}{b} \sin\frac{\pi x}{a}\right) \sin\frac{m\pi x}{a} \sin\frac{n\pi y}{b}\, dx\, dy$$

$$= \frac{2(-1)^n}{\alpha_{mn}can}$$

Thus, we have

$$v_{mn}(t) = \frac{2(-1)^n}{\alpha_{mn}can} \sin \alpha_{mn}ct$$

$$+ \frac{2(-1)^n}{\alpha_{mn}ca^3n(1-\alpha^2c^2)}(a^2 - c^2\pi^2)(\sin \alpha_{mn}ct - \alpha c \sin t)$$

The solution is therefore given by

$$u(x, y, t) = \frac{y}{b} \sin\frac{\pi x}{a} \sin t + \sum_{m=1}^{\infty} \sum_{n=1}^{\infty} v_{mn}(t) \sin\frac{m\pi x}{a} \sin\frac{n\pi y}{b}$$

## 9.13 EXERCISES

**1.** Solve the Dirichlet problem

$$\nabla^2 u = 0, \qquad 0 < x < a, \qquad 0 < y < b, \qquad 0 < z < c$$

$$u(0, y, z) = \sin\frac{\pi y}{b} \sin\frac{\pi z}{c}$$

$$u(a, y, z) = 0$$

$$u(x, 0, z) = 0$$

$$u(x, b, z) = 0$$

$$u(x, y, 0) = 0$$

$$u(x, y, c) = 0$$

**2.** Solve the Neumann problem

$$\nabla^2 u = 0, \qquad 0 < x < 1, \qquad 0 < y < 1, \qquad 0 < z < 1$$

$$u_x(0, y, z) = 0$$

$$u_x(1, y, z) = 0$$

$$u_y(x, 0, z) = 0$$

$$u_y(x, 1, z) = 0$$

$$u_z(x, y, 0) = \cos \pi x \cos \pi y$$

$$u_z(x, y, 1) = 0$$

3. Solve the Robin problem

$$\nabla^2 u = 0, \qquad 0 < x < \pi, \qquad 0 < y < \pi, \qquad 0 < z < \pi$$
$$u(0, y, z) = f(y, z)$$
$$u(\pi, y, z) = 0$$
$$u_y(x, 0, z) = 0$$
$$u_y(x, \pi, z) = 0$$
$$\left.\begin{aligned} u_z(x, y, 0) + hu(x, y, 0) &= 0, \\ u_z(x, y, \pi) + hu(x, y, \pi) &= 0 \end{aligned}\right. \qquad h = \text{constant}$$

4. Determine the solution of each of the following problems for a cylinder:

a.

$$\nabla^2 u = 0, \qquad r < a, \qquad 0 < \theta < 2\pi, \qquad 0 < z < l$$
$$u(a, \theta, z) = 0$$
$$u(r, \theta, l) = 0$$
$$u(r, \theta, 0) = f(r, \theta)$$

b.

$$\nabla^2 u = 0, \qquad r < a, \qquad 0 < \theta < 2\pi, \qquad 0 < z < l$$
$$u(a, \theta, z) = f(\theta, z)$$
$$u_z(r, \theta, 0) = 0$$
$$u_z(r, \theta, l) = 0$$

5. Find the solution of the Dirichlet problem for a sphere

$$\nabla^2 u = 0, \qquad r < a, \qquad 0 < \theta < \pi, \qquad 0 < \varphi < 2\pi$$
$$u(a, \theta, \varphi) = \cos^2 \theta$$

6. Solve the Dirichlet problem for a concentric sphere

$$\nabla^2 u = 0, \qquad a < r < b, \qquad 0 < \theta < \pi, \qquad 0 < \phi < 2\pi$$
$$u(a, \theta, \phi) = f(\theta, \phi)$$
$$u(b, \theta, \phi) = g(\theta, \phi)$$

7. Find the steady-state temperature distribution in a cylinder of radius $a$ if a constant flow of heat $T$ is supplied at the end $z = 0$, and the surface $r = a$ and the end $z = l$ are maintained at zero temperature.

8. Find the potential of the electrostatic field inside a cylinder of length $l$ and radius $a$ if both ends of the cylinder are each earthed, and the surface is charged to a potential $u_0$.

9. Determine the potential of the electric field inside a sphere of radius $a$ if the

upper half of the sphere is charged to a potential $u_1$ and the lower half to a potential $u_2$.

**10.** Solve the Dirichlet problem for a half cylinder

$$\nabla^2 u = 0, \qquad r < 1, \qquad 0 < \theta < \pi, \qquad 0 < z < 1$$

$$u(1, \theta, z) = 0$$

$$u(r, 0, z) = 0$$

$$u(r, \pi, z) = 0$$

$$u(r, \theta, 0) = 0$$

$$u(r, \theta, 1) = f(r, \theta)$$

**11.** Solve the Neumann problem for a sphere

$$\nabla^2 u = 0, \qquad r < 1, \qquad 0 < \theta < \pi, \qquad 0 < \varphi < 2\pi$$

$$u_r(1, \theta, \varphi) = f(\theta, \varphi)$$

where

$$\int_0^{2\pi} \int_0^{\pi} f(\theta, \varphi) \sin \theta \, d\theta \, d\varphi = 0$$

**12.** Find the solution of the initial-boundary value problem

$$u_{tt} = c^2 \nabla^2 u, \qquad 0 < x < 1, \qquad 0 < y < 1, \qquad t > 0$$

$$u(x, y, 0) = \sin^2 \pi x \sin \pi y, \qquad 0 \leqslant x \leqslant 1, \qquad 0 \leqslant y \leqslant 1$$

$$u_t(x, y, 0) = 0$$

$$u(0, y, t) = 0$$

$$u(1, y, t) = 0$$

$$u(x, 0, t) = 0$$

$$u(x, 1, t) = 0$$

**13.** Obtain the solution of

$$u_{tt} = c^2 \nabla^2 u, \qquad r < a, \qquad 0 < \theta < 2\pi, \qquad t > 0$$

$$u(r, \theta, 0) = f(r, \theta)$$

$$u_t(r, \theta, 0) = g(r, \theta)$$

$$u(a, \theta, t) = 0$$

**14.** Determine the temperature distribution in a rectangular plate with radiation from its surface. The temperature distribution is described by

$$u_t = k(u_{xx} + u_{yy}) - h(u - u_0), \qquad 0 < x < a, \qquad 0 < y < b, \qquad t > 0$$

$$u(x, y, 0) = f(x, y)$$
$$u(0, y, t) = 0$$
$$u(a, y, t) = 0$$
$$u(x, 0, t) = 0$$
$$u(x, b, t) = 0$$

where $u_0$ is a constant.

**15.** Solve the heat conduction problem in a circular plate

$$u_t = k\left(u_{rr} + \frac{1}{r}u_r + \frac{1}{r^2}u_{\theta\theta}\right), \qquad r < 1, \qquad 0 < \theta < 2\pi, \qquad t > 0$$

$$u(r, \theta, 0) = f(r, \theta)$$
$$u(1, \theta, t) = 0$$

**16.** Solve the initial-boundary value problem

$$u_{tt} = c^2\nabla^2 u, \qquad 0 < x < 1, \qquad 0 < y < 1, \qquad 0 < z < 1, \qquad t > 0$$

$$u(x, y, z, 0) = \sin \pi x \sin \pi y \sin \pi z$$
$$u_t(x, y, z, 0) = 0$$
$$u(0, y, z, t) = u(1, y, z, t) = 0$$
$$u(x, 0, z, t) = u(x, 1, z, t) = 0$$
$$u(x, y, 0, t) = u(x, y, 1, t) = 0$$

**17.** Solve

$$u_{tt} + ku_t = c^2\nabla^2 u, \qquad 0 < x < a, \qquad 0 < y < b, \qquad 0 < z < d, \qquad t > 0$$

$$u(x, y, z, 0) = f(x, y, z)$$
$$u_t(x, y, z, 0) = g(x, y, z)$$
$$u(0, y, z, t) = u(a, y, z, t) = 0$$
$$u(x, 0, z, t) = u(x, b, z, t) = 0$$
$$u(x, y, 0, t) = u(x, y, d, t) = 0$$

**18.** Obtain the solution of

$$u_{tt} = c^2\left(u_{rr} + \frac{1}{r}u_r + \frac{1}{r^2}u_{\theta\theta} + u_{zz}\right), \qquad r < a, \qquad 0 < z < l, \qquad t > 0$$

$$u(r, \theta, z, 0) = f(r, \theta, z)$$
$$u_t(r, \theta, z, 0) = q(r, \theta, z)$$
$$u(a, \theta, z, t) = 0$$
$$u(r, \theta, 0, t) = u(r, \theta, l, t) = 0$$

**19.** Determine the solution of the heat conduction problem

$$u_t = k\nabla^2 u, \qquad 0 < x < a, \qquad 0 < y < b, \qquad 0 < z < c, \qquad t > 0$$

$$u(x, y, z, 0) = f(x, y, z)$$

$$u_x(0, y, z, t) = u_x(a, y, z, t) = 0$$

$$u_y(x, 0, z, t) = u_y(x, b, z, t) = 0$$

$$u_z(x, y, 0, t) = u_z(x, y, c, t) = 0$$

**20.** Solve

$$u_t = k\nabla^2 u, \qquad r < a, \qquad 0 < z < l, \qquad t > 0$$

$$u(r, \theta, z, 0) = f(r, \theta, z)$$

$$u_r(a, \theta, z, t) = 0$$

$$u(r, \theta, 0, t) = u(r, \theta, l, t) = 0$$

**21.** Find the temperature distribution in the section of a sphere cut out by the cone $\theta = \theta_0$. The surface temperature is zero while the initial temperature is given by $f(r, \theta, \varphi)$.

**22.** Solve the initial-boundary value problem

$$u_{tt} = c^2\nabla^2 u + F(x, y, t), \qquad 0 < x < a, \qquad 0 < y < b, \qquad t > 0$$

$$u(x, y, 0) = f(x, y)$$

$$u_t(x, y, 0) = g(x, y)$$

$$u_x(0, y, t) = u_x(a, y, t) = 0 \quad \text{for all } t$$

$$u_y(x, 0, t) = u_y(x, b, t) = 0$$

**23.** Solve

$$u_{tt} = c^2\nabla^2 u + xy \sin t, \qquad 0 < x < \pi, \qquad 0 < y < \pi, \qquad t > 0$$

$$u(x, y, 0) = 0$$

$$u_t(x, y, 0) = 0$$

$$u(0, y, t) = u(\pi, y, t) = 0$$

$$u(x, 0, t) = u(x, \pi, t) = 0$$

**24.** Solve

$$u_t = k\nabla^2 u + F(x, y, z, t), \qquad 0 < x < a, \qquad 0 < y < b, \qquad 0 < z < c, \qquad t > 0$$

$$u(x, y, z, 0) = f(x, y, z)$$

$$u(0, y, z, t) = u(a, y, z, t) = 0$$

$$u(x, 0, z, t) = u(x, b, z, t) = 0$$

$$u_z(x, y, 0, t) = u_z(x, y, c, t) = 0$$

**25.** Solve

$$\begin{aligned}
&u_t = k\nabla^2 u + A, \qquad 0<x<\pi, \qquad 0<y<\pi, \qquad t>0 \\
&u(x, y, 0) = 0 \\
&u(0, y, t) = u(\pi, y, t) = 0 \\
&u_y(x, 0, t) + u(x, 0, t) = 0 \\
&u_y(x, \pi, t) + u(x, \pi, t) = 0
\end{aligned}$$

where $A$ is a constant.

**26.** Find the temperature distribution of the composite cylinder consisting of an inner cylinder $0 \leqslant r \leqslant r_0$ and an outer cylindrical tube $r_0 \leqslant r \leqslant a$. The surface temperature is maintained at zero degrees, and the initial temperature distribution is given by $f(r, \theta, z)$.

**27.** Solve the initial-boundary value problem

$$\begin{aligned}
&u_t - c^2\nabla^2 u = 0, \qquad 0<x<\pi, \qquad 0<y<\pi, \qquad t>0 \\
&u(x, y, 0) = 0 \\
&u(0, y, t) = u(\pi, y, t) = 0 \\
&u(x, 0, t) = x(x-\pi)\sin t, \qquad 0 \leqslant x \leqslant \pi, \qquad t \geqslant 0 \\
&u(x, \pi, t) = 0
\end{aligned}$$

**28.** Solve

$$\begin{aligned}
u_{tt} &= c^2\nabla^2 u, \qquad r<a, \qquad t>0 \\
u(r, \theta, 0) &= f(r, \theta) \\
u_t(r, \theta, 0) &= g(r, \theta) \\
u(a, \theta, t) &= p(\theta, t)
\end{aligned}$$

**29.** Solve

$$\begin{aligned}
u_t &= c^2\nabla^2 u, \qquad r<a, \qquad t>0 \\
u(r, \theta, 0) &= f(r, \theta) \\
u_r(a, \theta, t) &= g(\theta, t)
\end{aligned}$$

**30.** Determine the solution of the biharmonic equation

$$\nabla^4 u = q/D$$

with the boundary condition

$$u\left(-\frac{a}{2}, y\right) = u\left(\frac{a}{2}, y\right) = 0$$

$$u_{xx}\left(-\frac{a}{2}, y\right) = u_{xx}\left(\frac{a}{2}, y\right) = 0$$

$$u(x, 0) = u(x, b) = 0$$

$$u_{yy}(x, 0) = u_{yy}(x, b) = 0$$

where $q$ is the load per unit area and $D$ is the flexural rigidity of the plate. This is the problem of the deflection of a uniformly loaded plate, the sides of which are simply supported.

CHAPTER TEN

# Green's Function

## 10.1 THE DELTA FUNCTION

The application of Green's functions to boundary-value problems in ordinary differential equations was described earlier in Chapter 7. The Green's function method is applied here to boundary-value problems in partial differential equations. The method provides solutions in integral form and is applicable to a wide class of problems in applied mathematics and mathematical physics.

Before developing the method of Green's function, we will first define the *Dirac delta function* $\delta(x-\xi, y-\eta)$ in two dimensions by

a. $\delta(x-\xi, y-\eta)=0, \qquad x\neq\xi, \qquad y\neq\eta$ (10.1.1)

b. $\iint_{R_\varepsilon} \delta(x-\xi, y-\eta)\,dx\,dy=1, \qquad R_\varepsilon:(x-\xi)^2+(y-\eta)^2<\varepsilon^2$ (10.1.2)

c. $\iint_{R} F(x,y)\,\delta(x-\xi, y-\eta)\,dx\,dy=F(\xi,\eta)$ (10.1.3)

for arbitrary continuous function $F$ in the region $R$.

The delta function is not a function in the ordinary sense.[10] It is a symbolic function, and is often viewed as the limit of a distribution.

If $\delta(x-\xi)$ and $\delta(y-\eta)$ are one-dimensional delta functions, we have

$$\iint_{R} F(x,y)\,\delta(x-\xi)\,\delta(y-\eta)\,dx\,dy=F(\xi,\eta) \tag{10.1.4}$$

[10] For an elegant treatment of the delta function as a generalized function, see L. Schwartz, *Theory of Distributions*.

Since (10.1.3) and (10.1.4) hold for an arbitrary continuous function $F$, we conclude that

$$\delta(x-\xi, y-\eta) = \delta(x-\xi)\,\delta(y-\eta) \tag{10.1.5}$$

Thus, we may state that the two-dimensional delta function is the product of one-dimensional delta functions.

Higher dimensional delta functions can be defined in a similar manner.

## 10.2 GREEN'S FUNCTION

The solution of the Dirichlet problem

$$\begin{aligned} \nabla^2 u &= h(x, y) \quad \text{in } D \\ u &= f(x, y) \quad \text{on } B \end{aligned} \tag{10.2.1}$$

is given by[11]

$$u(x, y) = \iint_D G(x, y; \xi, \eta) h(\xi, \eta)\, d\xi\, d\eta + \int_B f \frac{\partial G}{\partial n}\, ds \tag{10.2.2}$$

where $G$ is the Green's function and $n$ denotes the outward normal to the boundary $B$ of the region $D$. It is rather obvious then that the solution $u(x, y)$ can be determined as soon as the Green's function $G$ is ascertained, so the problem in this technique is really to find the Green's function.

First we shall define the Green's function for the Dirichlet problem involving the Laplace operator. Then, the Green's function for the Dirichlet problem involving the Helmholtz operator may be defined in a completely analogous manner.

The Green's function for the Dirichlet problem involving the Laplace operator is the function which satisfies

a. $\nabla^2 G = \delta(x-\xi, y-\eta)$[12] in $D$ (10.2.3)

$G = 0$ on $B$ (10.2.4)

b. $G$ is symmetric, that is,

$$G(x, y; \xi, \eta) = G(\xi, \eta; x, y) \tag{10.2.5}$$

c. $G$ is continuous in $x$, $y$, $\xi$, $\eta$, but $\partial G/\partial n$ has a discontinuity at the point $(\xi, \eta)$ which is specified by the equation

$$\lim_{\varepsilon \to 0} \int_{C_\varepsilon} \frac{\partial G}{\partial n}\, ds = 1 \tag{10.2.6}$$

---

[11] The proof is given in Sec. 10.4.

[12] For other operators, see Greenberg [16].

where $n$ is the outward normal to the circle

$$C_\varepsilon : (x-\xi)^2 + (y-\eta)^2 = \varepsilon^2$$

The Green's function $G$ may be interpreted as the response of the system at a field point $(x, y)$ due to a $\delta$ function input at the source point $(\xi, \eta)$. $G$ is continuous everywhere in $D$, and its first and second derivatives are continuous in $D$ except at $(\xi, \eta)$. Thus, property (a) essentially states that $\nabla^2 G = 0$ everywhere except at the source point $(\xi, \eta)$.

We will now prove property (b).

***Theorem 10.2.1*** *The Green's function is symmetric.*

PROOF. Applying Green's second formula[13]

$$\iint_D (\phi \nabla^2 \psi - \psi \nabla^2 \phi)\, dS = \int_B \left( \phi \frac{\partial \psi}{\partial n} - \psi \frac{\partial \phi}{\partial n} \right) ds \qquad (10.2.7)$$

to the functions $\phi = G(x, y; \xi, \eta)$ and $\psi = G(x, y; \xi^*, \eta^*)$ we obtain

$$\begin{aligned}\iint_D &[G(x, y; \xi, \eta)\nabla^2 G(x, y; \xi^*, \eta^*) \\ &\quad - G(x, y; \xi^*, \eta^*)\nabla^2 G(x, y; \xi, \eta)]\, dx\, dy \\ &= \int_B \left[ G(x, y; \xi, \eta) \frac{\partial G}{\partial n}(x, y; \xi^*, \eta^*) - G(x, y; \xi^*, \eta^*) \frac{\partial G}{\partial n}(x, y; \xi, \eta) \right] ds\end{aligned}$$

Since $G(x, y; \xi, \eta)$ and hence $G(x, y; \xi^*, \eta^*)$ must vanish on $B$, we have

$$\begin{aligned}\iint_D &[G(x, y; \xi, \eta)\nabla^2 G(x, y; \xi^*, \eta^*) \\ &\qquad - G(x, y; \xi^*, \eta^*)\nabla^2 G(x, y; \xi, \eta)]\, dx\, dy = 0\end{aligned}$$

But

$$\nabla^2 G(x, y; \xi, \eta) = \delta(x - \xi, y - \eta)$$

and

$$\nabla^2 G(x, y; \xi^* \eta^*) = \delta(x - \xi^*, y - \eta^*)$$

Since

$$\iint_D G(x, y; \xi, \eta)\delta(x - \xi^*, y - \eta^*)\, dx\, dy = G(\xi^*, \eta^*; \xi, \eta)$$

[13] For the general second-order operator in two variables, see No. 1, Exercises 10.10.

and

$$\iint_D G(x, y; \xi^*, \eta^*)\delta(x-\xi, y-\eta)\, dx\, dy = G(\xi, \eta; \xi^*, \eta^*)$$

we obtain

$$G(\xi, \eta; \xi^*, \eta^*) = G(\xi^*, \eta^*; \xi, \eta) \qquad \square$$

***Theorem 10.2.2*** *$\partial G/\partial n$ is discontinuous at $(\xi, \eta)$; in particular,*

$$\lim_{\varepsilon\to 0} \int_{C_\varepsilon} \frac{\partial G}{\partial n}\, ds = 1, \qquad C_\varepsilon : (x-\xi)^2 + (y-\eta)^2 = \varepsilon^2$$

PROOF. Let $R_\varepsilon$ be the region bounded by $C_\varepsilon$. Then integrating both sides of Eq. (10.2.3), we obtain

$$\iint_{R_\varepsilon} \nabla^2 G\, dx\, dy = \iint_R \delta(x-\xi, y-\eta)\, dx\, dy = 1$$

It therefore follows that

$$\lim_{\varepsilon\to 0} \iint_{R_\varepsilon} \nabla^2 G\, dx\, dy = 1 \tag{10.2.8}$$

Thus, by the Divergence theorem,

$$\lim_{\varepsilon\to 0} \int_{C_\varepsilon} \frac{\partial G}{\partial n}\, ds = 1 \qquad \square$$

## 10.3 METHOD OF GREEN'S FUNCTION

It is often convenient to seek $G$ as the sum of a particular integral of the nonhomogeneous equation and the solution of the associated homogeneous equation. That is, $G$ may assume the form[14]

$$G(\xi, \eta; x, y) = F(\xi, \eta; x, y) + g(\xi, \eta; x, y) \tag{10.3.1}$$

where $F$, known as the free-space Green's function, satisfies

$$\nabla^2 F = \delta(\xi - x, \eta - y) \qquad \text{in } D \tag{10.3.2}$$

and $g$ satisfies

$$\nabla^2 g = 0 \qquad \text{in } D \tag{10.3.3}$$

so that by superposition $G = F + g$ satisfies Eq. (10.2.3). Also $G = 0$ on

[14] Hereafter, $(x, y)$ will denote the source point.

$B$ requires that

$$g = -F \qquad \text{on } B \tag{10.3.4}$$

Note that $F$ need not satisfy the boundary condition.

Before we determine the solution of a particular problem, let us first find $F$ for the Laplace and Helmholtz operators.

## (1) Laplace Operator

In this case $F$ must satisfy

$$\nabla^2 F = \delta(\xi - x, \eta - y) \qquad \text{in } D$$

Then for $r = [(\xi - x)^2 + (\eta - y)^2]^{\frac{1}{2}} > 0$, that is, for $\xi \neq x$, $\eta \neq y$, we have by taking $(x, y)$ as the center

$$\nabla^2 F = \frac{1}{r}\frac{\partial}{\partial r}\left(r\frac{\partial F}{\partial r}\right) = 0$$

since $F$ is independent of $\theta$. The solution, therefore, is

$$F = A + B \log r$$

Applying condition (10.2.6), we see that

$$\lim_{\varepsilon \to 0} \int_{C_\varepsilon} \frac{\partial F}{\partial n}\, ds = \lim_{\varepsilon \to 0} \int_0^{2\pi} \frac{B}{r} r\, d\theta = 1^{15}$$

Thus $B = 1/2\pi$ and $A$ is arbitrary. For simplicity we choose $A = 0$. Then $F$ takes the form

$$F = \frac{1}{2\pi} \log r \tag{10.3.5}$$

## (2) Helmholtz Operator

Here $F$ is required to satisfy

$$\nabla^2 F + \kappa^2 F = \delta(x - \xi, y - \eta)$$

Again for $r > 0$, we find

$$\frac{1}{r}\frac{\partial}{\partial r}\left(r\frac{\partial F}{\partial r}\right) + \kappa^2 F = 0$$

or

$$r^2 F_{rr} + r F_r + \kappa^2 r^2 F = 0$$

[15] This follows directly from Eq. (10.2.8) with $\nabla^2 g = 0$.

This is the Bessel equation of order zero, the solution of which is

$$F(\kappa r) = AJ_0(\kappa r) + BY_0(\kappa r)$$

Since the behavior of $J_0$ at $r = 0$ is not singular, we set $A = 0$. Thus, we have

$$F(\kappa r) = BY_0(\kappa r)$$

But for very small $r$,

$$Y_0(\kappa r) \sim \frac{2}{\pi} \log r$$

Applying condition (10.2.6), we obtain

$$\lim_{\varepsilon \to 0} \int_{C_\varepsilon} \frac{\partial F}{\partial n} ds = \lim_{\varepsilon \to 0} \int_{C_\varepsilon} B \frac{\partial Y_0}{\partial r} ds = 1$$

and hence $B = 1/4$. Thus $F(\kappa r)$ becomes

$$F(\kappa r) = \frac{1}{4} Y_0(\kappa r) \tag{10.3.6}$$

We may point out that, since

$$(\nabla^2 + \kappa^2) \text{ approaches } \nabla^2 \quad \text{as} \quad \kappa \to 0$$

it should (and does) follow that

$$\frac{1}{4} Y_0(\kappa r) \to \frac{1}{2\pi} \log r \quad \text{as} \quad \kappa \to 0+$$

## 10.4 DIRICHLET PROBLEM FOR THE LAPLACE OPERATOR

We are now in a position to determine the solution of the Dirichlet problem

$$\begin{aligned} \nabla^2 u &= h \qquad \text{in } D \\ u &= f \qquad \text{on } B \end{aligned} \tag{10.4.1}$$

by the method of Green's function.

By putting $\phi(\xi, \eta) = G(\xi, \eta; x, y)$ and $\psi(\xi, \eta) = u(\xi, \eta)$ in Eq. (10.2.7), we obtain

$$\iint_D [G(\xi, \eta; x, y)\nabla^2 u - u(\xi, \eta)\nabla^2 G] \, d\xi \, d\eta$$

$$= \int_B \left[ G(\xi, \eta; x, y) \frac{\partial u}{\partial n} - u(\xi, \eta) \frac{\partial G}{\partial n} \right] ds$$

But

$$\nabla^2 u = h(\xi, \eta)$$

and

$$\nabla^2 G = \delta(\xi - x, \eta - y)$$

in $D$. Thus, we have

$$\iint_D [G(\xi, \eta; x, y)h(\xi, \eta) - u(\xi, \eta)\delta(\xi - x, \eta - y)]\, d\xi\, d\eta$$

$$= \int_B \left[G(\xi, \eta; x, y)\frac{\partial u}{\partial n} - u(\xi, \eta)\frac{\partial G}{\partial n}\right] ds \quad (10.4.2)$$

Since $G = 0$ and $u = f$ on $B$, and noting that $G$ is symmetric, it follows that

$$u(x, y) = \iint_D G(x, y; \xi, \eta)h(\xi, \eta)\, d\xi\, d\eta + \int_B f\frac{\partial G}{\partial n}\, ds \quad (10.4.3)$$

which is the solution given in Sec. 10.2.

As a specific example, consider the Dirichlet problem for a unit circle. Then

$$\begin{aligned} \nabla^2 g = g_{\xi\xi} + g_{\eta\eta} &= 0 \qquad \text{in } D \\ g &= -F \qquad \text{on } B \end{aligned} \quad (10.4.4)$$

But we already have from Eq. (10.3.5) that $F = (1/2\pi) \log r$.

If we introduce the polar coordinates (see Fig. 10.4.1) $\rho$, $\theta$, $\sigma$, $\beta$ by means of the equations

$$\begin{aligned} x &= \rho \cos \theta, \qquad \xi\sigma \cos \beta \\ y &= \rho \sin \theta, \qquad \eta = \sigma \sin \beta \end{aligned} \quad (10.4.5)$$

then the solution of Eq. (10.4.4) is [see Sec. 8.4]

$$g = \frac{a_0}{2} + \sum_{n=1}^{\infty} \sigma^n (a_n \cos n\beta + b_n \sin n\beta)$$

where

$$g = -\frac{1}{4\pi} \log [1 + \rho^2 - 2\rho \cos (\beta - \theta)] \qquad \text{on } B$$

By using the relation

$$\log [1 + \rho^2 - 2\rho \cos (\beta - \theta)] = -2 \sum_{n=1}^{\infty} \frac{\rho^n \cos n(\beta - \theta)}{n}$$

and equating the coefficients of $\sin n\beta$ and $\cos n\beta$ to determine $a_n$ and $b_n$,

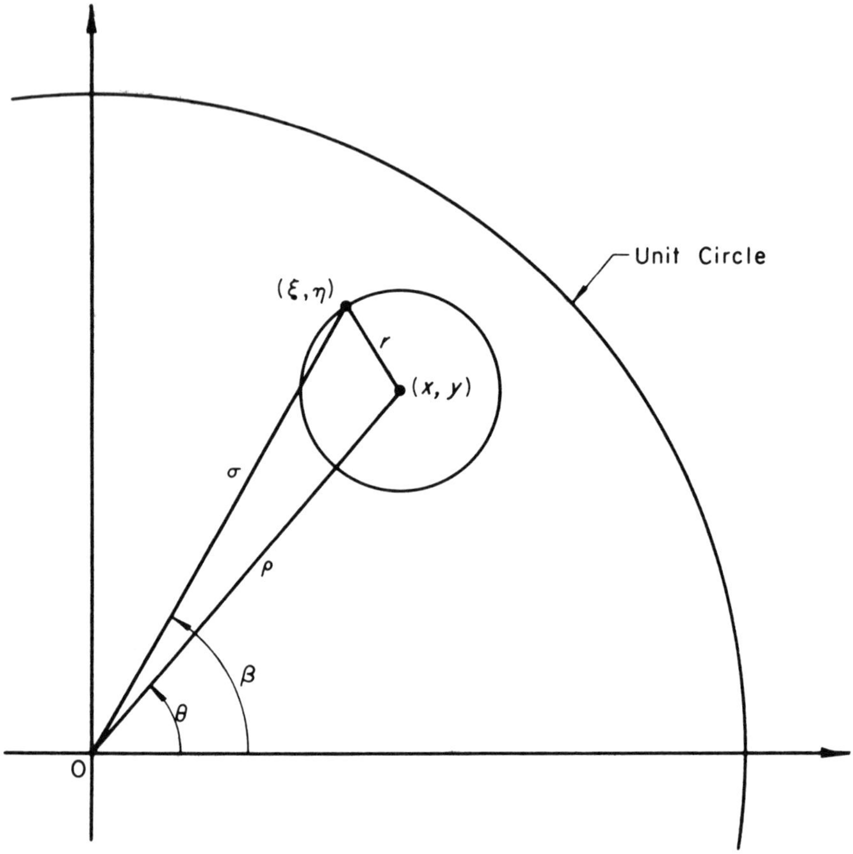


Figure 10.4.1

we find

$$a_n = \frac{\rho^n}{2\pi n} \cos n\theta$$

$$b_n = \frac{\rho^n}{2\pi n} \sin n\theta$$

It therefore follows that

$$g(\rho, \theta; \sigma, \beta) = \frac{1}{2\pi} \sum_{n=1}^{\infty} \frac{(\sigma\rho)^n}{n} \cos n(\beta - \theta)$$

$$= -\frac{1}{4\pi} \log \left[1 + (\sigma\rho)^2 - 2(\sigma\rho) \cos (\beta - \theta)\right]$$

Hence the Green's function for the problem is

$$G(\rho, \theta; \alpha, \beta) = \frac{1}{4\pi}\log\left[\sigma^2 + \rho^2 - 2\sigma\rho\cos(\beta - \theta)\right]$$

$$- \frac{1}{4\pi}\log\left[1 + (\sigma\rho)^2 - 2\sigma\rho\cos(\beta - \theta)\right] \quad (10.4.6)$$

from which we find

$$\left.\frac{\partial G}{\partial n}\right|_{\text{on } B} = \left(\frac{\partial G}{\partial \sigma}\right)_{\sigma=1} = \frac{1}{2\pi}\frac{1-\rho^2}{[1 + \rho^2 - 2\rho\cos(\beta - \theta)]}$$

If $h = 0$, then solution (10.4.3) reduces to the Poisson integral formula similar to (8.4.10) and assumes the form

$$u(\rho, \theta) = \frac{1}{2\pi}\int_0^{2\pi} \frac{1-\rho^2}{1 + \rho^2 - 2\rho\cos(\beta - \theta)} f(\beta)\, d\beta$$

## 10.5 DIRICHLET PROBLEM FOR THE HELMHOLTZ OPERATOR

We will now determine the Green's function solution of the Dirichlet problem involving the Helmholtz operator, namely,

$$\begin{aligned} \nabla^2 u + \kappa^2 u &= h \qquad \text{in } D \\ u &= f \qquad \text{on } B \end{aligned} \quad (10.5.1)$$

where $D$ is a circular domain of unit radius with boundary $B$. Then the Green's function must satisfy

$$\begin{aligned} \nabla^2 G + \kappa^2 G &= \delta(\xi - x, \eta - y) \qquad \text{in } D \\ G &= 0 \qquad \text{on } B \end{aligned} \quad (10.5.2)$$

Again, we seek the solution in the form

$$G(\xi, \eta; x, y) = F(\xi, \eta; x, y) + g(\xi, \eta; x, y)$$

From Eq. (10.3.6), we have

$$F = \frac{1}{4} Y_0(\kappa r) \quad (10.5.3)$$

where $r = [(\xi - x)^2 + (\eta - y)^2]^{\frac{1}{2}}$. The function $g$ must satisfy

$$\begin{aligned} \nabla^2 g + \kappa^2 g &= 0 \qquad \text{in } D \\ g &= -\frac{1}{4} Y_0(\kappa r) \qquad \text{on } B \end{aligned} \quad (10.5.4)$$

the solution of which can be easily determined by the method of separation of variables. Thus, the solution in the polar coordinates defined by Eq. (10.4.5) may be written in the form

$$g(\rho, \theta; \sigma, \beta) = \sum_{n=0}^{\infty} J_n(\kappa\sigma)[a_n \cos n\beta + b_n \sin n\beta] \tag{10.5.5}$$

where

$$a_0 = -\frac{1}{8\pi J_0(\kappa)} \int_{-\pi}^{\pi} Y_0[\kappa\sqrt{1+\rho^2 - 2\rho\cos(\beta-\theta)}]\, d\beta$$

$$\left.\begin{aligned} a_n &= -\frac{1}{4\pi J_n(\kappa)} \int_{-\pi}^{\pi} Y_0[\kappa\sqrt{1+\rho^2 - 2\rho\cos(\beta-\theta)}]\cos n\beta\, d\beta \\ b_n &= -\frac{1}{4\pi J_n(\kappa)} \int_{-\pi}^{\pi} Y_0[\kappa\sqrt{1+\rho^2 - 2\rho\cos(\beta-\theta)}]\sin n\beta\, d\beta \end{aligned}\right\} n = 1, 2, \ldots$$

To find the solution of the Dirichlet problem, we multiply both sides of the first equation of Eq. (10.5.1) by $G$ and integrate. Thus, we have

$$\iint_D (\nabla^2 u + \kappa^2 u) G(\xi, \eta; x, y)\, d\xi\, d\eta = \iint_D h(\xi, \eta) G(\xi, \eta; x, y)\, d\xi\, d\eta$$

We then apply Green's theorem on the left side of the preceding equation and obtain

$$\iint_D h(\xi, \eta) G(\xi, \eta; x, y)\, d\xi\, d\eta - \iint_D u(\nabla^2 G + \kappa^2 G)\, d\xi\, d\eta$$

$$= \int_B (G u_n - u G_n)\, ds$$

But $\nabla^2 G + \kappa^2 G = \delta(\xi - x, \eta - y)$ in $D$ and $G = 0$ on $B$. We therefore have

$$u(x, y) = \iint_D h(\xi, \eta) G(\xi, \eta; x, y)\, d\xi\, d\eta + \int_B f(\xi, \eta) G_n\, ds \tag{10.5.6}$$

where $G$ is given by Eqs. (10.5.3) and (10.5.5).

## 10.6 METHOD OF IMAGES

We shall describe another method of obtaining Green's function. This method, called the *method of images*, is based essentially on the construction of Green's function for a finite domain from that of an infinite domain. The disadvantage of this method is that it can be applied only to problems with simple boundary geometries.

As an illustration, we consider the same Dirichlet problem solved in Sec. 10.4.

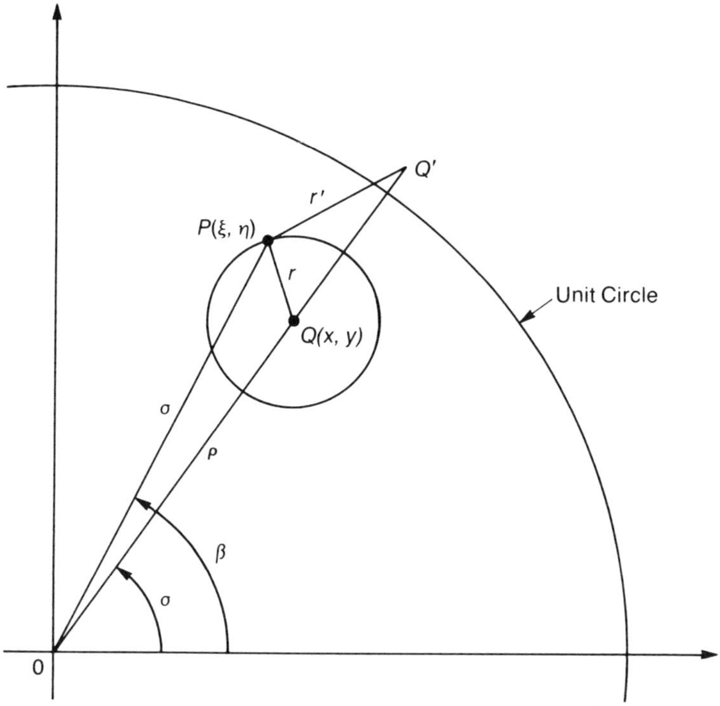


Figure 10.6.1

Let $P(\xi, \eta)$ be a point in the unit circle $D$, and let $Q(x, y)$ be the source point also in $D$. The distance between $P$ and $Q$ is $r$. Let $Q'$ be the *image* which lies outside of $D$ on the ray from the origin opposite to the source point $Q$ (as shown in Fig. 10.6.1) such that $OQ/\sigma = \sigma/OQ'$ where $\sigma$ is the radius of the circle passing through $P$ centered at the origin.

Since the two triangles $OPQ$ and $OPQ'$ are similar by virtue of the hypothesis $(OQ)(OQ') = \sigma^2$ and by possessing a common angle at $O$, we have

$$\frac{r'}{r} = \frac{\sigma}{\rho} \tag{10.6.1}$$

where $r' = PQ'$ and $\rho = OQ$.

If $\sigma = 1$, Eq. (10.6.1) becomes

$$\frac{r}{r'}\frac{1}{\rho} = 1$$

Then we can clearly see that the quantity

$$\frac{1}{2\pi}\log\left(\frac{r}{r'}\frac{1}{\rho}\right) = \frac{1}{2\pi}\log r - \frac{1}{2\pi}\log r' + \frac{1}{2\pi}\log\frac{1}{\rho} \tag{10.6.2}$$

which vanishes on the boundary $\sigma = 1$, is harmonic in $D$ except at $Q$, and satisfies Eq. (10.2.3). [Note the $\log r'$ is harmonic everywhere except at $Q'$, which is outside the domain $D$.] This suggests that we should choose the Green's function

$$G = \frac{1}{2\pi}\log r - \frac{1}{2\pi}\log r' + \frac{1}{2\pi}\log\frac{1}{\rho} \tag{10.6.3}$$

Noting that $Q'$ is at $(1/\rho, \theta)$, $G$ in polar coordinates takes the form

$$G(\rho, \theta; \sigma, \beta) = \frac{1}{4\pi}\log\left[\sigma^2 + \rho^2 - 2\sigma\rho\cos(\beta - \theta)\right]$$
$$- \frac{1}{4\pi}\log\left[\frac{1}{\sigma^2} + \rho^2 - 2\frac{\rho}{\sigma}\cos(\beta - \theta)\right] + \frac{1}{2\pi}\log\frac{1}{\sigma} \tag{10.6.4}$$

which is the same as $G$ given by (10.4.6).

It is quite interesting to observe the physical interpretation of the Green's function (10.6.3) or (10.6.4). The first term represents the potential due to a unit line charge at the source point, whereas the second term represents the potential due to a negative unit charge at the image point. The third term represents a uniform potential. The sum of these potentials makes up the potential field.

EXAMPLE 10.6.1. To illustrate an obvious and simple case, consider the semi-infinite plane $\eta > 0$. The problem is to solve

$$\nabla^2 u = h \quad \text{in} \quad \eta > 0$$
$$u = f \quad \text{on} \quad \eta = 0$$

The image point should be obvious by inspection. Thus, if we construct

$$G = \frac{1}{4\pi}\log\left[(\xi - x)^2 + (\eta - y)^2\right] - \frac{1}{4\pi}\log\left[(\xi - x)^2 + (\eta + y)^2\right] \tag{10.6.5}$$

the condition that $G = 0$ on $\eta = 0$ is clearly satisfied. It is also evident that $G$ is harmonic in $\eta > 0$ except at the source point, and that $G$ satisfies Eq. (10.2.3).

With $G_n|_B = [-G_\eta]_{\eta=0}$, the solution (10.4.3) is thus given by

$$u(x, y) = \frac{y}{\pi}\int_{-\infty}^{\infty}\frac{f(\xi)\,d\xi}{(\xi - x)^2 + y^2}$$
$$+ \frac{1}{4\pi}\int_0^{\infty}\int_{-\infty}^{\infty}\log\left[\frac{(\xi - x)^2 + (\eta - y)^2}{(\xi - x)^2 + (\eta + y)^2}\right]h(\xi, \eta)\,d\xi\,d\eta \tag{10.6.6}$$

EXAMPLE 10.6.2. Another example that illustrates the method of images well is the Robin's problem on the quarter infinite plane, namely

$$\begin{aligned} \nabla^2 u &= h(\xi, \eta) \quad \text{in} \quad \xi > 0, \quad \eta > 0 \\ u &= f(\eta) \quad \text{on} \quad \xi = 0 \\ u_n &= g(\xi) \quad \text{on} \quad \eta = 0 \end{aligned} \tag{10.6.7}$$

This is illustrated in Fig. 10.6.2.

Let $(-x, y)$, $(-x, -y)$, and $(x, -y)$ be the three image points of the source point $(x, y)$. Then, by inspection, we can immediately construct Green's function

$$G = \frac{1}{4\pi} \log \frac{[(\xi - x)^2 + (\eta - y)^2][(\xi - x)^2 + (\eta + y)^2]}{[(\xi + x)^2 + (\eta - y)^2][(\xi + x)^2 + (\eta + y)^2]} \tag{10.6.8}$$

This function satisfies $\nabla^2 G = 0$ except at the source point, and $G = 0$ on $\xi = 0$ and $G_\eta = 0$ on $\eta = 0$.

The solution from Eq. (10.4.2) is thus

$$\begin{aligned} u(x, y) &= \iint_D Gh \, d\xi \, d\eta + \int_B (Gu_n - uG_n) \, ds \\ &= \int_0^\infty \int_0^\infty Gh \, d\xi \, d\eta + \int_0^\infty g(\xi) G(\xi, 0; x, y) \, d\xi \\ &\quad + \int_0^\infty f(\eta) G_\xi(0, \eta; x, y) \, d\xi \end{aligned}$$

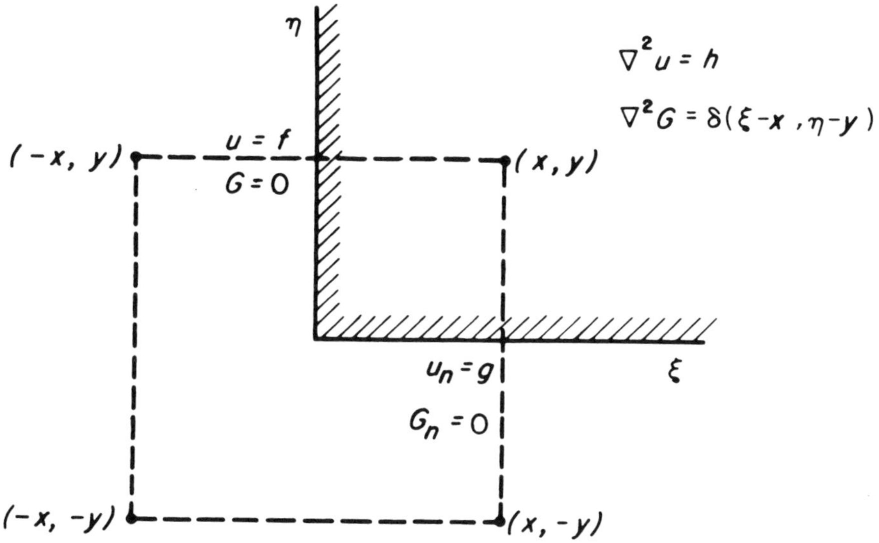


Figure 10.6.2

## 10.7 METHOD OF EIGENFUNCTIONS

In this section, we will apply the method of eigenfunctions, described in Chapter 9, to obtain Green's function.

We consider the boundary value problem

$$\begin{aligned} \nabla^2 u &= h \quad \text{in } D \\ u &= f \quad \text{on } B \end{aligned} \tag{10.7.1}$$

For this problem, $G$ must satisfy

$$\begin{aligned} \nabla^2 G &= \delta(\xi - x, \eta - y) \quad \text{in } D \\ G &= 0 \quad \text{on } B \end{aligned} \tag{10.7.2}$$

and hence the associated eigenvalue problem is

$$\begin{aligned} \nabla^2 \phi + \lambda \phi &= 0 \quad \text{in } D \\ \phi &= 0 \quad \text{on } B \end{aligned} \tag{10.7.3}$$

Let $\phi_{mn}$ be the eigenfunctions and $\lambda_{mn}$ be the corresponding eigenvalues. We then expand $G$ and $\delta$ in terms of the eigenfunctions $\phi_{mn}$. Consequently, we write

$$G(\xi, \eta; x, y) = \sum_m \sum_n a_{mn}(x, y) \phi_{mn}(\xi, \eta) \tag{10.7.4}$$

$$\delta(\xi - x, \eta - y) = \sum_m \sum_n b_{mn}(x, y) \phi_{mn}(\xi, \eta) \tag{10.7.5}$$

where

$$\begin{aligned} b_{mn} &= \frac{1}{\|\phi_{mn}\|^2} \iint_D \delta(\xi - x, \eta - y) \phi_{mn}(\xi, \eta)\, d\xi\, d\eta \\ &= \frac{\phi_{mn}(x, y)}{\|\phi_{mn}\|^2} \end{aligned} \tag{10.7.6}$$

in which

$$\|\phi_{mn}\|^2 = \iint_D \phi_{mn}^2 \, d\xi\, d\eta$$

Now substituting Eqs. (10.7.4) and (10.7.5) into Eq. (10.7.2) and using the relation from Eq. (10.7.3) that

$$\nabla^2 \phi_{mn} + \lambda_{mn} \phi_{mn} = 0$$

we obtain

$$-\sum_m \sum_n \lambda_{mn} a_{mn}(x, y) \phi_{mn}(\xi, \eta) = \sum_m \sum_n \frac{\phi_{mn}(x, y) \phi_{mn}(\xi, \eta)}{\|\phi_{mn}\|^2}$$

Hence

$$a_{mn}(x, y) = -\frac{\phi_{mn}(x, y)}{\lambda_{mn} \|\phi_{mn}\|^2} \tag{10.7.7}$$

and the Green's function is therefore given by

$$G(\xi, \eta; x, y) = -\sum_m \sum_n \frac{\phi_{mn}(x, y)\phi_{mn}(\xi, \eta)}{\lambda_{mn} \|\phi_{mn}\|^2} \tag{10.7.8}$$

EXAMPLE 10.7.1. As a particular example, consider the Dirichlet problem in a rectangular domain

$$\nabla^2 u = h \qquad \text{in } D$$
$$u = 0 \qquad \text{on } B$$

The eigenfunctions can be obtained explicitly by the method of separation of variables. We assume a solution in the form

$$u(\xi, \eta) = X(\xi)Y(\eta)$$

Substitution of this in

$$\nabla^2 u + \lambda u = 0 \qquad \text{in } D$$
$$u = 0 \qquad \text{on } B$$

yields, with $\alpha^2$ as separation constant,

$$X'' + \alpha^2 X = 0$$
$$Y'' + (\lambda - \alpha^2)Y = 0$$

With the homogeneous boundary conditions $X(0) = X(a) = 0$ and $Y(0) = Y(b) = 0$, $X$ and $Y$ are found to be

$$X_m(\xi) = A_m \sin\frac{m\pi\xi}{a}$$

$$Y_n(\eta) = B_n \sin\frac{n\pi\eta}{b}$$

We then have

$$\lambda_{mn} = \pi^2\left(\frac{m^2}{a^2} + \frac{n^2}{b^2}\right) \qquad \text{with} \qquad \alpha = \frac{m\pi}{a}$$

Thus, we obtain the eigenfunctions

$$\phi_{mn}(\xi, \eta) = \sin\frac{m\pi\xi}{a}\sin\frac{n\pi\eta}{b}$$

Knowing $\phi_{mn}$, we compute $\|\phi_{mn}\|$ and obtain

$$\|\phi_{mn}\|^2 = \int_0^a \int_0^b \sin^2 \frac{m\pi\xi}{a} \sin^2 \frac{n\pi\eta}{b} \, d\xi \, d\eta$$

$$= \frac{ab}{4}$$

We thus obtain from Eq. (10.7.8) the Green's function

$$G(\xi, \eta; x, y) = -\frac{4ab}{\pi^2} \sum_{m=1}^{\infty} \sum_{n=1}^{\infty} \frac{\sin \dfrac{m\pi x}{a} \sin \dfrac{n\pi y}{b} \sin \dfrac{m\pi\xi}{a} \sin \dfrac{n\pi\eta}{b}}{(m^2 b^2 + n^2 a^2)}$$

## 10.8 HIGHER DIMENSIONAL PROBLEMS

The Green's function method can be easily extended for applications in three and more dimensions. Since most of the problems encountered in the physical sciences are in three dimensions, we will illustrate some examples suitable for practical application.

First of all, let us extend our definition of Green's function in three dimensions.

The Green's function for the Dirichlet problem involving the Laplace operator is the function that satisfies

a.

$$\nabla^2 G = \delta(x - \xi, y - \eta, z - \zeta) \quad \text{in } R \tag{10.8.1}$$

$$G = 0 \quad \text{on } S \tag{10.8.2}$$

b.

$$G(x, y, z; \xi, \eta, \zeta) = G(\xi, \eta, \zeta; x, y, z) \tag{10.8.3}$$

c.

$$\lim_{\varepsilon \to 0} \iint_{S_\varepsilon} \frac{\partial G}{\partial n} \, dS = 1 \tag{10.8.4}$$

where $n$ is the outward unit normal to the surface

$$S_\varepsilon : (x - \xi)^2 + (y - \eta)^2 + (z - \zeta)^2 = \varepsilon^2$$

Proceeding as in the two-dimensional case, the solution of the Dirichlet problem

$$\begin{aligned} \nabla^2 u &= h \quad \text{in } R \\ u &= f \quad \text{on } S \end{aligned} \tag{10.8.5}$$

is

$$u(x, y, z) = \iiint_R Gh\, dR + \iint_S fG_n\, dS \tag{10.8.6}$$

Again we let

$$G(\xi, \eta, \zeta; x, y, z) = F(\xi, \eta, \zeta; x, y, z) + g(\xi, \eta, \zeta; x, y, z)$$

where

$$\nabla^2 F = \delta(x - \xi, y - \eta, z - \zeta) \qquad \text{in } R$$

and

$$\begin{aligned} \nabla^2 g &= 0 \qquad && \text{in } R \\ g &= -F && \text{on } S \end{aligned}$$

EXAMPLE 10.8.1. We consider a spherical domain with radius $a$. We must have

$$\nabla^2 F = 0$$

except at the source point. For

$$r = [(\xi - x)^2 + (\eta - y)^2 + (\zeta - z)^2]^{\frac{1}{2}} > 0$$

with $(x, y, z)$ as the origin, we have

$$\nabla^2 F = \frac{1}{r^2}\frac{d}{dr}\left(r^2 \frac{dF}{dr}\right) = 0$$

Integration then yields

$$F = A + \frac{B}{r} \quad \text{for} \quad r > 0$$

Applying the condition (10.8.4) we obtain

$$\lim_{\varepsilon \to 0} \iint_{S_\varepsilon} G_n\, dS = \lim_{\varepsilon \to 0} \iint_{S_\varepsilon} F_r\, dS = 1$$

Consequently, $B = -1/4\pi$ and $A$ is arbitrary. If we set $A = 0$ for convenience,[16] we have

$$F = -\frac{1}{4\pi r} \tag{10.8.7}$$

We apply the method of images to obtain the Green's function. If we

[16] This is the boundedness condition at infinity for exterior problems.

draw a three-dimensional diagram analogous to Fig. 10.6.1, we will have a relation similar to (10.6.1), namely,

$$r' = \frac{a}{\rho} r \tag{10.8.8}$$

where $r'$ and $\rho$ are measured in three-dimensional space. Thus, we seek Green's function

$$G = \frac{-1}{4\pi r} + \frac{a/\rho}{4\pi r'} \tag{10.8.9}$$

which is harmonic everywhere in $r$ except at the source point, and is zero on the surface $S$.

In terms of spherical coordinates

$$\begin{aligned} \xi &= \tau \cos \psi \sin \alpha, & x &= \rho \cos \phi \sin \theta \\ \eta &= \tau \sin \psi \sin \alpha, & y &= \rho \sin \phi \sin \theta \\ \zeta &= \tau \cos \alpha, & z &= \rho \cos \theta \end{aligned}$$

$G$ can be written in the form

$$G = \frac{-1}{4\pi(\tau^2 + \rho^2 - 2\tau\rho \cos \gamma)^{\frac{1}{2}}} + \frac{1}{4\pi \left[\dfrac{\tau^2 \rho^2}{a^2} + a^2 - 2\tau\rho \cos \gamma\right]^{\frac{1}{2}}} \tag{10.8.10}$$

where $\gamma$ is the angle between $r$ and $r'$. Now differentiating $G$, we have

$$\left[\frac{\partial G}{\partial \tau}\right]_{\tau=a} = \frac{a^2 - \rho^2}{4\pi a(a^2 + \rho^2 - 2a\rho \cos \gamma)^{\frac{3}{2}}}$$

Thus, the solution of the Dirichlet problem for $h = 0$ is

$$u(\rho, \theta, \phi) = \frac{a(a^2 - \rho^2)}{4\pi} \int_0^{2\pi} \int_0^{\pi} \frac{f(\alpha, \psi) \sin \alpha \, d\alpha \, d\psi}{(a^2 + \rho^2 - 2a\rho \cos \gamma)^{\frac{3}{2}}} \tag{10.8.11}$$

where $\cos \gamma = \cos \alpha \cos \theta + \sin \alpha \sin \theta \cos (\psi - \phi)$. This integral is called the *three-dimensional Poisson integral formula*.

For the exterior problem where the outward normal is radially inward towards the origin, the solution can be simply obtained by replacing $(a^2 - \rho^2)$ by $(\rho^2 - a^2)$ in Eq. (10.8.11).

EXAMPLE 10.8.2. Another example involving the Helmholtz operator is the three-dimensional radiation problem

$$\nabla^2 u + \kappa^2 u = 0$$

$$\lim_{r \to \infty} r(u_r + i\kappa u) = 0$$

where $i = \sqrt{-1}$; the limit condition is called the radiation condition, and $r$ is the field point distance.

In this case, the Green's function must satisfy

$$\nabla^2 G + \kappa^2 G = \delta(\xi - x, \eta - y, \zeta - z)$$

Since the point source solution is dependent only on $r$, we write the Helmholtz equation

$$G_{rr} + \frac{2}{r} G_r + \kappa^2 G = 0 \quad \text{for} \quad r > 0$$

Note that the source point is taken as the origin. If we write the above equation in the form

$$(Gr)_{rr} + \kappa^2 (Gr) = 0 \quad \text{for} \quad r > 0$$

then the solution can easily be seen to be

$$Gr = Ae^{i\kappa r} + Be^{-i\kappa r}$$

or

$$G = A \frac{e^{i\kappa r}}{r} + B \frac{e^{-i\kappa r}}{r}$$

In order for $G$ to satisfy the radiation condition

$$\lim_{r \to \infty} r(G_r + i\kappa G) = 0$$

$A = 0$ and $G$ thus takes the form

$$G = B \frac{e^{-i\kappa r}}{r}$$

To determine $B$ we have

$$\lim_{\varepsilon \to 0} \iint_{S_\varepsilon} \frac{\partial G}{\partial n} dS = -\lim_{\varepsilon \to 0} \iint_{S_\varepsilon} B \frac{e^{-i\kappa r}}{r} \left( \frac{1}{r} + i\kappa \right) dS = 1$$

from which we obtain $B = -1/4\pi$, and consequently, $G$ becomes

$$G = -\frac{e^{-i\kappa r}}{4\pi r}.$$

Note that this reduces to $-1/4\pi r$ when $\kappa = 0$.

## 10.9 NEUMANN PROBLEM

We have noted in the chapter on boundary-value problems that the Neumann problem requires more attention than Dirichlet's problem, because an additional condition is necessary for the existence of a solution of the Neumann problem.

Let us now consider the Neumann problem

$$\nabla^2 u + \kappa^2 u = h \qquad \text{in } R$$

$$\frac{\partial u}{\partial n} = 0 \qquad \text{on } S$$

By the divergence theorem, we have

$$\iiint_R \nabla^2 u \, dR = \iint_S \frac{\partial u}{\partial n} \, dS$$

Thus, if we integrate the Helmholtz equation and use the preceding result, we obtain

$$\kappa^2 \iiint_R u \, dR = \iiint_R h \, dR$$

In the case of Poisson's equation where $\kappa = 0$, this relation is satisfied only when

$$\iiint_R h \, dR = 0$$

If we consider a heat conduction problem, this condition may be interpreted as the requirement that the net generation of heat be zero. This is physically reasonable since the boundary is insulated in such a way that the net flux across it is zero.

If we define Green's function $G$, in this case, by

$$\nabla^2 G + \kappa^2 G = \delta(\xi - x, \eta - y, \zeta - z) \qquad \text{in } R$$

$$\frac{\partial G}{\partial n} = 0 \qquad \text{on } S$$

Then we must have

$$\kappa^2 \iiint_R G \, dR = 1$$

which cannot be satisfied for $\kappa = 0$. But, we know by physical reasoning that a solution exists if

$$\iiint_R h \, dR = 0$$

Hence we will modify the definition of Green's function so that

$$\frac{\partial G}{\partial n} = C \qquad \text{on } S$$

where $C$ is a constant. When integrating $\nabla^2 G = \delta$ over $R$, we obtain

$$C \iint_S dS = 1$$

It is not difficult to show that $G$ remains symmetric if

$$\iint_S G\, dS = 0$$

Thus, under this condition, if we take $C$ to be the reciprocal of the surface area, the solution of the Neumann problem for Poisson's equation is

$$u(x, y, z) = C^* + \iiint_R G(x, y, z; \xi, \eta, \zeta)\, h(\xi, \eta, \zeta)\, d\xi\, d\eta\, d\zeta$$

where $C^*$ is a constant.

We should remark here that the method of Green's functions provides the solution in integral form. This is made possible by replacing a problem involving nonhomogeneous boundary conditions with a problem of finding Green's function $G$ with homogeneous boundary conditions.

Regardless of methods employed, the Green's function of a problem with a nonhomogeneous equation and homogeneous boundary conditions is the same as the Green's function of a problem with a homogeneous equation and nonhomogeneous boundary conditions, since one problem can be transferred to the other without difficulty. To illustrate, consider the problem

$$\begin{aligned} Lu &= f && \text{in } R \\ u &= 0 && \text{on } \partial R \end{aligned}$$

where $\partial R$ denotes the boundary of $R$.

If we let $v = w - u$, where $w$ satisfies $Lw = f$ in $R$, then the problem becomes

$$\begin{aligned} Lv &= 0 && \text{in } R \\ v &= w && \text{on } \partial R \end{aligned}$$

Conversely, if we consider the problem

$$\begin{aligned} Lu &= 0 && \text{in } R \\ u &= g && \text{on } \partial R \end{aligned}$$

we can easily transform this problem into

$$\begin{aligned} Lv = Lw &\equiv w^* && \text{in } R \\ v &= 0 && \text{on } \partial R \end{aligned}$$

by putting $v = w - u$ and finding $w$ that satisfies $w = g$ on $\partial R$.

In fact, if we have

$$Lu = f \qquad \text{in } R$$

$$u = g \qquad \text{on } \partial R$$

we can transform this problem into either one of the above problems.

## 10.10 EXERCISES

**1.** If $L$ denotes the operator

$$Lu = Au_{xx} + Bu_{xy} + Cu_{yy} + Du_x + Eu_y + Fu$$

and if $M$ denotes the adjoint operator

$$Mv = (Av)_{xx} + (Bv)_{xy} + (Cv)_{yy} - (Dv)_x - (Ev)_y + Fv$$

show that

$$\iint_R (vLu - uMv)\, dx\, dy = \int_{\partial R} [U \cos(n, x) + V \cos(n, y)]\, ds$$

where

$$U = Avu_x - u(Av)_x - u(Bv)_y + Duv$$

$$V = Bvu_x + Cvu_y - u(Cv)_y + Euv$$

and $\partial R$ is the boundary of a region $R$.

**2.** Prove that the Green's function for a region, if it exists, is unique.

**3.** Determine the Green's function for the exterior Dirichlet problem for a unit circle

$$\nabla^2 u = 0 \quad \text{in} \quad r > 1$$

$$u = f \quad \text{on} \quad r = 1$$

**4.** Prove that for $x = x(\xi, \eta)$ and $y = y(\xi, \eta)$

$$\delta(x - x_0)\delta(y - y_0) = \frac{1}{|J|}\, \delta(\xi - \xi_0)\delta(\eta - \eta_0)$$

where $J$ is the Jacobian and $(x_0, y_0)$ corresponds to $(\xi_0, \eta_0)$. Hence show that for polar coordinates

$$\delta(x - x_0)\delta(y - y_0) = \frac{1}{r}\, \delta(r - r_0)\delta(\theta - \theta_0)$$

**5.** Determine, for an infinite wedge, the Green's function that satisfies

$$\nabla^2 G + \kappa^2 G = \frac{1}{r}\,\delta(r - r_0,\, \theta - \theta_0)$$

$$G = 0, \qquad \theta = 0, \quad \text{and} \quad \theta = \alpha$$

**6.** Determine, for the Poisson equation, the Green's function which vanishes on the boundary of a semicircular domain of radius $R$.

**7.** Find the solution of the Dirichlet problem

$$\nabla^2 u = 0, \qquad 0 < x < a, \qquad 0 < y < b$$

$$u(0, y) = u(a, y) = u(x, b) = 0$$

$$u(x, 0) = f(x)$$

**8.** Determine the solution of Dirichlet's problem

$$\nabla^2 u = f(r, \theta) \qquad \text{in } D$$

$$u = 0 \qquad \text{on } B$$

where $B$ is the boundary of a circle $D$ of radius $R$.

**9.** Determine the Green's function for the semi-infinite region $\zeta > 0$ for

$$\nabla^2 G + \kappa^2 G = \delta(\xi - x,\, \eta - y,\, \zeta - z)$$

$$G = 0 \quad \text{on} \quad \zeta = 0$$

**10.** Determine the Green's function for the semi-infinite region $\zeta > 0$ for

$$\nabla^2 G + \kappa^2 G = \delta(\xi - x,\, \eta - y,\, \zeta - z)$$

$$\frac{\partial G}{\partial n} = 0 \quad \text{on} \quad \zeta = 0$$

**11.** Find the Green's function in the quarter plane $\xi > 0$, $\eta > 0$ which satisfies

$$\nabla^2 G = \delta(\xi - x,\, \eta - y)$$

$$G = 0 \quad \text{on} \quad \xi = 0 \quad \text{and} \quad \eta = 0$$

**12.** Find the Green's function in the quarter plane $\xi > 0$, $\eta > 0$ which satisfies

$$\nabla^2 G = \delta(\xi - x,\, \eta - y)$$

$$G_\xi(0, \eta) = 0$$

$$G(\xi, 0) = 0$$

**13.** Find the Green's function in the plane $0 < x < \infty$, $-\infty < y < \infty$ for the problem

$$\nabla^2 u = f \qquad \text{in } R$$

$$u = 0 \quad \text{on} \quad x = 0$$

**14.** Determine the Green's function which satisfies

$$\nabla^2 G = \delta(x-\xi, y-\eta) \quad \text{in} \quad R: 0<x<a, \qquad 0<y<\infty$$
$$G=0 \quad \text{on} \quad \partial R: x=0, \qquad x=a, \qquad y=0$$

$G$ is bounded at infinity.

**15.** Find the Green's function that satisfies

$$\nabla^2 G = \frac{1}{r}\delta(r-\rho, \theta-\beta), \qquad 0<\theta<\frac{\pi}{3}, \qquad 0<r<1$$

$$G=0 \quad \text{on} \quad \theta=0 \quad \text{and} \quad \theta=\frac{\pi}{3}$$

$$\frac{\partial G}{\partial n}=0 \quad \text{on} \quad r=1$$

**16.** Solve the boundary-value problem

$$\frac{1}{r}\frac{\partial}{\partial r}\left(r\frac{\partial u}{\partial r}\right)+\frac{\partial^2 u}{\partial z^2}+\kappa^2 u=0, \qquad r\geqslant 0, \qquad z>0$$

$$\frac{\partial u}{\partial z}=\begin{cases} 0 & r>a, \quad z=0 \\ C & r<a, \quad z=0, \quad C=\text{constant}\end{cases}$$

**17.** Obtain the solution of

$$\nabla^2 u=0, \qquad 0<r<\infty, \qquad 0<\theta<2\pi$$
$$u(r, 0+)=u(r, 2\pi-)=0$$

**18.** Determine the Green's function for the equation

$$\nabla^2 u-\kappa^2 u=0$$

vanishing on all sides of the rectangle $0\leqslant x\leqslant a$, $0\leqslant y\leqslant b$.

**19.** Determine the Green's function of the Helmholtz equation

$$\nabla^2 u+\kappa^2 u=0, \qquad 0<x<a, \qquad -\infty<y<\infty$$

vanishing on $x=0$ and $x=a$.

**20.** Solve the exterior Dirichlet problem

$$\nabla^2 u=0 \quad \text{in} \quad r>1$$
$$u(1, \theta, \phi)=f(\theta, \phi)$$

**21.** By the method of images, determine the potential due to a point charge $q$ near a conducting sphere of radius $R$ with potential $V$.

**22.** By the method of images, show that the potential due to a conducting sphere of radius $R$ in a uniform electric field $E_0$ is given by

$$U = -E_0\left(r - \frac{R^2}{r^2}\right)\cos\theta$$

where $r$, $\theta$ are the polar coordinates with origin at the center of the sphere.

**23.** Determine the potential in a cylinder of radius $R$ and length $l$. The potential on the ends is zero while the potential on the cylindrical surface is prescribed to be $f(\theta, z)$.

CHAPTER ELEVEN

# Integral Transforms with Applications

Integral transform methods are found to be very useful for finding solutions of initial and/or boundary-value problems governed by partial differential equations. The aim of this chapter is to provide an introduction to the use of integral transform methods for students of applied mathematics, physics, and engineering. Since our major interest is in the application of integral transforms, no attempt will be made to discuss the basic results and theorems relating to transforms in their general forms. The present treatment is restricted to classes of functions which usually occur in physical and engineering applications.

## 11.1 FOURIER TRANSFORMS

We first give a formal definition of the Fourier transform by expressing the Fourier integral formula (5.12.9) in the form of a theorem.

***Theorem 11.1.1*** *Let $f(x)$ be continuous, piecewise smooth, and absolutely integrable. If*

$$F(\alpha) = \frac{1}{\sqrt{2\pi}} \int_{-\infty}^{\infty} f(t) e^{i\alpha t}\, dt, \tag{11.1.1}$$

*then for all $x$*

$$f(x) = \frac{1}{\sqrt{2\pi}} \int_{-\infty}^{\infty} F(\alpha) e^{-i\alpha x}\, d\alpha \tag{11.1.2}$$

The function $F(\alpha)$ is called the *Fourier transform* of $f(x)$, and $f(x)$ is called the *inverse Fourier transform* of $F(\alpha)$. One may note that the factor $1/2\pi$ has been split and placed in front of the integrals in (11.1.1) and (11.1.2). Often we find the factor $1/2\pi$ contained in only one of the relations (11.1.1) and (11.1.2). It is not uncommon to find the term $e^{i\alpha t}$ in (11.1.1) replaced by $e^{-i\alpha t}$, and, as a consequence, $e^{-i\alpha x}$ replaced by $e^{i\alpha x}$ in (11.1.2).

EXAMPLE 11.1.1. Find the Fourier transform of the function $f(x) = e^{-|x|}$.

$$\begin{aligned}
F(\alpha) &= \frac{1}{\sqrt{2\pi}} \int_{-\infty}^{\infty} e^{-|t|} e^{i\alpha t}\, dt \\
&= \frac{1}{\sqrt{2\pi}} \int_{-\infty}^{0} e^{t(1+i\alpha)}\, dt + \frac{1}{\sqrt{2\pi}} \int_{0}^{\infty} e^{-t(1-i\alpha)}\, dt \\
&= \frac{1}{\sqrt{2\pi}} \left( \frac{1}{1+i\alpha} + \frac{1}{1-i\alpha} \right) \\
&= \sqrt{\frac{2}{\pi}}\, \frac{1}{1+\alpha^2}
\end{aligned}$$

Analogous to Fourier sine and cosine series, there are Fourier sine and cosine transforms for odd and even functions respectively.

***Corollary 11.1.1*** *Let $f(x)$ be defined for $0 \leqslant x < \infty$. Let $f(x)$ be extended as an odd function in $(-\infty, \infty)$ satisfying the conditions of Fourier Integral Theorem 5.12.1. If, at the points of continuity,*

$$F_s(\alpha) = \sqrt{\frac{2}{\pi}} \int_0^{\infty} f(t) \sin \alpha t\, dt$$

*then*

$$f(x) = \sqrt{\frac{2}{\pi}} \int_0^{\infty} F_s(\alpha) \sin \alpha x\, d\alpha$$

PROOF. From Eq. (5.12.8), we have

$$f(x) = \frac{1}{\pi} \int_0^{\infty} \left[ \int_{-\infty}^{\infty} f(t)(\cos \alpha t \cos \alpha x + \sin \alpha t \sin \alpha x)\, dt \right] d\alpha$$

Since $f(-t) = -f(t)$ for all real $t$,

$$\int_{-\infty}^{\infty} f(t) \cos \alpha t\, dt = 0$$

$$\int_{-\infty}^{\infty} f(t) \sin \alpha t\, dt = 2\int_0^{\infty} f(t) \sin \alpha t\, dt$$

and consequently,

$$f(x) = \sqrt{\frac{2}{\pi}} \int_0^{\infty} \left[ \sqrt{\frac{2}{\pi}} \int_0^{\infty} f(t) \sin \alpha t\, dt \right] \sin \alpha x\, d\alpha$$

This is the assertion. □

***Corollary 11.1.2*** *Let $f(x)$ be defined for $0 \leqslant x < \infty$. Let $f(x)$ be extended as an even function in $(-\infty, \infty)$ satisfying the conditions of Fourier Integral Theorem* 5.12.1. *Then if, at the points of continuity,*

$$F_c(\alpha) = \sqrt{\frac{2}{\pi}} \int_0^\infty f(t) \cos \alpha t \, dt$$

*then*

$$f(x) = \sqrt{\frac{2}{\pi}} \int_0^\infty F_c(\alpha) \cos \alpha x \, d\alpha$$

PROOF. From Eq. (5.12.8), we have

$$f(x) = \frac{1}{\pi} \int_0^\infty \left[ \int_{-\infty}^\infty f(t)(\cos \alpha t \cos \alpha x + \sin \alpha t \sin \alpha x) \, dt \right] d\alpha$$

Since $f(-t) = f(t)$ for all real $t$,

$$\int_{-\infty}^\infty f(t) \cos \alpha t \, dt = 2 \int_0^\infty f(t) \cos \alpha t \, dt$$

$$\int_{-\infty}^\infty f(t) \sin \alpha t \, dt = 0$$

and consequently,

$$f(x) = \sqrt{\frac{2}{\pi}} \int_0^\infty \left[ \sqrt{\frac{2}{\pi}} \int_0^\infty f(t) \cos \alpha t \, dt \right] \cos \alpha x \, d\alpha$$

This completes the proof. □

EXAMPLE 11.1.2. Find the Fourier sine and cosine transforms of $f(x) = e^{-ax}$ for $a > 0$.[17]

$$\begin{aligned} F_s(\alpha) &= \sqrt{\frac{2}{\pi}} \int_0^\infty e^{-at} \sin \alpha t \, dt \\ &= \sqrt{\frac{2}{\pi}} \left[ -e^{-at} \frac{\cos \alpha t}{\alpha} \right]_0^\infty - \sqrt{\frac{2}{\pi}} \int_0^\infty \frac{a}{\alpha} e^{-at} \cos \alpha t \, dt \\ &= \sqrt{\frac{2}{\pi}} \frac{1}{\alpha} - \sqrt{\frac{2}{\pi}} \frac{a}{\alpha} \left[ e^{-at} \frac{\sin \alpha t}{\alpha} \right]_0^\infty - \sqrt{\frac{2}{\pi}} \frac{a^2}{\alpha^2} \int_0^\infty e^{-at} \sin \alpha t \, dt \end{aligned}$$

Hence

$$F_s\left(1 + \frac{a^2}{\alpha^2}\right) = \sqrt{\frac{2}{\pi}} \frac{1}{\alpha}$$

[17] Hereafter abbreviation $[\ ]_a^\infty$ will denote $\lim_{b \to \infty} [\ ]_a^b$.

and consequently,

$$F_s = \sqrt{\frac{2}{\pi}} \frac{\alpha}{a^2 + \alpha^2}$$

In a similar manner, $F_c(\alpha)$ is found to be

$$F_c = \sqrt{\frac{2}{\pi}} \frac{a}{a^2 + \alpha^2}$$

## 11.2 PROPERTIES OF FOURIER TRANSFORMS

***Theorem 11.2.1 (Linearity)*** *The Fourier transform is a linear integral transformation.*

PROOF. We have

$$\mathscr{F}[f] = \frac{1}{\sqrt{2\pi}} \int_{-\infty}^{\infty} f(t) e^{i\alpha t}\, dt$$

Then for any constants $a$ and $b$,

$$\begin{aligned} \mathscr{F}[af + bg] &= \frac{1}{\sqrt{2\pi}} \int_{-\infty}^{\infty} [af + bg] e^{i\alpha t}\, dt \\ &= \frac{a}{\sqrt{2\pi}} \int_{-\infty}^{\infty} f(t) e^{i\alpha t}\, dt + \frac{b}{\sqrt{2\pi}} \int_{-\infty}^{\infty} g(t) e^{i\alpha t}\, dt \\ &= a\mathscr{F}[f] + b\mathscr{F}[g] \end{aligned}$$

□

***Theorem 11.2.2 (Shifting)*** *Let $\mathscr{F}[f]$ be a Fourier transform of $f(t)$. Then $\mathscr{F}[f(t-c)] = e^{i\alpha c}\mathscr{F}[f(t)]$ where $c$ is a real constant.*

PROOF. From the definition, we have for $c > 0$

$$\begin{aligned} \mathscr{F}[f(t-c)] &= \frac{1}{\sqrt{2\pi}} \int_{-\infty}^{\infty} f(t-c) e^{i\alpha t}\, dt \\ &= \frac{1}{\sqrt{2\pi}} \int_{-\infty}^{\infty} f(\xi) e^{i\alpha(\xi + c)}\, d\xi \quad \text{where} \quad \xi = t - c \\ &= e^{i\alpha c}\mathscr{F}[f(t)] \end{aligned}$$

□

***Theorem 11.2.3 (Scaling)*** *If $\mathscr{F}[f]$ is the Fourier transform of $f$, then $\mathscr{F}[f(ct)] = (1/|c|)F(\alpha/c)$ where $c$ is a real constant and $c \neq 0$.*

PROOF. For $c \neq 0$,

$$\mathscr{F}[f(ct)] = \frac{1}{\sqrt{2\pi}} \int_{-\infty}^{\infty} f(ct) e^{i\alpha t}\, dt$$

If we let $\xi = ct$, then

$$\mathscr{F}[f(ct)] = \frac{1}{|c|}\frac{1}{\sqrt{2\pi}}\int_{-\infty}^{\infty} f(\xi)e^{i(\alpha/c)\xi}\, d\xi$$
$$= (1/|c|)F(\alpha/c) \qquad \square$$

***Theorem 11.2.4 (Differentiation)*** *Let $f$ be continuous and piecewise smooth in $(-\infty, \infty)$. Let $f(t)$ approach zero as $|t| \to \infty$. If $f$ and $f'$ are absolutely integrable, then*

$$\mathscr{F}[f'] = -i\alpha\mathscr{F}[f]$$

PROOF.

$$\mathscr{F}[f'(t)] = \frac{1}{\sqrt{2\pi}}\int_{-\infty}^{\infty} f'(t)e^{i\alpha t}\, dt$$
$$= \frac{1}{\sqrt{2\pi}}\left[f(t)e^{i\alpha t}\Big|_{-\infty}^{\infty} - i\alpha\int_{-\infty}^{\infty} f(t)e^{i\alpha t}\, dt\right]$$
$$= -i\alpha\mathscr{F}[f(t)] \qquad \square$$

This result can be easily extended. If $f$ and its first $(n-1)$ derivatives are continuous, and if its $n$th derivative is piecewise continuous, then

$$\mathscr{F}[f^{(n)}(t)] = (-i\alpha)^n\mathscr{F}[f(t)], \qquad n = 0, 1, 2, \ldots \tag{11.2.1}$$

provided $f$ and its derivatives are absolutely integrable. In addition, we assume that $f$ and its first $(n-1)$ derivatives tend to zero as $|t|$ tends to infinity.

EXAMPLE 11.2.1. Find the solution of the Dirichlet problem in the half-plane $y > 0$.

$$u_{xx} + u_{yy} = 0, \qquad -\infty < x < \infty, \qquad y > 0$$
$$u(x, 0) = f(x), \qquad -\infty < x < \infty$$
$$u \text{ is bounded as } y \to \infty$$
$$u \text{ and } u_x \text{ vanish as } |x| \to \infty$$

Let $U(\alpha, y)$ be the Fourier transform of $u(x, y)$ in the variable $x$. Then

$$U(\alpha, y) = \frac{1}{\sqrt{2\pi}}\int_{-\infty}^{\infty} u(x, y)e^{i\alpha x}\, dx$$

By the relation (11.2.1), we have

$$\mathscr{F}[u_{xx}] = (-i\alpha)^2\mathscr{F}[u]$$
$$= -\alpha^2 U(\alpha, y) \tag{11.2.2}$$

Note that

$$\mathscr{F}[u_{yy}] = \frac{1}{\sqrt{2\pi}} \int_{-\infty}^{\infty} u_{yy}\, e^{i\alpha x}\, dx$$
$$= \frac{\partial^2}{\partial y^2}\left[\frac{1}{\sqrt{2\pi}} \int_{-\infty}^{\infty} u(x, y) e^{i\alpha x}\, dx\right]$$
$$= \frac{\partial^2 U}{\partial y^2} \tag{11.2.3}$$

Transforming the Laplace equation, we obtain

$$\mathscr{F}[u_{xx} + u_{yy}] = 0$$

By Theorem 11.2.1, we have

$$\mathscr{F}[u_{xx}] + \mathscr{F}[u_{yy}] = 0 \tag{11.2.4}$$

Substitution of Eqs. (11.2.2) and (11.2.3) in Eq. (11.2.4) yields

$$U_{yy} - \alpha^2 U = 0$$

This is an ordinary differential equation in $y$, the solution of which is

$$U(\alpha, y) = A(\alpha)e^{\alpha y} + B(\alpha)e^{-\alpha y}$$

Since $u$ must be bounded as $y \to \infty$, $U(\alpha, y)$ must also be bounded as $y \to \infty$. Thus, for $\alpha > 0$, $A(\alpha)$ must vanish, and

$$U(\alpha, 0) = B(\alpha)$$

For $\alpha < 0$, $B(\alpha)$ must vanish, and

$$U(\alpha, 0) = A(\alpha)$$

Thus, for any $\alpha$, we have

$$U(\alpha, y) = U(\alpha, 0)e^{-|\alpha|y}$$

Observing that

$$U(\alpha, 0) = \mathscr{F}[u(x, 0)]$$
$$= \mathscr{F}[f(x)]$$

from the given condition, we find

$$U(\alpha, y) = \frac{1}{\sqrt{2\pi}} \int_{-\infty}^{\infty} f(x)e^{-|\alpha|y}e^{i\alpha x}\, dx$$

The inverse Fourier transform of $U(\alpha, y)$ is therefore given by

$$u(x, y) = \frac{1}{\sqrt{2\pi}} \int_{-\infty}^{\infty} \left[\frac{1}{\sqrt{2\pi}} \int_{-\infty}^{\infty} f(\xi)e^{-|\alpha|y}e^{i\alpha\xi}\, d\xi\right] e^{-i\alpha x}\, d\alpha$$
$$= \frac{1}{2\pi} \int_{-\infty}^{\infty} f(\xi)\, d\xi \int_{-\infty}^{\infty} e^{\alpha[i(\xi - x)] - |\alpha|y}\, d\alpha$$

It can be easily shown that

$$\int_{-\infty}^{\infty} e^{\alpha[i(\xi-x)]-|\alpha|y}\, d\alpha = \frac{2y}{(\xi-x)^2+y^2}$$

Hence the solution of the Dirichlet problem in the half-plane $y>0$ is

$$u(x, y) = \frac{y}{\pi}\int_{-\infty}^{\infty} \frac{f(\xi)}{(\xi-x)^2+y^2}\, d\xi$$

From this solution, we can readily deduce a solution of the Neumann problem in the half-plane $y>0$.

EXAMPLE 11.2.2. Find a solution of the Neumann problem in the half-plane $y>0$.

$$u_{xx}+u_{yy}=0, \qquad -\infty<x<\infty, \qquad y>0$$
$$u_y(x, 0)=g(x), \qquad -\infty<x<\infty$$
$$u \text{ is bounded as } y\to\infty$$
$$u \text{ and } u_x \text{ vanish as } |x|\to\infty$$

Let $v(x, y) = u_y(x, y)$ then

$$u(x, y) = \int_a^y v(x, \eta)\, d\eta + C, \quad C = \text{constant}$$

where $a$ is an arbitrary constant, and the Neumann problem becomes

$$\begin{aligned} \frac{\partial^2 v}{\partial x^2}+\frac{\partial^2 v}{\partial y^2} &= \frac{\partial^2 u_y}{\partial x^2}+\frac{\partial^2 u_y}{\partial y^2} \\ &= \frac{\partial}{\partial y}(u_{xx}+u_{yy}) \\ &= 0 \end{aligned}$$

$$v(x, 0) = u_y(x, 0) = g(x)$$

This is the Dirichlet problem, the solution of which is given by

$$v(x, y) = \frac{y}{\pi}\int_{-\infty}^{\infty} \frac{g(\xi)\, d\xi}{(\xi-x)^2+y^2}$$

Thus we have

$$\begin{aligned} u(x, y) &= \frac{1}{\pi}\int_a^y \eta \int_{-\infty}^{\infty} \frac{g(\xi)\, d\xi}{(\xi-x)^2+\eta^2}\, d\eta + C \\ &= \frac{1}{2\pi}\int_{-\infty}^{\infty} g(\xi)\log\left[\frac{(\xi-x)^2+y^2}{(\xi-x)^2+a^2}\right] d\xi + C \end{aligned}$$

## 11.3 CONVOLUTION THEOREM OF THE FOURIER TRANSFORM

The function

$$(f * g)(x) = \frac{1}{\sqrt{2\pi}} \int_{-\infty}^{\infty} f(x - \xi) g(\xi) \, d\xi \tag{11.3.1}$$

is called the *convolution* of the functions $f$ and $g$ over the interval $(-\infty, \infty)$.

***Theorem 11.3.1*** *If $F(\alpha)$ and $G(\alpha)$ are the Fourier transforms of $f(x)$ and $g(x)$ respectively, then the Fourier transform of the convolution $f * g$ is the product $F(\alpha)G(\alpha)$. Putting this statement in another way,*

$$\frac{1}{\sqrt{2\pi}} \int_{-\infty}^{\infty} F(\alpha) G(\alpha) e^{-i\alpha x} \, d\alpha = (f * g)(x)$$

$$= \frac{1}{\sqrt{2\pi}} \int_{-\infty}^{\infty} f(x - \xi) g(\xi) \, d\xi \tag{11.3.2}$$

PROOF. By definition

$$\mathscr{F}[f * g] = \frac{1}{2\pi} \int_{-\infty}^{\infty} e^{i\alpha x} \, dx \int_{-\infty}^{\infty} f(x - \xi) g(\xi) \, d\xi$$

$$= \frac{1}{2\pi} \int_{-\infty}^{\infty} g(\xi) e^{i\alpha\xi} \, d\xi \int_{-\infty}^{\infty} f(x - \xi) e^{i\alpha(x-\xi)} \, dx$$

With the change of variable $\eta = x - \xi$, we have

$$\mathscr{F}[f * g] = \frac{1}{\sqrt{2\pi}} \int_{-\infty}^{\infty} g(\xi) e^{i\alpha\xi} \, d\xi \frac{1}{\sqrt{2\pi}} \int_{-\infty}^{\infty} f(\eta) e^{i\alpha\eta} \, d\eta$$

$$= F(\alpha) G(\alpha)$$ □

The convolution has the following properties:

1. $f * g = g * f$ (commutative)
2. $f * (g * h) = (f * g) * h$ (associative)
3. $f * (g + h) = f * g + f * h$ (distributive)

***Theorem 11.3.2 (Parseval's formula)***

$$\int_{-\infty}^{\infty} |f(x)|^2 \, dx = \int_{-\infty}^{\infty} |F(k)|^2 \, dk \tag{11.3.3}$$

PROOF. The Convolution theorem 11.3.1 gives

$$\int_{-\infty}^{\infty} f(x) g(\xi - x) \, dx = \int_{-\infty}^{\infty} F(k) G(k) e^{-ik\xi} \, dk$$

which is, by putting $\xi = 0$,

$$\int_{-\infty}^{\infty} f(x)g(-x)\,dx = \int_{-\infty}^{\infty} F(k)G(k)\,dk \tag{11.3.4}$$

Setting $g(-x) = \overline{f(x)}$ so that

$$G(k) = \frac{1}{\sqrt{2\pi}} \int_{-\infty}^{\infty} g(x)e^{ikx}\,dx = \frac{1}{\sqrt{2\pi}} \int_{-\infty}^{\infty} \overline{f(-x)}e^{ikx}\,dx$$
$$= \overline{\frac{1}{\sqrt{2\pi}} \int_{-\infty}^{\infty} f(x)e^{ikx}\,dx} = \overline{F(k)}$$

where the bar denotes the complex conjugate.

Thus, result (11.3.4) becomes

$$\int_{-\infty}^{\infty} f(x)\overline{f(x)}\,dx = \int_{-\infty}^{\infty} F(k)\overline{F(k)}\,dk$$

or

$$\int_{-\infty}^{\infty} |f(x)|^2\,dx = \int_{-\infty}^{\infty} |F(k)|^2\,dk$$ □

In terms of the notation of the norms, this reads as

$$\|f\| = \|F\|$$

For physical systems, the quantity $|f|^2$ is a measure of energy, and $|F|^2$ represents the power spectrum of $f(x)$.

EXAMPLE 11.3.1. Obtain the solution of the initial-value problem of heat conduction in an infinite rod.

$$u_t = \kappa u_{xx}, \qquad -\infty < x < \infty, \quad t > 0 \tag{11.3.5}$$

$$u(x, 0) = f(x), \qquad -\infty < x < \infty \tag{11.3.6}$$

where $u(x, t)$ represents the temperature distribution and is bounded, and $\kappa$ is a constant of diffusivity.

The Fourier transform of $u(x, t)$ with respect to $x$ is defined by

$$U(\alpha, t) = \frac{1}{\sqrt{2\pi}} \int_{-\infty}^{\infty} e^{i\alpha x} u(x, t)\,dx$$

In view of this transformation, Eqs. (11.3.5)–(11.3.6) become

$$U_t + \kappa\alpha^2 U = 0 \tag{11.3.7}$$

$$U(\alpha, 0) = F(\alpha) \tag{11.3.8}$$

The solution of the transformed system is

$$U(\alpha, t) = F(\alpha)e^{-\alpha^2 \kappa t}$$

The inverse Fourier transformation gives

$$u(x, t) = \frac{1}{\sqrt{2\pi}} \int_{-\infty}^{\infty} F(\alpha) e^{-\alpha^2 \kappa t} e^{-i\alpha x} \, d\alpha$$

which is, by the Convolution theorem (11.3.1),

$$= \frac{1}{\sqrt{2\pi}} \int_{-\infty}^{\infty} f(\xi) g(x - \xi) \, d\xi$$

where $g(x)$ is the inverse transform of $G(\alpha) = e^{-\alpha^2 \kappa t}$ and has the form

$$g(x) = \frac{1}{\sqrt{2\pi}} \int_{-\infty}^{\infty} e^{-\alpha^2 \kappa t - i\alpha x} \, d\alpha = \frac{1}{\sqrt{2\kappa t}} e^{-x^2/4\kappa t}$$

Consequently, the final solution is

$$u(x, t) = \frac{1}{2\sqrt{\pi \kappa t}} \int_{-\infty}^{\infty} f(\xi) \exp\left[-\frac{(x - \xi)^2}{4\kappa t}\right] d\xi \tag{11.3.9}$$

Using the change of variable

$$\frac{\xi - x}{2\sqrt{\kappa t}} = \zeta, \qquad d\zeta = \frac{d\xi}{2\sqrt{\kappa t}}$$

We obtain

$$u(x, t) = \frac{1}{\sqrt{\pi}} \int_{-\infty}^{\infty} f(x + 2\sqrt{\kappa t}\, \zeta) e^{-\zeta^2} \, d\zeta \tag{11.3.10}$$

Integral (11.3.9) or (11.3.10) is called the *Poisson integral representation* of the temperature distribution. This integral is convergent for $t > 0$, and integrals obtained from it by differentiation under the integral sign with respect to $x$ and $t$ are uniformly convergent in the neighborhood of the point $(x, t)$. Hence $u(x, t)$ and its derivatives of all orders exist for $t > 0$.

Finally, it is noted that the integrand in (11.3.9) consists of the initial temperature distribution $f(x)$ and Green's function $G(x, t)$:

$$G(x, t) = \frac{1}{2\sqrt{\pi \kappa t}} \exp\left[-\frac{(x - \xi)^2}{4\kappa t}\right] \tag{11.3.11}$$

This represents the temperature response along the rod at time $t$ due to an initial unit impulse of heat at $x = \xi$. The physical meaning of the solution (11.3.9) is that the initial temperature distribution $f(x)$ is decomposed into a spectrum of impulses of magnitude $f(\xi)$ at each point $x = \xi$ to form the resulting temperature $f(\xi)\, G(x, t)$. Thus the resulting temperature is integrated to find the solution (11.3.9). This is the so-called principle of superposition.

In terms of $G(x, t)$, solution (11.3.9) can be written as

$$u(x, t) = \int_{-\infty}^{\infty} f(\xi) G(x, t)\, d\xi$$

so that in the limit $t \to 0+$ this becomes formally

$$u(x, 0) = f(x) = \int_{-\infty}^{\infty} f(\xi) \lim_{t \to 0+} G(x, t)\, d\xi$$

This limit represents the Dirac delta function

$$\delta(x - \xi) = \lim_{t \to 0+} \frac{1}{2\sqrt{\pi \kappa t}} e^{-(x-\xi)^2/4\kappa t} \tag{11.3.12}$$

Consider a special case where

$$f(x) = \begin{Bmatrix} 0, & x < 0 \\ a, & x > 0 \end{Bmatrix}$$

Then (11.3.9) gives

$$u(x, t) = \frac{a}{2\sqrt{\pi \kappa t}} \int_0^{\infty} \exp\left[ -\frac{(x - \xi)^2}{4\kappa t} \right] d\xi$$

If we introduce the new change of variable

$$\eta = \frac{\xi - x}{2\sqrt{\kappa t}}$$

then the above solution becomes

$$\begin{aligned} u(x, t) &= \frac{a}{\sqrt{\pi}} \int_{-x/2\sqrt{\kappa t}}^{\infty} e^{-\eta^2}\, d\eta \\ &= \frac{a}{\sqrt{\pi}} \left[ \int_{-x/2\sqrt{\kappa t}}^{0} e^{-\eta^2}\, d\eta + \int_0^{\infty} e^{-\eta^2}\, d\eta \right] \\ &= \frac{a}{\sqrt{\pi}} \left[ \int_0^{x/2\sqrt{\kappa t}} e^{-\eta^2}\, d\eta + \frac{\sqrt{\pi}}{2} \right] \\ &= \frac{a}{2} \left[ 1 + \operatorname{erf}\left( \frac{x}{2\sqrt{\kappa t}} \right) \right] \end{aligned}$$

where erf $(x)$ is called the *error function* defined by

$$\operatorname{erf}(x) = \frac{2}{\sqrt{\pi}} \int_0^x e^{-\eta^2}\, d\eta \tag{11.3.13}$$

This is a widely used and tabulated function.

## 11.4 THE FOURIER TRANSFORMS OF STEP AND IMPULSE FUNCTIONS

In this section, we shall determine the Fourier transforms of the step function and the impulse function, functions which occur frequently in applied mathematics and mathematical physics.

A *unit step function* is defined by

$$u_a(x) = \begin{cases} 0, & x < a \\ 1, & x \geqslant a \end{cases} \qquad a \geqslant 0 \tag{11.4.1}$$

as shown in Fig. 11.4.1.

The Fourier transform of the unit step function can be easily determined. We consider first

$$\begin{aligned}\mathscr{F}[u_a(x)] &= \frac{1}{\sqrt{2\pi}} \int_{-\infty}^{\infty} u_a(x) e^{i\alpha x}\, dx \\ &= \frac{1}{\sqrt{2\pi}} \int_{a}^{\infty} e^{i\alpha x}\, dx\end{aligned}$$

This integral does not exist. We avoid this procedure by defining a new function

$$u_a(x) e^{-\beta x} = \begin{cases} 0, & x < a \\ e^{-\beta x}, & x \geqslant a \end{cases}$$

This evidently is the unit step function as $\beta \rightarrow 0$. Thus, we find the

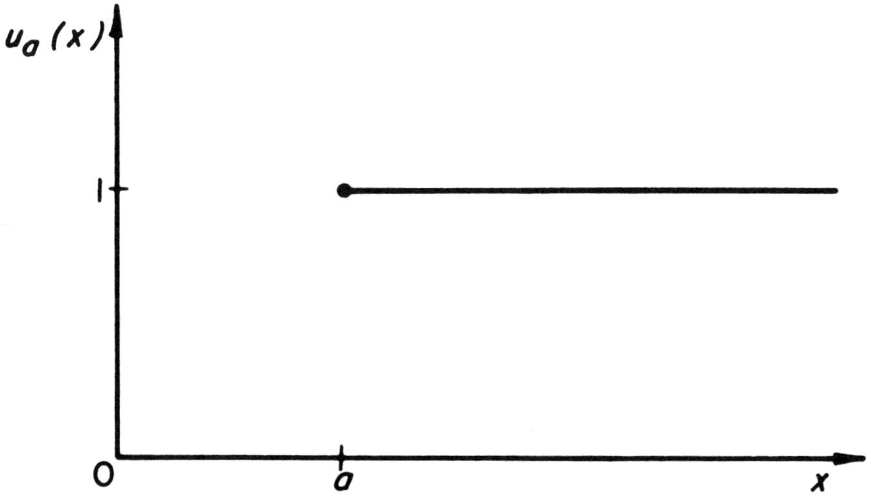


Figure 11.4.1

Fourier transform of the unit step function as

$$\begin{aligned}\mathscr{F}[u_a(x)] &= \lim_{\beta\to 0} \mathscr{F}[u_a(x)e^{-\beta x}] \\ &= \lim_{\beta\to 0} \frac{1}{\sqrt{2\pi}} \int_{-\infty}^{\infty} u_a(x) e^{-\beta x} e^{i\alpha x}\, dx \\ &= \lim_{\beta\to 0} \frac{1}{\sqrt{2\pi}} \int_{a}^{\infty} e^{-(\beta - i\alpha)x}\, dx \\ &= \frac{ie^{i\alpha a}}{\sqrt{2\pi}\,\alpha} \end{aligned} \tag{11.4.2}$$

For $a = 0$,

$$\mathscr{F}[u_0(x)] = (i/\sqrt{2\pi}\,\alpha) \tag{11.4.3}$$

An *impulse function* is defined by

$$p(x) = \begin{cases} h & a - \varepsilon < x < a + \varepsilon \\ 0 & x \leqslant a - \varepsilon, \quad x \geqslant a + \varepsilon \end{cases}$$

where $h$ is large and positive, $a > 0$, and $\varepsilon$ is a small positive constant, as shown in Fig. 11.4.2. This type of function appears in practical

Figure 11.4.2

applications; for instance, a force of large magnitude may act over a very short period of time.

The Fourier transform of the impulse function is

$$\begin{aligned}\mathscr{F}[p(x)] &= \frac{1}{\sqrt{2\pi}} \int_{-\infty}^{\infty} p(x)e^{i\alpha x}\, dx \\ &= \frac{1}{\sqrt{2\pi}} \int_{a-\varepsilon}^{a+\varepsilon} he^{i\alpha x}\, dx \\ &= \frac{h}{\sqrt{2\pi}} \frac{e^{i\alpha a}}{i\alpha} (e^{i\alpha\varepsilon} - e^{-i\alpha\varepsilon}) \\ &= \frac{2h\varepsilon}{\sqrt{2\pi}} e^{i\alpha a} \frac{\sin \alpha\varepsilon}{\alpha\varepsilon}\end{aligned}$$

Now if we choose the value of $h$ to be $1/2\varepsilon$, then the impulse defined by

$$I(\varepsilon) = \int_{-\infty}^{\infty} p(x)\, dx$$

becomes

$$\begin{aligned}I(\varepsilon) &= \int_{a-\varepsilon}^{a+\varepsilon} \frac{1}{2\varepsilon}\, dx \\ &= 1\end{aligned}$$

which is a constant independent of $\varepsilon$. In the limit $\varepsilon \to 0$, this particular function $p_\varepsilon(x)$ with $h = 1/2\varepsilon$ satisfies

$$\lim_{\varepsilon\to 0} p_\varepsilon(x) = 0, \qquad x \neq a$$

$$\lim_{\varepsilon\to 0} I(\varepsilon) = 1$$

Thus, we arrive at the result

$$\begin{gathered}\delta(x-a) = 0, \qquad x \neq a \\ \int_{-\infty}^{\infty} \delta(x-a)\, dx = 1\end{gathered} \tag{11.4.4}$$

This is the *Dirac delta function* which was defined earlier in Section 7.11.

We shall now define the Fourier transform of $\delta(x)$ as the limit of the transform of $p_\varepsilon(x)$. We consider then

$$\begin{aligned}\mathscr{F}[\delta(x-a)] &= \lim_{\varepsilon\to 0} \mathscr{F}[p_\varepsilon(x)] \\ &= \lim_{\varepsilon\to 0} \frac{e^{i\alpha a}}{\sqrt{2\pi}} \frac{\sin \alpha\varepsilon}{\alpha\varepsilon} \\ &= \frac{e^{i\alpha a}}{\sqrt{2\pi}}\end{aligned} \tag{11.4.5}$$

in which we note that, by L'Hospital's rule, $\lim_{\varepsilon\to 0} (\sin \alpha\varepsilon/\alpha\varepsilon) = 1$. When $a = 0$, we obtain

$$\mathscr{F}[\delta(x)] = 1/\sqrt{2\pi} \tag{11.4.6}$$

EXAMPLE 11.4.1. (Slowing-down of Neutrons).[18] Consider the problem

$$u_\tau = u_{xx} + \delta(x)\delta(\tau) \tag{11.4.7}$$

$$u(x, 0) = \delta(x) \tag{11.4.8}$$

$$\lim_{|x|\to\infty} u(x, \tau) = 0 \tag{11.4.9}$$

This is the problem of an infinite medium which slows neutrons, in which is located a source of neutrons. Here $u(x, \tau)$ represents the number of neutrons per unit volume per unit time which reach the age $\tau$, and $\delta(x)\delta(\tau)$ represents the source function.

Let $U(\alpha, \tau)$ be the Fourier transform of $u(x, \tau)$. Then the transformation of Eq. (11.4.7) yields

$$\frac{dU}{d\tau} + \alpha^2 U = \frac{1}{\sqrt{2\pi}}\delta(\tau)$$

The solution of this, after applying the condition $U(\alpha, 0) = 1/\sqrt{2\pi}$, is

$$U(\alpha, \tau) = \frac{1}{\sqrt{2\pi}} e^{-\alpha^2\tau}$$

Hence the inverse transform is given by

$$u(x, \tau) = \frac{1}{\sqrt{2\pi}} \int_{-\infty}^{\infty} e^{-\alpha^2\tau - i\alpha x}\, d\alpha$$

$$= \frac{1}{2}\frac{1}{\sqrt{\pi\tau}} e^{-x^2/4\tau}$$

## 11.5 FOURIER SINE AND COSINE TRANSFORMS

For semi-infinite regions, the Fourier sine and cosine transforms determined in Sec. 11.1 are particularly appropriate in solving boundary-value problems. Before we illustrate their applications, we must first prove the differentiation theorem.

---

[18] See I. N. Sneddon, *Fourier Transforms,* McGraw Hill, New York (1951), p. 215.

***Theorem 11.5.1*** *Let $f(x)$ and its first derivative vanish as $x \to \infty$. If $F_c(\alpha)$ is the Fourier cosine transform, then*

$$\mathscr{F}_c[f''(x)] = -\alpha^2 F_c(\alpha) - \sqrt{\frac{2}{\pi}} f'(0) \tag{11.5.1}$$

PROOF.

$$\begin{aligned}
\mathscr{F}_c[f''(x)] &= \sqrt{\frac{2}{\pi}} \int_0^\infty f''(x) \cos \alpha x \, dx \\
&= \sqrt{\frac{2}{\pi}} [f'(x) \cos \alpha x]_0^\infty + \sqrt{\frac{2}{\pi}} \alpha \int_0^\infty f'(x) \sin \alpha x \, dx \\
&= -\sqrt{\frac{2}{\pi}} f'(0) + \sqrt{\frac{2}{\pi}} \alpha [f(x) \sin \alpha x]_0^\infty \\
&\qquad - \sqrt{\frac{2}{\pi}} \alpha^2 \int_0^\infty f(x) \cos \alpha x \, dx \\
&= -\sqrt{\frac{2}{\pi}} f'(0) - \alpha^2 F_c(\alpha)
\end{aligned}$$

□

In a similar manner, transforms of higher-order derivatives can be obtained.

***Theorem 11.5.2*** *Let $f(x)$ and its first derivative vanish as $x \to \infty$. If $F_s(\alpha)$ is the Fourier sine transform, then*

$$\mathscr{F}_s[f''(x)] = \sqrt{\frac{2}{\pi}} \alpha f(0) - \alpha^2 F_s(\alpha) \tag{11.5.2}$$

The proof is left to the reader.

EXAMPLE 11.5.1. Find the temperature distribution in a semi-infinite rod for the following cases with zero initial temperature distribution:

(a) The heat supplied at the end $x = 0$ at the rate $g(t)$;
(b) The end $x = 0$ is kept at a constant temperature $T_0$.

The problem here is to solve the heat conduction equation

$$\begin{aligned}
u_t &= \kappa u_{xx}, \qquad x > 0, \quad t > 0 \\
u(x, 0) &= 0, \qquad x > 0
\end{aligned}$$

(a) $u_x(0, t) = g(t)$ and (*b*) $u(0, t) = T_0$, $t \geq 0$. Here we assume that $u(x, t)$ and $u_x(x, t)$ vanish as $x \to \infty$.

For case (a), let $U(\alpha, t)$ be the Fourier cosine transform of $u(x, t)$.

Then the transformation of the heat conduction equation yields

$$U_t + \kappa\alpha^2 U = -\sqrt{\frac{2}{\pi}}\, g(t)\kappa$$

The solution of this equation with $U(\alpha, 0) = 0$ is

$$U(\alpha, t) = -\kappa\sqrt{\frac{2}{\pi}} \int_0^t g(\tau) e^{-\alpha^2\kappa(t-\tau)}\, d\tau$$

Hence, the inverse transform gives

$$u(x, t) = \sqrt{\frac{2}{\pi}} \int_0^\infty U(\alpha, t) \cos \alpha x \, d\alpha$$

$$= -\frac{2\kappa}{\pi} \int_0^t g(\tau)\, d\tau \int_0^\infty e^{-\kappa\alpha^2(t-\tau)} \cos \alpha x \, d\alpha$$

The inner integral is given by (see Problem 6, Exercises 11.16)

$$\int_0^\infty e^{-\alpha^2\kappa(t-\tau)} \cos \alpha x \, d\alpha = \frac{1}{2}\sqrt{\frac{\pi}{\kappa(t-\tau)}} \exp\left[-\frac{x^2}{4\kappa(t-\tau)}\right]$$

The solution therefore is

$$u(x, t) = -\sqrt{\frac{\kappa}{\pi}} \int_0^t \frac{g(\tau)}{\sqrt{t-\tau}} e^{-x^2/4\kappa(t-\tau)}\, d\tau \tag{11.5.3}$$

For case (b), we apply the Fourier sine transform $U(\alpha, t)$ of $u(x, t)$ to obtain the transformed equation

$$U_t + \kappa\alpha^2 U = \sqrt{\frac{2}{\pi}}\, \alpha T_0 \kappa$$

The solution of this equation with zero initial condition is

$$U(\alpha, t) = T_0 \sqrt{\frac{2}{\pi}} \frac{(1 - e^{-\kappa t\alpha^2})}{\alpha}$$

Then the inverse transformation gives

$$u(x, t) = \frac{2T_0}{\pi} \int_0^\infty \frac{\sin \alpha x}{\alpha} (1 - e^{-\kappa t\alpha^2})\, d\alpha$$

Making use of the integral

$$\int_0^\infty e^{-a^2x^2} \frac{\sin \alpha x}{\alpha}\, d\alpha = \frac{\pi}{2} \operatorname{erf}\left(\frac{x}{2a}\right)$$

the solution is found to be

$$u(x, t) = \frac{2T_0}{\pi}\left[\frac{\pi}{2} - \frac{\pi}{2} \operatorname{erf}\left(\frac{x}{2\sqrt{\kappa t}}\right)\right]$$

$$= T_0 \operatorname{erfc}\left(\frac{x}{2\sqrt{\kappa t}}\right) \tag{11.5.4}$$

where $\operatorname{erfc}(y) = 1 - \operatorname{erf}(y)$ is the complementary error function defined by

$$\operatorname{erfc}(y) = \frac{2}{\sqrt{\pi}} \int_y^\infty e^{-x^2}\, dx$$

## 11.6 ASYMPTOTIC APPROXIMATION OF INTEGRALS BY THE KELVIN STATIONARY PHASE METHOD

Although definite integrals represent exact solutions for many physical problems, the physical meanings of the solutions are often difficult to determine. In many cases the exact evaluation of the integrals is a formidable task. It is then necessary to resort to asymptotic methods.

We consider the typical integral solution

$$u(x, t) = \int_a^b F(k) e^{it\theta(k)}\, dk \tag{11.6.1}$$

where $F(k)$ is called the spectral function determined by the initial or boundary data in $a < k < b$, and $\theta(k)$ is known as the phase function given by

$$\theta(k) \equiv k\frac{x}{t} - \omega(k), \qquad x > 0 \tag{11.6.1}$$

We examine the asymptotic behavior of (11.6.1) for both large $x$ and $t$; one of the interesting limits is $t \to \infty$ with $x/t$ held fixed. Integral (11.6.2) can be evaluated by the Kelvin stationary phase method for large $t$. As $t \to \infty$, the integrand of (11.6.1) oscillates very rapidly; consequently, the contributions to $u(x, t)$ from adjacent parts of the integrand cancel one another except in the neighborhood of the points, if any, at which the phase function $\theta(k)$ is stationary, that is, $\theta'(k) = 0$. Thus the main contribution to the integral for large $t$ comes from the neighborhood of the points $k = k_1$ which are determined by the solutions of

$$\theta'(k_1) = \frac{x}{t} - \omega'(k_1) = 0, \qquad a < k_1 < b \tag{11.6.3}$$

The points $k = k_1$ are known as *points of stationary phase,* or simply *stationary points.*

We expand both $F(k)$ and $\theta(k)$ in Taylor series about $k = k_1$ so that

$$u(x, t) = \int_a^b \left[F(k_1) + (k - k_1)F'(k_1) + \frac{1}{2}(k - k_1)^2 F''(k_1) + \ldots\right]$$
$$\times \exp\left\{it\left[\theta(k_1) + \frac{1}{2}(k - k_1)^2\theta''(k_1) + \frac{1}{6}(k - k_1)^3\theta'''(k_1) + \ldots\right]\right\} dk \tag{11.6.4}$$

provided $\theta''(k_1) \neq 0$.

Introducing the change of variable $k - k_1 = \varepsilon\alpha$ where

$$\varepsilon = \left\{\frac{2}{t\,|\theta''(k_1)|}\right\}^{\frac{1}{2}} \tag{11.6.5}$$

it turns out that the significant contribution to integral (11.6.4) is

$$u(x, t) \sim \varepsilon \int_{-(k_1-a)/\varepsilon}^{(b-k_1)/\varepsilon} \left[F(k_1) + \varepsilon\alpha F'(k_1) + \frac{1}{2}\varepsilon^2\alpha^2 F''(k_1) + \ldots\right]$$
$$\times \exp\left\{i\left[t\theta(k_1) + \alpha^2 \operatorname{sgn} \theta''(k_1) + \frac{1}{3}\varepsilon\left(\frac{\theta'''(k_1)}{|\theta''(k_1)|}\right)\alpha^3 + \ldots\right]\right\} d\alpha \tag{11.6.6}$$

where $\operatorname{sgn} x$ denotes the signum function defined by $\operatorname{sgn} x = 1,\ x > 0$ and $\operatorname{sgn} x = -1,\ x < 0$.

We then proceed to the limit $\varepsilon \to 0$ $(t \to \infty)$ and use the standard integral

$$\int_{-\infty}^{\infty} \exp(\pm i\alpha^2)\, d\alpha = \sqrt{\pi} \exp\left(\pm \frac{i\pi}{4}\right) \tag{11.6.7}$$

to obtain the asymptotic approximation, as $t \to \infty$,

$$u(x, t) \sim F(k_1)\left[\frac{2\pi}{t\,|\theta''(k_1)|}\right]^{\frac{1}{2}} \exp\left\{i\left[t\theta(k_1) + \frac{\pi}{4} \operatorname{sgn} \theta''(k_1)\right]\right\} + 0(\varepsilon^2) \tag{11.6.8}$$

If there is more than one stationary point, each one contributes a term similar to (11.6.8) and we obtain, for $n$ stationary points $k = k_r$, $r = 1, 2, \ldots n$;

$$u(x, t) \sim \sum_{r=1}^{n} F(k_r)\left\{\frac{2\pi}{t\,|\theta''(k_r)|}\right\}^{\frac{1}{2}} \exp\left\{i\left[t\theta(k_r) + \frac{\pi}{4} \operatorname{sgn} \theta''(k_r)\right]\right\},\ t \to \infty \tag{11.6.9}$$

If $\theta''(k_1) = 0$, but $\theta'''(k_1) \neq 0$, then asymptotic approximation (11.6.8) fails. This important special case can be handled in a similar fashion. The asymptotic approximation of (11.6.1) can be given by

$$u(x, t) = F(k_1) \exp\{it\theta(k_1)\} \int_{-\infty}^{\infty} \exp\left[\frac{i}{6} t\theta'''(k_1)(k - k_1)^3\right] dk$$
$$\sim \Gamma\left(\frac{4}{3}\right)\left[\frac{6}{t\,|\theta'''(k_1)|}\right]^{\frac{1}{3}} F(k_1) \exp\left[it\theta(k_1) + \frac{\pi i}{6}\right] + 0\left(\frac{1}{t^{\frac{2}{3}}}\right) \quad \text{as} \quad t \to \infty \tag{11.6.10}$$

For an elaborate treatment, see Copson [73].

## 11.7 LAPLACE TRANSFORMS

Due to simplicity, Laplace transforms are frequently used in determining solutions of a wide class of partial differential equations. Like other transforms, Laplace transforms are used to determine particular solutions. In solving partial differential equations, the general solutions are difficult, if not impossible, to obtain. Thus, the transform techniques sometimes offer a useful tool for attaining particular solutions.

The Laplace transform is closely related to the complex Fourier transform, so the Fourier integral formula (5.12.9) can be used to define the Laplace transform and its inverse. We replace $f(x)$ in (5.12.9) by $H(x)e^{-cx}f(x)$ to obtain

$$f(x)H(x)e^{-cx} = \frac{1}{2\pi}\int_{-\infty}^{\infty} e^{-i\alpha x}\, d\alpha \int_0^{\infty} f(t)e^{-t(c-i\alpha)}\, dt$$

or

$$f(x)H(x) = \frac{1}{2\pi}\int_{-\infty}^{\infty} e^{x(c-i\alpha)}\, d\alpha \int_0^{\infty} f(t)e^{-t(c-i\alpha)}\, dt$$

Substituting $s = c - i\alpha$ so that $ds = -i d\alpha$, we obtain

$$f(x)H(x) = \frac{1}{2\pi i}\int_{c-i\infty}^{c+i\infty} e^{xs}\, ds \int_0^{\infty} f(t)e^{-st}\, dt \tag{11.7.1}$$

Thus we give the following definition of the Laplace transform: If $f(t)$ is defined for all values of $t > 0$, then the Laplace transform of $f(t)$ is denoted by $F(s)$ or $\mathscr{L}\{f(t)\}$ and defined by the integral

$$F(s) = \mathscr{L}\{f(t)\} = \int_0^{\infty} e^{-st} f(t)\, dt \tag{11.7.2}$$

where $s$ is a positive real number or complex number with a positive real part so that the integral is convergent.

Hence (11.7.1) gives

$$f(x) = \mathscr{L}^{-1}\{F(s)\} = \frac{1}{2\pi i}\int_{c-i\infty}^{c+i\infty} e^{xs}F(s)\, ds, \qquad c > 0 \tag{11.7.3}$$

for $x > 0$ and zero for $x < 0$. This complex integral is used to define the *inverse Laplace transform* which is denoted by $\mathscr{L}^{-1}\{F(s)\} = f(x)$. It can easily be verified that both $\mathscr{L}$ and $\mathscr{L}^{-1}$ are linear integral operators.

Let us now find the Laplace transforms of some elementary functions.

1. Given $f(t) = c$, $c$ is a constant.

$$\begin{aligned}\mathscr{L}[c] &= \int_0^{\infty} e^{-st}c\, dt \\ &= \left[-\frac{ce^{-st}}{s}\right]_0^{\infty} \\ &= c/s\end{aligned}$$

2. Given $f(t) = e^{at}$, $a$ is a constant.

$$\mathscr{L}[e^{at}] = \int_0^\infty e^{-st}e^{at}\,dt$$

$$= \left[-\frac{e^{-(s-a)t}}{(s-a)}\right]_0^\infty$$

$$= \frac{1}{s-a}, \qquad s \geqslant a$$

3. Given $f(t) = t^2$. Then

$$\mathscr{L}[t^2] = \int_0^\infty e^{-st}t^2\,dt$$

Integration by parts yields

$$= \left[-\frac{t^2e^{-st}}{s}\right]_0^\infty + \int_0^\infty \frac{e^{-st}}{s}2t\,dt$$

Since $t^2e^{-st} \to 0$ as $t \to \infty$, we have

$$\mathscr{L}[t^2] = \frac{2}{s}\left[-\frac{e^{-st}}{s}t\right]_0^\infty + \frac{2}{s}\int_0^\infty \frac{e^{-st}}{s}\,dt$$

$$= 2/s^3$$

4. Given $f(t) = \sin \omega t$. Then

$$F(s) = \mathscr{L}[\sin \omega t] = \int_0^\infty e^{-st}\sin \omega t\,dt$$

$$= \left[-\frac{e^{-st}}{s}\sin \omega t\right]_0^\infty + \int_0^\infty \frac{e^{-st}}{s}\,\omega\cos \omega t\,dt$$

$$= \frac{\omega}{s}\left[-\frac{e^{-st}}{s}\cos \omega t\right]_0^\infty - \frac{\omega}{s}\int_0^\infty \frac{e^{-st}}{s}\,\omega\sin \omega t\,dt$$

$$F(s) = \frac{\omega}{s^2} - \frac{\omega^2}{s^2}F(s)$$

Thus, solving for $F(s)$, we obtain

$$\mathscr{L}[\sin \omega t] = \omega/(s^2 + \omega^2)$$

***Theorem 11.7.1*** *Let $f$ be piecewise continuous in the interval $[0, T]$ for every positive $T$, and let $f$ be of exponential order,*[19] *that is, $f(t) = O(e^{at})$ as $t \to \infty$ for some $a > 0$. Then the Laplace transform of $f(t)$ exists for $s > a$.*

---

[19] A function $f(t)$ is said to be of exponential order as $t \to \infty$ if there exists some $a > 0$ such that $e^{-at}|f(t)|$ is bounded for all $t > T$, that is, for $M > a$, $|f(t)| \leqslant Me^{at}$.

PROOF. Since $f$ is piecewise continuous and of exponential order, we have

$$|\mathscr{L}[f]| = \left|\int_0^\infty e^{-st} f(t)\, dt\right|$$
$$\leqslant \int_0^\infty e^{-st}\, |f(t)|\, dt$$
$$\leqslant \int_0^\infty e^{-st} M e^{at}\, dt$$
$$= M \int_0^\infty e^{-(s-a)t}\, dt$$
$$= M/(s-a), \quad s > a$$

Thus

$$\int_0^\infty e^{-st} f(t)\, dt$$

exists for $s > a$. □

## 11.8 PROPERTIES OF LAPLACE TRANSFORMS

***Theorem 11.8.1 (Linearity)*** *If $\mathscr{L}[f(t)]$ and $\mathscr{L}[g(t)]$ are the Laplace transforms of $f(t)$ and $g(t)$ respectively, then*

$$\mathscr{L}[af(t) + bg(t)] = a\mathscr{L}[f(t)] + b\mathscr{L}[g(t)]$$

*where $a$ and $b$ are constants.*

PROOF.

$$\mathscr{L}[af(t) + bg(t)] = \int_0^\infty [af(t) + bg(t)] e^{-st}\, dt$$
$$= a\int_0^\infty f(t) e^{-st}\, dt + b\int_0^\infty g(t) e^{-st}\, dt$$
$$= a\mathscr{L}[f(t)] + b\mathscr{L}[g(t)]$$ □

***Theorem 11.8.2 (Shifting)*** *If the Laplace transform of $f(t)$ is $F(s)$, then the Laplace transform of $e^{at}f(t)$ is $F(s-a)$.*

PROOF. *By definition*

$$\mathscr{L}[e^{at} f(t)] = \int_0^\infty e^{-st} e^{at} f(t)\, dt$$
$$= \int_0^\infty e^{-(s-a)t} f(t)\, dt$$
$$= F(s-a)$$ □

EXAMPLE 11.8.1.

a. If $\mathscr{L}[t^2] = 2/s^3$, then $\mathscr{L}[t^2 e^t] = 2/(s-1)^3$.
b. If $\mathscr{L}[\sin \omega t] = \omega/(s^2+\omega^2)$, then $\mathscr{L}[e^t \sin \omega t] = \omega/[(s-1)^2+\omega^2]$.

***Theorem 11.8.3 (Scaling)*** *If the Laplace transform of $f(t)$ is $F(s)$, then the Laplace transform of $f(ct)$ with $c>0$ is $(1/c)F(s/c)$.*

PROOF. *By definition*

$$\begin{aligned} \mathscr{L}[f(ct)] &= \int_0^\infty e^{-st} f(ct)\, dt \\ &= \int_0^\infty \frac{1}{c} e^{-(s\xi/c)} f(\xi)\, d\xi \qquad \text{(by substituting } \xi = ct\text{)} \\ &= (1/c)F(s/c) \end{aligned}$$

□

EXAMPLE 11.8.2.

a. If $\dfrac{s}{s+1} = \mathscr{L}[\cos t]$, then

$$\frac{1}{\omega}\frac{s/\omega}{(s/\omega)^2+1} = \frac{s}{s^2+\omega^2} = \mathscr{L}[\cos \omega t]$$

b. If $\dfrac{1}{s-1} = \mathscr{L}[e^t]$, then

$$\frac{1}{a}\frac{1}{\left(\dfrac{s}{a}-1\right)} = \mathscr{L}[e^{at}]$$

$$\frac{1}{s-a} = \mathscr{L}[e^{at}]$$

***Theorem 11.8.4 (Differentiation)*** *Let $f$ be continuous and $f'$ piecewise continuous, in $0 \leqslant t \leqslant T$ for all $T>0$. Let $f$ also be of exponential order at $t \to \infty$. Then the Laplace transform of $f'(t)$ exists and is given by*

$$\mathscr{L}[f'(t)] = s\mathscr{L}[f(t)] - f(0)$$

PROOF. Consider the integral

$$\begin{aligned} \int_0^T e^{-st} f'(t)\, dt &= [e^{-st} f(t)]_0^T + \int_0^T s e^{-st} f(t)\, dt \\ &= e^{-st} f(T) - f(0) + s\int_0^T e^{-st} f(t)\, dt \end{aligned}$$

Since $|f(t)| \leqslant Me^{at}$ for large $t$, with $a > 0$ and $M > 0$,

$$|e^{-st}f(T)| \leqslant Me^{-(s-a)T}$$

In the limit as $T \to \infty$, $e^{-st}f(T) \to 0$ whenever $s > a$. Hence

$$\mathscr{L}[f'(t)] = s\mathscr{L}[f(t)] - f(0) \qquad \Box$$

If $f'$ and $f''$ satisfy the same conditions imposed on $f$ and $f'$ respectively, then the Laplace transform of $f''(t)$ can be obtained immediately by applying the preceding theorem; that is

$$\begin{aligned}\mathscr{L}[f''(t)] &= s\mathscr{L}[f'(t)] - f'(0) \\ &= s\{s\mathscr{L}[f(t)] - f(0)\} - f'(0) \\ &= s^2\mathscr{L}[f(t)] - sf(0) - f'(0)\end{aligned}$$

Clearly, the Laplace transform of $f^{(n)}(t)$ can be obtained in a similar manner by successive application of the theorem. The result may be written as

$$\mathscr{L}[f^{(n)}(t)] = s^n\mathscr{L}[f(t)] - s^{n-1}f(0) - \ldots - sf^{(n-2)}(0) - f^{(n-1)}(0)$$

***Theorem 11.8.5 (Integration)*** *If $F(s)$ is the Laplace transform of $f(t)$, then*

$$\mathscr{L}\left[\int_0^t f(\tau)\,d\tau\right] = F(s)/s$$

PROOF.

$$\begin{aligned}\mathscr{L}\left[\int_0^t f(\tau)\,d\tau\right] &= \int_0^\infty \left[\int_0^t f(\tau)\,d\tau\right]e^{-st}\,dt \\ &= \left[-\frac{e^{-st}}{s}\int_0^t f(\tau)\,d\tau\right]_0^\infty + \frac{1}{s}\int_0^\infty f(t)e^{-st}\,dt \\ &= F(s)/s\end{aligned}$$

since $\int_0^t f(\tau)\,d\tau$ is of exponential order. $\Box$

In solving problems by this method, the difficulty arises in finding inverse transforms. Although the inversion formula exists, its use requires a knowledge of complex variables. However, for some problems of mathematical physics, we need not use this inversion formula. We can avoid its use by expanding a given transform by the method of partial fractions in terms of simple fractions in the transform variables. With these simple functions, one refers to the table of Laplace transforms[20]

[20] A short table of Laplace transforms is given in the Appendix. For extensive tables, see the references.

and obtains the inverse transforms. Here, we should note that we use the assumption that there is essentially a one-to-one correspondence between functions and their Laplace transforms. This may be stated as follows:

***Theorem 11.8.6 (Lerch)*** *Let $f$ and $g$ be piecewise continuous functions of exponential order. If there exists a constant $s_0$, such that $\mathscr{L}[f] = \mathscr{L}[g]$ for all $s > s_0$, then $f(t) = g(t)$ for all $t > 0$ except possibly at the points of discontinuity.*

For a proof, the reader is referred to Kreider et al. [24].

EXAMPLE 11.8.3. Consider the motion of a semi-infinite string with an external force $f(t)$ acting on it. One end is kept fixed while the other end is allowed to move freely in the vertical direction. If the string is initially at rest, the motion of the string is governed by

$$u_{tt} = c^2 u_{xx} + f(t), \qquad 0 < x < \infty, \qquad t > 0$$

$$u(x, 0) = 0$$

$$u_t(x, 0) = 0$$

$$u(0, t) = 0$$

$$u_x(x, t) \to 0 \qquad \text{as } x \to \infty$$

Let $U(x, s)$ be the Laplace transform of $u(x, t)$. Transforming the equation of motion and using the initial conditions, we obtain

$$U_{xx} - (s^2/c^2)U = -F(s)/c^2$$

where

$$F(s) = \int_0^\infty f(t) e^{-st}\, dt$$

The solution of this ordinary differential equation is

$$U(x, s) = Ae^{sx/c} + Be^{-sx/c} + [F(s)/s^2]$$

The transformed boundary conditions are

$$U(0, s) = 0$$

and

$$\lim_{x \to \infty} U_x(x, s) = 0$$

In view of the second condition, we have $A = 0$. Now applying the first condition, we obtain

$$U(0, s) = B + [F(s)/s^2] = 0$$

Hence

$$U(x, s) = [F(s)/s^2][1 - e^{-sx/c}]$$

a. When $f(t) = f_0$, a constant, then

$$U(x, s) = f_0\left(\frac{1}{s^3} - \frac{1}{s^3} e^{-sx/c}\right)$$

The inverse is given by

$$u(x, t) = \frac{f_0}{2}\left[t^2 - \left(t - \frac{x}{c}\right)^2\right] \quad \text{when} \quad t \geq x/c$$

$$= (f_0/2)t^2 \quad \text{when} \quad t \leq x/c$$

b. When $f(t) = \cos \omega t$, where $\omega$ is a constant, then

$$F(s) = \int_0^\infty \cos \omega t \, e^{-st} \, dt$$

$$= s/(\omega^2 + s^2)$$

Thus, we have

$$U(x, s) = \frac{1}{s(\omega^2 + s^2)} [1 - e^{-sx/c}] \tag{11.8.1}$$

By the method of partial fractions, we write

$$\frac{1}{s(s^2 + \omega^2)} = \frac{1}{\omega^2}\left[\frac{1}{s} - \frac{s}{s^2 + \omega^2}\right]$$

Hence

$$\mathscr{L}^{-1} \frac{1}{s(s^2 + \omega^2)} = \frac{1}{\omega^2}(1 - \cos \omega t) = \frac{2}{\omega^2} \sin^2\left(\frac{\omega t}{2}\right)$$

If we denote

$$\psi(t) = \sin^2\left(\frac{\omega t}{2}\right)$$

then the inverse of Eq. (11.8.1) may be written in the form

$$u(x, t) = \frac{2}{\omega^2}\left[\psi(t) - \psi\left(t - \frac{x}{c}\right)\right] \quad \text{when} \quad t \geq x/c$$

$$= \frac{2}{\omega^2} \psi(t) \quad \text{when} \quad t \leq x/c$$

## 11.9 CONVOLUTION THEOREM OF THE LAPLACE TRANSFORM

The function

$$(f * g)(t) = \int_0^t f(t - \xi) g(\xi) \, d\xi \tag{11.9.1}$$

is called the *convolution* of the functions $f$ and $g$.

***Theorem 11.9.1 (Convolution)*** *If $F(s)$ and $G(s)$ are the Laplace transforms of $f(t)$ and $g(t)$ respectively, then the Laplace transform of the convolution $f * g$ is the product $F(s)G(s)$.*

PROOF. *By definition*

$$\mathscr{L}[f * g] = \int_0^\infty e^{-st} \int_0^t f(t-\xi) g(\xi)\, d\xi\, dt$$
$$= \int_0^\infty \int_0^t e^{-st} f(t-\xi) g(\xi)\, d\xi\, dt$$

The region of integration is shown in Fig. 11.9.1. By reversing the order of integration, we have

$$\mathscr{L}[f * g] = \int_0^\infty \int_\xi^\infty e^{-st} f(t-\xi) g(\xi)\, dt\, d\xi$$
$$= \int_0^\infty g(\xi) \int_\xi^\infty e^{-st} f(t-\xi)\, dt\, d\xi$$

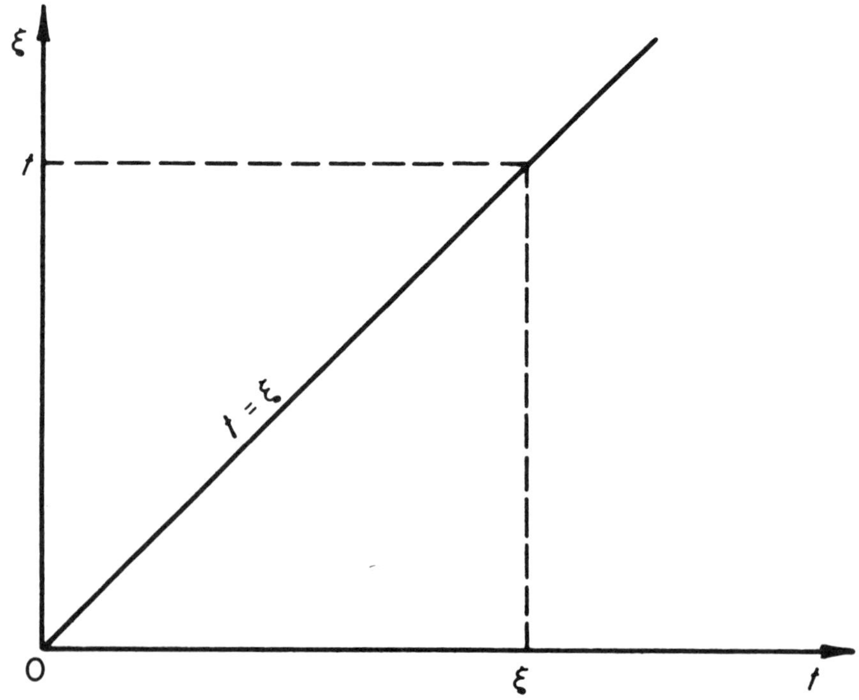


Figure 11.9.1

If we introduce the new variable $\eta = t - \xi$ in the inner integral, we obtain

$$\mathscr{L}[f * g] = \int_0^\infty g(\xi) \int_0^\infty e^{-s(\xi+\eta)} f(\eta)\, d\eta\, d\xi$$

$$= \int_0^\infty g(\xi) e^{-s\xi}\, d\xi \int_0^\infty f(\eta) e^{-s\eta}\, d\eta$$

$$= F(s)G(s) \tag{11.9.2}$$ □

The convolution has the following properties:

1. $f * g = g * f$ (commutative)
2. $f * (g * h) = (f * g) * h$ (associative)
3. $f * (g + h) = f * g + f * h$ (distributive)

EXAMPLE 11.9.1. Find the temperature distribution in a semi-infinite radiating rod. The temperature is kept constant at $x = 0$ while the other end is kept at zero temperature. If the initial temperature distribution is zero, the problem is governed by

$$u_t = k u_{xx} - hu, \qquad 0 < x < \infty, \qquad t > 0, \qquad h = \text{constant}$$

$$u(x, 0) = 0$$

$$u(0, t) = u_0, \qquad t > 0, \qquad u_0 = \text{constant}$$

$$u(x, t) \to 0 \quad \text{as} \quad x \to \infty$$

Let $U(x, s)$ be the Laplace transform of $u(x, t)$. Then the transformation with respect to $t$ yields

$$U_{xx} - \left(\frac{s+h}{k}\right) U = 0$$

$$U(0, s) = u_0/s$$

$$\lim_{x \to \infty} U(x, s) = 0$$

The solution of this equation is

$$U(x, s) = A e^{\sqrt{(s+h)}x/\sqrt{k}} + B e^{-\sqrt{(s+h)}x/\sqrt{k}}$$

The boundary condition at infinity requires that $A = 0$. Applying the remaining boundary condition, we have

$$U(0, s) = B = u_0/s$$

Hence the solution takes the form

$$U(x, s) = (u_0/s) e^{-\sqrt{(s+h)}x/\sqrt{k}}$$

We find that (by using Table III)

$$\mathscr{L}^{-1}\left[\frac{u_0}{s}\right] = u_0$$

and

$$\mathscr{L}^{-1}[e^{-\sqrt{(s+h)x/k}}] = \frac{xe^{-ht-(x^2/4kt)}}{2\sqrt{\pi kt^3}}$$

Thus the inverse Laplace transform of $U(x, s)$ is

$$u(x, t) = \mathscr{L}^{-1}\left[\frac{u_0}{s} e^{-\sqrt{(s+h)x/k}}\right]$$

By the Integration Theorem 11.8.5, we have

$$u(x, t) = \int_0^t \frac{u_0 x e^{-h\tau-(x^2/4k\tau)}}{2\sqrt{\pi k}\, \tau^{\frac{3}{2}}}\, d\tau$$

Substitution of the new variable $\eta = x/2\sqrt{k\tau}$ yields

$$u(x, t) = \frac{2u_0}{\sqrt{\pi}} \int_{x/2\sqrt{kt}}^{\infty} e^{-[\eta^2 + (hx^2/4k\eta^2)]}\, d\eta$$

For the case $h = 0$, $u(x, t)$ becomes

$$\begin{aligned}
u(x, t) &= \frac{2u_0}{\sqrt{\pi}} \int_{x/2\sqrt{kt}}^{\infty} e^{-\eta^2}\, d\eta \\
&= \frac{2u_0}{\sqrt{\pi}} \int_0^{\infty} e^{-\eta^2}\, d\eta - \frac{2u_0}{\sqrt{\pi}} \int_0^{x/2\sqrt{kt}} e^{-\eta^2}\, d\eta \\
&= u_0\left[1 - \operatorname{erf}\left(\frac{x}{2\sqrt{kt}}\right)\right] \\
&= u_0 \operatorname{erfc}\left(\frac{x}{2\sqrt{kt}}\right)
\end{aligned}$$

## 11.10 THE LAPLACE TRANSFORMS OF STEP AND IMPULSE FUNCTIONS

We have defined earlier the unit step function. Now, we will find the Laplace transform of it.

$$\begin{aligned}
\mathscr{L}[u_a(t)] &= \int_0^{\infty} e^{-st} u_a(t)\, dt \\
&= \int_a^{\infty} e^{-st}\, dt \qquad (11.10.1) \\
&= e^{-as}/s, \qquad s > 0
\end{aligned}$$

***Theorem 11.10.1 (Second Shifting)*** *If $F(s)$ is the Laplace transform of $f(t)$, then*

$$\mathscr{L}[u_a(t)f(t-a)] = e^{-as}F(s)$$

PROOF. *By definition*

$$\mathscr{L}[u_a(t)f(t-a)] = \int_0^\infty e^{-st}u_a(t)f(t-a)\,dt$$
$$= \int_a^\infty e^{-st}f(t-a)\,dt$$

Introducing the new variable $\xi = t - a$, we have

$$\mathscr{L}[u_a(t)f(t-a)] = \int_0^\infty e^{-(\xi+a)s}f(\xi)\,d\xi$$
$$= e^{-as}\int_0^\infty e^{-\xi s}f(\xi)\,d\xi$$
$$= e^{-as}F(s)$$

□

EXAMPLE 11.10.1. a. Given that

$$f(t) = \begin{cases} 0, & t < 2 \\ t-2, & t \geqslant 2 \end{cases}$$

find the Laplace transform of $f(t)$.

$$\mathscr{L}[f(t)] = \mathscr{L}[u_2(t)(t-2)]$$
$$= e^{-2s}\mathscr{L}[t]$$
$$= e^{-2s}/s^2$$

b. Find the inverse transform of

$$F(s) = \frac{1+e^{-2s}}{s^2}$$

$$\mathscr{L}^{-1}[F(s)] = \mathscr{L}^{-1}\left[\frac{1}{s^2} + \frac{e^{-2s}}{s^2}\right]$$
$$= \mathscr{L}^{-1}\left[\frac{1}{s^2}\right] + \mathscr{L}^{-1}\left[\frac{e^{-2s}}{s^2}\right]$$
$$= t + u_2(t)(t-2)$$
$$= \begin{cases} t, & 0 \leqslant t < 2 \\ 2(t-1), & t \geqslant 2 \end{cases}$$

The Laplace transform of the impulse function $p(t)$ is given by

$$
\begin{aligned}
\mathscr{L}[p(t)] &= \int_0^\infty e^{-st} p(t)\, dt \\
&= \int_{a-\varepsilon}^{a+\varepsilon} he^{-st}\, dt \\
&= h\left[-\frac{e^{-st}}{s}\right]_{a-\varepsilon}^{a+\varepsilon} \\
&= \frac{he^{-as}}{s}(e^{\varepsilon s} - e^{-\varepsilon s}) \\
&= 2\frac{he^{-as}}{s}\sinh \varepsilon s \qquad (11.10.2)
\end{aligned}
$$

If we choose the value of $h$ to be $1/2\varepsilon$, then the impulse is given by

$$
\begin{aligned}
I(\varepsilon) &= \int_{-\infty}^{\infty} p(t)\, dt \\
&= \int_{a-\varepsilon}^{a+\varepsilon} \frac{1}{2\varepsilon}\, dt \\
&= 1
\end{aligned}
$$

Thus, in the limit this particular impulse function satisfies

$$\lim_{\varepsilon \to 0} p_\varepsilon(t) = 0, \qquad t \neq a$$

$$\lim_{\varepsilon \to 0} I(\varepsilon) = 1$$

From this result, we obtain the Dirac delta function which satisfies

$$\delta(t-a) = 0, \qquad t \neq a$$

$$\int_{-\infty}^{\infty} \delta(t-a)\, dt = 1 \qquad (11.10.3)$$

Thus, we may define the Laplace transform of $\delta(t)$ as the limit of the transform of $p_\varepsilon(t)$.

$$
\begin{aligned}
\mathscr{L}[\delta(t-a)] &= \lim_{\varepsilon \to 0} \mathscr{L}[p_\varepsilon(t)] \\
&= \lim_{\varepsilon \to 0} e^{-as}\frac{\sinh \varepsilon s}{\varepsilon s} \qquad (11.10.4) \\
&= e^{-as}
\end{aligned}
$$

Hence if $a = 0$, we have

$$\mathscr{L}[\delta(t)] = 1 \qquad (11.10.5)$$

One of the very useful results that can be derived is the integral of the product of the delta function and any continuous function $f(t)$.

$$\int_{-\infty}^{\infty} \delta(t-a)f(t)\,dt = \lim_{\varepsilon\to 0}\int_{-\infty}^{\infty} p_\varepsilon(t)f(t)\,dt$$

$$= \lim_{\varepsilon\to 0}\int_{a-\varepsilon}^{a+\varepsilon} \frac{f(t)}{2\varepsilon}\,dt$$

$$= \lim_{\varepsilon\to 0}\frac{1}{2\varepsilon}\cdot 2\varepsilon f(t^*) \qquad a-\varepsilon < t^* < a+\varepsilon$$

$$= f(a) \tag{11.10.6}$$

Suppose that $f(t)$ is a periodic function with period $T$. Let $f$ be piecewise continuous on $[0, T]$. Then the Laplace transform of $f(t)$ is

$$\mathscr{L}[f(t)] = \int_0^{\infty} e^{-st}f(t)\,dt$$

$$= \sum_{n=0}^{\infty}\int_{nT}^{(n+1)T} e^{-st}f(t)\,dt$$

If we introduce the new variable $\xi = t - nT$, then

$$\mathscr{L}[f(t)] = \sum_{n=0}^{\infty} e^{-nTs}\int_0^{T} e^{-s\xi}f(\xi)\,d\xi$$

$$= \sum_{n=0}^{\infty} e^{-nTs}F_1(s)$$

where $F_1(s) = \int_0^T e^{-s\xi}f(\xi)\,d\xi$ is the transform of the function $f$ over the first period. Since the series is the geometric series, we obtain for the transform of the periodic function

$$\mathscr{L}[f(t)] = \frac{F_1(s)}{1-e^{-Ts}} \tag{11.10.7}$$

EXAMPLE 11.10.2. Find the Laplace transform of the square wave function

$$f(t) = \begin{cases} h, & 0<t<c \\ -h, & c<t<2c \end{cases}$$

and

$$f(t+2c) = f(t)$$

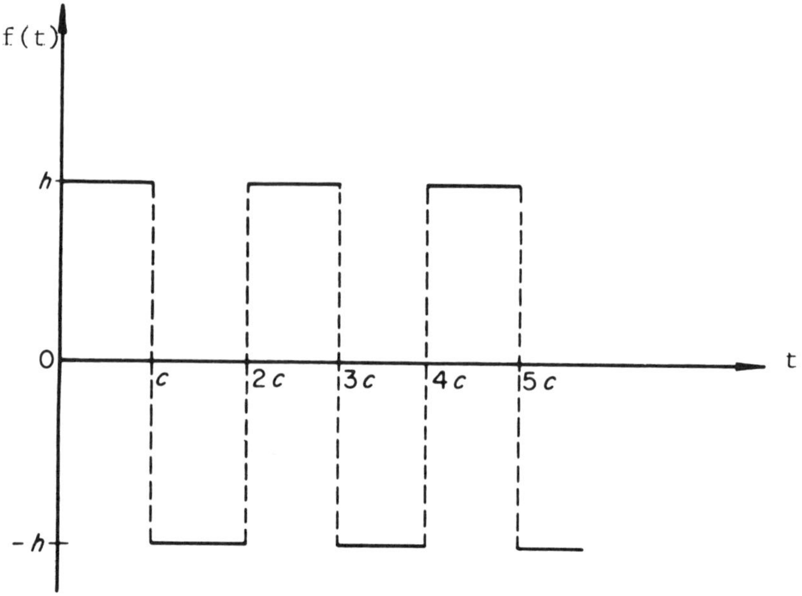


Figure 11.10.1

shown in Fig. 11.10.1.

$$F_1(s) = \int_0^{2c} e^{-s\xi} f(\xi)\, d\xi$$

$$= \int_0^{c} e^{-s\xi} h\, d\xi + \int_c^{2c} e^{-s\xi}(-h)\, d\xi$$

$$= \frac{h}{s}(1 - e^{-cs})^2$$

Thus, the Laplace transform of $f(t)$ is, by (11.10.7),

$$\mathscr{L}[f(t)] = \frac{F_1(s)}{1 - e^{-2cs}}$$

$$= \frac{h(1 - e^{-cs})^2}{s(1 - e^{-2cs})}$$

$$= \frac{h(1 - e^{-cs})}{s(1 + e^{-cs})}$$

$$= \frac{h}{s} \tanh \frac{cs}{2}$$

EXAMPLE 11.10.3. A uniform bar of length $l$ is fixed at one end. Let the force

$$f(t) = \begin{cases} f_0, & t > 0 \\ 0, & t < 0 \end{cases}$$

be suddenly applied at the end $x = l$. If the bar is initially at rest, find the longitudinal displacement for $t > 0$.

The motion of the bar is governed by

$$u_{tt} = a^2 u_{xx}, \qquad 0 < x < l, \qquad t > 0, \qquad a = \text{constant}$$

$$u(x, 0) = 0$$

$$u_t(x, 0) = 0$$

$$u(0, t) = 0$$

$$u_x(l, t) = f_0/E$$

where $E$ is a constant.

Let $U(x, s)$ be the Laplace transform of $u(x, t)$. Then $U(x, s)$ satisfies

$$U_{xx} - \frac{s^2}{a^2} U = 0$$

$$U(0, s) = 0$$

$$U_x(l, s) = f_0/Es$$

The solution of this differential equation is

$$U(x, s) = Ae^{xs/a} + Be^{-xs/a}$$

Applying the boundary conditions, we have

$$A + B = 0$$

$$\left(\frac{s}{a} e^{ls/a}\right) A + \left(-\frac{s}{a} e^{-ls/a}\right) B = f_0/Es$$

Solving for $A$ and $B$, we obtain

$$A = -B = \frac{af_0}{Es^2(e^{ls/a} + e^{-ls/a})}$$

Hence the transform is given by

$$U(x, s) = \frac{af_0(e^{xs/a} - e^{-xs/a})}{Es^2(e^{ls/a} + e^{-ls/a})}$$

Before finding the inverse transform of $U(x, s)$, multiply the numerator and denominator by $(e^{-ls/a} - e^{-3ls/a})$. Thus, we have

$$U(x, s) = \frac{af_0}{Es^2}\left[e^{-(l-x)s/a} - e^{-(l+x)s/a} - e^{-(3l-x)s/a} + e^{-(3l+x)s/a}\right]\frac{1}{1-e^{-4ls/a}}$$

Since the denominator has the term $(1 - e^{-4ls/a})$, the inverse transform $u(x, t)$ is periodic with period $4l/a$. Hence the final solution may be written in the form

$$u(x,t) = \begin{cases} 0, & 0 < t < \dfrac{l-x}{a} \\ \dfrac{af_0}{E}\left(t - \dfrac{l-x}{a}\right), & \dfrac{l-x}{a} < t < \dfrac{l+x}{a} \\ \dfrac{af_0}{E}\left[\left(t - \dfrac{l-x}{a}\right) - \left(t - \dfrac{l+x}{a}\right)\right], & \dfrac{l+x}{a} < t < \dfrac{3l-x}{a} \\ \dfrac{af_0}{E}\left[\left(t - \dfrac{l-x}{a}\right) - \left(t - \dfrac{l+x}{a}\right) - \left(t - \dfrac{3l-x}{a}\right)\right], & \dfrac{3l-x}{a} < t < \dfrac{3l+x}{a} \\ \dfrac{af_0}{E}\left[\left(t - \dfrac{l-x}{a}\right) - \left(t - \dfrac{l+x}{a}\right) - \left(t - \dfrac{3l-x}{a}\right) - \left(t - \dfrac{3l+x}{a}\right)\right], & \dfrac{3l+x}{a} < t < \dfrac{4l}{a} \end{cases}$$

which may be simplified to give

$$u(x,t) = \begin{cases} 0, & 0 < t < (l-x)/a \\ \dfrac{af_0}{E}\left(t - \dfrac{l-x}{a}\right), & (l-x)/a < t < (l+x)/a \\ \dfrac{af_0}{E}\left(\dfrac{2x}{a}\right), & (l+x)/a < t < (3l-x)/a \\ \dfrac{af_0}{E}\left(-t + \dfrac{3l+x}{a}\right), & (3l-x)/a < t < (3l+x)/a \\ 0, & (3l+x)/a < t < 4l/a \end{cases}$$

This result can clearly be seen in Fig. 11.10.2.

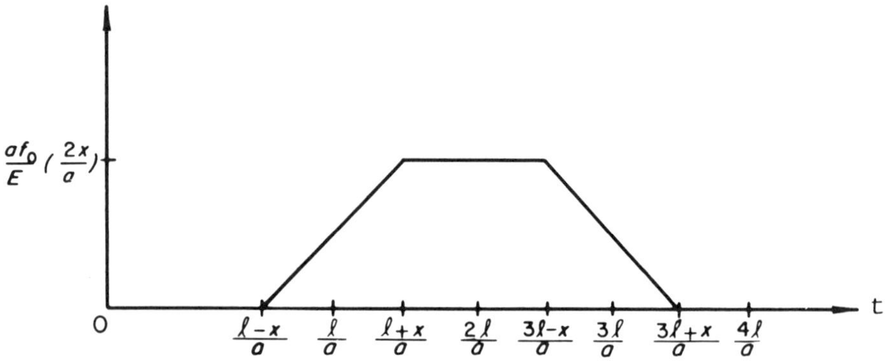


Figure 11.10.2

EXAMPLE 11.10.4. Consider a semi-infinite string fixed at the end $x = 0$. The string is initially at rest. Let there be an external force

$$f(x, t) = -f_0 \delta\left(t - \frac{x}{v}\right)$$

acting on the string. This is a concentrated force $f_0$ acting at the point $x = vt$.

The motion of the string is governed by

$$u_{tt} = c^2 u_{xx} - f_0 \delta\left(t - \frac{x}{v}\right)$$

$$u(x, 0) = 0$$

$$u_t(x, 0) = 0$$

$$u(0, t) = 0$$

$u(x, t)$ is bounded as $x \to \infty$

Let $U(x, s)$ be the Laplace transform of $u(x, t)$. Transforming the wave equation and using the initial conditions, we obtain

$$U_{xx} - \frac{s^2}{c^2} U = \frac{f_0}{c^2} e^{-xs/v}$$

The solution of this equation is

$$U(x, s) = Ae^{sx/c} + Be^{-sx/c} + \begin{cases} \dfrac{f_0 v^2 e^{-sx/v}}{(c^2 - v^2)s^2} & \text{for} \quad v \neq c \\ -\dfrac{f_0 x e^{-sx/v}}{2cs} & \text{for} \quad v = c \end{cases}$$

The condition that $u(x, t)$ must be bounded at infinity requires that $A = 0$. Application of the condition

$$U(0, s) = 0$$

yields

$$B = \begin{cases} \dfrac{-f_0 v^2}{(c^2 - v^2)s^2} & \text{for} \quad v \neq c \\ 0 & \text{for} \quad v = 0 \end{cases}$$

Hence the transform is given by

$$U(x, s) = \begin{cases} \dfrac{f_0 v^2 (e^{-xs/v} - e^{-xs/c})}{(c^2 - v^2)s^2} & \text{for} \quad v \neq c \\ -\dfrac{f_0 x e^{-xs/c}}{2cs} & \text{for} \quad v = c \end{cases}$$

The inverse transform is therefore given by

$$u(x,t)=\begin{cases}\dfrac{f_0v^2}{(c^2-v^2)}\left[\left(t-\dfrac{x}{v}\right)u\left(t-\dfrac{x}{v}\right)-\left(t-\dfrac{x}{c}\right)u\left(t-\dfrac{x}{c}\right)\right], & v\neq c\\ \dfrac{-f_0x}{2c}u\left(t-\dfrac{x}{c}\right) \quad \text{for} \quad v=c\end{cases}$$

EXAMPLE 11.10.5 (The Stokes Problem and the Rayleigh Problem in fluid dynamics). Solve the Stokes problem which is concerned with the unsteady boundary layer flows induced in a semi-infinite viscous fluid bounded by an infinite horizontal disk at $z=0$ due to oscillations of the disk in its own plane with a given frequency $\omega$.

We solve the boundary layer equation

$$u_t=\nu u_{zz}, \qquad z>0, \quad t>0$$

with the boundary and initial conditions

$$\begin{aligned} u(z,t)&=U_0e^{i\omega t} && \text{on} \quad z=0,\ t>0\\ u(z,t)&\to 0 && \text{as} \quad z\to\infty, t>0\\ u(z,0)&=0 && \text{at} \quad t\leqslant 0 \quad \text{for all} \quad z>0\end{aligned}$$

where $u(z,t)$ is the velocity of fluid of kinematic viscosity $\nu$ and $U_0$ is a constant.

The Laplace transform solution of the equation with the transformed boundary conditions is

$$U(z,s)=\frac{U_0}{(s-i\omega)}\exp\left(-z\sqrt{\frac{s}{\nu}}\right)$$

Using a standard table of the inverse Laplace transform, we obtain the solution

$$\begin{aligned} u(z,t)=\frac{U_0}{2}e^{i\omega t}[&\exp(-\lambda z)\operatorname{erfc}(\zeta-\sqrt{i\omega t})\\ &+\exp(\lambda z)\operatorname{erfc}(\zeta+\sqrt{i\omega t})]\end{aligned}$$

where $\zeta=z/2\sqrt{\nu t}$ is called the *similarity variable* of the viscous boundary layer theory and $\lambda=(i\omega/\nu)^{\frac{1}{2}}$. This result describes the unsteady boundary layer flow.

In view of the asymptotic formula for the complementary error function

$$\operatorname{erfc}(\zeta\mp\sqrt{i\omega t})\sim(2,0) \quad \text{as} \quad t\to\infty$$

the above solution for $u(z,t)$ has the asymptotic representation

$$\begin{aligned} u(z,t)&\sim U_0\exp(i\omega t-\lambda z)\\ &=U_0\exp\left[i\omega t-\left(\frac{\omega}{2\nu}\right)^{\frac{1}{2}}(1+i)z\right] \qquad (11.10.8)\end{aligned}$$

This is called the *Stokes steady-state solution.* This represents the propagation of shear waves which spread out from the oscillating disk with velocity $\omega/k = \sqrt{2\nu\omega}$ and exponentially decaying amplitude. The boundary layer associated with the solution has thickness of the order $(\nu/\omega)^{\frac{1}{2}}$ in which the shear oscillations imposed by the disk decay exponentially with distance $z$ from the disk. This boundary layer is called the *Stokes layer.* In other words, the thickness of the Stokes layer is equal to the depth of penetration of vorticity which is essentially confined to the immediate vicinity of the disk for high frequency $\omega$.

The Stokes problem with $\omega = 0$ becomes the Rayleigh problem. In other words, the motion is generated in the fluid from rest by moving the disk impulsively in its own plane with constant velocity $U_0$. In this case, the Laplace transformed solution is

$$U(z, s) = \frac{U_0}{s} \exp\left(-z\sqrt{\frac{s}{\nu}}\right)$$

so that the inversion gives the Rayleigh solution

$$u(z, t) = U_0 \operatorname{erfc}\left(\frac{z}{2\sqrt{\nu t}}\right) \tag{11.10.9}$$

This describes the growth of a boundary layer adjacent to the disk. The associated boundary layer is called the *Rayleigh layer* of thickness of the order $\delta \sim \sqrt{\nu t}$ which grows with increasing time. The rate of growth is of the order $d\delta/dt \sim \sqrt{\nu/t}$, which diminishes with increasing time.

The vorticity of the unsteady flow is given by

$$\frac{\partial u}{\partial z} = \frac{U_0}{\sqrt{\pi \nu t}} \exp(-\zeta^2) \tag{11.10.10}$$

which decays exponentially to zero as $z \gg \delta$.

Note that the vorticity is everywhere zero at $t = 0$ except at $z = 0$. This implies that it is generated at the disk and diffuses outward within the Rayleigh layer. The total viscous diffusion time $T_d \sim \delta^2/\nu$.

Another physical quantity related to the Stokes and Rayleigh problems is the *skin friction* on the disk defined by

$$\tau_0 = \mu\left(\frac{\partial u}{\partial z}\right)_{z=0} \tag{11.10.11}$$

where $\mu = \nu\rho$ is the dynamic viscosity and $\rho$ is the density of the fluid. The skin friction can readily be calculated from the flow field given by (11.10.8) or (11.10.9).

EXAMPLE 11.10.6. Obtain the d'Alembert solution of the Cauchy problem for a one-dimensional wave equation in $-\infty < x < \infty$:

$$u_{tt} = c^2 u_{xx}, \qquad t > 0$$

$$u(x, 0) = f(x), \qquad u_t(x, 0) = g(x)$$

We use the joint Laplace and Fourier transforms of $u(x, t)$ defined by

$$\bar{U}(\alpha, s) = \frac{1}{\sqrt{2\pi}} \int_{-\infty}^{\infty} e^{i\alpha x}\, dx \int_0^{\infty} e^{-st}\, u(x, t)\, dt$$

where $\bar{U}(\alpha, s)$ is the Laplace transform of $U(\alpha, t)$.

The transformed Cauchy problem has the solution in the form

$$\bar{U}(\alpha, s) = \frac{sF(\alpha) + G(\alpha)}{s^2 + c^2\alpha^2}$$

The joint inverse transformation gives the solution as

$$\begin{aligned}
u(x, t) &= \frac{1}{\sqrt{2\pi}} \int_{-\infty}^{\infty} e^{-i\alpha x} \mathscr{L}^{-1}\left[\frac{sF(\alpha) + G(\alpha)}{s^2 + c^2\alpha^2}\right] d\alpha \\
&= \frac{1}{\sqrt{2\pi}} \int_{-\infty}^{\infty} e^{-i\alpha x}\left[F(\alpha)\cos c\alpha t + \frac{G(\alpha)}{c\alpha}\sin c\alpha t\right] d\alpha \\
&= \frac{1}{\sqrt{2\pi}} \int_{-\infty}^{\infty} \frac{1}{2} F(\alpha) e^{-i\alpha x}(e^{ic\alpha t} + e^{-ic\alpha t})\, d\alpha \\
&\quad + \frac{1}{\sqrt{2\pi}} \int_{-\infty}^{\infty} \frac{1}{2ic\alpha} e^{-i\alpha x} G(\alpha)(e^{ic\alpha t} - e^{-ic\alpha t})\, d\alpha \\
&= \frac{1}{2}[f(x - ct) + f(x + ct)] + \frac{1}{\sqrt{2\pi}} \frac{1}{2c} \int_{-\infty}^{\infty} G(\alpha)\, d\alpha \int_{x-ct}^{x+ct} e^{-i\alpha\zeta}\, d\zeta \\
&= \frac{1}{2}[f(x - ct) + f(x + ct)] + \frac{1}{2c} \int_{x-ct}^{x+ct} g(\zeta)\, d\zeta \qquad (11.10.12)
\end{aligned}$$

This is the d'Alembert solution already obtained in Section 4.3.

## 11.11. HANKEL TRANSFORMS

Suppose $f(x, y)$ is a function of two independent variables $x$ and $y$ defined in $-\infty < (x, y) < \infty$. Then the double Fourier transform of $f(x, y)$ and its inverse are given by

$$F(\alpha, \beta) = \frac{1}{2\pi} \iint_{-\infty}^{\infty} e^{i(\alpha x + \beta y)} f(x, y)\, dx\, dy \qquad (11.11.1)$$

$$f(x, y) = \frac{1}{2\pi} \iint_{-\infty}^{\infty} e^{-i(\alpha x + \beta y)} F(\alpha, \beta)\, d\alpha\, d\beta \qquad (11.11.2)$$

provided these integrals exist.

In these integrals, we introduce the polar coordinates by writing $x = r\cos\theta$, $y = r\sin\theta$ and $\alpha = k\sin\phi$, $\beta = k\cos\phi$ so that

$$F(k, \phi) = \frac{1}{2\pi}\int_0^\infty r\,dr \int_{-\pi}^{\pi} e^{ikr\sin(\theta+\phi)} f(r, \theta)\, d\theta$$

$$f(r, \theta) = \frac{1}{2\pi}\int_0^\infty k\,dk \int_{-\pi}^{\pi} e^{-ikr\sin(\theta+\phi)} F(k, \phi)\, d\phi$$

We next write $f(r, \theta) = e^{-in\theta} f(r)$ and $F(k, \phi) = e^{in\phi}F(k)$ where $n$ is an integer or zero so that the above results become

$$e^{in\phi}F(k) = \frac{1}{2\pi}\int_0^\infty rf(r)\,dr \int_{-\pi}^{\pi} e^{-in\theta + ikr\sin(\theta+\phi)}\, d\theta$$

$$e^{-in\theta}f(r) = \frac{1}{2\pi}\int_0^\infty kF(k)\,dk \int_{-\pi}^{\pi} e^{in\phi - ikr\sin(\theta+\phi)}\, d\phi$$

Using the standard integral representation of the Bessel function $J_n(kr)$ in the form

$$e^{in\phi}J_n(kr) = \frac{1}{2\pi}\int_{-\pi}^{\pi} e^{-in\theta + ikr\sin(\theta+\phi)}\, d\theta \tag{11.11.3}$$

the above results assume the form

$$F_n(k) = H_n\{f(r)\} = \int_0^\infty rJ_n(kr)f(r)\,dr \tag{11.11.4}$$

$$f(r) = H_n^{-1}\{F(k)\} = \int_0^\infty kJ_n(kr)F_n(k)\,dk \tag{11.11.5}$$

where (11.11.5) is obtained by the complex conjugate of (11.11.3).

The function $F_n(k)$ defined by (11.11.4) is called the *Hankel transform* of $f(r)$ of order $n$. Instead of $F_n(k)$, we often simply write $F(k)$ for the Hankel transform specifying the order. Integral (11.11.4) exists for certain classes of functions which usually occur in practical applications. The *inverse Hankel transformation* is defined by the integral (11.11.5).

In particular, the Hankel transforms of the zero order ($n = 0$) and of the first order ($n = 1$) are widely used for solving problems involving Laplace's equation in cylindrical geometry.

EXAMPLE 11.11.1. Find the zero order Hankel transforms of (a) $(1/r)e^{-ar}$ and (b) $\delta(r)/r$.

(a) $F(k) = H_0\left\{\frac{1}{r}e^{-ar}\right\} = \int_0^\infty e^{-ar}J_0(kr)\,dr = \frac{1}{\sqrt{k^2 + a^2}}$

(b) $F(k) = H_0\left\{\frac{\delta(r)}{r}\right\} = \int_0^\infty \delta(r)J_0(kr)\,dr = 1.$

EXAMPLE 11.11.2. Find the first order Hankel transform of the function $f(r) = e^{-ar}$.

$$F(k) = H_1\{e^{-ar}\} = \int_0^\infty re^{-ar} J_1(kr)\, dr = \frac{k}{(a^2 + k^2)^{\frac{3}{2}}}$$

## 11.12. OPERATIONAL PROPERTIES OF HANKEL TRANSFORMS AND APPLICATIONS

We state the following properties of the Hankel transform:

(i) The Hankel transform is a linear integral operator, that is,

$$H_n\{af(r) + bg(r)\} = aH_n\{f(r)\} + bH_n\{g(r)\}$$

for all constants $a$ and $b$.

(ii) The Hankel transform satisfies the Parseval relation

$$\int_0^\infty rf(r)g(r)\, dr = \int_0^\infty kF(k)G(k)\, dk \tag{11.12.1}$$

where $F(k)$ and $G(k)$ are Hankel transforms of $f(r)$ and $g(r)$ respectively.

We proceed formally to obtain

$$\int_0^\infty kF(k)G(k)\, dk = \int_0^\infty kF(k)\, dk \int_0^\infty rJ_n(kr)g(r)\, dr$$

$$= \int_0^\infty rg(r)\, dr \int_0^\infty kJ_n(kr)F(k)\, dk$$

$$= \int_0^\infty rf(r)g(r)\, dr$$

(iii) $H_n\{f'(r)\} = \dfrac{k}{2n}[(n-1)F_{n+1}(k) - (n+1)F_{n-1}(k)]$

provided $rf(r)$ vanishes as $r \to 0$ and $r \to \infty$.

(iv) $$H_n\left\{\frac{1}{r}\frac{d}{dr}\left(r\frac{df}{dr}\right) - \frac{n^2}{r^2}f(r)\right\} = -k^2F_n(k) \tag{11.12.2}$$

provided both $\left(r\dfrac{df}{dr}\right)$ and $rf(r)$ vanish as $r \to 0$ and $r \to \infty$.

We have, by definition,

$$H_n\left\{\frac{1}{r}\frac{d}{dr}\left(r\frac{df}{dr}\right)-\frac{n^2}{r^2}f(r)\right\}$$

$$=\int_0^\infty \frac{d}{dr}\left(r\frac{df}{dr}\right)J_n(kr)\,dr-\int_0^\infty \frac{n^2}{r^2}rf(r)J_n(kr)\,dr$$

$$=\left[r\frac{df}{dr}J_n(kr)\right]_0^\infty-\int_0^\infty kJ_n'(kr)r\frac{df}{dr}\,dr$$

$$-\int_0^\infty \frac{n^2}{r^2}[rf(r)]J_n(kr)\,dr, \qquad \text{by partial integration}$$

$$=-[f(r)krJ_n'(kr)]_0^\infty+\int_0^\infty \frac{d}{dr}[krJ_n'(kr)]f(r)\,dr$$

$$-\int_0^\infty \frac{n^2}{r^2}rf(r)J_n(kr)\,dr, \qquad \text{by partial integration}$$

which is, by the given assumption and Bessel's equation (7.6.1),

$$=-\int_0^\infty \left(k^2-\frac{n^2}{r^2}\right)rf(r)J_n(kr)\,dr-\int_0^\infty \frac{n^2}{r^2}[rf(r)]J_n(kr)\,dr$$

$$=-k^2\int_0^\infty rf(r)J_n(kr)\,dr$$

$$=-k^2H_n\{f(r)\}=-k^2F_n(k)$$

This result is very widely used in solving partial differential equations in the axisymmetric cylindrical configurations. We illustrate this point by considering the following examples concerning applications:

EXAMPLE 11.12.1. Obtain the solution of the free vibration of a large circular membrane governed by the initial-value problem

$$\frac{\partial^2 u}{\partial r^2}+\frac{1}{r}\frac{\partial u}{\partial r}=\frac{1}{c^2}\frac{\partial^2 u}{\partial t^2}, \qquad 0<r<\infty, \qquad t>0$$

$$u(r,0)=f(r), \qquad u_t(r,0)=g(r), \qquad 0\leq r<\infty$$

where $c^2=T/\rho=$ constant, $T$ is the tension in the membrane, and $\rho$ is the surface density of the membrane.

Application of the Hankel transform of order zero

$$U(k,t)=\int_0^\infty ru(r,t)J_0(kr)\,dr$$

to the vibration problem gives

$$\frac{d^2U}{dt^2} + k^2c^2U = 0$$

$$U(k, 0) = F(k), \qquad U_t(k, 0) = G(k)$$

The general solution of this transformed system is

$$U(k, t) = F(k) \cos ckt + \frac{G(k)}{ck} \sin ckt$$

The inverse Hankel transformation gives

$$u(r, t) = \int_0^\infty kF(k) \cos ckt\, J_0(kr)\, dk + \frac{1}{c}\int_0^\infty G(k) \sin ckt\, J_0(kr)\, dr \tag{11.12.3}$$

This is the desired solution.

In particular, we consider

$$u(r, 0) = f(r) = \frac{A}{\left(1 + \frac{r^2}{a^2}\right)^{\frac{1}{2}}}, \qquad u_t(r, 0) = g(r) = 0$$

so that $G(k) = 0$ and

$$F(k) = Aa \int_0^\infty \frac{rJ_0(kr)\, dr}{(a^2 + r^2)^{\frac{1}{2}}} = \frac{Aa}{k} e^{-ak}$$

by means of Example 11.11.1.

Thus the solution (11.12.3) becomes

$$\begin{aligned} u(r, t) &= Aa \int_0^\infty e^{-ak} J_0(kr) \cos (ckt)\, dk \\ &= Aa \operatorname{Re} \int_0^\infty e^{-k(a+ict)} J_0(kr)\, dk \\ &= Aa \operatorname{Re} \{r^2 + (a + ict)^2\}^{-\frac{1}{2}} \end{aligned}$$

by Example 11.11.1.

EXAMPLE 11.12.2. Obtain the steady-state solution of the axisymmetric acoustic radiation problem governed by the wave equation in cylindrical polar coordinates $(r, \theta, z)$:

$$c^2\nabla^2 u = u_{tt}, \qquad 0 < r < \infty, \qquad z > 0, \qquad t > 0$$

$$u_z = f(r, t) \quad \text{on} \quad z = 0$$

where $f(r, t)$ is a given function and $c$ is a constant. We also assume that the solution is bounded and behaves as outgoing spherical waves. This is referred to as the *Sommerfeld radiation condition.*

We seek a steady-state solution of the acoustic radiation potential $u = e^{i\omega t}\phi(r, z)$ so that $\phi$ satisfies the Helmholtz equation

$$\phi_{rr} + \frac{1}{r}\phi_r + \phi_{zz} + \frac{\omega^2}{c^2}\phi = 0, \qquad 0 < r < \infty, \qquad z > 0$$

with the boundary condition representing the normal velocity prescribed on the $z = 0$ plane

$$\phi_z = f(r) \quad \text{on} \quad z = 0$$

where $f(r)$ is a known function of $r$.

We solve the problem by means of the zero order Hankel transformation

$$\Phi(k, z) = \int_0^\infty rJ_0(kr)\phi(r, z)\, dr$$

so that the given differential system becomes

$$\Phi_{zz} = \kappa^2\Phi, \qquad z > 0$$

$$\Phi_z = F(k) \quad \text{on} \quad z = 0$$

where $\kappa = (k^2 - \omega^2/c^2)^{\frac{1}{2}}$.

The solution of this system is

$$\Phi(k, z) = -\kappa^{-1}F(k)e^{-\kappa z}$$

where $\kappa$ is real and positive for $k > \omega/c$, and purely imaginary for $k < \omega/c$.

The inverse transformation yields the solution

$$\phi(r, z) = -\int_0^\infty \kappa^{-1}F(k)kJ_0(kr)e^{-\kappa z}\, dk$$

Since the exact evaluation of this integral is difficult, we choose a simple form of $f(r)$ as

$$f(r) = \begin{bmatrix} A, & r < a \\ 0, & r > a \end{bmatrix}$$

where $A$ is a constant so that

$$F(k) = \int_0^a kJ_0(ak)\, dk = \frac{a}{k}J_1(ak)$$

Then the solution for this special case is given by

$$\phi(r, z) = -Aa\int_0^\infty \kappa^{-1}J_1(ak)J_0(kr)e^{-\kappa z}\, dk \qquad (11.12.4)$$

For an asymptotic evaluation of this integral, we express it in terms of the spherical polar coordinates $(R, \theta, \phi)$ $(x = R \sin\theta \cos\phi,\ y = R \sin\theta \sin\phi,\ z = R\cos\theta)$, combined with the asymptotic result

$$J_0(kr) \sim \left(\frac{2}{\pi kr}\right)^{\frac{1}{2}} \cos\left(kr - \frac{\pi}{4}\right) \quad \text{as } r \to \infty$$

so that the acoustic potential $u = e^{i\omega t}\phi$ is

$$u \sim -\frac{Aa\sqrt{2}\, e^{i\omega t}}{\sqrt{\pi R \sin\theta}} \int_0^\infty \frac{\kappa^{-1}}{\sqrt{k}} J_1(ka) \cos\left(kR\sin\theta - \frac{\pi}{4}\right) e^{-\kappa z}\, dk$$

where $z = R\cos\theta$.

This integral can be evaluated asymptotically for $R \to \infty$ by using the stationary phase approximation formula (11.6.8) to obtain

$$u \sim -\frac{Aac}{\omega R \sin\theta} J_1(k_1 a) e^{i(\omega t - \omega R/c)} \qquad (11.12.5)$$

where $k_1 = \omega/c \sin\theta$ is the stationary point. This solution represents the outgoing spherical waves with constant velocity $c$ and decaying amplitude as $R \to \infty$.

## 11.13 THE MELLIN TRANSFORM AND ITS OPERATIONAL PROPERTIES

If $f(t)$ is not necessarily zero for $t < 0$, it is possible to define the *two-sided* (or *bilateral*) Laplace transform

$$F(p) = \int_{-\infty}^{\infty} e^{-pt} f(t)\, dt \qquad (11.13.1)$$

Then replacing $f(x)$ with $e^{-cx}f(x)$ in Fourier integral formula (5.12.9), we obtain

$$e^{-cx}f(x) = \frac{1}{2\pi}\int_{-\infty}^{\infty} e^{-i\alpha x}\, d\alpha \int_{-\infty}^{\infty} f(t) e^{-t(c - i\alpha)}\, dt$$

or

$$f(x) = \frac{1}{2\pi}\int_{-\infty}^{\infty} e^{x(c - i\alpha)}\, d\alpha \int_{-\infty}^{\infty} f(t) e^{-t(c - i\alpha)}\, dt$$

Making a change of variable $p = c - i\alpha$ and using result (11.13.1), we obtain the formal inverse transform of $f(t)$ as

$$f(x) = \frac{1}{2\pi i}\int_{c - i\infty}^{c + i\infty} e^{px} F(p)\, dp, \qquad c > 0 \qquad (11.13.2)$$

Putting $e^{-t}=x$ into (11.13.1) with $f(t)=g(x)$, it turns out that (11.13.1)–(11.13.2) become

$$G(p)=M\{g(x)\}=\int_0^\infty x^{p-1}g(x)\,dx \tag{11.13.3}$$

$$g(x)=M^{-1}\{G(p)\}=\frac{1}{2\pi i}\int_{c-i\infty}^{c+i\infty} x^{-p}G(p)\,dp \tag{11.13.4}$$

The function $G(p)$ is called the *Mellin transform* of $g(x)$. The *inverse Mellin* transformation is given by (11.13.4).

We state the following operational properties of the Mellin transform:

(i) Both $M$ and $M^{-1}$ are linear integral operators.

(ii) $M[f(ax)]=a^{-p}F(p)$

(iii) $M[x^a f(x)]=F(p+a)$

(iv) $M[f'(x)]=-(p-1)F(p-1)$ provided $[f(x)x^{p-1}]_0^\infty=0$

$M[f''(x)]=(p-1)(p-2)F(p-2)$

— — — —

$$M[f^{(n)}(x)]=\frac{(-1)^n\Gamma(p)}{\Gamma(p-n)}F(p-n) \qquad \text{provided}$$

$\lim_{x\to 0} x^{p-r-1}f^{(r)}(x)=0,\ r=0, 1, 2 \ldots (n-1)$

(v) $M\{xf'(x)\}=-pM\{f(x)\}=-pF(p)$ provided $[x^p f(x)]_0^\infty=0$

$M\{x^2 f''(x)\}=(p+p^2)F(p)$

— — — —

$$M\left\{\left(x\frac{d}{dx}\right)^n f(x)\right\}=(-1)^n p^n F(p),\ n=1,2,\ldots$$

(vi) Convolution Property

$$M\left[\int_0^\infty f(x\xi)g(\xi)\,d\xi\right]=F(p)G(1-p)$$

$$M\left[\int_0^\infty f\left(\frac{x}{\xi}\right)g(\xi)\frac{d\xi}{\xi}\right]=F(p)G(p)$$

(vii) If $F(p)=M(f(x))$ and $G(p)=M(g(x))$, then the following convolution result holds:

$$M[f(x)g(x)]=\frac{1}{2\pi i}\int_{c-i\infty}^{c+i\infty} F(s)G(p-s)\,ds$$

In particular, when $p=1$, we obtain the Parseval formula

$$\int_0^\infty f(x)g(x)\,dx=\frac{1}{2\pi i}\int_{c-i\infty}^{c+i\infty} F(s)G(1-s)\,ds$$

EXAMPLE 11.13.1. Show that the Mellin transform of $(1+x)^{-1}$ is $\pi \operatorname{cosec} \pi p$, $0 < \operatorname{Re} p < 1$

We consider the standard integral

$$\int_0^1 (1-t)^{m-1} t^{p-1}\, dt = \frac{\Gamma(m)\Gamma(p)}{\Gamma(m+p)}, \qquad \operatorname{Re} p > 0,\ \operatorname{Re} m > 0$$

and then change the variable $t = \dfrac{x}{1+x}$ to obtain

$$\int_0^\infty \frac{x^{p-1}\, dx}{(1+x)^{m+p}} = \frac{\Gamma(m)\Gamma(p)}{\Gamma(m+p)}$$

Replacing $m+p$ by $\alpha$, this gives

$$M[(1+x)^{-\alpha}] = \frac{\Gamma(p)\Gamma(\alpha - p)}{\Gamma(\alpha)}$$

Setting $\alpha = 1$ and using the result

$$\Gamma(p)\Gamma(1-p) = \pi \operatorname{cosec} \pi p, \qquad 0 < \operatorname{Re} p < 1$$

we obtain

$$M[(1+x)^{-1}] = \pi \operatorname{cosec} \pi p, \qquad 0 < \operatorname{Re} p < 1$$

EXAMPLE 11.13.2. Obtain the solution of the boundary-value problem

$$x^2 u_{xx} + x u_x + u_{yy} = 0, \qquad 0 \leqslant x < \infty, \quad 0 < y < 1$$

$$u(x, 0) = 0 \quad \text{and} \quad u(x, 1) = \begin{Bmatrix} A, & 0 \leqslant x \leqslant 1 \\ 0, & x > 1 \end{Bmatrix}$$

where $A$ is constant.

We apply the Mellin transform

$$U(p, y) = \int_0^\infty x^{p-1} u(x, y)\, dx$$

to reduce the system in the form

$$U_{yy} + p^2 U = 0, \qquad 0 < y < 1$$

$$U(p, 0) = 0 \quad \text{and} \quad U(p, 1) = A \int_0^1 x^{p-1}\, dx = \frac{A}{p}$$

The solution of this system is

$$U(p, y) = \frac{A}{p} \frac{\sin py}{\sin p}, \qquad 0 < \operatorname{Re} p < 1$$

The inverse Mellin transform gives

$$u(x, y) = \frac{A}{2\pi i} \int_{c-i\infty}^{c+i\infty} \frac{x^{-p}}{p} \frac{\sin py}{\sin p}\, dp$$

where $U(p, y)$ is analytic in a vertical strip $0 < \text{Re } p < \pi$ and hence $0 < c < \pi$. The integrand has simple poles at $p = r\pi$, $r = 1, 2, 3, \ldots$ which lie inside a semi-circular contour in the right-half plane. Application of the theory of residues gives the solution for $x > 1$

$$u(x, y) = \frac{A}{\pi} \sum_{r=1}^{\infty} \frac{(-1)^r x^{-r\pi}}{r} \sin r\pi y$$

EXAMPLE 11.13.3. Find the Mellin transform of the *Weyl fractional integral*

$$\omega(x, \alpha) = W_\alpha[f(\xi)] = \frac{1}{\Gamma(\alpha)} \int_x^\infty f(\xi)(\xi - x)^{\alpha-1}\, d\xi$$

We rewrite the Weyl integral by putting

$$k(x) = x^\alpha f(x), \qquad g(x) = \frac{1}{\Gamma(\alpha)}(1 - x)^{\alpha-1} H(1 - x)$$

so that

$$\omega(x, \alpha) = \int_0^\infty k(\xi) g\left(\frac{x}{\xi}\right) \frac{d\xi}{\xi}$$

The Mellin transform of this result is obtained by the convolution property (vi):

$$\Omega(p, \alpha) = K(p)G(p)$$

where $K(p) = F(p + \alpha)$ and

$$G(p) = \frac{1}{\Gamma(\alpha)} \int_0^1 (1 - x)^{\alpha-1} x^{p-1}\, dx = \frac{\Gamma(p)}{\Gamma(p + \alpha)}$$

Thus

$$\Omega(p, \alpha) = M[W_\alpha f(\xi)] = \frac{\Gamma(p)}{\Gamma(p + \alpha)} F(p + \alpha) \tag{11.13.5}$$

If $\alpha$ is complex with a positive real part such that $n - 1 < \text{Re } \alpha < n$ where $n$ is a positive integer, the *fractional derivative* of order $\alpha$ of a function $f(x)$ is defined by the formula

$$D_\infty^\alpha f(x) = \frac{d^n}{dx^n} W_{n-\alpha} f(x) = \frac{1}{\Gamma(n - \alpha)} \frac{d^n}{dx^n} \int_x^\infty f(\xi)(\xi - x)^{n-\alpha-1}\, d\xi \tag{11.13.6}$$

The Mellin transform of this fractional derivative can be given by using operational property (iv) and (11.13.5)

$$M[D_{\infty}^{\alpha}f(x)] = \frac{(-1)^n\Gamma(p)}{\Gamma(p-n)}\Omega(p-n, n-\alpha)$$
$$= \frac{(-1)^n\Gamma(p)}{\Gamma(p-\alpha)}F(p-\alpha)$$

This is an obvious generalization of the third result listed under (iv).

## 11.14 FINITE FOURIER TRANSFORMS AND APPLICATIONS

The finite Fourier transforms are often used in determining solutions of nonhomogeneous problems. These finite transforms, namely the sine and cosine transforms, follow immediately from the theory of Fourier series.

Let $f(x)$ be a piecewise continuous function in a finite interval, say, $(0, \pi)$. This interval is introduced for convenience, and the change of interval can be made without difficulty.

The finite Fourier sine transform denoted by $F_s(n)$ of the function $f(x)$ may be defined by

$$F_s(n) = \mathscr{F}[f(x)] = \frac{2}{\pi}\int_0^{\pi} f(x)\sin nx\,dx, \qquad n = 1, 2, 3, \ldots \tag{11.14.1}$$

and the inverse of the transform follows at once from the Fourier sine series; that is

$$f(x) = \sum_{n=1}^{\infty} F_s(n)\sin nx \tag{11.14.2}$$

The finite Fourier cosine transform $F_c(n)$ of $f(x)$ may be defined by

$$F_s(n) = \mathscr{F}[f(x)] = \frac{2}{\pi}\int_0^{\pi} f(x)\cos nx\,dx, \qquad n = 0, 1, 2, \ldots \tag{11.14.3}$$

The inverse of the transform is given by

$$f(x) = \frac{F_c(0)}{2} + \sum_{n=1}^{\infty} F_c(n)\cos nx \tag{11.14.4}$$

***Theorem 11.14.1*** *Let $f'(x)$ be continuous and $f''(x)$ be piecewise continuous in $[0, \pi]$. If $F_s(n)$ is the finite Fourier sine transform, then*

$$\mathscr{F}[f''(x)] = \frac{2n}{\pi}[f(0) - (-1)^n f(\pi)] - n^2F_s(n) \tag{11.14.5}$$

PROOF. *By definition*

$$\begin{aligned}\mathscr{F}[f''(x)] &= \frac{2}{\pi}\int_0^{\pi} f''(x)\sin nx\,dx \\ &= \frac{2}{\pi}[f'(x)\sin nx]_0^{\pi} - \frac{2n}{\pi}\int_0^{\pi} f'(x)\cos nx\,dx \\ &= -\frac{2n}{\pi}[f(x)\cos nx]_0^{\pi} - \frac{2n^2}{\pi}\int_0^{\pi} f(x)\sin nx\,dx \\ &= -\frac{2n}{\pi}\,[f(\pi)(-1)^n - f(0)] - n^2F_s(n)\end{aligned}$$

□

The transforms of higher-ordered derivatives can be derived in a similar manner.

***Theorem 11.14.2*** *Let $f'(x)$ be continuous and $f''(x)$ be piecewise continuous in $[0, \pi]$. If $F_c(n)$ is the finite Fourier cosine transform, then*

$$\mathscr{F}[f''(x)] = \frac{2}{\pi}[(-1)^n f'(\pi) - f'(0)] - n^2F_c(n) \qquad (11.14.6)$$

The proof is left to the reader.

EXAMPLE 11.14.1. Consider the motion of a string of length $\pi$ due to a force acting on it. Let the string be fixed at both ends. The motion is thus governed by

$$\begin{aligned} & u_{tt} = c^2u_{xx} + f(x, t), \qquad 0 < x < \pi, \qquad t > 0 \\ & u(x, 0) = 0 \\ & u_t(x, 0) = 0 \\ & u(0, t) = 0 \\ & u(\pi, t) = 0 \end{aligned} \qquad (11.14.7)$$

Transforming the equation of motion with respect to $x$, we obtain

$$\mathscr{F}[u_{tt} - c^2u_{xx} - f(x, t)] = 0$$

Due to its linearity (see Problem 52, Exercises 11.16), this can be written in the form

$$\mathscr{F}[u_{tt}] - c^2\mathscr{F}[u_{xx}] = \mathscr{F}[f(x, t)] \qquad (11.14.8)$$

let $U(n, t)$ be the finite Fourier sine transform of $u(x, t)$. Then we have

$$\begin{aligned}\mathscr{F}[u_{tt}] &= \frac{2}{\pi}\int_0^{\pi} u_{tt}\sin nx\,dx \\ &= \frac{d^2}{dt^2}\left[\frac{2}{\pi}\int_0^{\pi} u(x, t)\sin nx\,dx\right] \\ &= \frac{d^2U}{dt^2}\end{aligned}$$

We also have from Theorem 11.14.1

$$\mathscr{F}[u_{xx}] = \frac{2n}{\pi}[u(0, t) - (-1)^n u(\pi, t)] - n^2 U(n, t)$$

Because of the boundary conditions

$$u(0, t) = u(\pi, t) = 0$$

$\mathscr{F}[u_{xx}]$ becomes

$$\mathscr{F}[u_{xx}] = -n^2 U(n, t)$$

If we denote the finite Fourier sine transform of $f(x, t)$ by

$$F(n, t) = \frac{2}{\pi}\int_0^\pi f(x, t) \sin nx \, dx$$

then Eq. 11.14.8 takes the form

$$\frac{d^2U}{dt^2} + n^2c^2 U = F(n, t)$$

This is the second order ordinary differential equation, the solution of which is given by

$$U(n, t) = A \cos nct + B \sin nct + \frac{1}{nc}\int_0^t F(n, \tau) \sin nc(t - \tau) \, d\tau$$

Applying the initial conditions

$$\begin{aligned}\mathscr{F}[u(x, 0)] &= \frac{2}{\pi}\int_0^\pi u(x, 0) \sin nx \, dx \\ &= U(n, 0)\end{aligned}$$

and

$$\mathscr{F}[u_t(x, 0)] = \frac{dU}{dt}(n, 0)$$

we have

$$U(n, t) = \frac{1}{nc}\int_0^t F(n, \tau) \sin nc(t - \tau) \, d\tau$$

Thus the inverse transform of $U(n, t)$ is

$$\begin{aligned}u(x, t) &= \sum_{n=1}^{\infty} U(n, t) \sin nx \\ &= \sum_{n=1}^{\infty}\left[\frac{1}{nc}\int_0^t F(n, \tau) \sin nc(t - \tau) \, d\tau\right] \sin nx\end{aligned}$$

In the case when $f(x, t)$ is a constant, say $h$, then

$$\mathscr{F}[h] = \frac{2}{\pi}\int_0^\pi h \sin nx\, dx$$
$$= \frac{2h}{n\pi}[1-(-1)^n]$$

Now, we evaluate

$$U(n, t) = \frac{1}{nc}\int_0^t \frac{2h}{n\pi}[1-(-1)^n] \sin nc(t-\tau)\, d\tau$$
$$= \frac{2h}{n^3\pi c^2}[1-(-1)^n](1-\cos nct)$$

Hence the solution is given by

$$u(x, t) = \frac{2h}{\pi c^2}\sum_{n=1}^{\infty} \frac{[1-(-1)^n]}{n^3}(1-\cos nct) \sin nx$$

EXAMPLE 11.14.2. Find the temperature distribution in a rod of length $\pi$. The heat is generated in the rod at the rate $g(x, t)$ per unit time. The ends are insulated. The initial temperature distribution is given by $f(x)$.

The problem is to find the temperature function $u(x, t)$ of

$$u_t = u_{xx} + g(x, t), \qquad 0 < x < \pi, \qquad t > 0$$
$$u(x, 0) = f(x), \qquad 0 \leqslant x \leqslant \pi$$
$$u_x(0, t) = 0, \qquad t \geqslant 0$$
$$u_x(\pi, t) = 0, \qquad t \geqslant 0$$

Let $U(n, t)$ be the finite cosine transform of $u(x, t)$. As before, transformation of the heat equation with respect to $x$, using the boundary conditions, yields

$$\frac{dU}{dt} = -n^2 U + G(n, t)$$

where

$$G(n, t) = \frac{2}{\pi}\int_0^\pi g(x, t) \cos nx\, dx$$

Rewriting this equation, we obtain

$$\frac{d}{dt}(e^{n^2 t}U) = Ge^{n^2 t}$$

Thus, the solution is

$$U(n, t) = \int_0^t e^{-n^2(t-\tau)}G(n, \tau)\, d\tau + Ae^{-n^2 t}$$

Transformation of the initial condition gives

$$U(n, 0) = \frac{2}{\pi}\int_0^{\pi} u(x, 0)\cos nx\, dx$$
$$= \frac{2}{\pi}\int_0^{\pi} f(x)\cos nx\, dx$$

Hence $U(n, t)$ takes the form

$$U(n, t) = \int_0^t e^{-n^2(t-\tau)}G(n, \tau)\, d\tau + U(n, 0)e^{-n^2 t}$$

The solution $u(x, t)$, therefore, is

$$u(x, t) = \frac{U(0, 0)}{2} + \sum_{n=1}^{\infty} U(n, t)\cos nx$$

EXAMPLE 11.14.3. A rod with diffusion constant $\kappa$ contains a fuel which produces neutrons by fission. The ends of the rod are perfectly reflecting. If the initial neutron distribution is $f(x)$, find the neutron distribution $u(x, t)$ at any subsequent time $t$.

The problem is governed by

$$u_t = \kappa u_{xx} + bu, \qquad 0 < x < l, \qquad t > 0$$
$$u(x, 0) = f(x)$$
$$u_x(0, t) = u_x(l, t) = 0$$

If $U(n, t)$ is the cosine transform of $u(x, t)$, then by transforming the equation and using the boundary conditions, we obtain

$$U_t + (\kappa n^2 - b)U = 0$$

The solution of this equation is

$$U(n, t) = Ce^{-(\kappa n^2 - b)t}$$

where $C$ is a constant. Then applying the initial condition, we obtain

$$U(n, t) = U(n, 0)e^{-(\kappa n^2 - b)t}$$

where

$$U(n, 0) = \frac{2}{l}\int_0^l f(x)\cos nx\, dx$$

Thus, the solution takes the form

$$u(x, t) = \frac{U(0, 0)}{2} + \sum_{n=1}^{\infty} U(n, t)\cos nx$$

If for instance $f(x)=x$ in $0<x<\pi$ then

$$U(0, 0)=\pi$$

and

$$U(n, 0)=\frac{2}{n^2\pi}[(-1)^n-1], \qquad n=1, 2, 3, \ldots$$

Hence the solution is given by

$$u(x, t)=\frac{\pi}{2}+\sum_{n=1}^{\infty}\frac{2}{n^2\pi}[(-1)^n-1]e^{-(\kappa n^2-b)t}\cos nx$$

## 11.15. FINITE HANKEL TRANSFORMS AND APPLICATIONS

The Fourier-Bessel series representation of a function $f(r)$ defined in $0\leqslant r\leqslant a$ can be stated in the following theorem:

***Theorem 11.15.1*** *If $f(r)$ is defined in $0\leqslant r\leqslant a$ and*

$$F_n(k_i)=H_n\{f(r)\}=\int_0^a rf(r)J_n(rk_i)\,dr \tag{11.15.1}$$

*then*

$$f(r)=H_n^{-1}\{F_n(k_i)\}=\frac{2}{a^2}\sum_{i=1}^{\infty}F_n(k_i)\frac{J_n(rk_i)}{J_{n+1}^2(ak_i)} \tag{11.15.2}$$

*where $k_i$ $(0<k_1<k_2<\ldots)$ are the roots of the equation*

$$J_n(ak_i)=0$$

The function $F_n(k_i)$ defined by (11.15.1) is called the $n$th order *finite Hankel transform* of $f(r)$, and the *inverse Hankel transformation* is defined by (11.15.2). In particular, when $n=0$, the finite Hankel transform of order zero and its inverse are defined by the integral and series respectively

$$F(k_i)=H_0\{f(r)\}=\int_0^a rf(r)J_0(rk_i)\,dr \tag{11.15.3}$$

$$f(r)=H_0^{-1}\{F(k_i)\}=\frac{2}{a^2}\sum_{i=1}^{\infty}F(k_i)\frac{J_0(rk_i)}{J_1^2(ak_i)} \tag{11.15.4}$$

EXAMPLE 11.5.1. Find the $n$th order finite Hankel transform of $f(r)=r^n$.

We have the following result for the Bessel function

$$\int_0^a r^{n+1} J_n(k_i r)\, dr = \frac{a^{n+1}}{k_i} J_{n+1}(ak_i)$$

so that

$$H_n\{r^n\} = \frac{a^{n+1}}{k_i} J_{n+1}(ak_i)$$

In particular, when $n = 0$

$$H_0\{1\} = \frac{a}{k_i} J_1(ak_i)$$

EXAMPLE 11.15.2. Find $H_0\{(a^2 - r^2)\}$.

We have by definition

$$H_0\{(a^2 - r^2)\} = \int_0^a (a^2 - r^2) r J_0(k_i r)\, dr$$
$$= \frac{4a}{k_i^3} J_1(ak_i) - \frac{2a^2}{k_i^2} J_0(ak_i)$$

where $k_i$ is a root of $J_0(ax) = 0$. Hence

$$H_0\{(a^2 - r^2)\} = \frac{4a}{k_i^3} J_1(ak_i)$$

We state the following operational properties of the finite Hankel transform:

(i) $H_n\left\{\frac{df}{dr}\right\} = \frac{k_i}{2n}[(n-1)H_{n+1}\{f(r)\} - (n+1)H_{n-1}\{f(r)\}]$

(ii) $H_1\left\{\frac{df}{dr}\right\} = -k_i H_0\{f(r)\}$

(iii) $H_n\left[\frac{1}{r}\frac{d}{dr}\{rf'(r)\} - \frac{n^2}{r^2} f(r)\right] = -ak_i f(a) J_n'(ak_i) - k_i^2 H_n\{f(r)\}$ (11.15.5)

(iv) $H_0\left[f''(r) + \frac{1}{r} f'(r)\right] = ak_i f(a) J_1(ak_i) - k_i^2 H_0\{f(r)\}$ (11.15.6)

EXAMPLE 11.15.3. Find the solution of the axisymmetric heat conduction equation

$$u_t = \kappa\left(u_{rr} + \frac{1}{r} u_r\right), \qquad 0 < r < a, \quad t > 0$$

with the boundary and initial conditions

$$u(r,t)=f(t) \quad \text{on} \quad r=a, \qquad t\geqslant 0$$
$$u(r,0)=0, \qquad 0\leqslant r\leqslant a$$

where $u(r, t)$ represents the temperature distribution.

We apply the finite Hankel transform defined by

$$U(k,t)=H_0\{u(r,t)\}=\int_0^a rJ_0(k_ir)u(r,t)\,dr$$

so that the given equation with the boundary condition becomes

$$\frac{dU}{dt}+\kappa k_i^2 U=\kappa a k_i J_1(ak_i)f(t)$$

The solution of this equation with the transformed initial condition is

$$U(k,t)=a\kappa k_i J_1(ak_i)\int_0^t f(\tau)e^{-\kappa k_i^2(t-\tau)}\,d\tau$$

The inverse transformation gives the solution as

$$u(r,t)=\frac{2\kappa}{a}\sum_{i=1}^{\infty}\frac{k_iJ_0(rk_i)}{J_1(ak_i)}\int_0^t f(\tau)e^{-\kappa k_i^2(t-\tau)}\,d\tau \qquad (11.15.7)$$

In particular, if $f(t)=T_0=\text{constant}$, then this solution becomes

$$u(r,t)=\frac{2T_0}{a}\sum_{i=1}^{\infty}\frac{J_0(rk_i)}{k_iJ_1(ak_i)}(1-e^{-\kappa k_i^2 t}) \qquad (11.15.8)$$

In view of the inverse Hankel transform of one (see EXAMPLE 11.15.1), the above result becomes

$$u(r,t)=T_0\left[1-\frac{2}{a}\sum_{i=1}^{\infty}\frac{J_0(rk_i)}{k_iJ_1(ak_i)}e^{-\kappa k_i^2 t}\right] \qquad (11.15.9)$$

This solution consists of the steady-state term, and the transient term which tends to zero as $t\to\infty$. Consequently, the steady-state is attained in the limit $t\to\infty$.

EXAMPLE 11.15.4 (Unsteady Viscous Flow in a Rotating Cylinder). The axisymmetric unsteady motion of a viscous fluid in an infinitely long circular cylinder of radius $a$ is governed by

$$v_t=\nu\left(v_{rr}+\frac{1}{r}v_r-\frac{v}{r^2}\right), \qquad 0\leqslant r\leqslant a, \qquad t>0$$

where $v=v(r,t)$ is the tangential fluid velocity and $\nu$ is the kinematic viscosity of the fluid.

The cylinder is at rest until at $t = 0+$ it is caused to rotate so that the boundary and initial conditions are

$$v(r, t) = a\Omega f(t)H(t) \quad \text{on} \quad r = a$$
$$v(r, t) = 0 \quad \text{at} \quad t = 0 \quad \text{for} \quad r < a$$

where $f(t)$ is a physically realistic function of $t$.

We solve the problem by using the joint Laplace and finite Hankel transforms defined by

$$\bar{V}(k_m, s) = \int_0^a rJ_1(rk_m)\,dr \int_0^\infty e^{-st}v(r, t)\,dt$$

where $\bar{V}(k_m, s)$ is the Laplace transform of $V(k_m, t)$, and $k_m$ are the roots of $J_1(ak_m) = 0$.

Application of the transform yields

$$\frac{s}{\nu}\bar{V}(k_m, s) = -ak_m\bar{V}(a, s)J_1'(ak_m) - k_m^2\bar{V}(k_m, s)$$
$$\bar{V}(a, s) = a\Omega F(s)$$

where $F(s)$ is the Laplace transform of $f(t)$.

The solution of this system is

$$\bar{V}(k_m, s) = -\frac{a^2\nu k_m\Omega F(s)J_1'(ak_m)}{(s + \nu k_m^2)}$$

The joint inverse transformation gives

$$\begin{aligned} v(r, t) &= \frac{2}{a^2}\sum_{m=1}^\infty \frac{J_1(rk_m)}{[J_1'(ak_m)]^2}\frac{1}{2\pi i}\int_{c-i\infty}^{c+i\infty} e^{st}\bar{V}(k_m, s)\,ds \\ &= -2\nu\Omega\sum_{m=1}^\infty \frac{k_mJ_1(rk_m)}{J_1'(ak_m)}\frac{1}{2\pi i}\int_{c-i\infty}^{c+i\infty}\frac{e^{st}F(s)}{(s + \nu k_m^2)}\,ds \\ &= -2\nu\Omega\sum_{m=1}^\infty \frac{k_mJ_1(rk_m)}{J_1'(ak_m)}\int_0^t f(\tau)\exp[-\nu k_m^2(t-\tau)]\,d\tau \end{aligned}$$

by the Convolution theorem of the Laplace transform.

In particular, when $f(t) = \cos\omega t$, the velocity field becomes

$$\begin{aligned} v(r, t) &= -2\nu\Omega\sum_{m=1}^\infty \frac{k_mJ_1(rk_m)}{J_1'(ak_m)}\int_0^t \cos\omega\tau\exp[-\nu k_m^2(t-\tau)]\,d\tau \\ &= 2\nu\Omega\sum_{m=1}^\infty \frac{k_mJ_1(rk_m)}{J_1'(ak_m)}\left[\frac{\nu k_m^2\exp(-\nu tk_m^2) - (\omega\sin\omega t + \nu k_m^2\cos\omega t)}{(\omega^2 + \nu^2k_m^4)}\right] \\ &= v_{st}(r, t) + v_{tr}(r, t) \end{aligned} \tag{11.15.10}$$

where the steady-state flow field $v_{st}$ and the transient flow field $v_{tr}$ are given by

$$v_{st}(r, t) = -2\nu\Omega \sum_{m=1}^{\infty} \frac{k_m J_1(rk_m)(\omega \sin \omega t + \nu k_m^2 \cos \omega t)}{J_1'(ak_m)(\omega^2 + \nu^2 k_m^4)} \tag{11.15.11}$$

$$v_{tr}(r, t) = 2\nu^2\Omega \sum_{m=1}^{\infty} \frac{J_1(rk_m)k_m^3 e^{-\nu t k_m^2}}{J_1'(ak_m)(\omega^2 + \nu^2 k_m^4)} \tag{11.15.12}$$

Thus the solution consists of the steady-state and transient components. In the limit $t \to \infty$, the latter decays to zero, and the ultimate steady-state is attained and is given by (11.15.11), which has the form

$$v_{st}(r, t) = -2\nu\Omega \sum_{m=1}^{\infty} \frac{k_m J_1(rk_m) \cos(\omega t - \alpha)}{J_1'(ak_m)(\omega^2 + \nu^2 k_m^4)^{\frac{1}{2}}} \tag{11.15.13}$$

where $\tan \alpha = \omega / \nu k_m^2$. Thus we see that the steady solution suffers from a phase change of $\alpha + \pi$. The amplitude of the motion remains bounded for all values of $\omega$.

The frictional couple exerted on the fluid by unit length of the cylinder of radius $r = a$ is given by

$$C = \int_0^{2\pi} [P_{r\theta}]_{r=a} a^2 \, d\theta = 2\pi a^2 [P_{r\theta}]_{r=a}$$

where $P_{r\theta} = \mu r(d/dr)(v/r)$ with $\mu = \nu\rho$ calculated from (11.14.9).

Thus

$$C = 4\pi\mu\Omega\left[-a \sum_{m=1}^{\infty} \frac{\nu k_m J_1(ak_m)[\nu k_m^2 \exp(-\nu t k_m^2) - (\nu k_m^2 \cos \omega t + \omega \sin \omega t)]}{(\omega^2 + \nu^2 k_m^4) J_1'(ak_m)} \right.$$
$$\left. + a^2 \sum_{m=1}^{\infty} \frac{\nu k_m^2 [\nu k_m^2 \exp(-\nu t k_m^2) - (\nu k_m^2 \cos \omega t + \omega \sin \omega t)]}{(\omega^2 + \nu^2 k_m^4) J_1'(ak_m)}\right] \tag{11.15.14}$$

A subcase corresponding to $\omega = 0$ of this case is of special interest. The solution assumes the form

$$v(r, t) = -2\Omega \sum_{m=1}^{\infty} \frac{J_1(rk_m)[1 - e^{-\nu t k_m^2}]}{k_m J_1'(ak_m)} = v_{st} + v_{tr} \tag{11.15.15}$$

where $v_{st}$ and $v_{tr}$ represent the steady-state and the transient flow fields respectively given by

$$v_{st}(r, t) = -2\Omega \sum_{m=1}^{\infty} \frac{J_1(rk_m)}{k_m J_1'(ak_m)} \tag{11.15.16}$$

$$v_{tr}(r, t) = 2\Omega \sum_{m=1}^{\infty} \frac{J_1(rk_m)}{k_m J_1'(ak_m)} e^{-\nu t k_m^2} \tag{11.15.17}$$

Making use of the Fourier–Bessel series for $r$ as

$$r = \frac{2}{a^2} \sum_{m=1}^{\infty} \frac{J_1(rk_m) F(k_m)}{[J_1'(ak_m)]^2}$$

which is, by $F(k_m) = H_1\{r\} = \int_0^a r^2 J_1(rk_m)\,dr = (-a^2/k_m)J_1'(ak_m)$,

$$r = -2 \sum_{m=1}^{\infty} \frac{J_1(rk_m)}{k_m J_1'(ak_m)} \tag{11.15.18}$$

The steady-state solution has the closed form

$$v_{st}(r, t) = r\Omega$$

This represents the rigid body rotation of the fluid inside the cylinder. Thus the final form of (11.15.15) is

$$v(r, t) = r\Omega + 2\Omega \sum_{m=1}^{\infty} \frac{J_1(rk_m)e^{-\nu t k_m^2}}{k_m J_1'(ak_m)} \tag{11.15.19}$$

In the limit $t \to \infty$, the transients die out and the ultimate steady-state is attained as the rigid body rotation about the axis of the cylinder.

## 11.16 EXERCISES

**1.** Find the Fourier transform of $f(x) = e^{-ax^2}$ where $a$ is a constant.

**2.** Find the Fourier transform of

$$f(x) = \begin{cases} 1 & |x| < a \quad a \text{ is a positive constant} \\ 0 & |x| \geqslant a \end{cases}$$

**3.** Find the Fourier transform of

$$f(x) = \frac{1}{|x|}$$

**4.** Find the Fourier transform of

a. $f(x) = \sin(x^2)$
b. $f(x) = \cos(x^2)$

**5.** Show that

$$I = \int_0^{\infty} e^{-a^2x^2}\,dx = \sqrt{\pi}/2a, \qquad a > 0$$

by noting

$$I^2 = \int_0^{\infty}\int_0^{\infty} e^{-a^2(x^2+y^2)}\,dx\,dy = \int_0^{\pi/2}\int_0^{\infty} e^{-a^2r^2} r\,dr\,d\theta$$

**6.** Show that

$$\int_0^{\infty} e^{-a^2x^2}\cos bx\,dx = (\sqrt{\pi}/2a)e^{-b^2/4a^2}, \qquad a > 0$$

[Hint: Let $I(a, b) = \int_0^\infty e^{-a^2x^2} \cos bx \, dx$]

$$\therefore \frac{\partial I}{\partial b} = -\frac{b}{2a^2} I$$

$$\therefore I = Ce^{-b^2/4a^2}$$

Since

$$I(a, 0) = C = \int_0^\infty e^{-a^2x^2} \, dx = \frac{\sqrt{\pi}}{2a}$$

$$I(a, b) = \frac{\sqrt{\pi}}{2a} e^{-b^2/4a^2}$$

**7.** Show that

$$\mathscr{F}[f(at-b)] = \frac{1}{|a|} e^{i\alpha b/a} F(\alpha/a)$$

**8.** Prove that

$$\mathscr{F}[f^{(n)}(t)] = (-i\alpha)^n \mathscr{F}[f(t)]$$

**9.** Show that:

a. $f * 0 = 0 * f = 0$
b. $f * 1 \neq f$

**10.** Show that

$$\frac{1}{\sqrt{2\pi}} \int_{-\infty}^{\infty} e^{-\alpha^2 t - i\alpha x} \, d\alpha = \frac{1}{\sqrt{2t}} e^{-x^2/4t}$$

**11.** Determine the solution of

$$u_{tt} = c^2 u_{xx}, \qquad -\infty < x < \infty, \qquad t > 0$$
$$u(x, 0) = f(x)$$
$$u_t(x, 0) = g(x)$$

**12.** Solve

$$u_t = u_{xx}, \qquad x > 0, \qquad t > 0$$
$$u(x, 0) = f(x)$$
$$u(0, t) = 0$$

**13.** Solve

$$u_{tt} + c^2 u_{xxxx} = 0, \qquad -\infty < x < \infty, \qquad t > 0$$
$$u(x, 0) = f(x)$$
$$u_t(x, 0) = 0$$

**14.** Solve

$$\begin{aligned} u_{tt} + c^2 u_{xxxx} &= 0, \qquad x>0, \qquad t>0 \\ u(x,0) &= 0, \qquad x>0 \\ u_t(x,0) &= 0 \\ u(0,t) &= g(t), \qquad t>0 \\ u_{xx}(0,t) &= 0 \end{aligned}$$

**15.** Solve

$$\phi_{xx} + \phi_{yy} = 0, \qquad -a<x<a, \qquad 0<y<\infty$$

$$\phi_y(x,0) = \begin{cases} \delta_0, & 0<|x|<a \\ 0, & |x|>a \end{cases}$$

$$\phi(x,y) \to 0 \text{ uniformly in } x \text{ as } y\to\infty$$

**16.** Solve

$$u_t = u_{xx} + tu, \qquad -\infty<x<\infty, \qquad t>0$$

$$u(x,0) = f(x)$$

$$u(x,t) \text{ is bounded}$$

**17.** Solve

$$u_t - u_{xx} + hu = \delta(x)\delta(t), \qquad -\infty<x<\infty, \qquad t>0$$

$$u(x,t) \to 0 \text{ uniformly in } t \text{ as } |x|\to\infty$$

$$u(x,0) = 0$$

**18.** Solve

$$u_t - u_{xx} + h(t)u_x = \delta(x)\delta(t), \qquad 0<x<\infty, \qquad t>0$$

$$u(x,0) = 0$$

$$u_x(0,t) = 0$$

$$u(x,t) \to 0 \text{ uniformly in } t \text{ as } x\to\infty$$

**19.** Solve

$$\begin{aligned} u_{xx} + u_{yy} &= 0, \qquad 0<x<\infty, \qquad 0<y<\infty \\ u(x,0) &= f(x), \qquad 0\leq x<\infty \\ u_x(0,y) &= g(y), \qquad 0\leq y<\infty \end{aligned}$$

$$u(x,y) \to 0 \text{ uniformly in } x \text{ as } y\to\infty \text{ and uniformly in } y \text{ as } x\to\infty$$

**20.** Solve

$$\begin{aligned} u_{xx} + u_{yy} &= 0, \qquad -\infty < x < \infty, \qquad 0 < y < a \\ u(x, 0) &= f(x), \qquad -\infty < x < \infty \\ u(x, a) &= 0 \\ u(x, y) &\to 0 \quad \text{uniformly in } y \text{ as } |x| \to \infty \end{aligned}$$

**21.** Solve

$$\begin{aligned} u_t &= u_{xx}, \qquad x > 0, \qquad t > 0 \\ u(x, 0) &= 0, \qquad x > 0 \\ u(0, t) &= f(t), \qquad t > 0 \\ u(x, t) & \text{ is bounded} \end{aligned}$$

**22.** Solve

$$\begin{aligned} u_{xx} + u_{yy} &= 0, \qquad x > 0, \qquad 0 < y < 1 \\ u(x, 0) &= f(x) \\ u(x, 1) &= 0 \\ u(0, y) &= 0 \\ u(x, y) &\to 0 \text{ uniformly in } y \text{ as } x \to \infty \end{aligned}$$

**23.** Find the Laplace transform of each of the following functions:

a. $t^n$
b. $\cos \omega t$
c. $\sinh kt$
d. $\cosh kt$
e. $te^{at}$
f. $e^{at} \sin \omega t$
g. $e^{at} \cos \omega t$
h. $t \sinh kt$
i. $t \cosh kt$
j. $\sqrt{\dfrac{1}{t}}$
k. $\sqrt{t}$
l. $\dfrac{\sin at}{t}$

**24.** Find the inverse transform of each of the following functions:

a. $\dfrac{s}{(s^2+1)(s^2+2)}$
b. $\dfrac{1}{(s^2+1)(s^2+2)}$
c. $\dfrac{1}{(s-1)(s-2)}$
d. $\dfrac{1}{s(s+1)^2}$
e. $\dfrac{1}{s(s+1)}$
f. $\dfrac{s-4}{(s^2+4)^2}$

**25.** The velocity potential $\phi(x, z, t)$ and the free-surface evaluation $\eta(x, t)$ for surface waves in water of infinite depth satisfy the Laplace equation

$$\phi_{xx} + \phi_{zz} = 0, \qquad -\infty < x < \infty, \qquad -\infty < z \leq 0, \qquad t > 0$$

with the free-surface, boundary, and initial conditions

$$\phi_z = \eta_t \quad \text{on} \quad z = 0, \qquad t > 0$$

$$\phi_t + g\eta = 0 \quad \text{on} \quad z = 0, \qquad t > 0$$

$$\phi_z \to 0 \quad \text{as} \quad z \to -\infty$$

$\phi(x, 0, 0) = 0$ and $\eta(x, 0) = f(x)$, $-\infty < x < \infty$. Show that

$$\phi(x, z, t) = -\frac{\sqrt{g}}{\sqrt{2\pi}} \int_{-\infty}^{\infty} k^{-\frac{1}{2}} F(k) e^{|k|z - ikx} \sin(\sqrt{g\,|k|}\, t)\, dk$$

$$\eta(x, t) = \frac{1}{\sqrt{2\pi}} \int_{-\infty}^{\infty} F(k) e^{-ikx} \cos(\sqrt{g\,|k|}\, t)\, dk$$

where $k$ represents the Fourier transform variable.

Find the asymptotic solution for $\eta(x, t)$ as $t \to \infty$.

**26.** Use the Fourier transform method to show that the solution of the one-dimensional Schrödinger equation for a free particle of mass $m$

$$i\hbar\psi_t = -\frac{\hbar^2}{2m}\psi_{xx}, \qquad -\infty < x < \infty, \qquad t > 0$$

$$\psi(x, 0) = f(x), \qquad -\infty < x < \infty$$

where $\psi$ and $\psi_x$ tend to zero as $|x| \to \infty$, and $h = 2\pi\hbar$ is the Planck constant, is

$$\psi(x, t) = \frac{1}{\sqrt{2\pi}} \int_{-\infty}^{\infty} f(\xi) G(x - \xi)\, d\xi$$

where $G(x, t) = \dfrac{(1-i)}{2\sqrt{\gamma t}} \exp\left[-\dfrac{x^2}{4i\gamma t}\right]$ is the Green's function and $\gamma = \dfrac{\hbar}{2m}$.

**27.** Prove the following properties of convolution:

a. $f * g = g * f$

b. $f * (g * h) = (f * g) * h$

c. $f * (g + h) = f * g + f * h$

d. $f * 0 = 0 * f$

**28.** Obtain the solution of the problem

$$\begin{aligned} u_{tt} &= c^2 u_{xx}, \qquad 0<x<\infty, \qquad t>0 \\ u(x, 0) &= f(x) \\ u_t(x, 0) &= 0 \\ u(0, t) &= 0 \\ u(x, t) &\to 0 \text{ uniformly in } t \text{ as } x\to\infty \end{aligned}$$

**29.** Solve

$$\begin{aligned} u_{tt} &= c^2 u_{xx}, \qquad 0<x<1, \qquad t>0 \\ u(x, 0) &= 0 \\ u_t(x, 0) &= 0 \\ u(0, t) &= f(t), \qquad t\geqslant 0 \\ u(1, t) &= 0 \end{aligned}$$

**30.** Solve

$$\begin{aligned} u_t &= \kappa u_{xx}, \qquad 0<x<\infty, \qquad t>0 \\ u(x, 0) &= f_0, \qquad 0<x<\infty \\ u(0, t) &= f_1, \qquad t>0 \\ u(x, t) &\to f_0 \text{ uniformly in } t \text{ as } x\to\infty \end{aligned}$$

**31.** Solve

$$\begin{aligned} u_t &= \kappa u_{xx}, \qquad 0<x<\infty, \qquad t>0 \\ u(x, 0) &= x, \qquad x>0 \\ u(0, t) &= 0, \qquad t>0 \\ u(x, t) &\to x \text{ uniformly in } t \text{ as } x\to\infty \end{aligned}$$

**32.** Solve

$$\begin{aligned} u_t &= \kappa u_{xx}, \qquad 0<x<\infty, \qquad t>0 \\ u(x, 0) &= 0, \qquad 0<x<\infty \\ u(0, t) &= t^2, \qquad t\geqslant 0 \\ u(x, t) &\to 0 \text{ uniformly in } t \text{ as } x\to\infty \end{aligned}$$

**33.** Solve

$$\begin{aligned} u_t = \kappa u_{xx} - hu, \qquad 0<x<\infty, \qquad t>0, \qquad h = \text{constant} \\ u(x, 0) = f_0, \qquad x>0 \\ u(0, t) = 0, \qquad t>0 \\ u_x(0, t) \to 0 \text{ uniformly in } t \text{ as } x\to 0 \end{aligned}$$

**34.** Solve

$$\begin{aligned}
&u_t = \kappa u_{xx}, && 0<x<\infty, && t>0\\
&u(x, 0) = 0, && 0<x<\infty &&\\
&u(0, t) = f_0, && && t>0\\
&u(x, t) \to 0 \text{ uniformly in } t \text{ as } x\to\infty
\end{aligned}$$

**35.** Solve

$$\begin{aligned}
&u_{tt} = c^2 u_{xx}, && 0<x<\infty, && t>0\\
&u(x, 0) = 0, && 0<x<\infty &&\\
&u_t(x, 0) = f_0, && && \\
&u(0, t) = 0 && && t>0\\
&u_x(x, t) \to 0 \text{ uniformly in } t \text{ as } x\to\infty
\end{aligned}$$

**36.** Solve

$$\begin{aligned}
&u_{tt} = c^2 u_{xx}, && 0<x<\infty, && t>0\\
&u(x, 0) = f(x), && 0<x<\infty &&\\
&u_t(x, 0) = 0, && && \\
&u(0, t) = 0, && && t>0\\
&u_x(x, t) \to 0 \text{ uniformly in } t \text{ as } x\to\infty
\end{aligned}$$

**37.** A semi-infinite lossless transmission line has no initial current or potential. A time dependent EMF, $V_0(t)\ H(t)$ is applied at the end $x = 0$. Find the potential $V(x, t)$. Hence determine the potential for cases (i) $V_0(t) = V_0 =$ constant and (ii) $V_0(t) = V_0 \cos \omega t$.

**38.** Solve the Blasius problem of an unsteady boundary layer flow in a semi-infinite body of viscous fluid enclosed by an infinite horizontal disk at $z = 0$. The governing equation, boundary, and initial conditions are

$$\begin{aligned}
&\frac{\partial u}{\partial t} = \nu \frac{\partial^2 u}{\partial z^2}, && z>0, && t>0\\
&u(z, t) = Ut \quad \text{on} && z=0, && t>0\\
&u(z, t) \to 0 \quad \text{as} && z\to\infty, && t>0\\
&u(z, t) = 0 \quad \text{at} && t\leqslant 0, && z>0
\end{aligned}$$

Explain the implication of the solution.

**39.** The stress–strain relation and equation of motion for a viscoelastic rod in the absence of external force are

$$\frac{\partial e}{\partial t} = \frac{1}{E}\frac{\partial \sigma}{\partial t} + \frac{\sigma}{\eta}, \qquad \frac{\partial \sigma}{\partial x} = \rho \frac{\partial^2 u}{\partial t^2}$$

where $e$ is the strain, $\eta$ is the coefficient of viscosity, and the displacement $u(x, t)$ is related to the strain by $e = \partial u/\partial x$. Prove that the stress $\sigma(x, t)$ satisfies the equation

$$\frac{\partial^2 \sigma}{\partial x^2} - \frac{\rho}{\eta}\frac{\partial \sigma}{\partial t} = \frac{1}{c^2}\frac{\partial^2 \sigma}{\partial t^2}, \qquad c^2 = E/\rho$$

Show that the stress distribution in a semi-infinite viscoelastic rod subject to the boundary and initial conditions

$$\dot{u}(0, t) = UH(t), \qquad \sigma(x, t) \to 0 \quad \text{as} \quad x \to \infty, \qquad t > 0$$
$$\sigma(x, 0) = 0, \qquad \dot{u}(x, 0) = 0$$

is given by

$$\sigma(x, t) = -U\rho c \exp\left(-\frac{Et}{2\eta}\right) I_0\left[\frac{E}{2\eta}\left(t^2 - \frac{x^2}{c^2}\right)^{\frac{1}{2}}\right] H\left(t - \frac{x}{c}\right)$$

**40.** An elastic string is stretched between $x = 0$ and $x = l$ and is initially at rest in the equilibrium position. Show that the Laplace transform solution for the displacement field subject to the boundary conditions $y(0, t) = f(t)$ and $y(l, t) = 0$, $t > 0$ is

$$\bar{y}(x, s) = F(s)\frac{\sinh \frac{s}{c}(l - x)}{\sinh \frac{sl}{c}}$$

**41.** The end $x = 0$ of a semi-infinite submarine cable is maintained at a potential $V_0 H(t)$. If the cable has no initial current and potential, determine the potential $V(x, t)$ at a point $x$ and at time $t$.

**42.** Obtain the solution of the Stokes–Ekman problem of an unsteady boundary layer flow in a semi-infinite body of viscous fluid bounded by an infinite horizontal disk at $z = 0$, when both the fluid and the disk rotate with a uniform angular velocity $\Omega$ about the $z$-axis. The governing boundary layer equation, the boundary conditions, and the initial conditions are

$$\frac{\partial q}{\partial t} + 2\Omega i q = \nu \frac{\partial^2 q}{\partial z^2}, \quad z > 0, \qquad t > 0$$

$$q(z, t) = ae^{i\omega t} + be^{-i\omega t} \quad \text{on} \quad z = 0, \qquad t > 0$$
$$q(z, t) \to 0 \quad \text{as} \quad z \to \infty, \qquad t > 0$$
$$q(z, t) = 0 \quad \text{at} \quad t \leq 0 \quad \text{for all } z > 0$$

where $q = u + iv$, is the complex velocity field $\omega$ is the frequency of oscillations of the disk, and $a$ and $b$ are complex constants. Hence deduce the steady-state solution and determine the structure of the associated boundary layers.

**43.** Show that, when $\omega = 0$ in the Stokes–Ekman problem, the steady flow field is given by

$$q(z, t) \sim (a+b) \exp\left\{\left(-\frac{2i\Omega}{\nu}\right)^{\frac{1}{2}} z\right\}$$

Hence determine the thickness of the Ekman layer.

**44.** For problem 14 in Exercise 2.7, show that the potential and the current satisfy the partial differential equation

$$\left(\frac{\partial^2}{\partial t^2} + 2\kappa \frac{\partial}{\partial t} + \kappa^2\right)\binom{V}{I} = c^2 \frac{\partial^2}{\partial x^2}\binom{V}{I}$$

Find the solution for $V(x, t)$ with the boundary and initial data

$$V(x, t) = V_0(t) \quad \text{at} \quad x = 0, \qquad t > 0$$

$$V(x, t) \to 0 \quad \text{as} \quad x \to \infty, \qquad t > 0$$

$$V(x, 0) = V_t(x, 0) = 0 \quad \text{for} \quad 0 \leq x < \infty$$

**45.** Use the Laplace transform to solve the Abel integral equation

$$g(t) = \int_0^t f'(\tau)(t-\tau)^{-\alpha}\, d\tau, \qquad 0 < \alpha < 1$$

**46.** Solve Abel's problem of Tautochronous motion described in problem 17 of Exercise 13.10.

**47.** The velocity potential $\phi(r, z, t)$ and the free-surface elevation $\eta(r, t)$ for axisymmetric surface waves in water of infinite depth satisfy the equation

$$\phi_{rr} + \frac{1}{r}\phi_r + \phi_{zz} = 0, \qquad 0 \leq r < \infty, \qquad -\infty < z \leq 0, \qquad t > 0$$

with the free-surface, boundary, and initial conditions

$$\phi_z = \eta_t \quad \text{on} \quad z = 0, \qquad t > 0$$

$$\phi_t + g\eta = 0 \quad \text{on} \quad z = 0, \qquad t > 0$$

$$\phi_z \to 0 \quad \text{as} \quad z \to -\infty$$

$\phi(r, 0, 0) = 0$ and $\eta(r, 0) = f(r)$, $0 \leq r < \infty$, where $g$ is the acceleration due to gravity and $f(r)$ represents the initial elevation.

Show that

$$\phi(r, z, t) = -\frac{1}{\sqrt{g}} \int_0^\infty \sqrt{k}\, F(k) J_0(kr) e^{kz} \sin(\sqrt{gk}\, t)\, dk$$

$$\eta(r, t) = \int_0^\infty k F(k) J_0(kr) \cos(\sqrt{gk}\, t)\, dk$$

Derive the asymptotic solution

$$\eta(r, t) \sim \frac{gt^2}{2^{\frac{3}{2}} r^3} F\left(\frac{gt^2}{4r^2}\right) \cos\left(\frac{gt^2}{4r}\right) \quad \text{as} \quad t \to \infty$$

**48.** Write the solution for the Cauchy–Poisson problem where the initial elevation is concentrated in the neighborhood of the origin, that is, $f(r) = (a/2\pi r)\ \delta(r)$, $a$ is the total volume of the fluid displaced.

**49.** The steady temperature distribution $u(r, z)$ in a semi-infinite solid $z \geqslant 0$ is governed by the system

$$u_{rr} + \frac{1}{r} u_r + u_{zz} = -Aq(r), \qquad 0 < r < \infty, \qquad z > 0$$

$$u(r, 0) = 0$$

where $A$ is a constant and $q(r)$ represents the steady heat source. Show that the solution is given by

$$u(r, z) = A \int_0^\infty Q(k) J_0(kr) k^{-1} (1 - e^{-kz})\, dk$$

where $Q(k)$ is the zero order Hankel transform of $q(r)$.

**50.** Find the solution for the small deflection $u(r)$ of an elastic membrane subjected to a concentrated loading distribution which is gov·rned by

$$u_{rr} + \frac{1}{r} u_r - \kappa^2 u = \frac{1}{2\pi} \frac{\delta(r)}{r}, \qquad 0 \leqslant r < \infty$$

where $u$ and its derivatives vanish as $r \to \infty$.

**51.** Obtain the solution for the potential $v(r, z)$ due to a flat electrified disk of radius unity with the center of the disk as the origin and the axis along the $z$-axis. The function $v(r, z)$ satisfies the Laplace equation

$$v_{rr} + \frac{1}{r} v_r + v_{zz} = 0, \qquad 0 < r < \infty, \qquad z > 0$$

with the boundary conditions

$$v(r, 0) = v_0, \qquad 0 \leqslant r < 1$$

$$v_z(r, 0) = 0, \qquad r > 1$$

**52.** Prove that the Fourier sine and cosine transforms are linear.

**53.** If $F_s(n)$ is the Fourier sine transform of $f(x)$ on $0 \leqslant x \leqslant l$, show that

$$\mathscr{F}[f''(x)] = \frac{2n\pi}{l^2} [f(0) - (-)^n f(l)] - \left(\frac{n\pi}{l}\right)^2 F_s(n)$$

**54.** If $F_c(n)$ is the Fourier cosine transform of $f(x)$ on $0 \leqslant x \leqslant l$, show that

$$\mathscr{F}[f''(x)] = \frac{2}{l} [(-1)^n f'(l) - f'(0)] - \left(\frac{n\pi}{l}\right)^2 F_c(n)$$

When $l=\pi$, show that

$$\mathscr{F}[f''(x)] = \frac{2}{\pi}[(-1)^n f'(\pi) - f'(0)] - n^2 F_c(n)$$

**55.** By the transform method, solve

$$\begin{aligned} u_t &= u_{xx} + g(x, t), && 0<x<\pi, && t>0 \\ u(x, 0) &= f(x), && 0 \leq x \leq \pi \\ u(0, t) &= 0, && && t>0 \\ u(\pi, t) &= 0, \end{aligned}$$

**56.** By the transform method, solve

$$\begin{aligned} u_t &= u_{xx} + g(x, t), && 0<x<\pi, && t>0 \\ u(x, 0) &= 0, && 0<x<\pi \\ u(0, t) &= 0 && && t>0 \\ u_x(\pi, t) + hu(\pi, t) &= 0, \end{aligned}$$

**57.** By the transform method, solve

$$\begin{aligned} u_t &= u_{xx} + g(x, t), && 0<x<\pi, && t>0 \\ u(x, 0) &= 0 \\ u(0, t) &= 0 && && t>0 \\ u_x(\pi, t) &= 0 \end{aligned}$$

**58.** By the transform method, solve

$$\begin{aligned} u_t &= u_{xx} - hu, && 0<x<\pi, && t>0 \\ u(x, 0) &= \sin x, && 0 \leq x \leq \pi \\ u(0, t) &= 0 && && t>0 \\ u(\pi, t) &= 0 \end{aligned}$$

**59.** By the transform method, solve

$$\begin{aligned} u_{tt} &= u_{xx} + h, && 0<x<\pi, && t>0, && h = \text{constant} \\ u(x, 0) &= 0 \\ u_t(x, 0) &= 0 \\ u_x(0, t) &= 0 \\ u_x(\pi, t) &= 0 \end{aligned}$$

**60.** By the transform method, solve

$$u_{tt} = u_{xx} + g(x), \qquad 0 < x < \pi, \qquad t > 0$$
$$u(x, 0) = 0$$
$$u_t(x, 0) = 0$$
$$u(0, t) = 0$$
$$u(\pi, t) = 0$$

**61.** By the transform method, solve

$$u_{tt} + c^2 u_{xxxx} = 0, \qquad 0 < x < \pi, \qquad t > 0$$
$$u(x, 0) = 0$$
$$u_t(x, 0) = 0$$
$$u(0, t) = 0$$
$$u(\pi, t) = 0$$
$$u_{xx}(0, t) = 0$$
$$u_{xx}(\pi, t) = \sin t, \qquad t \geq 0$$

**62.** Find the temperature distribution $u(r, t)$ in a long cylinder of radius $a$ when the initial temperature is constant, $u_0$ and radiation occurs at the surface into a medium with zero temperature. Here $u(r, t)$ satisfies the initial-boundary problem

$$u_t = \kappa\left(u_{rr} + \frac{1}{r} u_r\right), \qquad 0 \leq r < a, \qquad t > 0$$
$$u_r + \alpha u = 0 \quad \text{at} \quad r = a, \qquad t > 0$$
$$u(r, 0) = u_0 \quad \text{for} \quad 0 \leq r < a$$

**63.** Apply the finite Fourier sine transform to solve the longitudinal displacement field in a uniform bar of length $l$ and cross section $A$ subjected to an external force $FA$ applied at the end $x = l$. The governing equation and boundary and initial conditions are

$$c^2 u_{xx} = u_{tt}, \qquad \left(c^2 = \frac{E}{\rho}\right), \qquad 0 < x < l, \qquad t > 0$$
$$u(0, t) = 0, \qquad t > 0$$
$$E\,u(l, t) = F, \qquad t > 0$$
$$u(x, 0) = u_t(x, 0) = 0, \qquad 0 < x < l$$

where $E$ is the constant Young's modulus, $\rho$ is the density, and $F$ is constant.

**64.** Use the finite Fourier cosine transform to solve the heat conduction problem

$$u_t = \kappa u_{xx}, \qquad 0 < x < l, \qquad t > 0$$

$$u_x(x, t) = 0 \quad \text{at} \quad x = 0 \quad \text{and} \quad x = l, \qquad t > 0$$

$$u(x, 0) = u_0 \quad \text{for} \quad 0 < x < l$$

where $u_0$ and $\kappa$ are constant.

**65.** Use Mellin's transform to find the solution of the integral equation

$$\int_0^\infty f(x)k(xt)\,dx = g(t), \qquad t > 0$$

**66.** Use the Mellin transform to show the following results:

a. $\displaystyle\sum_{n=1}^{\infty} f(n) = \frac{1}{2\pi i}\int_{c-i\infty}^{c+i\infty} \zeta(p)\,F(p)\,dp$

b. $\displaystyle\sum_{n=1}^{\infty} f(nx) = M^{-1}[\zeta(p)\,F(p)]$

where $\zeta(s)$ is the Riemann zeta function defined by

$$\zeta(s) = \sum_{n=1}^{\infty} n^{-s}, \qquad \operatorname{Re} s > 1.$$

**67.** Show that the solution of the boundary-value problem

$$u_{rr} + \frac{1}{r}u_r + u_{zz} = 0, \qquad r \geqslant 0, \qquad z > 0$$

$$u(r, 0) = u_0 \quad \text{for} \quad 0 \leqslant r \leqslant a$$

$$u(r, z) \to 0 \quad \text{as} \quad z \to \infty$$

is

$$u(r, z) = au_0 \int_0^\infty J_1(ak)J_0(kr)e^{-kz}\,dk.$$

**68.** Show that the asymptotic representation of the Bessel function $J_n(kr)$ for large $kr$ is

$$J_n(kr) = \frac{1}{\pi}\int_0^\pi \cos(n\theta - kr\sin\theta)\,d\theta$$

$$\sim \left(\frac{2}{\pi kr}\right)^{\frac{1}{2}} \cos\left(kr - \frac{n\pi}{2} - \frac{\pi}{4}\right)$$

CHAPTER TWELVE

# Nonlinear Partial Differential Equations with Applications

## 12.1 INTRODUCTION

The linear wave equation

$$u_{tt} = c^2 \nabla^2 u \tag{12.1.1}$$

arises in the areas of elasticity, fluid dynamics, acoustics, magnetohydrodynamics, and electromagnetism.

The general solution of the one-dimensional form (12.1.1) is

$$u(x, t) = \phi(x - ct) + \psi(x + ct) \tag{12.1.2}$$

where $\phi$ and $\psi$ are determined by the initial or boundary conditions. Physically, $\phi$ and $\psi$ represent waves moving with constant speed $c$ and without change of shape, along the positive and the negative directions of $x$ respectively.

The solutions $\phi$ and $\psi$ correspond to the two factors when the one-dimensional Eq. (12.1.1) is written in the form

$$\left(\frac{\partial}{\partial t} + c\frac{\partial}{\partial x}\right)\left(\frac{\partial}{\partial t} - c\frac{\partial}{\partial x}\right)u = 0 \tag{12.1.3}$$

Obviously, the simplest linear wave equation is

$$u_t + cu_x = 0 \tag{12.1.4}$$

with the solution $u = \phi(x - ct)$.

## 12.2 ONE-DIMENSIONAL NONLINEAR WAVE EQUATION AND METHOD OF CHARACTERISTICS

The simplest first order nonlinear wave equation is given by

$$u_t + c(u)u_x = 0, \qquad -\infty < x < \infty, \qquad t > 0 \tag{12.2.1}$$

where $c(u)$ is a given function of $u$.

We solve this nonlinear equation subject to the initial condition

$$u(x, 0) = f(x), \qquad -\infty < x < \infty \tag{12.2.2}$$

Before we embark on the method of solution, the following comments are in order. First, unlike the linear differential equations, the principle of superposition cannot be applied to find the general solution of nonlinear partial differential equations. Second, the effect of nonlinearity can change the entire nature of the solution. Third, a study of the above initial-value problem reveals most of the important ideas for nonlinear hyperbolic waves. Finally, a large number of physical and engineering problems are governed by the above nonlinear system or an extension of it.

Although the nonlinear system governed by (12.2.1)–(12.2.2) looks simple, it poses nontrivial problems in applied mathematics and it leads surprisingly to new phenomena. We solve the system by the method of characteristics.

In order to construct continuous solutions, we consider the total derivative $du$ given by

$$du = \frac{\partial u}{\partial t}\, dt + \frac{\partial u}{\partial x}\, dx \tag{12.2.3}$$

so that the points $(x, t)$ are assumed to lie on a curve $\Gamma$. Then, $dx/dt$ represents the slope of the curve $\Gamma$ at any point $P$ on $\Gamma$. Thus, (12.2.3) becomes

$$\frac{du}{dt} = u_t + \left(\frac{dx}{dt}\right) u_x \tag{12.2.4}$$

It follows from this result that (12.2.1) can be regarded as the ordinary differential equation

$$\frac{du}{dt} = 0 \tag{12.2.5}$$

along any member of the family of curves $\Gamma$ which are the solution curves of

$$\frac{dx}{dt} = c(u) \tag{12.2.6}$$

These curves $\Gamma$ are called the *characteristic curves* of the main equation (12.2.1). Thus, the solution of (12.2.1) has been reduced to the solution of a pair of simultaneous ordinary differential equations (12.2.5) and (12.2.6).

Equation (12.2.5) implies that $u = \text{constant}$ along each characteristic curve $\Gamma$, and each $c(u)$ remains constant on $\Gamma$. Therefore, (12.2.6) shows that the characteristic curves of (12.2.1) form a family of straight lines in the $(x, t)$ plane with slope $c(u)$. This indicates that the general solution of (12.2.1) depends on finding the family of lines. Also, each line with slope $c(u)$ corresponds to the value of $u$ on it. If the initial point on the characteristic curve $\Gamma$ is denoted by $\xi$ and if one of the curves $\Gamma$ intersects $t = 0$ at $x = \xi$, then $u(0) = f(\xi)$ on the whole of that curve $\Gamma$ as shown in Fig. 12.2.1.

Thus we have the following characteristic form on $\Gamma$:

$$\frac{dx}{dt} = c(u), \qquad x(0) = \xi \tag{12.2.7}$$

$$\frac{du}{dt} = 0, \qquad u(0) = f(\xi) \tag{12.2.8}$$

These constitute a pair of coupled ordinary differential equations on $\Gamma$. Equation (12.2.7) cannot be solved independently because $c$ is a function of $u$. However, (12.2.8) can readily be solved to obtain $u = \text{constant}$ on $\Gamma$ and hence $u = f(\xi)$ on the whole of $\Gamma$. Thus (12.2.7) leads to

$$\frac{dx}{dt} = F(\xi), \qquad x(0) = \xi \tag{12.2.9}$$

where

$$F(\xi) = c(f(\xi)) \tag{12.2.10}$$

Integration of (12.2.9) gives

$$x = tF(\xi) + \xi \tag{12.2.11}$$

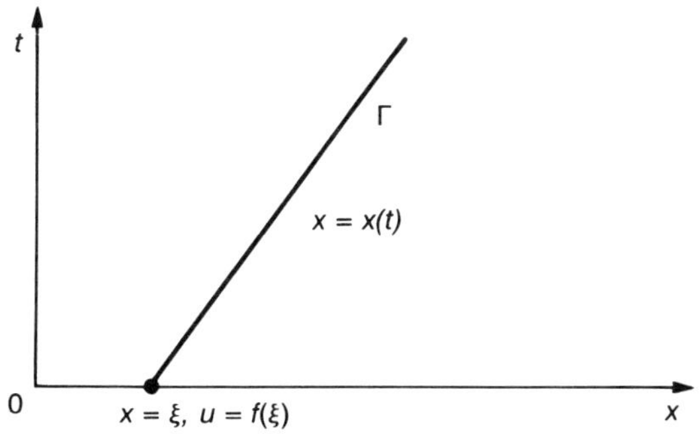


Figure 12.2.1

This represents the characteristic curve which is a straight line whose slope is not a constant, but depends on $\xi$.

Combining these results we obtain the solution of the initial-value problem in parametric form

$$\left.\begin{aligned} u(x, t) &= f(\xi) \\ x &= \xi + tF(\xi) \end{aligned}\right\} \tag{12.2.12ab}$$

where

$$F(\xi) = c(f(\xi))$$

We next verify that this final form represents an analytic expression of the solution. Differentiating (12.2.12ab) w.r.t $x$ and $t$, we obtain

$$u_x = f'(\xi)\xi_x, \qquad u_t = f'(\xi)\xi_t$$
$$1 = \{1 + tF'(\xi)\}\xi_x$$
$$0 = F(\xi) + \{1 + tF'(\xi)\}\xi_t$$

Elimination of $\xi_x$ and $\xi_t$ gives

$$u_x = \frac{f'(\xi)}{1 + tF'(\xi)}, \qquad u_t = -\frac{F(\xi)f'(\xi)}{1 + tF'(\xi)} \tag{12.2.13ab}$$

Since $F(\xi) = c(f(\xi))$, Eq. (12.2.1) is satisfied provided $1 + tF'(\xi) \neq 0$.

The solution (12.2.12ab) also satisfies the initial condition at $t = 0$ since $\xi = x$, and the solution (12.2.12ab) is unique.

Suppose $u(x, t)$ and $v(x, t)$ are two solutions. Then, on $x = \xi + tF(\xi)$,

$$u(x, t) = u(\xi, 0) = f(\xi) = v(x, t)$$

Thus we have proved the following:

***Theorem 12.2.1*** *The nonlinear initial-value problem*

$$u_t + c(u)u_x = 0, \qquad -\infty < x < \infty, \qquad t > 0$$
$$u(x, t) = f(x) \quad \text{at} \quad t = 0, \qquad -\infty < x < \infty$$

*has a unique solution provided* $1 + tF'(\xi) \neq 0$, $f$ *and* $c$ *are* $C^1(R)$ *functions where* $F(\xi) = c(f(\xi))$.

The solution is given in the parametric form:

$$u(x, t) = f(\xi)$$
$$x = \xi + tF(\xi)$$

Remark: When $c(u) = \text{constant} = c > 0$, Eq. (12.2.1) becomes the linear wave equation (12.1.4). The characteristic curves are $x = ct + \xi$ and the solution $u$ is given by

$$u(x, t) = f(\xi) = f(x - ct)$$

## Physical Significance of (12.2.12ab)

We assume $c(u) > 0$. The graph of $u$ at $t = 0$ is the graph of $f$. In view of the fact

$$u(x, t) = u(\xi + tF(\xi), t) = f(\xi)$$

the point $(\xi, f(\xi))$ moves parallel to the $x$-axis in the positive direction through a distance $tF(\xi) = ct$, and the distance moved $(x = \xi + ct)$ depends on $\xi$. This is a typical nonlinear phenomenon. In the linear case, the curve moves parallel to the $x$-axis with constant velocity $c$, and the solution represents waves travelling without change of shape. Thus, there is a striking difference between the linear and the nonlinear solutions.

Theorem 12.2.1 asserts that the solution of the nonlinear initial-value problem exists provided

$$1 + tF'(\xi) \neq 0, \qquad x = \xi + tF(\xi) \tag{12.2.14ab}$$

However, the former condition is always satisfied for sufficiently small time $t$. By a solution of the problem, we mean a differentiable function $u(x, t)$. It follows from results (12.2.13ab) that both $u_x$ and $u_t$ tend to infinity as $1 + tF'(\xi) \to 0$. This means that the solution develops a singularity (discontinuity) when $1 + tF'(\xi) = 0$. We consider a point $(x, t) = (\xi, 0)$ so that this condition is satisfied on the characteristics through the point $(\xi, 0)$ at a time $t$ such that

$$t = -\frac{1}{F'(\xi)} \tag{12.2.15}$$

which is positive provided $F'(\xi) = c'(f)f'(\xi) < 0$. If we assume $c'(f) > 0$, the above inequality implies that $f'(\xi) < 0$. Hence the solution (12.2.12ab) ceases to exist for all time if the initial data is such that $f'(\xi) < 0$ for some value of $\xi$. Suppose $t = \tau$ is the time when the solution first develops a singularity (discontinuity) for some value of $\xi$. Then

$$\tau = -\frac{1}{\min\limits_{-\infty<\xi<\infty} \{c'(f)f'(\xi)\}} > 0$$

We draw the graphs of the nonlinear solution $u(x, t) = f(\xi)$ below for different values of $t = 0, \tau, 2\tau, \ldots$. The shape of the initial curve for $u(x, t)$ changes with increasing values of $t$, and the solution becomes multiple-valued for $t \geqslant \tau$. Therefore, the solution breaks down when $F'(\xi) < 0$ for some $\xi$, and such breaking is a typical nonlinear phenomenon. In linear theory, such breaking will never occur.

More precisely, the development of a singularity in the solution for $t \geqslant \tau$ can be seen by the following consideration. If $f'(\xi) < 0$, we can find two values of $\xi = \xi_1$, $\xi_2$ $(\xi_1 < \xi_2)$ on the initial line such that the characteristics through them have different slopes $1/c(u_1)$ and $1/c(u_2)$

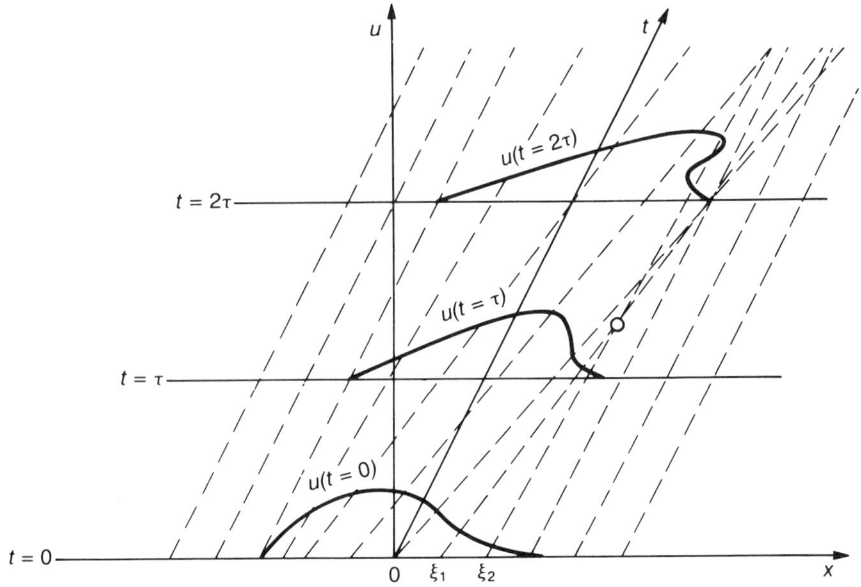


Figure 12.2.2. The solution $u(x, t)$ for different times $t=0$, $\tau$ and $2\tau$; the characteristics are shown by the dotted lines with two of them from $x=\xi_1$ and $x=\xi_2$ intersect at $t>\tau$.

where $u_1=f(\xi_1)$ and $u_2=f(\xi_2)$ and $c(u_2)<c(u_1)$. Thus, these two characteristics will intersect at a point in the $(x, t)$ plane for some $t>0$. Since the characteristics carry constant values of $u$, the solution ceases to be single-valued at their point of intersection. Figure 12.2.2 shows that the wave profile progressively distorts itself, and at any instant of time there exists an interval on the $x$-axis where $u$ assumes three values for a given $x$. The end result is the development of a non-unique solution and this leads to breaking.

Therefore, when conditions (12.2.14ab) are violated, the solution develops a discontinuity known as a *shock*. The analysis of shock involves extension of a solution to allow for discontinuities. Also, it is necessary to impose on the solution certain restrictions to be satisfied across its discontinuity. This point will be discussed further in a subsequent section.

## 12.3 LINEAR DISPERSIVE WAVES

We consider a single linear partial differential equation with constant coefficients in the form

$$P\left(\frac{\partial}{\partial t}, \frac{\partial}{\partial x}, \frac{\partial}{\partial y}, \frac{\partial}{\partial z}\right)u(\mathbf{x}, t)=0 \tag{12.3.1}$$

where $P$ is a polynomial in partial derivatives and $\mathbf{x}=(x, y, z)$.

We seek an elementary plane wave solution of (12.3.1) in the form

$$u(\mathbf{x}, t) = ae^{i(\boldsymbol{\kappa}\cdot\mathbf{x}-\omega t)} \tag{12.3.2}$$

where $a$ is the amplitude, $\boldsymbol{\kappa} = (k, l, m)$ is the wavenumber vector, $\omega$ is the frequency and $a$, $\boldsymbol{\kappa}$, $\omega$ are constants. When this plane wave solution is substituted in the equation, $\partial/\partial t$, $\partial/\partial x$, $\partial/\partial y$, and $\partial/\partial z$ produce factors $-i\omega$, $ik$, $il$, and $im$ respectively, and the solution exists provided $\omega$ and $\boldsymbol{\kappa}$ are related by an equation

$$P(-i\omega, ik, il, im) = 0 \tag{12.3.3}$$

This equation is known as the *dispersion relation*. Evidently, we have a direct correspondence between Eq. (12.3.1) and the dispersion relation (12.3.3) through the correspondence

$$\frac{\partial}{\partial t} \leftrightarrow -i\omega, \qquad \left(\frac{\partial}{\partial x}, \frac{\partial}{\partial y}, \frac{\partial}{\partial z}\right) \leftrightarrow i(k, l, m) \tag{12.3.4}$$

Equation (12.3.1) and the corresponding dispersion relation (12.3.3) indicate that the former can be derived from the latter and vice-versa by using (12.3.4). The dispersion relation characterizes the plane wave motion. In many problems, the dispersion relation can be written in the explicit form

$$\omega = W(k, l, m) \tag{12.3.5}$$

The phase and the group velocities of the waves are defined by

$$\mathbf{C}_p(\kappa) = \frac{\omega}{\kappa}\,\hat{\kappa} \tag{12.3.6}$$

$$\mathbf{C}_g(\kappa) = \nabla_{\boldsymbol{\kappa}}\omega \tag{12.3.7}$$

where $\hat{\kappa}$ is the unit vector in the direction of wave vector $\boldsymbol{\kappa}$.

In the one-dimensional case, (12.3.5)–(12.3.7) become

$$\omega = W(k), \qquad C_p = \frac{\omega}{k}, \qquad C_g = \frac{d\omega}{dk} \tag{12.3.8abc}$$

The one-dimensional waves given by (12.3.2) are called *dispersive* if the group velocity $C_g \equiv \omega'(k)$ is not constant, that is, $\omega''(k) \neq 0$. Physically, as time progresses, the different waves disperse in the medium with the result that a single hump breaks into wavetrains.

EXAMPLE 12.3.1

(i) Linearized one-dimensional wave equation

$$u_{tt} - c^2 u_{xx} = 0, \qquad \omega = \pm ck \tag{12.3.9ab}$$

(ii) Linearized Korteweg and deVries (KdV) equation for long water waves

$$u_t + \alpha u_x + \beta u_{xxx} = 0, \qquad \omega = \alpha k - \beta k^3 \tag{12.3.10ab}$$

(iii) Klein-Gordon equation

$$u_{tt} - c^2 u_{xx} + \alpha^2 u = 0, \qquad \omega = \pm(c^2k^2 + \alpha^2)^{\frac{1}{2}} \quad (12.3.11\text{ab})$$

(iv) Schrödinger Equation in quantum mechanics and de Broglie waves

$$i\hbar\psi_t - \left(V - \frac{\hbar^2}{2m}\nabla^2\right)\psi = 0, \qquad \hbar\omega = \frac{\hbar^2\kappa^2}{2m} + V \quad (12.3.12\text{ab})$$

where $V$ is a constant potential energy, and $h = 2\pi\hbar$ is the Planck constant.

The group velocity of de Broglie wave is $\hbar\boldsymbol{\kappa}/m$, and through the correspondence principle, $\hbar\omega$ is to be interpreted as the total energy, $\hbar^2\kappa^2/2m$ as the kinetic energy, and $\hbar\boldsymbol{\kappa}$ as the particle momentum. Hence the group velocity is the classical particle velocity.

(v) Equation for vibration of a beam

$$u_{tt} + \alpha^2 u_{xxxx} = 0, \qquad \omega = \pm\alpha k^2 \qquad (12.3.13\text{ab})$$

(vi) The dispersion relation for water waves in an ocean of depth $h$ is

$$\omega^2 = gk \tanh kh \qquad (12.3.14)$$

where $g$ is the acceleration due to gravity

(vii) The Boussinesq equation

$$u_{tt} - \alpha^2\nabla^2 u - \beta^2\nabla^2 u_{tt} = 0, \qquad \omega \pm \frac{\alpha\kappa}{\sqrt{1 + \beta^2\kappa^2}} \quad (12.3.15\text{ab})$$

This equation arises in elasticity for longitudinal waves in bars, long water waves, and plasma waves.

(viii) Electromagnetic waves in dielectrics

$$(u_{tt} + \omega_0^2 u)(u_{tt} - c_0^2 u_{xx}) - \omega_p^2 u_{tt} = 0,$$
$$(\omega^2 - \omega_0^2)(\omega^2 - c_0^2k^2) - \omega_p^2\omega^2 = 0 \quad (12.3.16\text{ab})$$

where $\omega_0$ is the natural frequency of the oscillator, $c_0$ is the speed of light, and $\omega_p$ is the plasma frequency.

In view of the superposition principle, the general solution can be obtained from (12.3.2) with the dispersion solution (12.3.3). For the one-dimensional case, the general solution has the Fourier integral representation

$$u(x, t) = \int_{-\infty}^{\infty} F(k)e^{i[kx - tW(k)]}\,dk \qquad (12.3.17)$$

where $F(k)$ is chosen to satisfy the initial or boundary data provided the data are physically realistic enough to have Fourier transforms.

In many cases, as cited in examples 12.3.1, there are two modes

$\omega = \pm W(k)$ so that the solution (12.3.17) has the form

$$u(x, t) = \int_{-\infty}^{\infty} F_1(k) e^{i[kx - tW(k)]} \, dk + \int_{-\infty}^{\infty} F_2(k) e^{i[kx + tW(k)]} \, dk \quad (12.3.18)$$

with the initial data at $t = 0$

$$u(x, t) = \phi(x), \qquad u_t(x, t) = \psi(x) \qquad (12.3.19ab)$$

The initial conditions give

$$\phi(x) = \int_{-\infty}^{\infty} [F_1(k) + F_2(k)] e^{ikx} \, dk$$

$$\psi(x) = -i \int_{-\infty}^{\infty} [F_1(k) - F_2(k)] W(k) e^{ikx} \, dk$$

Applying the Fourier inverse transformations, we have

$$F_1(k) + F_2(k) = \Phi(k) = \frac{1}{\sqrt{2\pi}} \int_{-\infty}^{\infty} \phi(x) e^{-ikx} \, dx$$

$$-iW(k)[F_1(k) - F_2(k)] = \Psi(k) = \frac{1}{\sqrt{2\pi}} \int_{-\infty}^{\infty} \psi(x) e^{-ikx} \, dx$$

so that

$$[F_1(k), F_2(k)] = \frac{1}{2}\left[\Phi(k) \pm \frac{i\Psi(k)}{W(k)}\right] \qquad (12.3.20ab)$$

The asymptotic behavior of $u(x, t)$ for large $t$ with fixed $x/t$ can be obtained by the Kelvin stationary phase approximation. For real $\phi(x)$, $\psi(x)$, $\Phi(-k) = \Phi^*(k)$ and $\Psi(-k) = \Psi^*(k)$ where the asterisk denotes a complex conjugate. It follows from (12.3.20ab) that, for $W(k)$ even,

$$[F_1(-k), F_2(-k)] = [F_2^*(k), F_1^*(k)] \qquad (12.3.21)$$

and for $W(k)$ odd,

$$[F_1(-k), F_2(-k)] = [F_1^*(k), F_2^*(k)] \qquad (12.3.22)$$

In particular, when $\phi(x) = \delta(x)$ and $\psi(x) \equiv 0$, then $F_1(k) = F_2(k) = 1/\sqrt{8\pi}$, and the solution (12.3.18) reduces to the form

$$u(x, t) = \sqrt{\frac{2}{\pi}} \int_{0}^{\infty} \cos kx \cos \{tW(k)\} \, dk \qquad (12.3.23)$$

In order to obtain the asymptotic approximation by the Kelvin stationary phase method (see § 11.6) for $t \to \infty$, we consider both cases when $W(k)$ is even ($W'(k)$ is odd) and when $W(k)$ is odd ($W'(k)$ is even) and make an extra reasonable assumption that $W'(k)$ is monotonic and positive for $k > 0$. It turns out that the asymptotic solution for $t \to \infty$ is

$$u(x, t) \sim 2\,\mathrm{Re}\left\{F_1(k)\left\{\frac{2\pi}{t\,|W''(k)|}\right\}^{\frac{1}{2}} \exp\left[i\left\{\theta(x, t) - \frac{\pi}{4}\,\mathrm{sgn}\,W''(k)\right\}\right]\right\} + O\left(\frac{1}{t}\right)$$

$$= \mathrm{Re}\,[a(x, t)e^{i\theta(x,t)}] \tag{12.3.24}$$

where $k(x, t)$ is the positive root of

$$W'(k) = \frac{x}{t}, \qquad \omega = W(k), \qquad \frac{x}{t} > 0 \tag{12.3.25ab}$$

$$\theta(x, t) = xk(x, t) - t\omega(x, t) \tag{12.3.26}$$

and

$$a(x, t) = 2F_1(k)\left\{\frac{2\pi}{t\,|W''(k)|}\right\}^{\frac{1}{2}} \exp\left\{-\frac{i\pi}{4}\,\mathrm{sgn}\,W''(k)\right\} \tag{12.3.27}$$

It is important to point out that solution (12.3.24) has a form similar to that of the elementary plane wave solution, but $k$, $\omega$, and $a$ are no longer constants and they are functions of space variable $x$ and time $t$. The solution still represents an oscillatory wavetrain with the phase function $\theta(x, t)$ describing the variations between local maxima and minima. Unlike the elementary plane wavetrain, the present asymptotic result (12.3.24) represents a non-uniform wavetrain in the sense that the amplitude, the distance, and the time between successive maxima are not constant.

It also follows from (12.3.25a) that

$$\frac{k_t}{k} = -\frac{W'(k)}{kW''(k)}\frac{1}{t} \sim O\left(\frac{1}{t}\right) \tag{12.3.28}$$

$$\frac{k_x}{k} = \frac{1}{kW''(k)}\frac{1}{t} \sim O\left(\frac{1}{t}\right) \tag{12.3.29}$$

These results indicate the $k(x, t)$ is a slowly varying function of $x$ and $t$ as $t \to \infty$. Applying a similar argument to $\omega$ and $a$, we conclude that $k(x, t)$, $\omega(x, t)$, and $a(x, t)$ are all slowly varying functions of $x$ and $t$ as $t \to \infty$.

Finally, all these results seem to provide an important clue for natural generalization of the concept of nonlinear and non-uniform wavetrains.

## 12.4 NONLINEAR DISPERSIVE WAVES AND WHITHAM'S EQUATIONS

To describe a slowly varying nonlinear and non-uniform oscillatory wavetrain in a medium, we assume the existence of a solution in the form (12.3.24) so that

$$u(x, t) = a(x, t)e^{i\theta(x,t)} + \text{c.c.} \tag{12.4.1}$$

where c.c. stands for the complex congugate, $a(x, t)$ is the complex amplitude given by (12.3.27), the phase function $\theta(x, t)$ is

$$\theta(x, t) = xk(x, t) - t\omega(x, t) \tag{12.4.2}$$

and $k$, $\omega$, and $a$ are all slowly varying function of $x$ and $t$.

Due to slow variations of $k$ and $\omega$, it is reasonable to assume that these quantities still satisfy the dispersion relation

$$\omega = W(k) \tag{12.4.3}$$

Differentiating (12.4.2) with respect to $x$ and $t$ respectively, we obtain

$$\theta_x = k + \{x - tW'(k)\}k_x \tag{12.4.4}$$

$$\theta_t = -W(k) + \{x - tW'(k)\}k_t \tag{12.4.5}$$

In the neighborhood of stationary points defined by (12.3.25a), these results become

$$\theta_x = k(x, t), \qquad \theta_t = -\omega(x, t) \tag{12.4.6ab}$$

These results can be used as a definition of *local wavenumber* and *local frequency* of the slowly varying nonlinear wavetrain.

In view of (12.4.6ab), (12.4.3) gives a nonlinear partial differential equation for the phase $\theta$ in the form

$$\frac{\partial \theta}{\partial t} + W\left(\frac{\partial \theta}{\partial x}\right) = 0 \tag{12.4.7}$$

The solution of this equation determines the geometry of the wave pattern.

However, it is convenient to eliminate $\theta$ from (12.4.6ab) to obtain

$$\frac{\partial k}{\partial t} + \frac{\partial \omega}{\partial x} = 0 \tag{12.4.8}$$

This is known as the *Whitham equation* for the conservation of waves, where $k$ represents the density of waves and $\omega$ is the flux of waves.

Using the dispersion relation (12.4.3), we obtain

$$\frac{\partial k}{\partial t} + C_g(k)\frac{\partial k}{\partial x} = 0 \tag{12.4.9}$$

where $C_g(k) = W'(k)$ is the group velocity. This represents the simplest nonlinear wave (hyperbolic) equation for the propagation of $k$ with the group velocity $C_g(k)$.

Since Eq. (12.4.9) is similar to (12.2.1), we can use the analysis of § 12.1 to find the general solution of (12.4.9) with the initial condition $k = f(x)$ at $t = 0$. In this case, the solution has the form

$$k(x, t) = f(\xi), \qquad x = \xi + tF(\xi) \tag{12.4.10ab}$$

where $F(\xi) = C_g(f(\xi))$. This further confirms the propagation of $k$ with the velocity $C_g$. Some physical interpretations of this kind of solution have already been discussed in § 12.2.

Equations (12.4.9) and (12.4.3) reveal that $\omega$ also satisfies the nonlinear wave (hyperbolic) equation

$$\frac{\partial \omega}{\partial t} + W'(k)\frac{\partial \omega}{\partial x} = 0 \tag{12.4.11}$$

It follows from Eqs. (12.4.9) and (12.4.11) that both $k$ and $\omega$ remain constant on the characteristic curves defined by

$$\frac{dx}{dt} = W'(k) = C_g(k) \tag{12.4.12}$$

in the $(x, t)$ plane. Since $k$ or $\omega$ is constant on each curve, the characteristic curves are straight lines with slope $C_g(k)$. The solution for $k$ is given by (12.4.10ab).

Finally, it follows from the above analysis that any constant value of the phase $\theta$ propagates according to $\theta(x, t) = \text{constant}$, and hence

$$\theta_t + \left(\frac{dx}{dt}\right)\theta_x = 0 \tag{12.4.13}$$

which gives, by (12.4.6ab),

$$\frac{dx}{dt} = -\frac{\theta_t}{\theta_x} = \frac{\omega}{k} = C_p \tag{12.4.14}$$

Thus the phase of the waves propagates with the phase speed $C_p$. On the other hand, (12.4.9) ensures that the wavenumber $k$ propagates with the group velocity $C_g(k) = d\omega/dk = W'(k)$.

We next investigate how the wave energy propagates in the dispersive medium. We consider the following integral involving the square of the wave amplitude (energy) given by (12.3.24) between any two points $x = x_1$ and $x = x_2$ $(0 < x_1 < x_2)$

$$Q(t) = \int_{x_1}^{x_2} |a|^2\,dx = \int_{x_1}^{x_2} aa^*\,dx \tag{12.4.15}$$

$$= 8\pi \int_{x_1}^{x_2} \frac{F_1(k)F_1^*(k)}{t\,|W''(k)|}\,dx \tag{12.4.16}$$

which is, due to a change of variable $x = tW'(k)$,

$$= 8\pi \int_{k_1}^{k_2} F_1(k)F_1^*(k)\,dk \tag{12.4.17}$$

where $k_r = tW'(k_r)$, $r = 1, 2$.

When $k_r$ is kept fixed as $t$ varies, $Q(t)$ remains constant so that

$$0=\frac{dQ}{dt}=\frac{d}{dt}\int_{x_1}^{x_2}|a|^2\,dx$$

$$=\int_{x_1}^{x_2}\frac{\partial}{\partial t}|a|^2\,dx+|a|_2^2\,W'(k_2)-|a|_1^2\,W'(k_1) \qquad (12.4.18)$$

In the limit $x_2-x_1\to 0$, this result reduces to the partial differential equation

$$\frac{\partial}{\partial t}|a|^2+\frac{\partial}{\partial x}[W'(k)\,|a|^2]=0 \qquad (12.4.19)$$

This represents the equation for the conservation of wave energy where $|a|^2$ and $|a|^2\,W'(k)$ are the *energy density* and *energy flux* respectively. It also follows that the energy propagates with the group velocity $W'(k)$. It has been shown that the wavenumber $k$ also propagates with the group velocity. Thus, the group velocity plays a double role.

The above analysis reveals another important fact, (12.4.3), (12.4.8), and (12.4.19) constitute a closed set of equations for the three quantities $k$, $\omega$, and $a$. Indeed, these are the fundamental equations for nonlinear dispersive waves, and are known as *Whitham's equations.*

## 12.5 NONLINEAR INSTABILITY

For infinitesimal waves, the wave amplitude ($ak \ll 1$) is very small, so that nonlinear effects can be neglected altogether. However, for finite amplitude waves the terms involving $a^2$ cannot be neglected, and hence the effects of nonlinearity becomes important. In the theory of water waves, Stokes first obtained the connection due to inherent nonlinearity on the wave-profile and the frequency of a steady periodic wave system. According to the Stokes theory, the remarkable change of nonlinearity is the dependence of $\omega$ on $a$ which couples (12.4.8) to (12.4.19). This leads to a new nonlinear phenomenon.

For finite amplitude waves, $\omega$ has the Stokes expansion

$$\omega=\omega_0(k)+a^2\omega_2(k)+\ldots=\omega(k,a^2) \qquad (12.5.1)$$

This can be regarded as the nonlinear dispersion relation which depends on both $k$ and $a^2$. In the linear case, $a\to 0$, (12.5.1) gives the linear dispersion relation (12.4.3).

In order to discuss nonlinear instability, we substitute (12.5.1) into (12.4.8) and retain (12.4.19) in the linear approximation to obtain the following coupled system:

$$\frac{\partial k}{\partial t}+\frac{\partial}{\partial x}\{\omega_0(k)+\omega_2(k)a^2\}=0 \tag{12.5.2}$$

$$\frac{\partial a^2}{\partial t}+\frac{\partial}{\partial x}\{\omega_0'(k)a^2\}=0 \tag{12.5.3}$$

where $W(k)\equiv\omega_0(k)$.

These equations can be further approximated to obtain

$$\frac{\partial k}{\partial t}+\omega_0'\frac{\partial k}{\partial x}+\omega_2\frac{\partial a^2}{\partial x}=0(a^2) \tag{12.5.4}$$

$$\frac{\partial a^2}{\partial t}+\omega_0'\frac{\partial a^2}{\partial x}+\omega_0''a^2\frac{\partial k}{\partial x}=0 \tag{12.5.5}$$

In matrix form, these equations read

$$\begin{pmatrix}\omega_0' & \omega_2\\ \omega_0''a^2 & \omega_0'\end{pmatrix}\begin{pmatrix}\dfrac{\partial k}{\partial x}\\ \dfrac{\partial a^2}{\partial x}\end{pmatrix}+\begin{pmatrix}1 & 0\\ 0 & 1\end{pmatrix}\begin{pmatrix}\dfrac{\partial k}{\partial t}\\ \dfrac{\partial a^2}{\partial t}\end{pmatrix}=0 \tag{12.5.6}$$

Hence the eigenvalues $\lambda$ are the roots of the determinant equation

$$|a_{ij}-\lambda b_{ij}|=\begin{vmatrix}\omega_0'-\lambda & \omega_2\\ \omega_0''a^2 & \omega_0'-\lambda\end{vmatrix}=0 \tag{12.5.7}$$

where $a_{ij}$ and $b_{ij}$ are the coefficient matrices of (12.5.6). This equation gives the characteristic velocities

$$\lambda=\frac{dx}{dt}=\omega_0'\pm(\omega_2\omega_0'')^{\frac{1}{2}}a+O(a^2) \tag{12.5.8ab}$$

If $\omega_2\omega_0''>0$, the characteristics are real and the system is hyperbolic. The double characteristic velocity splits into two separate real velocities. This provides a new extension of the group velocity to nonlinear problems. If the disturbance is initially finite in extent, it would eventually split into two disturbances. In general, any initial disturbance or modulating source would introduce disturbances on both families of characteristics. In the hyperbolic case, compressive modulation will progressively distort and steepen so that the question of breaking will arise. These results are remarkably different from those found in linear theory, where there is only one characteristic velocity and any hump may distort, due to the dependence of $\omega_0'(k)$ on $k$, but would never split.

On the other hand, if $\omega_2\omega_0''<0$, the characteristics are not real and the system is elliptic. This leads to ill-posed problems. Any small perturbations in $k$ and $a$ will be given by the solutions of the form $\exp[i\alpha(x-\lambda t)]$

where $\lambda$ is calculated from (12.5.8ab) for unperturbed values of $k$ and $a$. In this elliptic case, $\lambda$ is complex, and the perturbation will grow as $t \to \infty$. Hence the original wavetrain will become unstable. In the linear theory, the elliptic case does not arise at all.

EXAMPLE 12.5.1. For Stokes waves in deep water, the dispersion relation is

$$\omega = (gk)^{\frac{1}{2}}\left[1 + \frac{1}{2}k^2a^2\right] \tag{12.5.9}$$

so that $\omega_0(k) = (gk)^{\frac{1}{2}}$ and $\omega_2(k) = \frac{1}{2}\sqrt{g}\,k^{\frac{5}{2}}$.

In this case, $\omega_0'' = -\frac{1}{4}\sqrt{g}\,k^{-\frac{3}{2}}$ so that $\omega_0''\omega_2 = -\frac{g}{8} < 0$. The conclusion is that the Stokes waves in deep water are unstable. This is the most remarkable discovery in the theory of nonlinear water waves developed during the 1960's.

## 12.6 TRAFFIC FLOW MODEL

We consider the flow of cars on a long highway under the assumptions that cars do not enter or leave the highway at any one of its points. We take the $x$-axis along the highway and assume that the traffic flows in the positive direction. Suppose $\rho(x, t)$ is the density representing the number of cars per unit length at the point $x$ of the highway at time $t$, and $q(x, t)$ is the flow of cars per unit time.

We assume a conservation law which states that the change in the total amount of a physical quantity contained in any region of space must be equal to the flux of that quantity across the boundary of that region. In this case, the time rate of change of the total number of cars in any segment $x_1 \leqslant x \leqslant x_2$ of the highway is given by

$$\frac{d}{dt}\int_{x_1}^{x_2} \rho(x, t)\,dx = \int_{x_1}^{x_2} \frac{\partial \rho}{\partial t}\,dx \tag{12.6.1}$$

This rate of change must be equal to the net flux across $x_1$ and $x_2$ given by

$$q(x_1, t) - q(x_2, t) \tag{12.6.2}$$

which measures the flow of cars entering the segment at $x_1$ minus the flow of cars leaving the segment at $x_2$. Thus we have the conservation equation

$$\frac{d}{dt}\int_{x_1}^{x_2} \rho(x, t)\,dx = q(x_1, t) - q(x_2, t) \tag{12.6.3}$$

or

$$\int_{x_1}^{x_2} \frac{\partial \rho}{\partial t}\, dx = -\int_{x_1}^{x_2} \frac{\partial q}{\partial x}\, dx$$

or

$$\int_{x_1}^{x_2} \left(\frac{\partial \rho}{\partial t} + \frac{\partial q}{\partial x}\right) dx = 0 \tag{12.6.4}$$

Since the integrand in (12.6.4) is continuous, and (12.6.4) holds for every segment $[x_1, x_2]$, it follows that the integrand must vanish so that we have the partial differential equation

$$\frac{\partial \rho}{\partial t} + \frac{\partial q}{\partial x} = 0 \tag{12.6.5}$$

We now introduce an additional assumption which is supported by both theoretical and experimental findings. According to this assumption, the flow rate $q$ depends on $x$ and $t$ only through the density, that is, $q = Q(\rho)$ for some function $Q$. This assumption seems to be reasonable in the sense that the density of cars surrounding a given car indeed controls the speed of that car. The functional relation between $q$ and $\rho$ depends on many factors, including speed limits, weather conditions, and road characteristics. Several specific relations are suggested by Haight [93].

We consider here a particular relation $q = \rho v$ where $v$ is the average local velocity of cars. We assume $v$ is a function of $\rho$ to a first approximation. In view of this relation, (12.6.5) reduces to the nonlinear hyperbolic equation

$$\frac{\partial \rho}{\partial t} + c(\rho)\frac{\partial \rho}{\partial x} = 0 \tag{12.6.6}$$

where

$$c(\rho) = q'(\rho) = v(\rho) + \rho v'(\rho) \tag{12.6.7}$$

In general, the local velocity $v(\rho)$ is a decreasing function of $\rho$ so that $v(\rho)$ has a finite maximum value $v_{\max}$ at $\rho = 0$ and decreases to zero at $\rho = \rho_{\max} = \rho_m$. For the value of $\rho = \rho_m$, the cars are bumper to bumper. Since $q = \rho v$, $q(\rho) = 0$ when $\rho = 0$ and $\rho = \rho_m$. This means that $q$ is an increasing function of $\rho$ until it attains a maximum value $q_{\max} = q_M$ for some $\rho = \rho_M$ and then decreases to zero at $\rho = \rho_m$. Both $q(\rho)$ and $v(\rho)$ are shown in Fig. 12.6.1.

Equation (12.6.6) is similar to (12.2.1) with the wave propagation velocity $c(\rho) = v(\rho) + \rho v'(\rho)$. Since $v'(\rho) < 0$, $c(\rho) < v(\rho)$, that is, the propagation velocity is less than the car velocity. In other words, waves propagate backwards through the stream of cars, and drivers are warned of disturbances ahead. It follows from Fig. 12.6.1a that $q(\rho)$ is an increasing function in $[0, \rho_M]$, a decreasing function in $[\rho_M, \rho_m]$, and

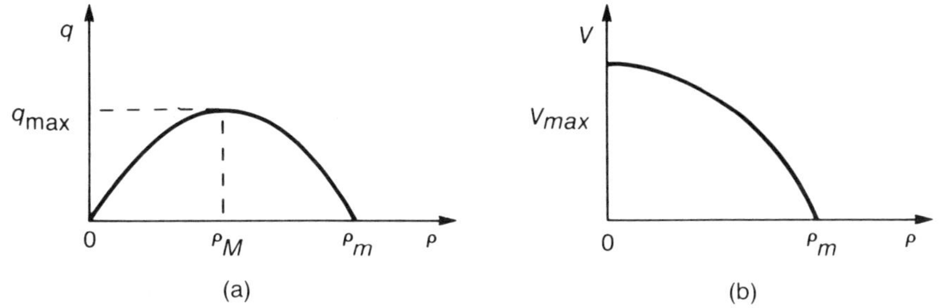


Figure 12.6.1

attains a maximum at $\rho_M$. Hence $c(\rho) = q'(\rho)$ is positive in $[0, \rho_M]$, zero at $\rho_M$, and negative in $[\rho_M, \rho_m]$. All these mean that waves propagate forward relative to the highway in $[0, \rho_M]$, are stationary at $\rho_M$, and then travel backward in $[\rho_M, \rho_m]$.

We use § 12.1 to solve the initial-value problem for the nonlinear equation (12.6.6) with the initial condition $\rho(x, 0) = f(x)$. The solution is

$$\rho(x, t) = f(\xi), \qquad x = \xi + tF(\xi) \tag{12.6.8ab}$$

where

$$F(\xi) = c(f(\xi))$$

Since $c'(\rho) = q''(\rho) < 0$ ($q(\rho)$ is convex), $c(\rho)$ is a decreasing function of $\rho$. This means that breaking occurs at the left due to formation of shock at the back. Waves propagate slower than the cars, so drivers enter such a local density increase from behind; they must decelerate rapidly through the shock but speed up slowly as they get out from the crowded area. These conclusions are in accord with observational results.

Actual observational data of traffic flow indicate that a typical result on a single lane highway is $\rho_m \approx 225$ cars per mile, $\rho_M \approx 80$ cars per mile, and $q_M \approx 1590$ cars per hour. Thus the maximum flow rate $q_M$ occurs at a low velocity $v = q_M/\rho_M \approx 20$ miles per hour.

## 12.7 FLOOD WAVES IN RIVERS

We consider flood waves in a long rectangular river of constant breadth. We take the $x$-axis along the river which flows in the positive $x$-direction and assume that the disturbance is approximately the same across the breadth. In this problem, the depth $h(x, t)$ of the river plays the role of density involved in the traffic flow model discussed in § 12.6. Let $q(x, t)$ be the flow per unit breadth and per unit time. According to the Conservation Law, the rate of change of the mass of the fluid in any section $x_1 \leqslant x \leqslant x_2$ must be balanced by the net flux across $x_2$ and $x_1$ so that the conservation equation is

$$\frac{d}{dt}\int_{x_1}^{x_2} h(x,t)\,dx + q(x_2,t) - q(x_1,t) = 0 \tag{12.7.1}$$

An argument similar to the previous section gives

$$\frac{\partial h}{\partial t} + \frac{\partial q}{\partial x} = 0 \tag{12.7.2}$$

Although the fluid flow is extremely complicated, we assume a simple functional relation $q = Q(h)$ as a first approximation to express the increase in flow as the water level rises. Thus, Eq. (12.7.2) becomes

$$h_t + c(h)h_x = 0 \tag{12.7.3}$$

where $c(h) = Q'(h)$ and $Q(h)$ is determined from the balance between the gravitational force and the frictional force of the river bed. This equation is similar to (12.2.1) and the solution has already been obtained in § 12.2.

Here we discuss the velocity of wave propagation for some particular values of $Q(h)$. One such result is given by the Chezy result as

$$Q(h) = hv \tag{12.7.4}$$

where $v = \alpha\sqrt{h}$ is the velocity of fluid flow and $\alpha$ is a constant so that the propagation velocity of flood waves is given by

$$c(h) = Q'(h) = \frac{3}{2}\alpha\sqrt{h} = \frac{3}{2}v \tag{12.7.5}$$

Thus, the flood waves propagate one and a half times faster than the stream velocity.

For a general case where $v = \alpha h^n$ so that

$$Q(h) = hv = \alpha h^{n+1} \tag{12.7.6}$$

so the propagation velocity of flood waves is

$$c(h) = Q'(h) = (n+1)v \tag{12.7.7}$$

This result also indicates that flood waves propagate faster than the fluid.

## 12.8 RIEMANN'S SIMPLE WAVES OF FINITE AMPLITUDE

We consider a one-dimensional unsteady isentropic flow of gas of density $\rho$ and pressure $p$ with the direction of motion along the $x$-axis. Suppose $u(x, t)$ is the $x$-component of the velocity at time $t$ and $A$ is an area-element of the $y-z$ plane. The volume of the rectangular cylinder of height $dx$ standing on the element $A$ is then $A\,dx$ and its mass $A\rho\,dx$. The change in this mass during the time $dt$ or the quantity $A\rho_t\,dx\,dt$ is determined by the mass entering it, which is equal to $-A(\partial/\partial x)(\rho u)\,dx\,dt$. Its acceleration is $u_t + uu_x$ and the force impelling it in the positive $x$-direction is $-p_x A\,dx = -c^2\rho_x A\,dx$ where $p = f(\rho)$ and $c^2 = f'(\rho)$. These results lead to two coupled non-linear partial differential equations

$$\frac{\partial \rho}{\partial t} + \frac{\partial}{\partial x}(\rho u) = 0 \tag{12.8.1}$$

$$(u_t + uu_x) + \frac{c^2}{\rho}\rho_x = 0 \tag{12.8.2}$$

In matrix form, this system is

$$A\frac{\partial U}{\partial x} + I\frac{\partial U}{\partial t} = 0 \tag{12.8.3}$$

where $U$, $A$ and $I$ are matrices given by

$$U = \begin{pmatrix} \rho \\ u \end{pmatrix}, \qquad A = \begin{pmatrix} u & \rho \\ c^2/\rho & u \end{pmatrix} \quad \text{and} \quad I = \begin{pmatrix} 1 & 0 \\ 0 & 1 \end{pmatrix} \tag{12.8.4}$$

The concept of characteristic curves introduced briefly in § 12.2 requires generalization if it is to be applied to quasilinear systems of first order partial differential equations (12.8.1)–(12.8.2).

It is of interest to determine how a solution evolves with time $t$. Hence, we leave the time variable unchanged and replace the space variable $x$ by some arbitrary curvilinear coordinate $\xi$ so that the semi-curvilinear coordinate transformation from $(x, t)$ to $(\xi, t')$ can be introduced by

$$\xi = \xi(x, t), \qquad t' = t \tag{12.8.5}$$

If the Jacobian of this transformation is non-zero, we can transform (12.8.3) by the following correspondence rule:

$$\frac{\partial}{\partial t} \equiv \frac{\partial \xi}{\partial t}\frac{\partial}{\partial \xi} + \frac{\partial t'}{\partial t}\cdot\frac{\partial}{\partial t'} = \frac{\partial \xi}{\partial t}\frac{\partial}{\partial \xi} + \frac{\partial}{\partial t'}$$

$$\frac{\partial}{\partial x} \equiv \frac{\partial \xi}{\partial x}\frac{\partial}{\partial \xi} + \frac{\partial t'}{\partial x}\frac{\partial}{\partial t'} = \frac{\partial \xi}{\partial x}\frac{\partial}{\partial \xi}$$

This rule transforms (12.8.3) into the form

$$I\frac{\partial U}{\partial t'} + \left(\frac{\partial \xi}{\partial t}I + \frac{\partial \xi}{\partial x}A\right)\frac{\partial U}{\partial \xi} = 0 \tag{12.8.6}$$

This equation can be used to determine $\partial U/\partial \xi$ provided the determinant of its coefficient matrix is non-zero. Obviously, this condition depends on the nature of the curvilinear coordinate curves $\xi(x, t) =$ constant, which has been kept arbitrary. We assume now that the determinant vanishes for the particular choice $\xi = \eta$ so that

$$\left| \frac{\partial \eta}{\partial t} I + \frac{\partial \eta}{\partial x} A \right| = 0 \tag{12.8.7}$$

In view of this, $\partial U/\partial \eta$ will become indeterminate on the family of curves $\eta =$ constant, and hence $\partial U/\partial \eta$ may be discontinuous across the curves $\eta =$ constant. This implies that each element of $\partial U/\partial \eta$ will be discontinuous across any of the curves $\eta =$ constant. It is then necessary to find out how these discontinuities in the elements of $\partial U/\partial \eta$ are related across the curve $\eta =$ constant. We next consider the solutions $U$ which are everywhere continuous with discontinuous derivatives $\partial U/\partial \eta$ across the particular curve $\eta =$ constant $= \eta_0$. Since $U$ is continuous, elements of the matrix $A$ are not discontinuous across $\eta = \eta_0$ so that $A$ can be determined in the neighborhood of a point $P$ on $\eta = \eta_0$. And since $\partial/\partial t'$ represents differentiation along the curves $\eta =$ constant, $\partial U/\partial t'$ is continuous everywhere and hence it is continuous across the curve $\eta = \eta_0$ at $P$.

In view of all the above facts, it follows that differential equation (12.8.6) across the curve $\xi = \eta = \eta_0$ at $P$ gives

$$\left( \frac{\partial \eta}{\partial t} I + \frac{\partial \eta}{\partial x} A \right)_P \left[ \frac{\partial U}{\partial \eta} \right]_P = 0 \tag{12.8.8}$$

where $[f]_P = f(P+) - f(P-)$ denotes the discontinuous jump in the quantity $f$ across the curve $\eta = \eta_0$, and $f(P-)$ and $f(P+)$ represent the values to the immediate left and immediate right of the curve at $P$. Since $P$ is any arbitrary point on the curve, $\partial/\partial \eta$ denotes the differentiation normal to the curves $\eta =$ constant so that Eq. (12.8.8) can be regarded as the compatibility conditions satisfied by $\partial U/\partial \eta$ on either side of and normal to these curves in the $(x, t)$ plane.

Obviously, (12.8.8) is a homogeneous system of equations for the two jump quantities $[\partial U/\partial \eta]$. Therefore, for the existence of a non-trivial solution, the coefficient determinant must vanish, that is,

$$\left| \frac{\partial \eta}{\partial t} I + \frac{\partial \eta}{\partial x} A \right| = 0 \tag{12.8.9}$$

However, along the curves $\eta =$ constant, we have

$$0 = d\eta = \eta_t + \left( \frac{dx}{dt} \right) \eta_x \tag{12.8.10}$$

so that these curves have the slope

$$\frac{dx}{dt} = -\frac{\eta_t}{\eta_x} = \lambda \text{ (say)} \tag{12.8.11}$$

Consequently, Eqs. (12.8.9) and (12.8.8) can be expressed in terms of $\lambda$ in the form

$$|A-\lambda I| = 0 \tag{12.8.12}$$

$$(A-\lambda I)\left[\frac{\partial U}{\partial \eta}\right] = 0 \tag{12.8.13}$$

where $\lambda$ represents the eigenvalues of the matrix $A$, and $[\partial U/\partial \eta]$ are proportional to the corresponding right eigenvectors of $A$.

Since $A$ is a $2\times 2$ matrix, it must have two eigenvalues. If these are real and distinct, integration of (12.8.11) leads to two distinct families of real curves $\Gamma_1$ and $\Gamma_2$ in the $(x, t)$ plane:

$$\Gamma_r\colon \frac{dx}{dt} = \lambda_r, \qquad r = 1, 2 \tag{12.8.14ab}$$

The families of curves $\Gamma_r$ are called the *characteristic curves* of the system (12.8.3). Any one of these families of curves $\Gamma_r$ may be chosen for the curvilinear coordinate curves $\eta = \text{constant}$. The eigenvalues $\lambda_r$ have the dimensions of a velocity, and the $\lambda_r$ associated with each family will then be the velocity of propagation of the matrix column vector $[\partial U/\partial \eta]$ along the curves $\Gamma_r$ belonging to that family.

In this particular case, the eigenvalues $\lambda$ are determined by (12.8.12), that is,

$$\begin{vmatrix} u-\lambda & \rho \\ c^2/\rho & u-\lambda \end{vmatrix} = 0 \tag{12.8.15}$$

so that

$$\lambda = \lambda_r = u \pm c, \qquad r = 1, 2 \tag{12.8.16ab}$$

Consequently, the families of the characteristic curves $\Gamma_r$ $(r = 1, 2)$ defined by (12.8.14ab) become

$$\Gamma_1\colon \frac{dx}{dt} = u + c, \quad \text{and} \quad \Gamma_2\colon \frac{dx}{dt} = u - c \tag{12.8.17ab}$$

In physical terms, these results indicate that disturbances propagate with the sum of the velocities of the fluid and sound along the family of curves $\Gamma_1$. In the second family $\Gamma_2$, they propagate with the difference of the fluid velocity, $u$ and the sound velocity, $c$.

The right eigenvectors $\mu_r \equiv \begin{pmatrix} \mu_r^{(1)} \\ \mu_r^{(2)} \end{pmatrix}$ are solutions of the equations

$$(A - \lambda_r I)\mu_r = 0, \qquad r = 1, 2 \tag{12.8.18ab}$$

or

$$\begin{pmatrix} u-\lambda_r & \rho \\ c^2/\rho & u-\lambda_r \end{pmatrix}\begin{pmatrix} \mu_r^{(1)} \\ \mu_r^{(2)} \end{pmatrix} = 0, \qquad r = 1, 2 \tag{12.8.19ab}$$

This result combined with (12.8.13) gives

$$\begin{pmatrix}[\rho_\eta] \\ [u_\eta]\end{pmatrix} = \begin{pmatrix}\mu_r^{(1)} \\ \mu_r^{(2)}\end{pmatrix} = \alpha\begin{pmatrix}1 \\ \pm c/\rho\end{pmatrix}, \qquad r = 1, 2 \tag{12.8.20}$$

where $\alpha$ is a constant.

In other words, across a wavefront in the $\Gamma_1$ family of characteristic curves

$$\frac{[\partial \rho/\partial \eta]}{1} = \frac{[\partial u/\partial \eta]}{c/\rho} \tag{12.8.21}$$

and across a wavefront in the $\Gamma_2$ family of characteristic curves

$$\frac{[\partial \rho/\partial \eta]}{1} = \frac{[\partial u/\partial \eta]}{-c/\rho} \tag{12.8.22}$$

where $c$ and $\rho$ have values appropriate to the wavefront.

The above method of characteristics can be applied to a more general system

$$\frac{\partial U}{\partial t} + A\frac{\partial U}{\partial x} = 0 \tag{12.8.23}$$

where $U$ is an $n \times 1$ matrix with elements $u_1, u_2, \ldots, u_n$ and $A$ is an $n \times n$ matrix with elements $a_{ij}$. An argument similar to that given above leads to $n$ eigenvalues of (12.8.13). If these eigenvalues are real and distinct, integration of equations (12.8.14ab) with $r = 1, 2, \ldots, n$ gives $n$ distinct families of real curves $\Gamma_r$ in the $(x, t)$ plane so that

$$\Gamma_r\colon\ \frac{dx}{dt} = \lambda_r, \qquad r = 1, 2, \ldots, n \tag{12.8.24}$$

When the eigenvalues $\lambda_r$ of $A$ are all real and distinct, there are $n$ distinct linearly independent right eigenvectors $\mu_r$ of $A$ satisfying the equation

$$A\mu_r = \lambda_r\mu_r$$

where $\mu_r$ is an $n \times 1$ matrix with elements $\mu_r^{(1)}, \mu_r^{(2)}, \ldots, \mu_r^{(n)}$. Then across a wavefront belonging to the $\Gamma_r$ family of characteristics, it turns out that

$$\frac{[\partial u_1/\partial \eta]}{\mu_r^{(1)}} = \frac{[\partial u_2/\partial \eta]}{\mu_r^{(2)}} = \cdots = \frac{[\partial u_n/\partial \eta]}{\mu_r^{(n)}} \tag{12.8.25}$$

where the elements of $\mu_r$ are known on the wavefront.

In order to introduce the Riemann invariants, we form the linear combination of the eigenvectors $(\pm c/\rho, 1)$ with Eqs. (12.8.1)–(12.8.2) to

obtain

$$\pm \frac{c}{\rho}(\rho_t + \rho u_x + u\rho_x) + \left(u_t + uu_x + \frac{c^2}{\rho}\rho_x\right) = 0 \qquad (12.8.26)$$

We use $\partial u/\partial \rho = \pm c/\rho$ from (12.8.21)–(12.8.22) and rewrite (12.8.26) as

$$\pm \frac{c}{\rho}[\rho_t + (u \pm c)\rho_x] + [u_t + (u \pm c)u_x] = 0 \qquad (12.8.27)$$

In view of (12.8.17ab), (12.8.27) becomes

$$du \pm \frac{c}{\rho} d\rho = 0 \quad \text{on} \quad \Gamma_r, \qquad r = 1, 2 \qquad (12.8.28)$$

or

$$d[F(\rho) \pm u] = 0 \quad \text{on} \quad \Gamma_r \qquad (12.8.29)$$

where

$$F(\rho) = \int_{\rho_0}^{\rho} \frac{c(\rho)}{\rho} d\rho \qquad (12.8.30)$$

Integration of (12.8.29) gives

$$F(\rho) + u = 2r \quad \text{on} \quad \Gamma_1; \qquad F(\rho) - u = 2s \quad \text{on} \quad \Gamma_2 \quad (12.8.31\text{ab})$$

where $2r$ and $2s$ are constants of integration on $\Gamma_1$ and $\Gamma_2$ respectively.

The quantities $r$ and $s$ are called the *Riemann invariants*. As stated above, $r$ is an arbitrary constant on characteristics $\Gamma_1$ and hence, in general, $r$ will vary on each $\Gamma_2$. Similarly, $s$ is constant on each $\Gamma_2$ but will vary on $\Gamma_1$. It is natural to introduce $r$ and $s$ as new curvilinear coordinates. Since $r$ is constant on $\Gamma_1$, $s$ can be treated as the parameter on $\Gamma_1$. Similarly, $r$ can be regarded as the parameter on $\Gamma_2$. Then $dx = (u \pm c)\, dt$ on $\Gamma_r$ implies

$$\frac{dx}{ds} = (u + c)\frac{dt}{ds} \quad \text{on} \quad \Gamma_1 \qquad (12.8.32)$$

$$\frac{dx}{dr} = (u - c)\frac{dt}{dr} \quad \text{on} \quad \Gamma_2 \qquad (12.8.33)$$

In fact, $r$ is a constant on $\Gamma_1$ and $s$ is a constant on $\Gamma_2$. Therefore, the derivatives in these equations are really partial derivations with respect to $s$ and $r$ so that we can rewrite them as

$$\frac{\partial x}{\partial s} = (u + c)\frac{\partial t}{\partial s} \qquad (12.8.34)$$

$$\frac{\partial x}{\partial r} = (u - c)\frac{\partial t}{\partial r} \qquad (12.8.35)$$

These two first order PDE's can, in general, be solved for $x = x(r, s)$,

$t = t(r, s)$ and then, by inversion, $r$ and $s$ as functions $x$ and $t$ can be obtained. Once this is done, we use (12.8.31ab) to determine $u(x, t)$ and $\rho(x, t)$ in terms of $r$ and $s$ as

$$u(x, t) = r - s, \qquad F(\rho) = r + s \tag{12.8.36ab}$$

When one of the Riemann invariants $r$ or $s$ is constant throughout the flow, the corresponding solution is tremendously simplified. The solutions are known as *simple wave motions* representing simple waves in one direction only. The generating mechanisms of simple waves with their propagation laws can be illustrated by the so called *piston problem* in gas dynamics.

EXAMPLE 12.8.1. Determine the Riemann invariants for a polytropic gas characterized by the law $p = k\rho^{\gamma}$ where $k$ and $\gamma$ are constants.

In this case

$$c^2 = \frac{dp}{d\rho} = k\gamma\rho^{\gamma-1}, \qquad F(\rho) = \int_0^{\rho} \frac{c(\rho)}{\rho}\, d\rho = \frac{2c}{\gamma - 1}$$

Hence the Riemann invariants are

$$\frac{2}{\gamma - 1} c \pm u = (2r, 2s) \quad \text{on} \quad \Gamma_r \tag{12.8.37ab}$$

It is also possible to express the dependent variables $u$ and $c$ in terms of the Riemann invariants. It turns out that

$$u = r - s, \qquad c = \frac{\gamma - 1}{2}(r + s) \tag{12.8.38ab}$$

EXAMPLE 12.8.2. (The piston problem in a polytropic gas). The problem is to determine how a simple wave is produced by the prescribed motion of a piston in the closed end of a semi-infinite tube filled with gas.

This is a one dimensional unsteady problem in gas dynamics. We assume that the gas is initially at rest with a uniform state $u = 0$, $\rho = \rho_0$, and $c = c_0$. The piston starts from rest at the origin and is allowed to withdraw from the tube with a variable velocity for a time $t_1$, after which the velocity of withdrawal remains constant. The piston path is shown by a dotted curve in Fig. 12.8.1. In the $(x, t)$ plane, the path of the piston is given by $x = X(t)$ with $X(0) = 0$. The fluid velocity $u$ is equal to the piston velocity $\dot{X}(t)$ on the piston $x = X(t)$ which will be used as the boundary condition on the piston.

The initial state of the gas is given by $u = u_0$, $\rho = \rho_0$, $c = c_0$ at $t = 0$ in $x \geqslant 0$. The characteristic line $\Gamma_0$ that bounds it and passes through the

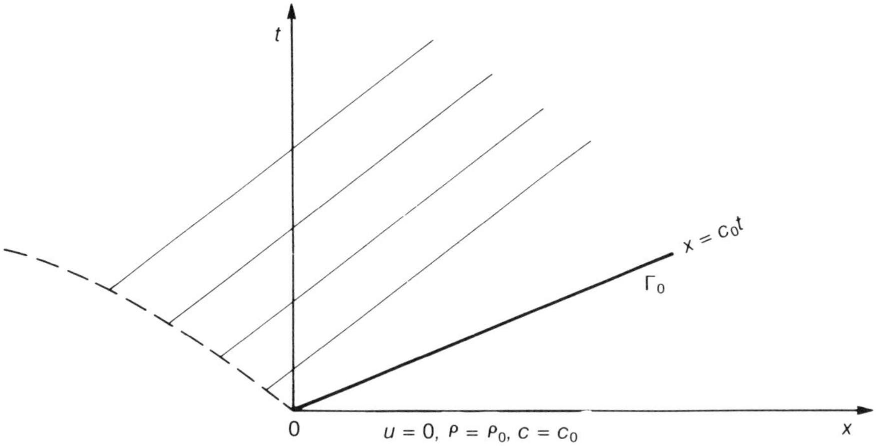


Figure 12.8.1. Simple waves generated by the motion of a piston.

origin is determined by the equation

$$\frac{dx}{dt} = (u + c)_{t=0} = c_0$$

so that the equation of $\Gamma_0$ is $x = c_0 t$.

In view of the uniform initial state, all the $\Gamma_2$ characteristics start on the $x$-axis so that the Riemann invariants $s$ in (12.8.37b) must be constant and of the form

$$\frac{2c}{\gamma - 1} - u = \frac{2c_0}{\gamma - 1} \tag{12.8.39}$$

or

$$u = \frac{2(c - c_0)}{\gamma - 1}, \qquad c = c_0 + \frac{\gamma - 1}{2} u \tag{12.8.40ab}$$

The characteristics $\Gamma_1$ meeting the piston are given by

$$\frac{2c}{\gamma - 1} + u = 2r \qquad \text{on each } \Gamma_1\text{: } \frac{dx}{dt} = u + c \tag{12.8.41}$$

which is, by (12.8.40) which holds everywhere,

$$u = \text{constant on } \Gamma_1\text{: } \frac{dx}{dt} = c_0 + \frac{1}{2}(\gamma + 1)u \tag{12.8.42}$$

Since the flow is continuous with no shocks, $u = 0$ and $c = c_0$ ahead of and on $\Gamma_0$ which separates those $\Gamma_1$ meeting the $x$-axis from those meeting the piston. The family of lines $\Gamma_1$ through the origin has the equation $dx/dt = \xi$ where $\xi$ is a parameter with $\xi = c_0$ on $\Gamma_0$. The $\Gamma_1$ characteristics are also defined by $dx/dt = u + c$ so that $\xi = u + c$. Hence elimination of $c$ from (12.8.40b) gives

$$u = \left(\frac{2}{\gamma+1}\right)(\xi - c_0) \tag{12.8.43}$$

Substituting this value of $u$ in (12.8.40b) we obtain

$$c = \left(\frac{\gamma-1}{\gamma+1}\right)\xi + \frac{2c_0}{\gamma+1} \tag{12.8.44}$$

It follows from $c^2 = \gamma k \rho^{\gamma-1}$ and (12.8.40b) with the initial data, $\rho = \rho_0$, $c = c_0$ that

$$\rho = \rho_0\left[1 + \frac{\gamma-1}{2c_0}u\right]^{2/(\gamma-1)} \tag{12.8.45}$$

With $\xi = x/t$, results (12.8.43) through (12.8.45) give the complete solution of the piston problem in terms of $x$ and $t$.

Finally, the equation of the characteristic line $\Gamma_1$ is found by integrating the second equation (12.8.42) and using the boundary condition on the piston. When a $\Gamma_1$ intersects the piston path at time $t = \tau$, then $u = \dot{X}(\tau)$ along it, and the equation becomes

$$x = X(\tau) + \left\{c_0 + \frac{\gamma+1}{2}\dot{X}(\tau)\right\}(t - \tau) \tag{12.8.46}$$

It is noted that the family $\Gamma_1$ represents straight lines with slope $dx/dt$ increasing with velocity $u$. Consequently, the characteristics are likely to overlap on the piston, that is $\ddot{X}(\tau) > 0$ for any $\tau$. If $u$ increases, so do $c$, $\rho$, and $p$ so that instability develops. It shows that shocks will be formed in the compressive part of the disturbance.

## 12.9 THE DISCONTINUOUS SOLUTIONS AND SHOCK WAVES

The development of a non-unique solution of a nonlinear hyperbolic equation has already been discussed in connection with several different problems. In real physical situations, this non-uniqueness usually manifests itself in the formation of discontinuous solutions which propagate in the medium. Such discontinuous solutions across some surface are called *shock waves*. These waves are indeed found to occur widely in high speed flows in gas dynamics.

In order to investigate the nature of discontinuous solutions, we reconsider the nonlinear conservation equation (12.6.5), that is,

$$\frac{\partial \rho}{\partial t} + \frac{\partial q}{\partial x} = 0 \tag{12.9.1}$$

This equation has been solved under two basic assumptions: (i) There exists a functional relation between $q$ and $\rho$, that is, $q = Q(\rho)$; (ii) $\rho$ and $q$ are continuously differentiable. In some physical situations, the solution of (12.9.1) leads to breaking phenomenon. When breaking occurs, questions arise on the validity of these assumptions. To examine the formation of discontinuities, we consider the following: (a) we assume the relation $q = Q(\rho)$ but allow jump discontinuity for $\rho$ and $q$; (b) in addition to the fact that $\rho$ and $q$ are continuously differentiable, we assume that $q$ is a function of $\rho$ and $\rho_x$. One of the simplest forms is

$$q = Q(\rho) - \nu\rho_x, \qquad \nu > 0 \tag{12.9.2}$$

In case (a), we assume our conservation equation (12.6.1) still holds and has the form

$$\frac{d}{dt}\int_{x_1}^{x_2} \rho(x, t)\, dx + q(x_2, t) - q(x_1, t) = 0 \tag{12.9.3}$$

We now assume that there is a discontinuity at $x = s(t)$ where $s$ is a continuously differentiable function of $t$, and $x_1$ and $x_2$ are chosen so that $x_2 > s(t) > x_1$, and $U(t) = \dot{s}(t)$. Equation (12.9.3) can be written as

$$\frac{d}{dt}\left[\int_{x_1}^{s^-} \rho\, dx + \int_{s^+}^{x_2} \rho\, dx\right] + q(x_2, t) - q(x_1, t) = 0$$

which implies

$$\int_{x_1}^{s^-} \rho_t\, dx + \dot{s}\rho(s^-, t) + \int_{s^+}^{x_2} \rho_t\, dx - \dot{s}\rho(s^+, t) + q(x_2, t) - q(x_1, t) = 0 \tag{12.9.4}$$

where $\rho(s^-, t)$, $\rho(s^+, t)$ are the values of $\rho(x, t)$ as $x \to s$ from below and above respectively. Since $\rho_t$ is bounded in each of the intervals separately, the integrals tend to zero as $x_1 \to s^-$ and $x_2 \to s^+$. Thus, in the limit,

$$q(s^+, t) - q(s^-, t) = U\{\rho(s^+, t) - \rho(s^-, t)\} \tag{12.9.5}$$

In conventional notation of shock dynamics, this can be written as

$$q_2 - q_1 = U(\rho_2 - \rho_1) \tag{12.9.6}$$

or

$$-U[\rho] + [q] = 0 \tag{12.9.7}$$

where subscripts 1 and 2 are used to denote the values behind and ahead of the shock respectively, and [ ] denotes the discontinuous jump in the quantity involved. Equation (12.9.7) is called the *shock condition*. Thus the basic problem can be written as

$$\frac{\partial \rho}{\partial t} + \frac{\partial q}{\partial x} = 0 \quad \text{at points on continuity} \tag{12.9.8}$$

$$-U[\rho] + [q] = 0 \quad \text{at points of discontinuity} \tag{12.9.9}$$

Therefore, we have a nice correspondence

$$\frac{\partial}{\partial t} \leftrightarrow -U[\ ], \qquad \frac{\partial}{\partial x} \leftrightarrow [\ ] \tag{12.9.10ab}$$

between the differential equation and the shock condition.

It is now possible to find discontinuous solutions of (12.9.3). In any continuous part of the solution, Eq. (12.9.1) is still satisfied and the assumption $q = Q(\rho)$ remains valid. But we have $q_1 = Q(\rho_1)$ and $q_2 = Q(\rho_2)$ on the two sides of any shock, and the shock condition (12.9.6) has the form

$$U(\rho_2 - \rho_1) = Q(\rho_2) - Q(\rho_1) \tag{12.9.11}$$

EXAMPLE 12.9.1. The simplest example in which breaking occurs is

$$\rho_t + c(\rho)\rho_x = 0$$

with discontinuous initial data at $t = 0$

$$\rho = \begin{Bmatrix} \rho_2, & x < 0 \\ \rho_1, & x > 0 \end{Bmatrix} \tag{12.9.12}$$

and

$$F(x) = \begin{Bmatrix} c_2 = c_2(\rho), & x < 0 \\ c_1 = c_1(\rho), & x > 0 \end{Bmatrix} \tag{12.9.13}$$

where

$$\rho_2 > \rho_1 \quad \text{and} \quad c_2 > c_1$$

In this case, breaking will occur immediately and this can be seen from Fig. 12.9.1ab with $c'(\rho) > 0$. The multivalued region begins at the origin $\xi = 0$ and is bounded by the characteristics $x = c_1 t$ and $x = c_2 t$ with $c_1 < c_2$. This corresponds to a centered compression wave with overlapping characteristics in the $(x, t)$ plane.

On the other hand, if the initial condition is expansive with $c_2 < c_1$, there is a continuous solution obtained from (12.2.12ab) in which all values of $F(x)$ in $[c_2, c_1]$ are taken on characteristics through the origin $\xi = 0$. This corresponds to a centered fan of characteristics $x = ct$, $c_2 \leqslant c \leqslant c_1$ in the $(x, t)$ plane so that the solution has the explicit form $c = x/t$, $c_2 < x/t < c_1$. The density distribution and the expansion wave are shown in Fig. 12.9.2ab.

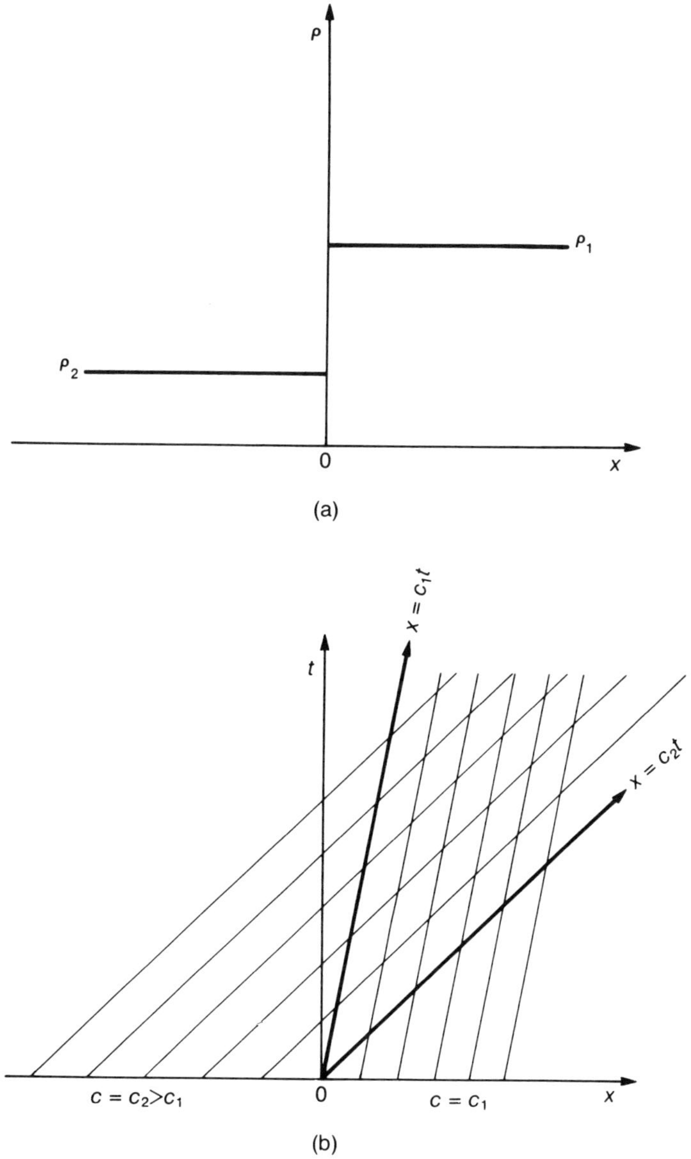


Figure 12.9.1ab. Density distribution and centered compression wave with overlapping characteristics.

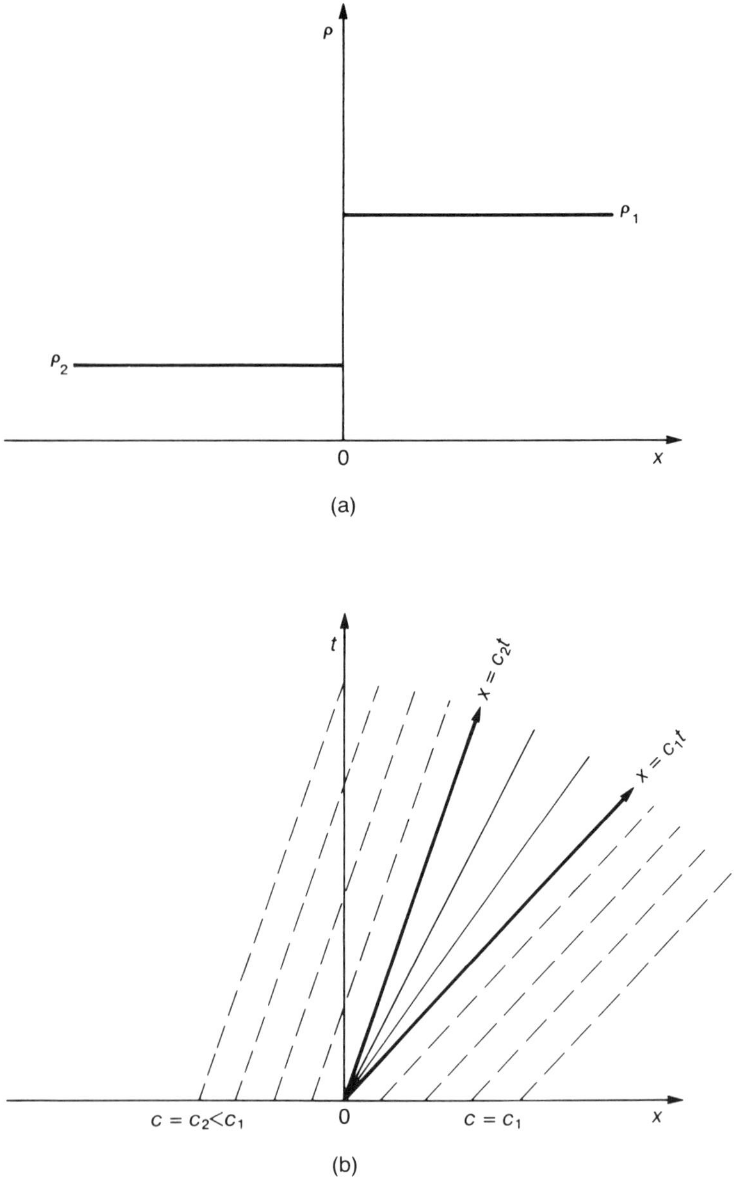


Figure 12.9.2ab. Density distribution and centered expansion wave.

In this case, the complete solution is given by

$$c = \begin{cases} c_2, & x \leqslant c_2 t \\ \dfrac{x}{t}, & c_2 t < x < c_1 t \\ c_1, & x \geqslant c_1 t \end{cases} \tag{12.9.14}$$

## 12.10 SHOCK-WAVE STRUCTURE AND BURGERS' EQUATION

In order to resolve breaking we assumed a functional relation in $\rho$ and $q$ with appropriate shock conditions. Now we investigate the case when

$$q = Q(\rho) - \nu \rho_x, \qquad \nu > 0 \tag{12.10.1}$$

It is noted that near breaking where $\rho_x$ is large, (12.10.1) gives a better approximation. With (12.10.1), the basic equation (12.9.1) becomes

$$\rho_t + c(\rho)\rho_x = \nu \rho_{xx} \tag{12.10.2}$$

where $c(\rho) = Q'(\rho)$, the second and the third terms represent the effects of nonlinearity and diffusion.

We first solve (12.10.2) for two simple cases: (i) $c(\rho) = \text{constant} = c$, and (ii) $c(\rho) \equiv 0$. In the first case, Eq. (12.10.1) becomes linear and we seek a plane wave solution

$$u(x, t) = a \exp\{i(kx - \omega t)\} \tag{12.10.3}$$

Substituting this solution into the linear equation (12.10.2), we have the dispersion relation

$$\omega = ck - i\nu k^2 \tag{12.10.4}$$

where

$$\operatorname{Im} \omega = -\nu k^2 < 0, \quad \text{since} \quad \nu > 0$$

Thus the wave profile has the form

$$u(x, t) = a e^{-\nu k^2 t} \exp[ik(x - ct)] \tag{12.10.5}$$

which represents a diffusive wave ($\operatorname{Im} \omega < 0$) with wavenumbers $k$ and phase velocity $c$ whose amplitude decays exponentially with time $t$. The decay time is given by $t_0 = (\nu k^2)^{-1}$ which becomes smaller and smaller as $k$ increases with fixed $\nu$. Thus, the waves of smaller wavelengths decay faster than waves of longer wavelengths. On the other hand, for a fixed wavenumber $k$, $t_0$ decreases as $\nu$ increases so that waves of a given wavelength attenuate faster in a medium with larger $\nu$. The quantity $\nu$ may be regarded as a measure of diffusion. Finally, after a sufficiently long time ($t \gg t_0$) only disturbances of long wavelength will survive, while all short wavelength disturbances will decay rapidly.

In the second case, (12.10.2) reduces to the linear diffusion equation

$$\rho_t = \nu\rho_{xx} \tag{12.10.6}$$

This equation with the initial data at $t = 0$

$$\rho = \begin{bmatrix} \rho_1, & x < 0 \\ \rho_2, & x > 0 \end{bmatrix}, \qquad \rho_1 > \rho_2 \tag{12.10.7}$$

can readily be solved, and the solution for $t > 0$ is

$$u(x, t) = \frac{\rho_1}{2\sqrt{\pi\nu t}} \int_{-\infty}^{0} e^{-(x-\xi)^2/4\nu t}\, d\xi + \frac{\rho_2}{2\sqrt{\pi\nu t}} \int_{0}^{\infty} e^{-(x-\xi)^2/4\nu t}\, d\xi \tag{12.10.8}$$

After some manipulation involving changes of variables of integration $(x - \xi/2\sqrt{\nu t} = \eta)$, the solution is simplified to the form

$$u(x, t) = \frac{1}{2}(\rho_1 + \rho_2) + (\rho_2 - \rho_1)\frac{1}{\sqrt{\pi}} \int_0^{x/2\sqrt{\nu t}} e^{-\eta^2}\, d\eta \tag{12.10.9}$$

$$= \frac{1}{2}(\rho_1 + \rho_2) + \frac{1}{2}(\rho_2 - \rho_1)\,\mathrm{erf}\left(\frac{x}{2\sqrt{\nu t}}\right) \tag{12.10.10}$$

This shows that the effect of the term $\nu\rho_{xx}$ is to smooth out the initial distribution like $(\nu t)^{-\frac{1}{2}}$. The solution tends to values $\rho_1$, $\rho_2$ as $x \to \mp\infty$. The absence of the term $\nu\rho_{xx}$ in (12.10.2) leads to nonlinear steepening and breaking. Indeed, Eq. (12.10.2) combines the two opposite effects of breaking and diffusion. It is also noted that the sign of $\nu$ is important. Indeed, solutions are stable or unstable according as $\nu > 0$ or $< 0$.

In order to investigate the solutions having the balance between steepening and diffusion, we seek solutions of (12.10.2) in the form

$$\rho = \rho(X), \qquad X = x - Ut \tag{12.10.11ab}$$

where $U$ is a constant to be determined.

It follows from (12.10.2) that

$$[c(\rho) - U]\rho_X = \nu\rho_{XX} \tag{12.10.12}$$

Integration of this equation gives

$$Q(\rho) - U\rho + A = \nu\rho_X \tag{12.10.13}$$

where $A$ is a constant of integration.

Integrating (12.10.13) w.r.t $X$ gives an implicit relation for $\rho(X)$ in the form

$$\frac{X}{\nu} = \int \frac{d\rho}{Q(\rho) - U\rho + A} \tag{12.10.14}$$

We would like to have a solution which tends to $\rho_1$, $\rho_2$ as $X \to \mp\infty$. If such a solution exists with $\rho_X \to 0$ as $|X| \to \infty$, the quantities $U$ and $A$ must satisfy

$$Q(\rho_1) - U\rho_1 + A = Q(\rho_2) - U\rho_2 + A = 0 \qquad (12.10.15)$$

which implies

$$U = \frac{Q(\rho_1) - Q(\rho_2)}{\rho_1 - \rho_2} \qquad (12.10.16)$$

This is exactly the same as the shock velocity obtained before.

Result (12.10.15) shows that $\rho_1$, $\rho_2$ are zeros of $Q(\rho) - U\rho + A$. In the limit $\rho \to \rho_1$ or $\rho_2$, the integral (12.10.14) diverges and so $X \to \pm\infty$. If $c'(\rho) > 0$, then $Q(\rho) - U\rho + A \leqslant 0$ in $\rho_2 \leqslant \rho \leqslant \rho_1$ and then $\rho_X \leqslant 0$ because of (12.10.11). Thus $\rho$ is decreasing monotonically from $\rho_1$ at $X = \infty$ to $\rho_2$ at $X = -\infty$ as shown in Fig. 12.10.1.

Physically, a continuous waveform carrying an increase in $\rho$ will progressively distort itself and eventually break forward and require a shock with $\rho_1 > \rho_2$ provided $c'(\rho) > 0$. It will break backward and require a shock with $\rho_1 < \rho_2$ and $c'(\rho) < 0$.

EXAMPLE 12.10.1. Obtain the solution of (12.10.2) with the initial data (12.10.7) and $Q(\rho) = \alpha\rho^2 + \beta\rho + \gamma$, $\alpha > 0$.

We write

$$Q(\rho) - U\rho + A = -\alpha(\rho_1 - \rho)(\rho - \rho_2)$$

where

$$U = \beta + \alpha(\rho_1 + \rho_2) \quad \text{and} \quad A = \alpha\rho_1\rho_2 - \gamma$$

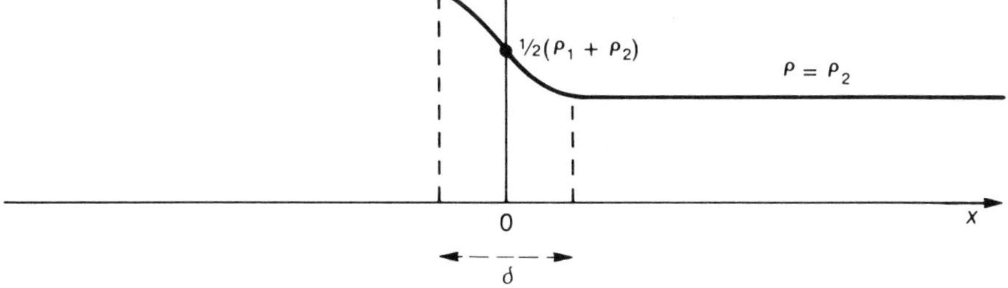


Figure 12.10.1. Shock structure and shock thickness.

Integral (12.10.14) becomes

$$\frac{X}{\nu} = -\frac{1}{\alpha}\int \frac{d\rho}{(\rho-\rho_2)(\rho_1-\rho)} = \frac{1}{\alpha(\rho_1-\rho_2)}\log\left(\frac{\rho_1-\rho}{\rho-\rho_2}\right)$$

which gives the solution

$$\rho(X) = \rho_2 + (\rho_1-\rho_2)\frac{\exp\left[\frac{\alpha X}{\nu}(\rho_2-\rho_1)\right]}{1+\exp\left[\frac{\alpha X}{\nu}(\rho_2-\rho_1)\right]} \tag{12.10.17}$$

The exponential factor in the solution indicates the existence of a transition layer of thickness $\delta$ of the order of $\nu/[\alpha(\rho_1-\rho_2)]$. This can also be referred to as the *shock thickness*. Hence, the thickness $\delta$ increases as $\rho_1 \to \rho_2$ for a fixed $\nu$. It tends to zero as $\nu \to 0$ for fixed $\rho_1$ and $\rho_2$.

In this case, the shock velocity (12.10.16) becomes

$$U = \alpha(\rho_1+\rho_2) + \beta = \frac{1}{2}(c_1+c_2) \tag{12.10.18}$$

where $c(\rho) = Q'(\rho)$, $c_1 = c(\rho_1)$, and $c_2 = c(\rho_2)$.

We multiply (12.10.2) by $c'(\rho)$ and simplify to deduce

$$ct + cc_x = \nu c_{xx} - \nu c''(\rho)\rho_x^2 \tag{12.10.19}$$

If $Q(\rho)$ is a quadratic expression in $\rho$, then $c(\rho) = Q'(\rho)$ becomes linear in $\rho$ and $c''(\rho) = 0$. Thus, (12.10.19) leads to *Burgers' equation* replacing $c$ with $u$

$$u_t + uu_x = \nu u_{xx} \tag{12.10.20}$$

This equation incorporates the combined effects of nonlinearity and diffusion. It is the simplest nonlinear model equation for diffusive waves in fluid dynamics. Using the Cole–Hopf transformation

$$u = -2\nu\frac{\phi_x}{\phi} \tag{12.10.21}$$

Burgers' equation can be solved exactly and the opposite effects of nonlinearity and diffusion can be investigated in some detail.

We introduce the transformation in two steps. First, we write $u = \psi_x$ so that (12.10.20) can readily be integrated to obtain

$$\psi_t + \frac{1}{2}\psi_x^2 = \nu\psi_{xx} \tag{12.10.22}$$

The next step is to introduce $\psi = -2\nu \log\phi$ to transform this equation into the so called diffusion equation

$$\phi_t = \nu\phi_{xx} \tag{12.10.23}$$

This equation was solved in earlier chapters. We simply quote the solution of the initial-value problem of (12.10.23) with the initial data

$$\phi(x, 0) = \Phi(x), \qquad -\infty < x < \infty \tag{12.10.24}$$

The solution for $\phi$ is

$$\phi(x, t) = \frac{1}{2\sqrt{\pi\nu t}} \int_{-\infty}^{\infty} \Phi(\zeta) \exp\left[-\frac{(x-\zeta)^2}{4\nu t}\right] d\zeta \tag{12.10.25}$$

where $\Phi(\zeta)$ can be written in terms of the initial value $u(x, 0) = F(x)$ by using (12.10.21). It turns out that, at $t = 0$,

$$\phi(x) = \Phi(x) = \exp\left\{-\frac{1}{2\nu}\int_0^x F(\alpha)\, d\alpha\right\} \tag{12.10.26}$$

It is then convenient to write down $\phi(x, t)$ in the form

$$\phi(x, t) = \frac{1}{2\sqrt{\pi\nu t}} \int_{-\infty}^{\infty} \exp\left(-\frac{f}{2\nu}\right) d\zeta \tag{12.10.27}$$

where

$$f(\zeta, x, t) = \int_0^{\zeta} F(\alpha)\, d\alpha + \frac{(x-\zeta)^2}{2t} \tag{12.10.28}$$

Consequently

$$\phi_x(x, t) = -\frac{1}{4\nu\sqrt{\pi\nu t}} \int_{-\infty}^{\infty} \frac{(x-\zeta)}{t} \exp\left(-\frac{f}{2\nu}\right) d\zeta \tag{12.10.29}$$

Therefore, the solution for $u$ follows from (12.10.21) and has the form

$$u(x, t) = \frac{\displaystyle\int_{-\infty}^{\infty} \left(\frac{x-\zeta}{t}\right) \exp\left(-\frac{f}{2\nu}\right) d\zeta}{\displaystyle\int_{-\infty}^{\infty} \exp\left(-\frac{f}{2\nu}\right) d\zeta} \tag{12.10.30}$$

Although this is the exact solution of Burgers' equation, physical interpretation can hardly be given unless a suitable simple form of $F(x)$ is specified. Even in many cases, finding an exact evaluation of the integrals in (12.10.30) is almost a formidable task. It is then necessary to resort to asymptotic methods. Before we deal with asymptotic analysis, the following example may be considered for an investigation of shock formation.

EXAMPLE 12.10.2. Find the solution of Burgers' equation with physical significance for the case

$$F(x) = \begin{bmatrix} A\,\delta(x), & x<0 \\ 0, & x>0 \end{bmatrix}$$

We first find

$$f(\zeta, x, t) = A \int_{0+}^{\zeta} \delta(\alpha)\, d\alpha + \frac{(x-\zeta)^2}{2t}$$

$$= \begin{bmatrix} \dfrac{(x-\zeta)^2}{2t} - A, & \zeta<0 \\ \dfrac{(x-\zeta)^2}{2t}, & \zeta>0 \end{bmatrix}$$

Thus

$$\int_{-\infty}^{\infty} \frac{x-\zeta}{t} \exp\left(-\frac{f}{2\nu}\right) d\zeta = \int_{-\infty}^{0} \frac{x-\zeta}{t} \exp\left[\frac{A}{2\nu} - \frac{(x-\zeta)^2}{4\nu t}\right] d\zeta$$

$$+ \int_{0}^{\infty} \frac{x-\zeta}{t} \exp\left[-\frac{(x-\zeta)^2}{4\nu t}\right] d\zeta$$

$$= 2\nu(e^{A/2\nu} - 1) \exp\left(-\frac{x^2}{4\nu t}\right)$$

which is obtained by substitution

$$\frac{x-\zeta}{2\sqrt{\nu t}} = \alpha,$$

Similarly,

$$\int_{-\infty}^{\infty} \exp\left(-\frac{f}{2\nu}\right) d\zeta = 2\sqrt{\nu t}\left[\sqrt{\pi} + (e^{A/2\nu} - 1)\,\mathrm{erfc}\left(\frac{x}{2\sqrt{\nu t}}\right)\right]$$

where erfc $(\alpha)$ is the complementary error function defined by

$$\mathrm{erfc}\,(x) = \frac{2}{\sqrt{\pi}} \int_{x}^{\infty} e^{-\eta^2}\, d\eta \tag{12.10.31}$$

Therefore, the solution for $u(x, t)$ is

$$u(x, t) = \sqrt{\frac{\nu}{t}} \frac{(e^{A/2\nu} - 1) \exp\left(-\dfrac{x^2}{4\nu t}\right)}{\sqrt{\pi} + (e^{A/2\nu} - 1)(\sqrt{\pi}/2)\,\mathrm{erfc}\left(\dfrac{x}{2\sqrt{\nu t}}\right)} \tag{12.10.32}$$

In the limit $\nu \to \infty$, the effect of diffusion would be more significant than that of nonlinearity. Since

$$\mathrm{erfc}\left(\frac{x}{2\sqrt{\nu t}}\right) \to 0, \qquad e^{A/2\nu} \sim 1 + \frac{A}{2\nu} \quad \text{as} \quad \nu \to \infty,$$

the solution (12.10.32) tends to the limiting value

$$u(x, t) \sim \frac{A}{2\sqrt{\pi \nu t}} \exp\left(-\frac{x^2}{4\nu t}\right) \tag{12.10.33}$$

This represents the well known source solution of the classical linear heat equation $u_t = \nu u_{xx}$.

On the other hand, when $\nu \to 0$ nonlinearity would dominate over diffusion. It is expected that solution (12.10.32) tends to that of Burgers' equation as $\nu \to 0$. We next introduce the similarity variable $\eta = x/\sqrt{2At}$ to rewrite (12.10.32) in the form

$$u(x \cdot t) = \left(\frac{\nu}{t}\right)^{\frac{1}{2}} \frac{(e^{A/2\nu} - 1)\exp\left(-\frac{A\eta^2}{2\nu}\right)}{\sqrt{\pi} + (e^{A/2\nu} - 1)(\sqrt{\pi}/2)\operatorname{erfc}\left(\sqrt{\frac{A}{2\nu}}\,\eta\right)} \tag{12.10.34}$$

$$\sim \left(\frac{\nu}{t}\right)^{\frac{1}{2}} \frac{\exp\left\{\frac{A}{2\nu}(1-\eta^2)\right\}}{\sqrt{\pi} + (\sqrt{\pi}/2)\exp\left(\frac{A}{2\nu}\right)\operatorname{erfc}\left(\sqrt{\frac{A}{2\nu}}\,\eta\right)} \quad \text{as } \nu \to 0 \text{ for all } \eta \tag{12.10.35}$$

$$\sim 0 \quad \text{as} \quad \nu \to 0, \quad \text{for} \quad \eta < 0 \quad \text{and} \quad \eta > 1 \tag{12.10.36}$$

Invoking the asymptotic result

$$\operatorname{erfc}(x) \sim (2/\sqrt{\pi})\frac{e^{-x^2}}{2x} \quad \text{as} \quad x \to \infty \tag{12.10.37}$$

the solution (12.10.34) has the form, for $0 < \eta < 1$; $\nu \to 0$,

$$u(x, t) \sim \left(\frac{\nu}{t}\right)^{\frac{1}{2}} \frac{2\eta\left(\frac{A}{2\nu}\right)^{\frac{1}{2}} \exp\left\{\frac{A}{2\nu}(1-\eta^2)\right\}}{2\eta\left(\frac{A\pi}{2\nu}\right)^{\frac{1}{2}} + \exp\left\{\frac{A}{2\nu}(1-\eta^2)\right\}}$$

$$= \left(\frac{2A}{t}\right)^{\frac{1}{2}} \frac{\eta}{1 + 2\eta\left(\frac{A\pi}{2\nu}\right)^{\frac{1}{2}} \exp\left\{\frac{A}{2\nu}(\eta^2 - 1)\right\}}$$

$$\sim \left(\frac{2A}{t}\right)^{\frac{1}{2}} \eta \quad \text{as} \quad \nu \to 0$$

The final asymptotic solution as $\nu \to 0$ is

$$u(x, t) \sim \begin{bmatrix} \frac{x}{t}, & 0 < x < (2At)^{\frac{1}{2}} \\ 0, & \text{otherwise} \end{bmatrix} \tag{12.10.38}$$

This result represents a shock at $x = (2At)^{\frac{1}{2}}$ with the velocity $U = (A/2t)^{\frac{1}{2}}$. This solution $u$ has a jump from 0 to $x/t = (2A/t)^{\frac{1}{2}}$ so that the shock condition is fulfilled.

## Asymptotic Behavior of Burgers' Solution as $\nu \to 0$

We use the Kelvin stationary phase approximation method discussed in § 11.6 to examine the asymptotic behavior of Burgers' solution (12.10.30). According to this method, the significant contribution to the integrals involved in (12.10.30) comes from points of stationary phase for fixed $x$ and $t$, that is, from the roots of the equation

$$\frac{\partial f}{\partial \zeta} = F(\zeta) - \frac{(x-\zeta)}{t} = 0 \tag{12.10.39}$$

Suppose $\zeta = \xi(x, t)$ is a root. According to result (11.12.8), the integral in (12.10.30) as $\nu \to 0$ gives

$$\int_{-\infty}^{\infty} \frac{x-\zeta}{t} \exp\left(-\frac{f}{2\nu}\right) d\zeta \sim \frac{x-\xi}{t} \left\{\frac{4\pi\nu}{|f''(\xi)|}\right\}^{\frac{1}{2}} \exp\left\{-\frac{f(\xi)}{2\nu}\right\}$$

$$\int_{-\infty}^{\infty} \exp\left(-\frac{f}{2\nu}\right) d\zeta \sim \left\{\frac{4\pi\nu}{|f''(\xi)|}\right\}^{\frac{1}{2}} \exp\left\{-\frac{f(\xi)}{2\nu}\right\}$$

Therefore, the final asymptotic solution is

$$u(x, t) \sim \frac{x-\xi}{t} \tag{12.10.40}$$

where $\xi$ satisfies (12.10.39). In other words, the solution can be rewritten in the form

$$\left.\begin{aligned} u &= F(\xi) \\ x &= \xi + tF(\xi) \end{aligned}\right\} \tag{12.10.41ab}$$

This is identical with the solution of (12.2.12ab) which was obtained in § 12.2. In this case, the stationary point $\xi$ represents the characteristic variable.

Although the exact solution of Burgers' equation is a single-valued and continuous function for all time $t$, the asymptotic solution (12.10.41ab) exhibits instability. It has already been shown that (12.10.41ab) progressively distorts itself and becomes multiple-valued after sufficiently long time. Eventually, breaking will occur.

It follows from the analysis of Burgers' equation that the nonlinear and diffusion terms show opposite effects. The former introduces steepening in the solution profile, whereas the latter tends to diffuse (spread) the sharp discontinuities into a smooth profile. In view of this property, the solution represents the diffusive wave. In the context of fluid flows, $\nu$ denotes the kinematic viscosity which measures the viscous dissipation.

Finally, Burgers' equation arises in many physical problems, including one-dimensional turbulence (where this equation had its origin), sound waves in viscous media, shock waves in viscous media, waves in fluid-filled viscous elastic pipes, and magnetohydrodynamic waves in media with finite conductivity.

## 12.11 THE KORTEWEG–DE VRIES EQUATION AND SOLITONS

The dispersion relation (12.3.14) for dispersive surface waves on water of constant depth $h_0$ is

$$\begin{aligned} \omega &= (gk \tanh kh_0)^{\frac{1}{2}} \\ &= c_0 k\left(1 - \frac{1}{3}k^2h_0^2\right)^{\frac{1}{2}}, \qquad c_0 = (gh_0)^{\frac{1}{2}} \\ &\approx c_0 k\left(1 - \frac{1}{6}k^2h_0^2\right) \end{aligned} \tag{12.11.1}$$

In many physical problems, wave motions with small dispersion exhibit such a $k^2$ term in the departure of the linearized theory term from a value $c_0k$. An equation for the free surface elevation $\eta(x, t)$ with this dispersion relation is given by

$$\eta_t + c_0\eta_x + \sigma\eta_{xxx} = 0 \tag{12.11.2}$$

where $\sigma = \frac{1}{6}c_0h_0^2$ is a constant for fairly long waves. This equation is called the *linearized Korteweg–de Vries (KdV) equation* for fairly long waves moving to the positive $x$ direction only. The phase and group velocities of the waves are found from (12.11.1) and they are

$$C_p = \frac{\omega}{k} = c_0 - \sigma k^2 \tag{12.11.3}$$

$$C_g = \frac{d\omega}{dk} = c_0 - 3\sigma k^2 \tag{12.11.4}$$

It is noted that $C_g < C_p$ and the dispersion comes from the term involving $k^3$ in the dispersion relation (12.11.1), and hence from the term $\sigma\eta_{xxx}$. For sufficiently long waves ($k \to 0$), $C_p = C_g = c_0$, and waves are non-dispersive.

In 1895, Korteweg–de Vries derived the nonlinear equation for long water waves in a channel of depth $h_0$ which has the remarkable form

$$\eta_t + c_0\left(1 + \frac{3}{2}\frac{\eta}{h_0}\right)\eta_x + \sigma\eta_{xxx} = 0 \tag{12.11.5}$$

This is the simplest nonlinear model equation for dispersion waves, and combines nonlinearity and dispersion. This KdV equation arises in many physical problems, which include water waves of long wavelengths, plasma waves, and magnetohydynamic waves. Like Burgers' equation, the nonlinearity and dispersion have opposite effects on the KdV equation. The former introduces steepening of the wave profile while the latter counteract waveform steepening. The most remarkable feature is that the dispersive term in the KdV equation does allow the solitary and periodic waves which are not found in the shallow water wave theory. In Burgers' equation the nonlinear term leads to steepening which produces a shock wave, on the other hand, in the KdV equation the steepening process is balanced by dispersion to give rise to a steady solitary wave.

We now seek the steady solution of the KdV equation (12.11.5) in the form

$$\eta(x, t) = h_0 f(X), \qquad X = x - Ut \tag{12.11.6}$$

for some function $f$ and constant wave velocity $U$. We determine $f$ and $U$ by substitution of the form (12.11.6) into (12.11.5). This gives, with $\sigma = \frac{1}{6}c_0 h_0^2$,

$$\frac{1}{6}h_0^2 f''' + \frac{3}{2}ff' + \left(1 - \frac{U}{c_0}\right)f' = 0 \tag{12.11.7}$$

and then integration leads to

$$\frac{1}{6}h_0^2 f'' + \frac{3}{4}f^2 + \left(1 - \frac{U}{c_0}\right)f + A = 0$$

where $A$ is an integrating constant.

We next multiply this equation by $f'$ and integrate again to obtain

$$\frac{1}{3}h_0^2 f'^2 + f^3 + 2\left(1 - \frac{U}{c_0}\right)f^2 + 4Af + B = 0 \tag{12.11.8}$$

where $B$ is a constant of integration.

We first seek a solitary wave solution under the boundary conditions $f, f', f'' \to 0$ as $|X| \to \infty$. Therefore $A, B = 0$ and (12.11.8) assumes the form

$$\frac{1}{3}h_0^2 f'^2 + f^2(f - \alpha) = 0 \tag{12.11.9}$$

where

$$\alpha = 2\left(\frac{U}{c_0} - 1\right) \tag{12.11.10}$$

Finally, we have

$$X = \int_0^f \frac{df}{f'} = \left(\frac{h_0^2}{3}\right)^{\frac{1}{2}} \int_0^f \frac{df}{f\sqrt{(\alpha - f)}}$$

which is, by the substitution $f = \alpha \operatorname{sech}^2 \theta$,

$$X - X_0 = \left(\frac{4h_0^2}{3\alpha}\right)^{\frac{1}{2}} \theta \tag{12.11.11}$$

for some constant $X_0$ of integration.

Therefore, the solution for $f(X)$ is

$$f(X) = \alpha \operatorname{sech}^2 \left[\left(\frac{3\alpha}{4h_0^2}\right)^{\frac{1}{2}} (X - X_0)\right] \tag{12.11.12}$$

The solution $f(X)$ increases from $f = 0$ as $X \to \infty$ so that it attains a maximum value $f = f_{max} = \alpha$ at $X = 0$, and then decreases symmetrically to $f = 0$ as $X \to -\infty$ as shown in Fig. 12.11.1. These also imply that $X_0 = 0$, so that (12.11.12) becomes

$$f(X) = \alpha \operatorname{sech}^2 \left[\left(\frac{3\alpha}{4h_0^2}\right)^{\frac{1}{2}} X\right] \tag{12.11.13}$$

Therefore, the final solution is

$$\eta(x, t) = \eta_0 \operatorname{sech}^2 \left[\left(\frac{3\eta_0}{4h_0^3}\right)^{\frac{1}{2}} (x - Ut)\right] \tag{12.11.14}$$

where $\eta_0 = \alpha h_0$. This is called the *solitary wave* solution of the KdV equation for any positive constant $\eta_0$. However, it has come to be known as *soliton* since Zabusky and Kruskal coined the term in 1965. Since $\eta > 0$ for all $X$, the soliton is a wave of elevation which is symmetrical about $X = 0$. It propagates in the medium without change of shape with velocity

$$U = c_0\left(1 + \frac{\alpha}{2}\right) = c_0\left(1 + \frac{1}{2}\frac{\eta_0}{h_0}\right) \tag{12.11.15}$$

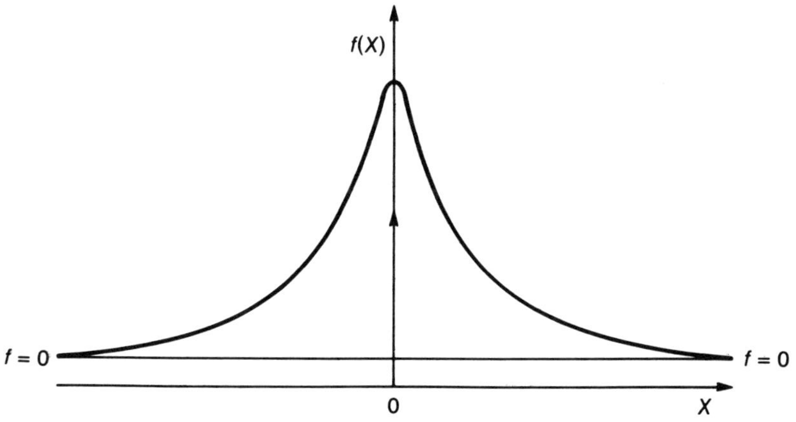


Figure 12.11.1. A soliton.

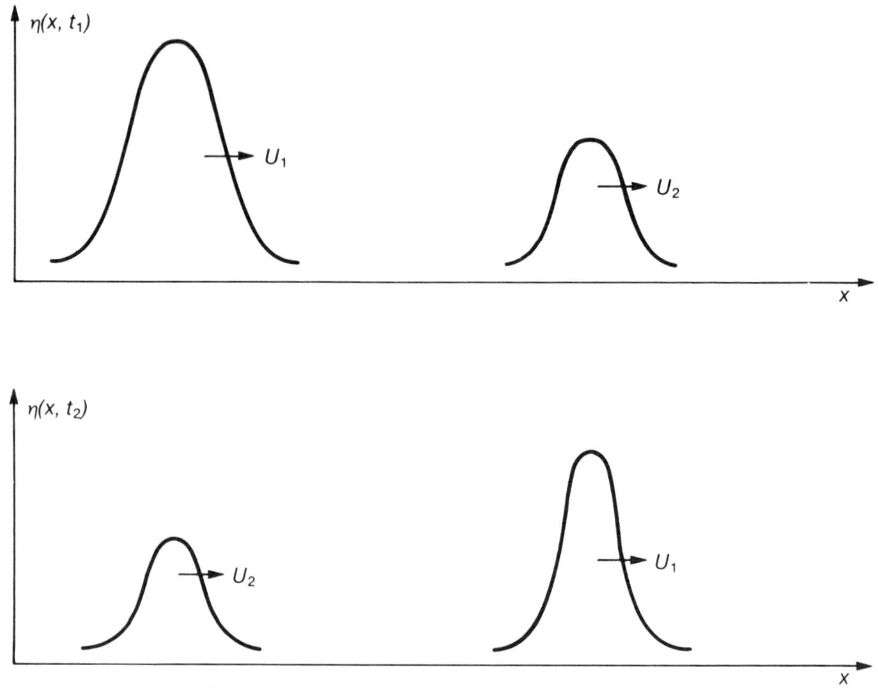


Figure 12.11.2. Interaction of two solitons ($U_1 > U_2$, $t_2 > t_1$).

which is directly proportional to the amplitude $\eta_0$. The width, $(3\eta_0/4h_0^3)^{-\frac{1}{2}}$ is inversely proportional to $\sqrt{\eta_0}$. In other words, the solitary wave propagates to the right with the velocity $U$ which is directly proportional to the amplitude, and has the width that is inversely proportional to the square root of the amplitude. Therefore, taller solitons travel faster and are narrower than the shorter (or slower) ones. They can overtake the shorter ones, and surprisingly they emerge from the interaction without change of shape as shown in Fig. 12.11.2. Indeed, solitons behave like particles involved in mutual interaction.

## General Waves of Permanent Form

We now consider the general case given by (12.11.8) which can be written

$$\frac{h_0^2}{3}f'^2 = -f^3 + \alpha f^2 - 4Af - B \equiv F(f)$$

We seek real bounded solutions for $f(X)$. Therefore $f'^2 \geqslant 0$ and varies monotonically until $f'$ is zero. Hence the zeros of the cubic $F(f)$ are crucial. For bounded solutions, all the three zeros $f_1$, $f_2$, $f_3$ must be real. Without loss of generality, we choose $f_1 = 0$ and $f_2 = \alpha$. The third zero must be negative so we set $f_3 = \alpha - \beta$ with $0 < \alpha < \beta$. Therefore, the equation for $f(X)$ is

$$\frac{1}{3} h_0^2 \left(\frac{df}{dX}\right)^2 = f(\alpha - f)(f - \alpha + \beta) \tag{12.11.16}$$

or

$$\sqrt{\frac{3}{h_0^2}}\, dX = -\frac{df}{[f(\alpha - f)(f - \alpha + \beta)]^{\frac{1}{2}}} \tag{12.11.17}$$

where

$$U = c_0\left(1 + \frac{2\alpha - \beta}{2}\right) \tag{12.11.18}$$

We put $\alpha - f = p^2$ in (12.11.17) to obtain

$$\left(\frac{3}{4h_0^2}\right)^{\frac{1}{2}} dX = \frac{dp}{[(\alpha - p^2)(\beta - p^2)]^{\frac{1}{2}}} \tag{12.11.19}$$

We next substitute $p = \sqrt{\alpha}\, q$ into (12.11.19) to transform it into the standard form

$$\left(\frac{3\beta}{4h_0^2}\right)^{\frac{1}{2}} X = \int_0^q \frac{dq}{[(1 - q^2)(1 - m^2 q^2)]^{\frac{1}{2}}} \tag{12.11.20}$$

where $m = (\alpha/\beta)^{\frac{1}{2}}$.

The right hand side is an elliptic integral of the first kind, and hence $q$, can be expressed in terms of the Jacobian $sn$ function (see Dutta and Debnath [74])

$$q = sn\left[\left(\frac{3\beta}{4h_0^2}\right)^{\frac{1}{2}} X,\, m\right] \tag{12.11.21}$$

where $m$ is the modulus of the Jacobian elliptic function $sn(z, m)$. Therefore,

$$\begin{aligned} f(X) &= \alpha\left[1 - sn^2\left\{\left(\frac{3\beta}{4h_0^2}\right)^{\frac{1}{2}} X\right\}\right] \\ &= \alpha\, cn^2\left[\left(\frac{3\beta}{4h_0^2}\right)^{\frac{1}{2}} X\right] \end{aligned} \tag{12.11.22}$$

where $cn(z, m)$ is also the Jacobian elliptic function of modulus $m$ and $cn^2(z) = 1 - sn^2(z)$.

From (12.11.20), the period $P$ is given by

$$P = 2\left(\frac{4h_0^2}{3\beta}\right)^{\frac{1}{2}} \int_0^1 \frac{dq}{[(1 - q^2)(1 - m^2 q^2)]^{\frac{1}{2}}} \tag{12.11.23}$$

$$= \frac{4h_0}{\sqrt{3\beta}} K(m) \equiv \lambda \tag{12.11.24}$$

where $K(m)$ is the complete elliptic integral of the first kind defined by

$$K(m) = \int_0^{\pi/2} (1 - m \sin^2 \theta)^{-\frac{1}{2}} \, d\theta \tag{12.11.25}$$

and $\lambda$ denotes the wavelength of the cnoidal wave.

Two limiting cases are of special interest: (i) $m \to 1$ and (ii) $m \to 0$.

When $m \to 1$ $(\alpha \to \beta)$, it is easy to show that $cn(z) \to \operatorname{sech} z$. Hence the cnoidal wave solution (12.11.22) tends to the solitary wave with the wavelength $\lambda$, given by (12.11.24) which approaches infinity because $K(1) = \infty$, $K(0) = \pi/2$.

In the other limit $m \to 0$ $(\alpha \to 0)$ $snz \to \sin z$ and $cnz \to \cos z$ so that the solution (12.11.22) becomes

$$f(X) = \alpha \cos^2 \left[ \left( \frac{3\beta}{4h_0^2} \right)^{\frac{1}{2}} X \right] \tag{12.11.26}$$

where

$$U = c_0 \left( 1 - \frac{\beta}{2} \right) \tag{12.11.27}$$

Using $\cos 2\theta = 2\cos^2 \theta - 1$, we can rewrite (12.11.26) in the form

$$f(X) = \frac{\alpha}{2} \left[ 1 + \cos \left( \frac{\sqrt{3\beta}}{h_0} \right) X \right] \tag{12.11.28}$$

We next introduce $k = \sqrt{3\beta}/h_0$ (or $\beta = \frac{1}{3}k^2h_0^2$) to simplify (12.11.28) as

$$f(X) = \frac{\alpha}{2} [1 + \cos (kx - \omega t)] \tag{12.11.29}$$

where

$$\omega = Uk = c_0 k \left( 1 - \frac{1}{6} k^2 h_0^2 \right) \tag{12.11.30}$$

This is in agreement with the classical result of the linear theory.

Remark: It is important to point out that the phase velocity (12.11.3) becomes negative for $k^2 > c_0/\sigma$ which indicates that waves propagate in the negative $x$ direction. This is in contradiction to the original assumption of forward travelling waves. Moreover, the group velocity given by (12.11.4) assumes large negative values for large $k$ so that the fine-scale features of the solution are propagated in the negative $x$ direction. The solution of (12.11.2) involves the Airy function which shows fiercely oscillatory character for large negative arguments. Consequently, this leads to a lack of continuity and a tendency to emphasize short wave components which is in contradiction to the KdV model representing fairly long waves. In order to eliminate these physically undesirable features of the KdV equation, Benjamin, Bona, and Mahony

[88] proposed a new nonlinear model equation in the form

$$\eta_t + \eta_x + \eta\eta_x - \eta_{xxt} = 0 \tag{12.11.31}$$

This is known as the *Benjamin, Bona and Mahony* (*BBM*) *equation*. The advantage of this model over the KdV equation becomes apparent when we examine their linearized forms and the corresponding solutions. The linearized form (12.11.31) gives the dispersion relation

$$\omega = \frac{k}{1+k^2} \tag{12.11.32}$$

which shows that both the phase velocity and the group velocity are bounded for all $k$, and both velocities tend to zero for large $k$. In other words, the model has the desirable feature of responding very insignificantly to short wave components that may be introduced in the initial wave form. Thus, the BBM model seems to be a preferable long wave model of physical interest. However, whether the BBM equation is a better model than the KdV equation has not yet been established.

Another important property of the KdV equation is that it satisfies the conservation law of the form

$$T_t + X_x = 0 \tag{12.11.33}$$

where $T$ is called the *density* and $X$ is called the *flux*.

If $T$ and $X$ are integrable in $-\infty < x < \infty$, and $X \to 0$ as $|x| \to \infty$, then

$$\frac{d}{dt}\int_{-\infty}^{\infty} T\,dx = -[X]_{-\infty}^{\infty} = 0$$

Therefore

$$\int_{-\infty}^{\infty} T\,dx = \text{constant}$$

so that the density is conserved.

The canonical form of the KdV equation

$$u_t - 6uu_x + u_{xxx} = 0 \tag{12.11.34}$$

can be written as

$$(u)_t + (-3u^2 + u_{xx})_x = 0$$

so that

$$T = u \quad \text{and} \quad X = -3u^2 + u_{xx} \tag{12.11.35ab}$$

If we assume that $u$ is periodic or that $u$ and its derivatives decay very rapidly as $|x| \to \infty$, then

$$\frac{d}{dt}\int_{-\infty}^{\infty} u\,dx = 0$$

This leads to the conservation of mass, that is,

$$\int_{-\infty}^{\infty} u\,dx = \text{constant} \tag{12.11.36}$$

This is often called the *time invariant function* of the solutions of the KdV equation.

The second conservation law for (12.11.33) can be obtained by multiplying it by $u$ so that

$$\left(\frac{1}{2}u^2\right)_t + \left(-2u^3 + uu_X - \frac{1}{2}u_x^2\right)_x = 0 \tag{12.11.37}$$

This gives

$$\int_{-\infty}^{\infty} \frac{1}{2}u^2\,dx = \text{constant} \tag{12.11.38}$$

This is the principle of conservation of energy.

It is well known that the KdV equation has an infinite number of polynomial conservation laws. It is generally believed that the existence of a soliton solution to the KdV equation is closely related to the existence of an infinite number of conservation laws.

## 12.12 THE NONLINEAR SCHRÖDINGER EQUATION, SOLITARY WAVES, AND THE INSTABILITY OF WATER WAVES

We first derive the one-dimensional linear Schrödinger equation from the Fourier integral representation of the plane wave solution

$$\phi(x, t) = \int_{-\infty}^{\infty} F(k)\exp[i(kx - \omega t)]\,dk \tag{12.12.1}$$

where the spectrum function $F(k)$ is determined from the given initial or boundary conditions.

We assume that the wave is slowly modulated as it propagates in a dispersive medium. For such a modulated wave, most of the energy is confined in the neighborhood of $k = k_0$ so that the dispersion relation $\omega = \omega(k)$ can be expanded about the point $k = k_0$ as

$$\omega = \omega(k) = \omega_0 + (k - k_0)\omega_0' + \frac{1}{2}(k - k_0)^2\omega_0'' + \ldots \tag{12.12.2}$$

where $\omega_0 \equiv \omega(k_0)$, $\omega_0' \equiv \omega'(k_0)$ and $\omega_0'' \equiv \omega''(k_0)$.

In view of (12.12.2), we can rewrite (12.12.1) as

$$\phi(x, t) = \psi(x, t)\exp[i(k_0 x - \omega_0 t)] \tag{12.12.3}$$

where the amplitude $\psi(x, t)$ is

$$\psi(x, t) = \int_{-\infty}^{\infty} F(k) \exp\left[i(k - k_0)x - i\left\{(k - k_0)\omega_0' + \frac{1}{2}(k - k_0)^2\omega_0''\right\}t\right] dk \tag{12.12.4}$$

Evidently, this represents the slowly varying (or modulated) part of the basic wave. A simple computation of $\psi_t$, $\psi_x$, $\psi_{xx}$ gives

$$\psi_t = -i\left\{(k - k_0)\omega_0' + \frac{1}{2}(k - k_0)^2\omega_0''\right\}\psi$$

$$\psi_x = i(k - k_0)\psi$$

$$\psi_{xx} = -(k - k_0)^2\psi$$

so that

$$i(\psi_t + \omega_0'\psi_x) + \frac{1}{2}\omega_0''\psi_{xx} = 0 \tag{12.12.5}$$

The dispersion relation associated with this equation is

$$\omega = k\omega_0' + \frac{1}{2}k^2\omega_0'' \tag{12.12.6}$$

If we choose a frame of reference moving with the linear group velocity, that is, $x^* = x - \omega_0' t$, $t^* = t$, the term involving $\psi_x$ is dropped and $\psi$ satisfies the linear Schrödinger equation, dropping the asterisks,

$$i\psi_t + \frac{1}{2}\omega_0''\psi_{xx} = 0 \tag{12.12.7}$$

We next derive the nonlinear Schrödinger equation from the nonlinear dispersion relation involving both frequency and amplitude in the most general form

$$\omega = \omega(k, a^2) \tag{12.12.8}$$

We first expand $\omega$ in a Taylor series about $k = k_0$ and $|a|^2 = 0$ in the form

$$\omega = \omega_0 + (k - k_0)\left(\frac{\partial\omega}{\partial k}\right)_{k=k_0} + \frac{1}{2}(k - k_0)^2\left(\frac{\partial^2\omega}{\partial k^2}\right)_{k=k_0} + \left(\frac{\partial\omega}{\partial |a|^2}\right)_{|a|^2=0} |a|^2 \tag{12.12.9}$$

where $\omega_0 \equiv \omega(k_0)$.

If we now replace $\omega - \omega_0$ with $i(\partial/\partial t)$ and $k - k_0$ with $-i(\partial/\partial x)$, and assume that the resulting operators act on $a$, it turns out that

$$i(a_t + \omega_0' a_x) + \frac{1}{2}\omega_0'' a_{xx} + \gamma\, |a|^2 a = 0 \tag{12.12.10}$$

where

$$\omega_0' \equiv \omega'(k_0),\ \omega_0'' = \omega''(k_0) \quad \text{and} \quad \gamma \equiv -\left(\frac{\partial \omega}{\partial |a|^2}\right)_{|a|^2=0} \quad \text{is a constant.}$$

Equation (12.12.10) is known as the *nonlinear Schrödinger* (*NLS*) *equation.* If we choose a frame of reference moving with the linear group velocity $\omega_0'$, that is, $x^* = x - \omega_0' t$ and $t^* = t$, the term involving $a_x$ will drop out from (12.12.10), and the amplitude $a(x, t)$ satisfies the normalized NLS equation, dropping the asterisks,

$$ia_t + \frac{1}{2}\omega_0'' a_{xx} + \gamma |a|^2 a = 0 \tag{12.12.11}$$

The corresponding dispersion relation is

$$\omega = \frac{1}{2}\omega_0'' k^2 - \gamma a^2 \tag{12.12.12}$$

According to the stability criterion established in § 12.5, the wave modulation is stable or unstable if $\gamma\omega_0'' < 0$ or $> 0$.

To study the solitary wave solution, it is convenient to use the NLS equation in the standard form

$$i\psi_t + \psi_{xx} + \gamma |\psi|^2 \psi = 0, \qquad -\infty < x < \infty, \qquad t \geq 0 \tag{12.12.13}$$

We seek waves of permanent form by trying the solution

$$\psi = f(X)e^{i(mX - nt)}, \qquad X = x - Ut \tag{12.12.14}$$

for some functions $f$ and constant wave speed $U$ to be determined, and $m$, $n$ are constants.

Substitution of (12.12.14) into (12.12.13) gives

$$f'' + i(2m - U)f' + (n - m^2)f + \gamma |f|^2 f = 0 \tag{12.12.15}$$

We eliminate $f'$ by setting $2m - U = 0$, and then write $n = m^2 - \alpha$ so that $f$ can be assumed to be real. Thus (12.12.15) becomes

$$f'' - \alpha f + \gamma f^3 = 0 \tag{12.12.16}$$

Multiplying this equation by $2f'$ and integrating, we find

$$f'^2 = A + \alpha f^2 - \frac{\gamma}{2} f^4 \equiv F(f) \tag{12.12.17}$$

where $F(f) \equiv (\alpha_1 - \alpha_2 f^2)(\beta_1 - \beta_2 f^2)$ so that $A = \alpha_1\beta_1$, $\alpha = -(\alpha_1\beta_2 + \alpha_2\beta_1)$, $\gamma = -2(\alpha_2\beta_2)$ and $\alpha$'s and $\beta$'s are assumed to be real and distinct.

Evidently

$$X = \int_0^f \frac{df}{\sqrt{(\alpha_1 - \alpha_2 f^2)(\beta_1 - \beta_2 f^2)}} \tag{12.12.18}$$

Putting $(\alpha_2/\alpha_1)^{\frac{1}{2}} f = u$ in this integral, we deduce the following elliptic integral of the first kind (see Dutta and Debnath [74]):

$$\sigma X = \int_0^u \frac{du}{\sqrt{(1-u^2)(1-\kappa^2 u^2)}} \tag{12.12.19}$$

where $\sigma = (\alpha_2 \beta_1)^{\frac{1}{2}}$ and $\kappa = (\alpha_1 \beta_2)/(\beta_1 \alpha_2)$

Thus, the final solution can be expressed in terms of the Jacobian *sn* function

$$u = sn(\sigma X, \kappa)$$

or

$$f(X) = \left(\frac{\alpha_1}{\alpha_2}\right)^{\frac{1}{2}} sn(\sigma X, \kappa) \tag{12.12.20}$$

In particular, when $A = 0$, $\alpha > 0$, and $\gamma > 0$, we obtain a solitary wave solution. In this case, (12.12.17) can be rewritten

$$\sqrt{\alpha}\, X = \int_0^f \frac{df}{f\left(1 - \frac{\gamma}{2\alpha} f^2\right)^{\frac{1}{2}}} \tag{12.12.21}$$

Substitution of $(\gamma/2\alpha)^{\frac{1}{2}} f = \operatorname{sech} \theta$ in this integral gives the exact solution

$$f(X) = \left(\frac{2\alpha}{\gamma}\right)^{\frac{1}{2}} \operatorname{sech}\left[\sqrt{\alpha}\,(x - Ut)\right] \tag{12.12.22}$$

This represents a solitary wave which propagates without change of shape with constant velocity $U$. Unlike the solution of the KdV equation, the amplitude and the velocity of the wave are independent parameters. It is noted that the solitary wave exists only for the unstable case ($\gamma > 0$). This means that small modulations of the unstable wavetrain lead to a series of solitary waves.

The nonlinear dispersion relation for deep water waves is

$$\omega = \sqrt{gk}\left(1 + \frac{1}{2} a^2 k^2\right) \tag{12.12.23}$$

Therefore,

$$\omega_0' = \frac{\omega_0}{2k_0}, \qquad \omega_0'' = -\frac{\omega_0}{4k_0^2} \quad \text{and} \quad \gamma = -\frac{1}{2}\omega_0 k_0^2 \tag{12.12.24}$$

and the NLS equation for deep water waves is obtained from (12.12.10) in the form

$$i\left(a_t + \frac{\omega_0}{2k_0} a_x\right) - \frac{\omega_0}{8k_0^2} a_{xx} - \frac{1}{2}\omega_0 k_0^2 |a|^2 a = 0 \tag{12.12.25}$$

The normalized form of this equation in a frame of reference moving with the linear group velocity $\omega_0'$ is

$$ia_t - \frac{\omega_0}{8k_0^2} a_{xx} = \frac{1}{2} \omega_0 k_0^2 |a|^2 a \tag{12.12.26}$$

Since $\gamma\omega_0'' = \omega_0^2/8 > 0$, this equation confirms the instability of deep water waves. This is one of the most remarkable recent results in the theory of water waves.

We next discuss the uniform solution and the solitary wave solution of (12.12.26). We look for solutions in the form

$$a(x, t) = A(X) \exp(i\gamma^2 t), \qquad X = x - \omega_0' t \tag{12.12.27}$$

and substitute this into equation (12.12.26) to obtain

$$A_{XX} = -\frac{8k_0^2}{\omega_0}\left(\gamma^2 A + \frac{1}{2}\omega_0 k_0^2 A^3\right) \tag{12.12.28}$$

We multiply this equation by $2A_X$ and then integrate to find

$$\begin{aligned} A_X^2 &= -\left(A_0^4 m'^2 + \frac{8}{\omega_0}\gamma^2 k_0^2 A^2 + 2k_0^4 A^4\right) \\ &= (A_0^2 - A^2)(A^2 - m'^2 A_0^2) \end{aligned} \tag{12.12.29}$$

where $A_0^4 m'^2$ is an integrating constant and $2k_0^4 = 1$, $m'^2 = 1 - m^2$, and $A_0^2 = 4\gamma^2/\omega_0 k_0^2(m^2 - 2)$ which relates $A_0$, $\gamma$, and $m$.

Finally, we rewrite (12.12.29) as

$$A_0^2\, dX = \frac{dA}{\left[\left(1 - \frac{A^2}{A_0^2}\right)\left(\frac{A^2}{A_0^2} - m'^2\right)\right]^{\frac{1}{2}}} \tag{12.12.30}$$

or

$$A_0(X - X_0) = \int^{s} \frac{ds}{[(1 - s^2)(s^2 - m'^2)]^{\frac{1}{2}}}, \qquad s = A/A_0$$

This can readily be expressed in terms of the Jacobian *dn* function (see Dutta and Debnath [74])

$$A = A_0 dn[A_0(X - X_0), m] \tag{12.12.31}$$

where $m$ is the modulus of the *dn* function.

In the limit $m \to 0$, $dn\, z \to 1$ and $\gamma^2 \to -\frac{1}{2}\omega_0 k_0^2 A_0^2$. Hence the solution is

$$a(x, t) = A(t) = A_0 \exp\left\{-\frac{1}{2} i\omega_0 k_0^2 A_0^2 t\right\} \tag{12.12.32}$$

On the other hand, when $m \to 1$, $dn\, z \to \operatorname{sech} z$ and $\gamma^2 \to -\frac{1}{4}\omega_0 k_0^2 A_0^2$.

Therefore, the solitary wave solution is

$$a(x, t) = A_0 \operatorname{sech}[A_0(x - \omega_0' t - X_0)] \exp\left[-\frac{i}{4}\omega_0 k_0^2 A_0^2 t\right] \tag{12.12.33}$$

We next use the NLS equation (12.12.26) to discuss the instability of deep water waves, which is known as the *Benjamin and Feir instability.* We consider a perturbation of (12.12.32) and write

$$a(X, t) = A(t)[1 + B(X, t)] \tag{12.12.34}$$

where $A(t)$ is the uniform solution given by (12.12.32).
Substituting (12.12.34) into (12.12.26) we obtain

$$iA_t(1 + B) + iA(t)B_t - \frac{\omega_0}{8k_0^2}A(t)B_{XX}$$
$$= \frac{1}{2}\omega_0 k_0^2 A_0^2[(1 + B) + BB^*(1 + B) + (B + B^*)B + (B + B^*)]A$$

where $B^*$ is the complex conjugate of $B$.
Neglecting squares of $B$, it follows that

$$iB_t - \frac{\omega_0}{8k_0^2}B_{XX} = \frac{1}{2}\omega_0 k_0^2 A_0^2(B + B^*) \tag{12.12.35}$$

We now seek a solution of the form

$$B(X, t) = B_1 e^{\Omega t + i\kappa X} + B_2 e^{\Omega t - i\kappa X} \tag{12.12.36}$$

where $B_1$, $B_2$ are complex constants, $\kappa$ is a real wavenumber, and $\Omega$ is a growth rate (possibly complex) to be determined.
Substitution of $B$ into (12.12.35) leads to the pair of coupled equations

$$\left(i\Omega + \frac{\omega_0\kappa^2}{8k_0^2}\right)B_1 - \frac{1}{2}\omega_0 k_0^2 A_0^2(B_1 + B_2^*) = 0 \tag{12.12.37}$$

$$\left(i\Omega + \frac{\omega_0\kappa^2}{8k_0^2}\right)B_2 - \frac{1}{2}\omega_0 k_0^2 A_0^2(B_1^* + B_2) = 0 \tag{12.12.38}$$

It is convenient to take the complex conjugate of (12.12.38) so that it assumes the form

$$\left(-i\Omega + \frac{\omega_0\kappa^2}{8k_0^2}\right)B_2^* - \frac{1}{2}\omega_0 k_0^2 A_0^2(B_1 + B_2^*) = 0 \tag{12.12.39}$$

The pair of linear homogeneous equations (12.12.37) and (12.12.39) for $B_1$ and $B_2^*$ admits a nontrivial solution provided

$$\begin{vmatrix} i\Omega + \dfrac{\omega_0\kappa^2}{8k_0^2} - \dfrac{1}{2}\,\omega_0 k_0^2 A_0^2 & -\dfrac{1}{2}\,\omega_0 k_0^2 A_0^2 \\ -\dfrac{1}{2}\,\omega_0 k_0^2 A_0^2 & -i\Omega + \dfrac{\omega_0\kappa^2}{8k_0^2} - \dfrac{1}{2}\,\omega_0 k_0^2 A_0^2 \end{vmatrix} = 0$$

or

$$\Omega^2 = \frac{\omega_0^2\kappa^2}{8k_0^2}\left(k_0^2A_0^2 - \frac{\kappa^2}{8k_0^2}\right) \tag{12.12.40}$$

The growth rate $\Omega$ is purely imaginary or real and positive depending on whether $\kappa^2/k_0^2 >$ or $< 8k_0^2A_0^2$. The former case corresponds to an oscillatory solution for $B$, and the latter case represents the Benjamin and Feir instability with the criterion

$$\kappa^{*2} < 8k_0^2A_0^2 \tag{12.12.41}$$

where $\kappa^* = \kappa/k_0$ is the non-dimensional wavenumber. The range of instability

$$0 < \kappa^* < \kappa_c^* = 2\sqrt{2}\,k_0A_0 \tag{12.12.42}$$

Since $\Omega$ is a function of $\kappa^*$, the maximum instability occurs at $\kappa^* = \kappa_{\max}^* = 2k_0A_0$ with a maximum growth rate given by

$$(\text{Re}\,\Omega)_{\max} = \frac{1}{2}\,\omega_0 k_0^2 A_0^2 \tag{12.12.43}$$

In 1967, Benjamin and Feir (see Whitham [99]) confirmed these remarkable results both theoretically and experimentally.

## Conservation Laws for the NLS Equation

Zakharov and Shabat [101] proved that Eq. (12.12.13) has an infinite number of polynomial conservation laws. They have the form of an integral, with respect to $x$, of a polynomial expression in terms of the function $\psi(x, t)$ and its derivatives with respect to $x$. These laws are somewhat similar to those already proved for the KdV equation. Therefore, the proofs of the conservation laws are based on similar assumptions used in the context of the KdV equation.

We prove here three conservation laws for the nonlinear Schrödinger equation (12.12.13):

$$\int_{-\infty}^{\infty} |\psi|^2\,dx = \text{constant} = C_1 \tag{12.12.44}$$

$$\int_{-\infty}^{\infty} i(\psi\bar{\psi}_x - \bar{\psi}\psi_x)\,dx = \text{constant} = C_2 \tag{12.12.45}$$

$$\int_{-\infty}^{\infty} \left(|\psi_x|^2 - \frac{1}{2}\gamma\,|\psi|^4\right) dx = \text{constant} = C_3 \tag{12.12.46}$$

We multiply (12.12.13) by $\bar{\psi}$ and its complex conjugate by $\psi$ and subtract the latter from the former to obtain

$$i\frac{d}{dt}(\psi\bar{\psi}) + \frac{d}{dx}(\psi_x\bar{\psi} - \bar{\psi}_x\psi) = 0 \tag{12.12.47}$$

where the bar denotes the complex conjugate.

Integration with respect to $x$ in $-\infty < x < \infty$ gives

$$i\frac{d}{dt}\int_{-\infty}^{\infty} |\psi|^2\,dx = 0$$

This proves result (12.12.44).

We multiply (12.12.13) by $\bar{\psi}_x$ and its complex conjugate by $\psi_x$ and then add them to obtain

$$i(\psi_t\bar{\psi}_x - \bar{\psi}_t\psi_x) + (\psi_{xx}\bar{\psi}_x + \bar{\psi}_{xx}\psi_x) + \nu\,|\psi|^2(\psi\bar{\psi}_x + \bar{\psi}\psi_x) = 0 \tag{12.12.48}$$

We differentiate (12.12.13) and its complex conjugate with respect to $x$, and multiply the former by $\bar{\psi}$ and the latter by $\psi$ and then add them together. This leads to the result

$$i(\bar{\psi}\psi_{xt} - \psi\bar{\psi}_{xt}) + (\psi_{xxx}\bar{\psi} + \bar{\psi}_{xxx}\psi) + \nu[\bar{\psi}(|\psi|^2\,\psi)_x + \psi(|\psi|^2\,\bar{\psi})_x] = 0 \tag{12.12.49}$$

If we subtract (12.12.49) from (12.12.48) and then simplify, it turns out that

$$\begin{aligned} i\frac{d}{dt}&(\bar{\psi}\psi_x - \psi\bar{\psi}_x) \\ &= \frac{d}{dx}(\psi_x\bar{\psi}_x) + \frac{d}{dx}(\bar{\psi}\psi_{xx} + \psi\bar{\psi}_{xx}) - \frac{d}{dx}(\psi_x\bar{\psi}_x) + \nu\frac{d}{dx}(\psi\bar{\psi})^2 \\ &= \frac{d}{dx}(\bar{\psi}\psi_{xx} + \psi\bar{\psi}_{xx}) + \nu\frac{d}{dx}|\psi|^4 \end{aligned}$$

Integrating this result with respect to $x$, we obtain

$$\frac{d}{dt}\int_{-\infty}^{\infty} i(\bar{\psi}\psi_x - \psi\bar{\psi}_x)\,dx = 0$$

This proves the second result.

We multiply (12.12.13) by $\bar{\psi}_t$ and its complex conjugate by $\psi_t$ and add the resulting equations to derive

$$(\bar{\psi}_t\psi_{xx} + \bar{\psi}_{xx}\psi_t) + \nu(\psi^2\bar{\psi}\bar{\psi}_t + \bar{\psi}^2\psi\psi_t) = 0$$

or

$$\frac{d}{dx}(\bar{\psi}_t\psi_x + \psi_t\bar{\psi}_x) - \frac{d}{dt}(\psi_x\bar{\psi}_x) + \frac{\gamma}{2}\frac{d}{dt}(\psi^2\bar{\psi}^2) = 0$$

Integrating this with respect to $x$, we obtain

$$\frac{d}{dt}\int_{-\infty}^{\infty}\left(|\psi_x|^2-\frac{\gamma}{2}|\psi|^4\right)dx=\int_{-\infty}^{\infty}\frac{d}{dx}(\bar{\psi}_t\psi_x+\psi_t\bar{\psi}_x)\,dx=0 \quad (12.12.50)$$

This gives (12.12.46).

The above three conservation integrals have a simple physical meaning. In fact, the constants of motion $C_1$, $C_2$, and $C_3$ are related to the number of particles, the momentum, and the energy of a system governed by the nonlinear Schrödinger equation.

An analysis of this section reveals several remarkable features of the nonlinear Schrödinger equation. This equation can also be used to investigate instability phenomena in many other physical systems. Like the various forms of the KdV equation, the NLS equation arises in many physical problems, including nonlinear water waves and ocean waves, waves in plasmas, propagation of heat pulses in a solid, self-trapping phenomena in nonlinear optics, nonlinear waves in a fluid filled visco-elastic tube, and various nonlinear instability phenomena in fluids and plasmas.

## 12.13 EXERCISES

**1.** For the flow density relation $q=v\rho(1-\rho/\rho_1)$, find the solution of the traffic flow problem with the initial condition $\rho(x,0)=f(x)$ for all $x$.

If

$$f(x)=\rho_0\begin{bmatrix}\frac{1}{3}, & x\leqslant 0\\ \frac{1}{3}+\frac{5}{12}x, & 0\leqslant x\leqslant 1\\ \frac{3}{4}, & x\geqslant 1\end{bmatrix}$$

Show that

i. $\rho=\frac{1}{3}\rho_0$ along the characteristic lines $ct=3(x-x_0)$, $x_0\leqslant 0$

ii. $\rho=\frac{3}{4}\rho_0$ along the characteristic lines $ct=2(x_0-x)$, $x_0\geqslant 1$

iii. $\rho=\frac{1}{3}\rho_0+\frac{5}{12}x_0\rho_0$ along $ct\left(\frac{2}{5}-x_0\right)=\frac{6}{5}(x-x_0)$, $0\leqslant x_0\leqslant 1$

Discuss what happens at the intersection of the two lines $ct=3x$ and $ct=2(1-x)$. Draw the characteristic lines $ct$ versus $x$.

**2.** A mountain of height $h(x, t)$ is vulnerable to erosion if its slope $h_x$ is very large. If $h_t$ and $h_x$ satisfy the functional relation $h_t = -Q(h_x)$, show that $u = h_x$ is governed by the nonlinear wave equation

$$u_t + c(u)u_x = 0, \qquad c(u) = Q'(u)$$

**3.** Consider the flow of water in a river carrying some particles through the solid bed. During the sedimentation process, some particles will be deposited in the bed. Assuming that $v$ is the constant velocity of water and $\rho = \rho_f + \rho_b$ is the density where $\rho_f$ is the density of the particles carried in the fluid and $\rho_b$ is the density of the material deposited on the solid bed, the conservation law is

$$\frac{\partial \rho}{\partial t} + \frac{\partial q}{\partial x} = 0, \qquad q = v\rho_f$$

a. Show that $\rho_b$ satisfies the equation

$$\frac{\partial \rho_f}{\partial t} + c(\rho_f)\frac{\partial \rho_f}{\partial x} = 0, \qquad c(\rho_f) = \frac{v}{1 + Q'(\rho_f)}$$

where $\rho_b = Q(\rho_f)$.

b. This problem also arises in chemical engineering with a second relation between $\rho_f$ and $\rho_b$ in the form

$$\frac{\partial \rho_b}{\partial t} = k_1(\alpha - \rho_b)\rho_f - k_2(\beta - \rho_f)\rho_b$$

where $k_1$, $k_2$ represent constant reaction rates and $\alpha$, $\beta$ are constant values of the saturation levels of the particles in the solid bed and fluid respectively. Show that the propagation speed $c$ is

$$c = \frac{k_2 \beta v}{k_1 \alpha + k_2 \beta}$$

provided the densities are small.

**4.** Show that a steady solution $u(x, t) = f(\zeta)$, $\zeta = x - ct$ of Burgers' equation with the boundary conditions $f \to u_\infty^-$ or $u_\infty^+$ as $\zeta \to -\infty$ or $+\infty$ is

$$u(x, t) = \left[c - \frac{1}{2}(u_\infty^- - u_\infty^+)\tanh\left\{\frac{(u_\infty^- - u_\infty^+)}{4v}(x - ct)\right\}\right]$$

where $u_\pm^\infty = c \mp (c^2 + 2A)^{\frac{1}{2}}$ are the two roots of $f^2 - 2cf - 2A = 0$ and $A$ is a constant of integration.

**5.** Show that the transformations $x \to \gamma^{\frac{1}{3}}x$, $u \to -6\gamma^{\frac{1}{3}}u$, $t \to t$ reduce the KdV equation $u_t + uu_x + \gamma u_{xxx} = 0$ into the canonical form $u_t - 6uu_x + u_{xxx} = 0$.

Hence or otherwise, prove that the solution of the canonical equation with the boundary conditions on $u(x, t)$ and its derivatives tend to zero as $|x| \to \infty$ is

$$u(x, t) = -\frac{a^2}{2}\operatorname{sech}^2\left\{\frac{a}{2}(x - a^2 t)\right\}$$

**6.** Verify that the Riccati transformation $u = v^2 + v_x$ transforms the KdV equation so that $v$ satisfies the associated KdV equation

$$v_t - 6v^2 v_x + v_{xxx} = 0$$

For a given $u$, show that the Riccati equation $v_x + v^2 = u$ becomes the linear Schrödinger equation, $\psi_{xx} = u\psi$ (without the energy-level term) under the transformation $v = \psi_x/\psi$.

**7.** Apply the method of characteristics to solve the pair of equations

$$\frac{\partial u}{\partial t} + \frac{\partial v}{\partial x} = 0, \qquad \frac{\partial v}{\partial t} + \frac{\partial u}{\partial x} = 0$$

with the initial data

$$u(x, 0) = e^x, \qquad v(x, 0) = e^{-x}.$$

Show that the Riemann invariants are

$$2r(\alpha) = 2\cosh \alpha, \qquad 2s(\beta) = 2\sinh \beta.$$

Also, show that solutions are $(u, v) = \cosh(x - t) \pm \sinh(x + t)$.

**8.** For an anisentropic flow, the Euler equations are

$$\rho_t + (\rho u)_x = 0$$

$$u_t + uu_x + \frac{1}{\rho} p_x = 0$$

$$S_t + uS_x = 0$$

where $\rho$ is the density, $u$ is the velocity in the $x$ direction, $p$ is the pressure and $S$ is the entropy.

Show that this system has three families of characteristics.

$$\Gamma_0 : \frac{dx}{dt} = u, \qquad \Gamma_\pm : \frac{dx}{dt} = u \pm c$$

where $c^2 = \left(\frac{\partial p}{\partial \rho}\right)_S$ = constant.

Hence derive the following full set of characteristic equations

$$\frac{dS}{dt} = 0 \quad \text{on} \quad \Gamma_0$$

$$\frac{dp}{dt} \pm \rho c \frac{du}{dt} = 0 \quad \text{on} \quad \Gamma_\pm.$$

In particular, when the flow is isentropic ($S$ = constant everywhere), show that the two characteristic equations are

$$\int \frac{c(\rho)}{\rho} d\rho \pm u = \text{constant on } \Gamma_\pm : \frac{dx}{dt} = u \pm c.$$

**9.** a. Using equations (12.8.32)–(12.8.36ab), deduce a second order equation

$$t_{rs} + \phi(r+s)(t_r + t_s) = 0, \qquad \phi(r+s) = \frac{1}{2c}\left(1 + \frac{\rho}{c}\frac{dc}{d\rho}\right).$$

b. For a polytropic gas, show that

$$\phi(r+s) = \frac{\gamma+1}{4c} = \frac{\alpha}{F(\rho)}, \qquad \alpha \equiv \frac{1}{2}\left(\frac{\gamma+1}{\gamma-1}\right), \quad \text{and} \quad t_{rs} + \frac{\alpha}{r+s}(t_r + t_s) = 0$$

c. With $F(\rho) = \dfrac{2c}{\gamma - 1} = r + s$ and $u = r - s$, show that the differential equation in 9b reduces to the *Euler–Poisson–Darboux equation*

$$t_{uu} - \left(\frac{\gamma-1}{2}\right)^2\left(t_{cc} + \frac{2\alpha}{c}t_c\right) = 0$$

where $u$ and $c$ are independent variables.

**10.** Show that the KdV equation

$$u_t - 6uu_x + u_{xxx} = 0$$

satisfies the conservation law in the form $U_t + V_x = 0$ where (i) $U = u$, $V = -3u^2 + u_{xx}$, and (ii) $U = \frac{1}{2}u^2$, $V = -2u^3 + uu_x - \frac{1}{2}u_x^2$.

**11.** Show that the KdV equation

$$u_t + 6uu_x + u_{xxx} = 0$$

satisfies the conservation law

$$U_t + V_x = 0$$

where

(i) $U = u$, $V = 3u^2 + u_{xx}$

(ii) $U = \dfrac{1}{2}u^2$, $V = 2u^3 + uu_{xx} - \dfrac{1}{2}u_x^2$

(iii) $U = u^3 - \dfrac{1}{2}u_x^2$, $V = \dfrac{9}{2}u^4 + 3u^2u_{xx} + \dfrac{1}{2}u_{xx}^2 + u_xu_t$

**12.** Show that the conservation laws for the associated KdV equation

$$v_t - 6v^2v_x + v_{xxx} = 0$$

are

(i) $(v)_t + \left(\dfrac{1}{3}v^3 + v_{xx}\right)_x = 0$

(ii) $\left(\dfrac{1}{2}v^2\right)_t + \left(\dfrac{1}{4}v^4 + vv_{xx} - \dfrac{1}{2}v_x^2\right)_x = 0$

**13.** For the nonlinear Schrödinger equation (12.12.13), prove that

$$\int_{-\infty}^{\infty} T\,dx = \text{constant}$$

where (i) $T \equiv i(\bar{\psi}\psi_x - \psi\bar{\psi}_x)$ and (ii) $T \equiv |\psi_x|^2 - \frac{1}{2}\gamma\,|\psi|^4$.

**14.** Show that the conservation law for Burgers' equation (12.10.20) is

$$(u)_t + \left(\frac{1}{2}u^2 + \nu u_x\right)_x = 0$$

**15.** Show that the conservation laws for the equation

$$u_t - uu_x - u_{xxt} = 0$$

are

(i) $(u)_t - \left(\dfrac{1}{2}u^2 + u_{xt}\right)_x = 0$

(ii) $\dfrac{1}{2}\left(u^2 + u_x^2\right)_t - \left(\dfrac{1}{3}u^3 + uu_{xt}\right)_x = 0$

**16.** Show that the linear Schrödinger system

$$i\psi_t + \psi_{xx} = 0, \qquad -\infty < x < \infty, \qquad t > 0$$

$$\psi \to 0 \quad \text{as} \quad |x| \to \infty$$

$$\psi(x, 0) = \psi(x) \quad \text{with} \quad \int_{-\infty}^{\infty} |\psi|^2\,dx = 1$$

has the conservation law

$$(i\,|\psi|^2)_t + (\psi^*\psi_x - \psi\psi_x^*)_x = 0$$

and the energy integral

$$\int_{-\infty}^{\infty} |\psi|^2\,dx = 1$$

CHAPTER THIRTEEN

# Numerical and Approximation Methods

## 13.1 INTRODUCTION

The preceding chapters have been devoted to the analytical treatment of linear and nonlinear partial differential equations. Several analytical methods to find the exact analytical solution of these equations within simple domains have been discussed. The boundary and initial conditions in these problems were also relatively simple, and were expressible in simple mathematical form. In dealing with many equations arising from the modelling of physical problems, the determination of such exact solutions in a simple domain is almost a formidable task even when the boundary and/or initial data are simple. It is then necessary to resort to numerical or approximation methods in order to deal with the problems that cannot be solved analytically. In view of the wide-spread accessibility of today's high speed electronic computers, numerical and approximation methods are becoming increasingly important and useful in applications.

In this chapter some of the major numerical and approximation approaches to the solution of partial differential equations are discussed in some detail. These include numerical methods based on finite difference approximations, variational methods, and the Rayleigh–Ritz, Galerkin, and Kantorovich methods of approximation. The chapter also contain a large section on analytical treatment of variational methods and the Euler–Lagrange equations and their applications.

## 13.2 FINITE DIFFERENCE APPROXIMATIONS, ERRORS, CONSISTENCY, CONVERGENCE, AND STABILITY

The Taylor series expansion of a function $u(x, y)$ of two independent variables $x$ and $y$ is

$$u(x_i \pm h, y_j) = u_{i\pm1,j} = u_{i,j} \pm h(u_x)_{i,j} + \frac{h^2}{2!}(u_{xx})_{i,j} \pm \frac{h^3}{3!}(u_{xxx})_{i,j} + \ldots \qquad (13.2.1\text{ab})$$

$$u(x_i, y_j \pm k) = u_{i,j\pm1} = u_{i,j} \pm k(u_y)_{i,j} + \frac{k^2}{2!}(u_{yy})_{i,j} \pm \frac{k^3}{3!}(u_{yyy})_{i,j} + \ldots \qquad (13.2.2\text{ab})$$

where $u_{i,j} = u(x, y)$, $u_{i\pm1,j} = u(x \pm h, y)$, and $u_{i,j\pm1} = u(x, y \pm k)$.

We choose a set of uniformly spaced rectangles with vertices at $P_{i,j}$ with coordinates $(ih, jk)$ where $i$, $j$ are positive or negative integers or zero, as shown in Fig. 13.2.1. We denote $u(ih, jk)$ by $u_{i,j}$.

Using the above Taylor Series expansion, we write approximate expressions for $u_x$ at $P_{i,j}$ in terms of $u_{i,j}u_{i\pm1,j}$:

$$u_x = \frac{1}{h}[u(x+h, y) - u(x, y)] \sim \frac{1}{h}(u_{i+1,j} - u_{i,j}) + O(h) \qquad (13.2.3)$$

$$u_x = \frac{1}{h}[u(x, y) - u(x-h, y)] \sim \frac{1}{h}(u_{i,j} - u_{i-1,j}) + O(h) \qquad (13.2.4)$$

$$u_x = \frac{1}{2h}[u(x+h, y) - u(x-h, y)] \sim \frac{1}{2h}(u_{i+1,j} - u_{i-1,j}) + O(h^2) \qquad (13.2.5)$$

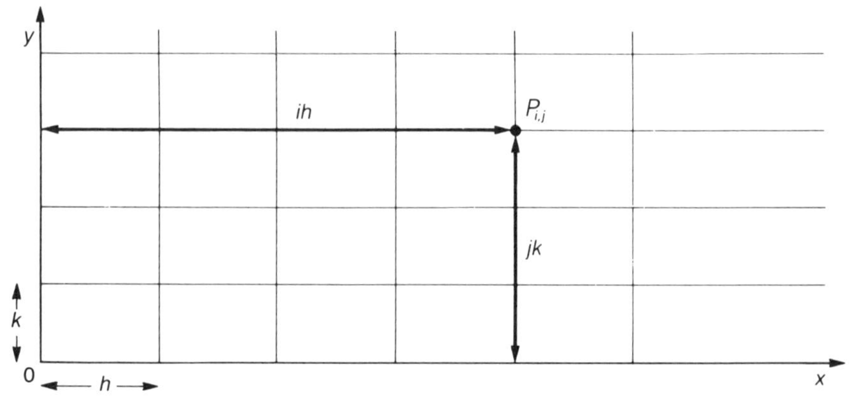


Figure 13.2.1

These expressions are called the *forward first difference, backward first difference,* and *central first difference* of $u_x$ respectively. The quantity $O(h)$ or $O(h^2)$ is known as the *truncation error* in this discretization process.

A similar approximate result for $u_{xx}$ at $P_{i,j}$ is

$$u_{xx} = \frac{1}{h^2}[u(x+h, y) - 2u(x, y) + u(x-h, y)]$$

$$\sim \frac{1}{h^2}[u_{i+1,j} - 2u_{i,j} + u_{i-1,j}] + O(h^2) \qquad (13.2.6)$$

Similarly, the approximate formulas for $u_y$ and $u_{yy}$ at $P_{i,j}$ are

$$u_y = \frac{1}{k}[u(x, y+k) - u(x, y)] \sim \frac{1}{k}(u_{i,j+1} - u_{i,j}) + O(k) \qquad (13.2.7)$$

$$u_y = \frac{1}{k}[u(x, y) - u(x, y-k)] \sim \frac{1}{k}(u_{i,j} - u_{i,j-1}) + O(k) \qquad (13.2.8)$$

$$u_y = \frac{1}{2k}[u(x, y+k) - u(x, y-k)] \sim \frac{1}{2k}(u_{i,j+1} - u_{i,j-1}) + O(k^2) \qquad (13.2.9)$$

$$u_{yy} = \frac{1}{k^2}[u(x, y+k) - 2u(x, y) + u(x, y-k)]$$

$$\sim \frac{1}{k^2}[u_{i,j+1} - 2u_{i,j} + u_{i,j-1}] + O(k^2) \qquad (13.2.10)$$

All these difference formulas are extremely useful in finding numerical solutions of first or second order partial differential equations.

Suppose $U(x, y)$ represents the exact solution of a partial differential equation $L(U) = 0$ with independent variables $x$ and $y$, and $u_{i,j}$ is the exact solution of the corresponding finite difference equation $F(u_{i,j}) = 0$. Then the finite difference scheme is said to be *convergent* if $u_{i,j}$ tends to $U$ as $h$ and $k$ tend to zero. The difference, $d_{i,j} \equiv U_{i,j} - u_{i,j}$ is called the *cummulative truncation* (or *discretization*) *error.*

This error can generally be minimized by decreasing the grid sizes $h$ and $k$. However, this error depends not only on $h$ and $k$, but also on the number of terms in the truncated series which is used to approximate each partial derivative.

Another kind of error is introduced when a partial differential equation is approximated by a finite difference equation. If the exact finite difference solution $u_{i,j}$ is replaced by the exact solution $U_{i,j}$ of the partial differential equation at the grid points $P_{i,j}$, then the value $F(U_{i,j})$ is called the *local truncation error* at $P_{i,j}$. The finite difference scheme and the partial differential equation are said to be *consistent* if $F(U_{i,j})$ tends to zero as $h$ and $k$ tend to zero.

In general, finite difference equations cannot be solved exactly because the numerical computation is carried out only up to a finite number of decimal places. Consequently, another kind of error is introduced in the finite difference solution during the actual process of computation. This kind of error is called the *round-off error,* and also depends upon the type of computer used. In practice, the actual computational solution is $u_{i,j}^*$, but not $u_{i,j}$, so that the difference $r_{i,j} = (u_{i,j} - u_{i,j}^*)$ is the round-off error at the grid point $P_{i,j}$. In fact, this error is introduced into the solution of the finite difference equation by round-off errors. In reality, the round-off depends mainly on the actual computational process and the finite difference itself. In contrast to the cummulative truncation error, the round-off error cannot be made small by allowing $h$ and $k$ to tend to zero.

Thus the total error involved in the finite difference analysis at the point $P_{i,j}$ is

$$U_{i,j} - u_{i,j}^* = (U_{i,j} - u_{i,j}) + (u_{i,j} - u_{i,j}^*) = d_{i,j} - r_{i,j} \tag{13.2.11}$$

Usually the discretization error $d_{i,j}$ is bounded when $u_{i,j}$ is bounded because the value of $U_{i,j}$ is fixed for a given partial differential equation with the prescribed boundary and initial data. This fact is used or assumed in order to introduce the concept of stability. The finite difference algorithm is said to be *stable* if the round-off errors are sufficiently small for all $i$ as $j \to \infty$, that is, the growth of $r_{i,j}$ can be controlled. It should be pointed out again that the round-off error depends not only on the actual computational process and the type of computer used, but also on the finite difference equation itself. Lax (1954) proved a remarkable theorem which establishes the relationship between consistency, stability, and convergence for the finite difference algorithm.

***Theorem 13.2.1 (Lax's Equivalence Theorem).*** *Given a properly posed linear initial-value problem and a finite difference approximation to it that satisfies the consistency criterion, stability is the necessary and sufficient condition for convergence.*

## Von Neumann's Stability Method

This method is essentially based upon a finite Fourier series. It expresses the initial errors on the line $t = 0$ in terms of a finite Fourier series and then examines the propagation of errors as $t \to \infty$. It is convenient to denote the error function by $e_{r,s}$ instead of $e_{i,j}$ so that $e_{r,s}$ gives the initial values $e_{r,0} = e(rh) = e_r$ on the line $t = 0$ between $x = 0$ and $x = l$ where $r = 0, 1, 2, \ldots, N$ and $Nh = l$. The finite Fourier series expansion of $e_r$ is

$$e_r = \sum_{n=0}^{N} A_n \exp(in\pi x/l) = \sum_{n=0}^{N} A_n \exp(i\alpha_n rh) \tag{13.2.12}$$

where $\alpha_n = n\pi/l$, $x = rh$, and $A_n$ are the Fourier coefficients which are determined from the $(N+1)$ equations (13.2.12).

Since we are concerned with the linear finite difference scheme, errors form an additive system so that the total error can be found by the superposition principle. Thus it is sufficient to consider a single term $\exp(i\alpha rh)$ in the Fourier series (13.2.12). Following the method of separation of variables commonly used for finding the analytical solution of a partial differential equation, we seek a separable solution of the finite difference equation for $e_{r,s}$ in the form

$$e_{r,s} = \exp(i\alpha rh + \beta sk) = \exp(i\alpha rh)p^s \tag{13.2.13}$$

which reduces to $\exp(i\alpha rh)$ at $s = 0$ $(t = sk = 0)$, where $p = \exp(\beta k)$, and $\beta$ is a complex constant. This shows that the error is bounded as $s \to \infty$ $(t \to \infty)$ provided that

$$|p| \leqslant 1 \tag{13.2.14}$$

is satisfied. This condition is found to be necessary and sufficient for the stability of the finite difference algorithm.

## 13.3 LAX–WENDROFF EXPLICIT METHOD

To describe this method, we consider the first order conservation equation

$$\frac{\partial u}{\partial t} + c\frac{\partial u}{\partial x} = 0 \tag{13.3.1}$$

where $u \equiv u(x, t)$ is some physical function of space variable $x$ and time $t$. This equation occurs frequently in applied mathematics.

Lax–Wendroff used the Taylor series expansion in $t$ in the form

$$u_{i,j+1} = u_{i,j} + k(u_t)_{i,j} + \frac{k^2}{2!}(u_{tt})_{i,j} + \frac{k^3}{3!}(u_{ttt})_{i,j} + \ldots \tag{13.3.2}$$

where $k \equiv \delta t$.

The partial derivatives in $t$ in (13.3.2) can easily be eliminated by using $u_t = -cu_x$ so that (13.3.2) becomes

$$u_{i,j+1} = u_{i,j} - ck(u_x)_{i,j} + \frac{c^2k^2}{2}(u_{xx})_{i,j} - \ldots \tag{13.3.3}$$

Replacing $u_x$, $u_{xx}$ by the central difference formulas, (13.3.3) becomes

$$u_{i,j+1} = u_{i,j} - \frac{ck}{2h}(u_{i+1,j} - u_{i-1,j}) + \frac{1}{2}\left(\frac{ck}{h}\right)^2(u_{i+1,j} - 2u_{i,j} + u_{i-1,j})$$

or

$$u_{i,j+1} = (1 - \varepsilon^2)u_{i,j} + \frac{\varepsilon}{2}(1 + \varepsilon)u_{i-1,j} - \frac{\varepsilon}{2}(1 - \varepsilon)u_{i+1,j} + O(\varepsilon^3) \tag{13.3.4}$$

where $\varepsilon = ck/h$. This is called the *Lax–Wendroff second order finite difference scheme*, and has widely been used to solve first order hyperbolic equations.

Von Neumann criterion (13.2.14) can be applied to investigate the stability of the Lax–Wendroff scheme. It is noted that the error function $e_{r,s}$ given by (13.2.13) satisfies the finite difference equation (13.3.4). We then substitute (13.2.13) into (13.3.4) and cancel common factors to obtain

$$p = (1-\varepsilon^2) + \frac{\varepsilon}{2}[(1+\varepsilon)e^{-i\alpha h} - (1-\varepsilon)e^{i\alpha h}]$$

$$= 1 - 2\,\varepsilon^2 \sin^2\frac{\alpha h}{2} - 2i\varepsilon \sin\frac{\alpha h}{2}\cos\frac{\alpha h}{2}$$

so that

$$|p|^2 = 1 - 4\varepsilon^2(1-\varepsilon^2)\sin^4\frac{\alpha h}{2} \tag{13.3.5}$$

According to the von Neumann criterion, the Lax–Wendroff scheme (13.3.4) is stable as $t\to\infty$ if $|p|\leqslant 1$, which gives $4\varepsilon^2(1-\varepsilon^2)\geqslant 0$, that is, $0<\varepsilon\leqslant 1$.

The local truncation error of the Lax–Wendroff equation (13.3.4) at $P_{i,j}$ is

$$T_{i,j} = \frac{1}{k}(u_{i,j+1} - u_{i-1,j})$$

which is, by (13.2.2a) and (13.2.1b) with $ck = h$ ($\varepsilon = 1$),

$$= (u_t + cu_x)_{i,j} + \frac{k}{2}(u_{tt} - c^2 u_{xxx})_{i,j} + \frac{k^2}{6}(u_{ttt} + c^3 u_{xxx}) + O((ck)^3) \tag{13.3.6}$$

The first two terms on the right side of (13.3.6) vanish by Eq. (13.3.1) so that the local truncation error becomes

$$T_{i,j} = \frac{1}{6}(k^2 u_{ttt} + ch^2 u_{xxx})_{i,j} \tag{13.3.7}$$

Another approximation to (13.3.1) with first order accuracy is

$$\frac{1}{k}(u_{i,j+1} - u_{i,j}) + \frac{c}{h}(u_{i,j} - u_{1-i,j}) = 0 \tag{13.3.8}$$

A final explicit scheme for (13.3.1) is based upon the central difference approximation. This scheme is called the *leap frog algorithm*. In this method, the finite difference approximation to (13.3.1) is

$$\frac{1}{2k}(u_{i,j+1} - u_{i,j-1}) + \frac{c}{2h}(u_{i+1,j} - u_{i-1,j}) = 0$$

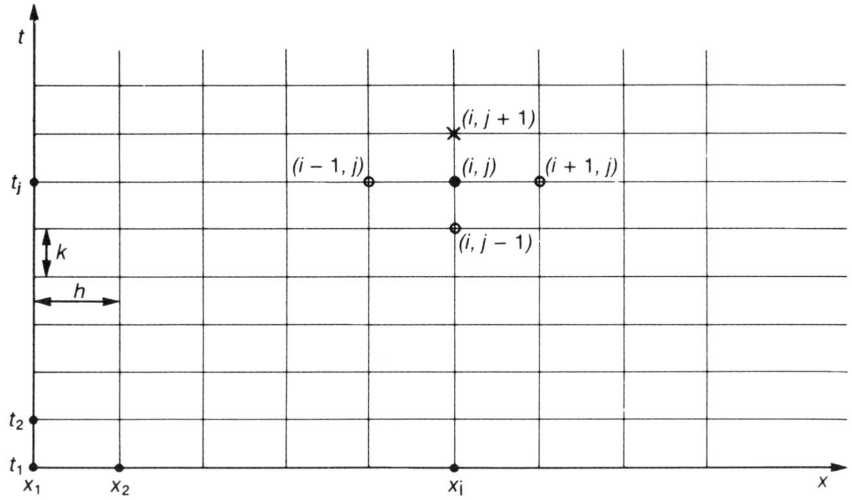


Figure 13.3.1. Grid system for the leap frog algorithm.

or

$$u_{i,j+1} = u_{i,j-1} - \varepsilon(u_{i+1,j} - u_{i-1,j}) \tag{13.3.9}$$

As shown in Fig. 13.3.1, this equation shows that the value of $u$ at $P_{i,j+1}$ is computed from the previously computed values at three grid points at two previous time steps.

## 13.4 EXPLICIT FINITE DIFFERENCE METHODS

### (A) Wave Equation and the Courant–Friedrichs-Lewy Convergence Criterion

The method of characteristics provides the most convenient and accurate procedure for solving the Cauchy problems involving hyperbolic equations. One of the main advantages of this method is that discontinuities of the initial data propagate into the solution domain along the characteristics. However, when the initial data are discontinuous, the finite difference algorithm for the hyperbolic systems is not very convenient. Problems concerning hyperbolic equations with continuous initial data can be solved successfully by finite difference methods with rectangular grid systems.

A commonly cited problem is the propagation of a one-dimensional wave governed by the system

$$u_{tt} = c^2 u_{xx}, \qquad -\infty < x < \infty, \quad t > 0 \tag{13.4.1}$$

$$u(x, 0) = f(x), \qquad u_t(x, 0) = g(x) \qquad \text{for all } x \tag{13.4.2ab}$$

Using a rectangular grid system with $h=\delta x$, $k=\delta t$, $u_{i,j}=u(ih, jk)$, $-\infty<x<\infty$, and $0\leqslant j<\infty$, the central difference approximation to equation (13.4.1) gives

$$\frac{1}{k^2}(u_{i,j+1}-2u_{i,j}+u_{i,j-1})=\frac{c^2}{h^2}(u_{i+1,j}-2u_{i,j}+u_{i-1,j})$$

or

$$u_{i,j+1}=\varepsilon^2(u_{i+1,j}+u_{i-1,j})+2(1-\varepsilon^2)u_{i,j}-u_{i,j-1} \tag{13.4.3}$$

where $\varepsilon\equiv ck/h$, and is often called the *Courant parameter*. This explicit formula allows us to determine the approximate values at the grid points on the line $t=2k, 3k, 4k, \ldots$ when the grid values at $t=k$ have been obtained.

The approximate values of the initial data on the line $t=0$ are

$$u_{i,0}=f_i, \qquad \frac{1}{2k}(u_{i,1}-u_{i,-1})=g_{i,0} \tag{13.4.4ab}$$

so that the second result gives

$$u_{i,-1}=u_{i,1}-2kg_{i,0} \tag{13.4.5}$$

When $j=0$ in (13.4.3) and (13.4.5) is used, we obtain

$$u_{i,1}=\frac{1}{2}\varepsilon^2(f_{i-1}+f_{i+1})+(1-\varepsilon^2)f_i+kg_{i,0} \tag{13.4.6}$$

This result determines grid values on the line $t=k$.

The value of $u$ at $P_{i,j+1}$ is obtained in terms of its previously calculated values at $P_{i\pm1,j}$, $P_{i,j}$, and $P_{i,j-1}$, which are determined from previously computed values on the lines $t=(j-1)k$, $(j-2)k$, $(j-3)k$. Thus, the computation from the lines $t=0$ and $t=k$ suggests that $u$ at $P_{i,j}$ will represent a function of the values of $u$ within the domain bounded by the lines drawn backwards $t=0$ from $P$ whose gradients are $\pm\varepsilon$ as shown in Fig. 13.4.1. Thus the triangular regions $PAB$, $PCD$ represent the domains of dependence at $P$ of the solutions of the finite difference equation (13.4.3) and the differential equation (13.4.1). In analogy with real characteristic lines $PC$ and $PD$ of the differential equation, the straight lines $PA$ and $PB$ are called the *numerical characteristics*. Thus it follows from Fig. 13.4.1 that $\Delta PAB$ lies inside $\Delta PCD$, which means that the solution of the finite difference system at $P$ would remain unchanged even when the initial data along $PA$ and $PB$ are changed. Courant, Friedrichs, and Lewy (CFL) proved that the solution of the finite difference system converges to that of the differential equation system as $h$ and $k$ tend to zero provided the domain of dependence of the difference equation lies inside that of the partial differential equation. This condition for convergence is known as the CFL condition, which means $1/c\geqslant k/h$, that is, $0<\varepsilon\leqslant1$.

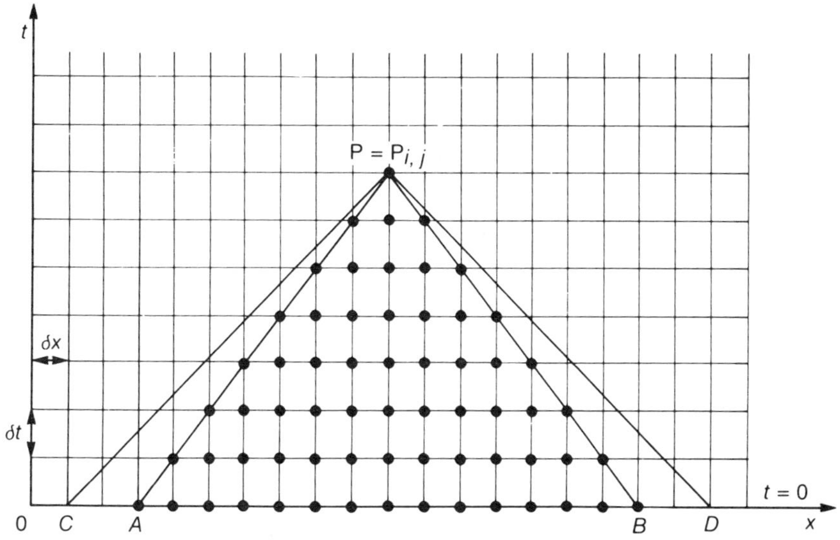


Figure 13.4.1. Computational grid systems and characteristics.

If the Courant parameter is $\varepsilon = 1$, (13.4.3) reduces to a simple form

$$u_{i,j+1} = u_{i+1,j} + u_{i-1,j} - u_{i,j-1} \tag{13.4.7}$$

As shown in Fig. 13.3.1, this equation shows that the value of $u$ at $P_{i,j+1}$ is computed from the previously computed values at three grid points at two previous time steps. This is the leap frog algorithm.

From Eq. (4.3.4) in Chapter 4 we know that the solution of the Cauchy problem for the wave equation has the form

$$u(x, t) = \phi(x + ct) + \psi(x - ct)$$

where the functions $\phi$ and $\psi$ represent waves propagating without changing shape along the negative and positive $x$ directions with a constant speed $c$. The lines of slope $dt/dx = \pm 1/c$ in the $x - t$ plane, which trace the progress of the waves, are known as the *characteristics* of the wave equation.

In terms of a grid point $(x_i, t_j)$, the above solution has the form

$$u_{i,j} = \phi(x_i + ct_j) + \psi(x_i - ct_j) \tag{13.4.8}$$

It follows from Fig. 13.3.1 that $x_i = x_1 + (i - 1)h$ and $t_j = t_1 + (j - 1)k$ so that (13.4.8) takes the form

$$u_{i,j} = \phi(\alpha + ih + jck) + \psi(\beta + ih - jck) \tag{13.4.9}$$

where

$$\alpha = (x_1 - h) + c(t_1 - k), \qquad \beta = (x_1 - h) - c(t_1 - k).$$

Since $\varepsilon = 1$, $ck = h$, (13.4.9) becomes

$$u_{i,j} = \phi(\alpha + (i+j)h) + \psi(\beta + (i-j)h).$$

This satisfies the equation (13.4.7). It follows that the leap frog method gives the exact solution of the partial differential equation (13.4.1).

We apply the von Neumann stability analysis to investigate the stability of the above numerical method for the wave equation. We seek a separable solution of the error function $e_{r,s}$ as

$$e_{r,s} = \exp(i\alpha rh)p^s \tag{13.4.10}$$

where $p = \exp(\beta k)$. This function satisfies the finite difference equation (13.4.3). Substituting (13.4.10) into (13.4.3) and cancelling the common factors, we obtain a quadratic equation for $p$

$$p^2 - 2bp + 1 = 0 \tag{13.4.11}$$

where $b = 1 - 2\varepsilon^2 \sin^2 \alpha h/2$, $\varepsilon = ck/h$, and $b \leqslant 1$ for all real $\varepsilon$ and $\alpha$. This quadratic equation has two complex roots $p_1$ and $p_2$ if $b^2 < 1$. Since $p_1 \cdot p_2 = 1$, it follows that one of the roots will always have modulus greater than one unless $|p_1| = |p_2| = 1$. Thus the scheme is unstable as $s \to \infty$ if the modulus of one of the roots exceeds unity.

On the other hand, when $-1 \leqslant b \leqslant 1$, $b^2 \leqslant 1$, then $|p_1| = |p_2| = 1$. Thus the finite difference scheme is stable provided $-1 \leqslant b \leqslant 1$, which leads to the useful condition for stability as $b \geqslant -1$ that

$$\varepsilon^2 \leqslant \operatorname{cosec}^2\left(\frac{\alpha h}{2}\right) \tag{13.4.12}$$

This shows the dependence of the stability limit on the space-grid size $h$. However, this stability condition is always true if $\varepsilon^2 \leqslant 1$.

EXAMPLE 13.4.1. Find the explicit finite difference solution of the wave equation

$$u_{tt} - u_{xx} = 0, \qquad 0 < x < 1, \qquad t > 0$$

with the boundary conditions

$$u(0, t) = u(1, t) = 0, \qquad t \geqslant 0$$

and the initial conditions

$$u(x, 0) = \sin \pi x, \qquad u_t(x, 0) = 0, \qquad 0 \leqslant x \leqslant 1$$

Compare the numerical solution with the analytical solution $u(x, t) = \cos \pi t \sin \pi x$ at several points.

The explicit finite difference approximation to the wave equation with $\varepsilon = k/h = 1$ is found from (13.4.3) in the form

$$u_{i,j+1} = u_{i-1,j} + u_{i+1,j} - u_{i,j-1}, \qquad j \geqslant 1$$

The problem is symmetric with respect to $x = \frac{1}{2}$, so we need to calculate the solution only for $0 \leqslant x \leqslant \frac{1}{2}$. We take $h = k = \frac{1}{10} = 0.1$. The boundary conditions give $u_{0,j} = 0$ for $j = 0, 1, 2, 3, 4, 5$. The initial condition $u_t(x, 0) = 0$ yields

$$u_t(x, 0) = \frac{1}{2k}(u_{i,1} - u_{i,-1}) = 0$$

or

$$u_{i,1} = u_{i,-1}$$

The explicit formula with $j = 0$ gives

$$u_{i,1} = \frac{1}{2}(u_{i-1,0} + u_{i+1,0}), \qquad i = 1, 2, 3, 4, 5$$

Thus

$$u_{1,1} = \frac{1}{2}(u_{0,0} + u_{2,0}) = \frac{1}{2}u_{2,0} = \frac{1}{2}\sin(0.2\pi) = 0.2939$$

Similarly,

$$u_{2,1} = 0.5590, \qquad u_{3,1} = 0.7695, \qquad u_{4,1} = 0.9045, \qquad u_{5,1} = 0.9511$$

We next use the basic explicit formula to compute

$$u_{1,2} = u_{0,1} + u_{2,1} - u_{1,0} = 0 + 0.5590 - 0.3090 = 0.2500$$
$$u_{2,2} = u_{1,1} + u_{3,1} - u_{2,0} = 0.2939 + 0.7695 - 0.5878 = 0.4756$$

Similarly, we can compute other values for $u_{i,j}$ which are shown in Table 13.4.1.

Table 13.4.1

| $j$ \ $i$ | $t$ \ $x$ | 0<br>0.0 | 1<br>0.1 | 2<br>0.2 | 3<br>0.3 | 4<br>0.4 | 5<br>0.5 |
|---|---|---|---|---|---|---|---|
| 1 | 0.1 | 0 | 0.2939 | 0.5590 | 0.7695 | 0.9045 | 0.9511 |
| 2 | 0.2 | 0 | 0.2500 | 0.4756 | 0.6545 | 0.7695 | 0.8090 |
| 3 | 0.3 | 0 | 0.1817 | 0.3455 | 0.4756 | 0.5590 | 0.5878 |
| 4 | 0.4 | 0 | 0.9045 | 0.7695 | 0.2500 | 0.2939 | 0.3090 |
| 5 | 0.5 | 0 | 0 | 0 | 0 | 0 | 0 |

The analytical solutions at $(x, t) = (0.1, 0.1)$ and $(0.2, 0.3)$ are

$$u(0.1, 0.1) = \cos(0.1\pi)\sin(0.1\pi) = (0.9511)(0.3090) = 0.2939$$
$$u(0.2, 0.2) = \cos(0.2\pi)\sin(0.2\pi) = (0.8090)(0.5878) = 0.4755$$
$$u(0.3, 0.3) = \cos(0.3\pi)\sin(0.3\pi) = (0.5878)(0.5878) = 0.3455.$$

Comparison of the analytical solutions with the above finite difference solutions shows that the latter results are very accurate.

## (B) Parabolic Equations

As a prototype diffusion problem, we consider

$$u_t = \kappa u_{xx}, \qquad 0<x<1, \qquad t>0 \tag{13.4.13}$$

$$u(0, t) = u(1, t) = 0 \qquad \text{for all } t \tag{13.4.14ab}$$

$$u(x, 0) = f(x) \qquad \text{for all } x \text{ in } (0, 1) \tag{13.4.15}$$

where $f(x)$ is a given function.

The explicit finite difference approximation to (13.4.13) is

$$\frac{1}{k}(u_{i,j+1} - u_{i,j}) = \frac{\kappa}{h^2}(u_{i+1,j} - 2u_{i,j} + u_{i-1,j}) \tag{13.4.16}$$

or

$$u_{i,j+1} = \varepsilon(u_{i+1,j} + u_{i-1,j}) + (1-2\varepsilon)u_{i,j} \tag{13.4.17}$$

where $\varepsilon = \kappa k/h^2$.

This explicit finite difference formula gives approximate values of $u$ on $t=(j+1)k$ in terms of values on $t = jk$ with given $u_{i,0} = f_i$. Thus $u_{i,j}$ can be obtained for all $j$ by successive use of (13.4.17).

The problems of stability and convergence of the parabolic equation are similar to those of the wave equation. It can be shown that the solution of the finite difference equation converges to that of the differential equation system (13.4.13)–(13.4.15) as $h$ and $k$ tend to zero provided $\varepsilon \leqslant \frac{1}{2}$.

In particular, when $\varepsilon = \frac{1}{2}$, (13.4.17) takes a simple form

$$u_{i,j+1} = \frac{1}{2}(u_{i+1,j} + u_{i-1,j}) \tag{13.4.18}$$

This is called the *Bender–Schmidt explicit formula* which determines the solution at $(x_i, t_{j+1})$ as the mean of the values at the grid points $(i \pm 1, j)$. However, more accurate results can be found from (13.4.17) for $\varepsilon < \frac{1}{2}$.

To investigate the stability of the numerical scheme, we assume the error function

$$e_{r,s} = \exp(i\alpha rh)p^s \tag{13.4.19}$$

where $p = e^{\beta k}$. The error function satisfies the same difference equation as $u_{r,s}$. Hence we substitute (13.4.19) into (13.4.17) to obtain

$$p = 1 - 4\varepsilon \sin^2 \frac{\alpha h}{2} \tag{13.4.20}$$

Clearly, $p$ is always less than 1 because $\varepsilon > 0$. If $p \geqslant 0$, the function given by (13.4.19) will decay steadily as $s = j \to \infty$. If $-1 < p < 0$, then the solution will have a decaying amplitude as $s \to \infty$. Therefore, the finite difference scheme will be stable if $p > -1$, that is,

$$0 < \varepsilon \leqslant \frac{1}{2} \operatorname{cosec}^2 \frac{\alpha h}{2} \tag{13.4.21}$$

This shows that the stability limit depends on $h$. However, in view of the inequality

$$\varepsilon \leqslant \frac{1}{2} \leqslant \frac{1}{2} \operatorname{cosec}^2 \frac{\alpha h}{2}$$

we conclude that the stability condition is $\varepsilon \leqslant \frac{1}{2}$. Finally, if $p < -1$, the solution oscillates with increasing amplitude as $s \to \infty$, and hence the scheme will be unstable for $\varepsilon > \frac{1}{2}$.

EXAMPLE 13.4.2. Show that the Richardson explicit finite difference scheme for (13.4.13) is unconditionally unstable.

The Richardson finite difference approximation to (13.4.13) is

$$\frac{1}{2k}(u_{i,j+1} - u_{i,j-1}) = \frac{\kappa}{h^2}(u_{i+1,j} - 2u_{i,j} + u_{i-1,j}) \tag{13.4.22}$$

To establish the instability of this equation, we use the Fourier method and assume

$$e_{r,s} = \exp(i\alpha rh)p^s, \qquad p = e^{\beta k}$$

This function satisfies the Richardson difference equation as $u_{r,s}$. Consequently,

$$p - \frac{1}{p} = -8\varepsilon \sin^2 \frac{\alpha h}{2}$$

or

$$p^2 + 8p\varepsilon \sin^2 \frac{\alpha h}{2} - 1 = 0$$

This quadratic equation gives two roots

$$p_1, p_2 = -4\varepsilon \sin^2 \frac{\alpha h}{2} \pm \left(1 + 16\varepsilon^2 \sin^4 \frac{\alpha h}{2}\right)^{\frac{1}{2}} \tag{13.4.23ab}$$

or

$$p_1, p_2 = \pm 1 - 4\varepsilon \sin^2 \frac{\alpha h}{2}\left(1 \mp 2\varepsilon \sin^2 \frac{\alpha h}{2}\right) + O(\varepsilon^4)$$

This gives $|p_1| \leqslant 1$ and

$$|p_2| > 1 + 4\varepsilon \sin^2 \frac{\alpha h}{2} > 1$$

for all positive $\varepsilon$ and, consequently, the Richardson scheme is always unstable for all values of $\varepsilon$.

The unstable feature of the Richardson scheme can be eliminated by replacing $u_{i,j}$ with $\frac{1}{2}(u_{i,j+1}+u_{i,j-1})$ in (13.4.22), which now becomes

$$(1+2\varepsilon)u_{i,j+1} = 2\varepsilon(u_{i+1,j}+u_{i-1,j}) + (1-2\varepsilon)u_{i,j-1} \qquad (13.4.24)$$

This is called the *Du Fort-Frankel explicit algorithm* and can be shown to be stable for all $\varepsilon$.

EXAMPLE 13.4.3. Prove that the solution of the finite difference equation for the diffusion equation (13.4.13) in $-\infty<x<\infty$ with the initial condition $u(x, 0)=e^{i\alpha x}$ converges to the exact solution of (13.4.13) as $h$ and $k$ tend to zero.

We obtain the exact solution of (13.4.13) by seeking a separable form

$$u(x, t) = e^{i\alpha x}v(t)$$

where $v(t)$ is a function of $t$ alone, which is to be determined.

Substituting this solution into (13.4.13) gives

$$\frac{dv}{dt} + \kappa\alpha^2 v = 0$$

which admits solutions

$$v(t) = Ae^{-\kappa\alpha^2 t}$$

where $A$ is an integrating constant. The initial condition $v(0)=1$ gives $A=1$. Hence

$$u(x, t) = \exp(i\alpha x - \alpha^2\kappa t) \qquad (13.4.25)$$

We now solve the corresponding finite difference equation (13.4.17) by replacing $i$, $j$ with $r$, $s$. We seek a separable solution of the difference equation

$$u_{r,s} = e^{i\alpha rh}v_s$$

with the initial condition

$$u_{r,0} = e^{i\alpha rh}v_0 \qquad \text{so that } v_0 = 1$$

Substituting this solution into the finite difference equation (13.4.7) yields

$$v_{s+1} = \left(1 - 4\varepsilon\sin^2\frac{\alpha h}{2}\right)v_s, \qquad v_0 = 1$$

so that the solution can be obtained by a simple inspection as

$$v_s = \left(1 - 4\varepsilon\sin^2\frac{\alpha h}{2}\right)^s \qquad (13.4.26)$$

$$u_{r,s} = e^{i\alpha rh}\left(1 - 4\varepsilon \sin^2 \frac{\alpha h}{2}\right)^s \tag{13.4.27}$$

where $\varepsilon = \kappa k/h^2$.

For small $h$, $1 - 4\varepsilon \sin^2 \alpha h/2 \sim 1 - \varepsilon\alpha^2 h^2$ so that $1 - \varepsilon\alpha^2 h^2 \approx \exp(-\varepsilon\alpha^2 h^2)$ for small $\varepsilon\alpha^2 h^2 = \kappa\alpha^2 k$. Consequently, the final solution becomes

$$u_{r,s} \sim e^{i\alpha rh - \kappa\alpha^2 ks} \quad \text{as} \quad h, k \to 0 \tag{13.4.28}$$

This is identical with the exact solution of the differential equation (13.4.13) with $rh = x$ and $sk = t$. This example shows that the finite difference approximation is reasonably good.

EXAMPLE 13.4.4. Calculate a finite difference solution of the initial boundary-value problem

$$u_t = u_{xx}, \qquad 0 < x < 1, \qquad t > 0$$

with the boundary conditions

$$u(0, t) = u(1, t) = 0, \qquad t \geqslant 0$$

and the initial condition

$$u(x, 0) = x(1 - x), \qquad 0 \leqslant x \leqslant 1$$

Compare the numerical solution with the exact analytical solution at $x = 0.04$ and $t = 0.02$.

The explicit finite difference approximation to the parabolic equation is

$$u_{i,j+1} = \varepsilon u_{i-1,j} + (1 - 2\varepsilon)u_{i,j} + \varepsilon u_{i+1,j}$$

where $\varepsilon = k/h^2$. This gives the unknown value $u_{i,j+1}$ at the $(i, j+1)$th grid point in terms of given values of $u$ along the $j$th time row.

We set $h = \frac{1}{5}$ and $k = \frac{1}{100}$ so that $\varepsilon = k/h^2 = \frac{1}{4}$ and the above formula becomes

$$u_{i,j+1} = \frac{1}{4}(u_{i-1,j} + 2u_{i,j} + u_{i+1,j})$$

With the notation $u_{i,0} = u(ih, 0)$, the initial condition gives

$$u_{4,0} = 0.16, \quad \text{and} \quad u_{5,0} = 0$$

The boundary conditions yield $u_{0,j} = u(0, jk) = 0$ and $u_{5,j} = u(5h, jk) = u(1, jk) = 0$, for all $j = 0, 1, 2, \ldots$.

Using these initial and boundary data, we calculate $u_{i,j}$ as follows:

$$u_{1,1}=\frac{1}{4}(u_{0,0}+2u_{1,0}+u_{2,0})=0.14, \qquad u_{1,2}=\frac{1}{4}(u_{0,1}+2u_{1,1}+u_{2,1})=0.125$$

$$u_{2,1}=\frac{1}{4}(u_{1,0}+2u_{2,0}+u_{3,0})=0.22, \qquad u_{2,2}=\frac{1}{4}(u_{1,1}+2u_{2,1}+u_{3,1})=0.200$$

$$u_{3,1}=\frac{1}{4}(u_{2,0}+2u_{3,0}+u_{4,0})=0.22, \qquad u_{3,2}=\frac{1}{4}(u_{2,1}+2u_{3,1}+u_{4,1})=0.200$$

$$u_{4,1}=\frac{1}{4}(u_{3,0}+2u_{4,0}+u_{5,0})=0.14, \qquad u_{4,2}=\frac{1}{4}(u_{3,1}+2u_{4,1}+u_{5,1})=0.125$$

$$u_{1,3}=\frac{1}{4}(u_{0,2}+2u_{1,2}+u_{2,2})=0.1125, \qquad u_{1,4}=\frac{1}{4}(u_{0,3}+2u_{1,3}+u_{2,3})=0.1016$$

$$u_{2,3}=\frac{1}{4}(u_{1,2}+2u_{2,2}+u_{3,2})=0.1813, \qquad u_{2,4}=\frac{1}{4}(u_{1,3}+2u_{2,3}+u_{3,3})=0.1641$$

$$u_{3,3}=\frac{1}{4}(u_{2,2}+2u_{3,2}+u_{4,2})=0.1813, \qquad u_{3,4}=\frac{1}{4}(u_{2,3}+2u_{3,3}+u_{4,3})=0.1641$$

$$u_{4,3}=\frac{1}{4}(u_{3,2}+2u_{4,2}+u_{5,2})=0.1125, \qquad u_{4,4}=\frac{1}{4}(u_{3,3}+2u_{4,3}+u_{5,3})=0.1016$$

The method of separation of variables gives the analytical solution of the problem as

$$u(x, t)=\frac{8}{\pi^3}\sum_{n=0}^{\infty}\frac{1}{(2n+1)^3}\exp\left[-(2n+1)^2\pi^2 t\right]\sin(2n+1)\pi x$$

This exact solution $u(x, t)$ at $x=0.4$ $(i=2)$ and $t=0.02$ $(j=2)$ gives

$$u\sim\frac{8}{3}\left[\frac{1}{1^3}\exp(-0.02\pi^2)\sin(0.4)\pi+\frac{1}{3^3}\exp(-0.18\pi^2)\sin(1.2)\pi\right]=0.2000$$

The analytical solution is seen to be identical with the numerical value.

EXAMPLE 13.4.5. Obtain the numerical solution of the initial boundary problem

$$\begin{aligned} &u_t=\kappa u_{xx}, && 0\le x\le 1, && t>0\\ &u(0, t)=1, && u(1, t)=0, && t\ge 0\\ &u(x, 0)=0, && 0\le x\le 1 \end{aligned}$$

We use the explicit finite-difference formula (13.4.17)

$$u_{i,j+1}=\varepsilon(u_{i+1,j}+u_{i-1,j})+(1-2\varepsilon)u_{i,j}$$

where $\varepsilon=\kappa k/h^2$.

We set $h = 0.25 = \frac{1}{4}$ and $\varepsilon = \frac{2}{5} = 0.4$ to compute $u_{i,j}$ for $i, j = 0, 1, 2, 3, 4$ as follows:

| $j$ \ $i$ | 0 | 1 | 2 | 3 | 4 |
|---|---|---|---|---|---|
| 0 | 1.000 | 0.000 | 0.000 | 0.000 | 0.000 |
| 1 | 1.000 | 0.400 | 0.000 | 0.000 | 0.000 |
| 2 | 1.000 | 0.480 | 0.160 | 0.000 | 0.000 |
| 3 | 1.000 | 0.560 | 0.224 | 0.064 | 0.000 |
| 4 | 1.000 | 0.602 | 0.295 | 0.103 | 0.000 |

## (C) Elliptic Equations

As a prototype boundary-value problem, we consider the Dirichlet problem for the Laplace equation

$$\nabla^2 u \equiv u_{xx} + u_{yy} = 0, \qquad 0 \leqslant x \leqslant a, \qquad 0 \leqslant y \leqslant b \qquad (13.4.29)$$

where the value of $u(x, y)$ is prescribed everywhere on the boundary of the rectangular domain.

The rectangular grid system is the most common and convenient system for this problem. We choose the vertices of the rectangular domain as the nodal points and set $h = a/m$ and $k = b/n$ where $m$ and $n$ are positive integers so that the domain is divided into $mn$ subrectangles.

The finite difference approximation to the Laplace equation (13.4.29) is

$$\frac{1}{h^2}(u_{i+1,j} - 2u_{i,j} + u_{i-1,j}) + \frac{1}{k^2}(u_{i,j+1} - 2u_{i,j} + u_{i,j-1}) = 0 \quad (13.4.30)$$

or

$$2(h^2 + k^2)u_{i,j} = k^2(u_{i+1,j} + u_{i-1,j}) + h^2(u_{i,j+1} + u_{i,j-1}) \quad (13.4.31)$$

where $1 \leqslant i \leqslant m - 1$ and $1 \leqslant j \leqslant n - 1$.

The prescribed conditions on the boundary of the rectangular domain determines the values $u_{0,j}$, $u_{m,j}$, $u_{i,0}$, and $u_{i,n}$ for a square grid system $(k = h)$ so that (13.4.30) becomes

$$u_{i,j} = \frac{1}{4}(u_{i+1,j} + u_{i-1,j} + u_{i,j+1} + u_{i,j-1}) \qquad (13.4.32)$$

This means that the value of $u$ at an interior point is equal to the average of the value of $u$ at four adjacent points. This is the well known *mean value theorem* for harmonic functions that satisfy the Laplace equation.

As $i$ and $j$ vary, the present numerical scheme reduces to a set of

$(m-1)(n-1)$ linear non-homogeneous algebraic equations for $(m-1)(n-1)$ unknown values of $u$ at interior grid points. It can be shown the solution of the finite difference equation (13.4.31) converges to the exact solution of the problem as $h, k \to 0$. The proof of the existence of a solution and its convergence to the exact solution as $h$ and $k$ tend to zero is essentially based on the *Maximum Modulus Principle.* It follows from the finite difference equation (13.4.30) or (13.4.31) that the value of $|u|$ at any interior grid point does not exceed its value at any of the four adjoining nodal points. In other words, the value of $u$ at $P_{i,j}$ cannot exceed its values at the four adjoining points $P_{i\pm1,j}$ and $P_{i,j\pm1}$. The successive application of this argument at all interior grid points leads to the conclusion that $|u|$ at the interior grid points cannot be greater than the maximum value of $|u|$ on the boundary. This may be recognized as the finite difference analogue of the Maximum Modulus Principle discussed in § 8.2. Thus, the success of the numerical method is directly associated with the existence of the Maximum Modulus Principle.

Clearly, the present numerical algorithm deals with a large number of algebraic equations. Even though numerical accuracy can be improved by making $h$ and $k$ sufficiently small, there is a major computational difficulty involved in the numerical solution for large number equations. It is possible to handle such a large number of algebraic equations by direct methods or by iterative methods, but it would be very difficult to obtain the numerical solution with sufficient accuracy. It is therefore necessary to develop some alternative methods of solution that can be conveniently and efficiently carried out on a computer.

In order to eliminate some of the drawbacks stated above, one of the numerical schemes, the so called *Liebmann's iterative method,* is found to be useful. In this method values of $u$ are first guessed for all interior grid points in addition to those given as the boundary points on the edges of the given domain. These values are denoted by $u_{i,j}^{(0)}$ where the superscript 0 indicates the zeroth iteration. It is convenient to choose the square grid so that the simplified finite difference equation (13.4.31) can be used. The values of $u$ are calculated for the next iteration by using (13.4.31) at every interior point based on the values of $u$ at the present iteration. The sequence of computation starts from the interior grid point located at the lowest left corner, proceeds upward until reaching the top, and then goes to the bottom of the next vertical line on the right. This process is repeated until the new value of $u$ at the last interior grid point at the upper right corner has been obtained.

At the starting point, formula (13.4.32) gives

$$u_{2,2}^{(1)} = \frac{1}{4}\left[u_{3,2}^{(0)} + u_{1,2}^{(0)} + u_{2,3}^{(0)} + u_{2,1}^{(0)}\right] \tag{13.4.33}$$

where $u_{1,2}^{(0)}$ and $u_{2,1}^{(0)}$ are boundary values which remain constant during the iteration process. They may be replaced, respectively, with $u_{1,2}^{(0)}$ $u_{2,1}^{(0)}$ in (13.4.33). The computation at the next step involves $u_{2,2}^{(0)}$. Since an improved value $u_{2,2}^{(1)}$ is available at this time, it will be utilized instead. Hence

$$u_{2,3}^{(1)} = \frac{1}{4}\left[u_{3,3}^{(0)} + u_{1,3}^{(1)} + u_{2,4}^{(0)} + u_{2,2}^{(1)}\right] \tag{13.4.34}$$

where $u_{1,3}^{(1)}$ is used to replace the constant boundary value $u_{1,3}^{(0)}$.

We repeat this argument to obtain a general iteration formula for computation of $u$ at the $(n+1)$th step

$$u_{i,j}^{(n+1)} = \frac{1}{4}\left[u_{i+1,j}^{(n)} + u_{i-1,j}^{(n+1)} + u_{i,j+1}^{(n)} + u_{i,j-1}^{(n+1)}\right], \tag{13.4.35}$$

This result is valid for any interior point, whether it is next to some boundary point or not. If $P_{i,j}$ is a true interior point, the second and fourth terms on the right side of (13.4.35) represent, respectively, the values of $u$ at the grid points to the left of and below that point. These values have already been recomputed according to our scheme, and therefore carry the superscript $(n+1)$. Result (13.4.35) is known as the *Liebmann* $(n+1)$*th iteration formula.* It can be proved that $u_{i,j}^{(n)}$ converges to $u_{i,j}$ as $n \to \infty$.

Another iteration scheme similar to (13.4.35) is given by

$$u_{i,j}^{(n+1)} = \frac{1}{4}\left[u_{i+1,j}^{(n)} + u_{i-1,j}^{(n)} + u_{i,j+1}^{(n)} + u_{i,j-1}^{(n)}\right] \tag{13.4.36}$$

This is called the *Richardson iteration formula,* and is also found to be useful. However, this scheme converges slower than (13.4.35).

One of the major difficulties of the above methods is the slow rate of convergence. An improved numerical method, the so called *Successive Over-Relaxation* (*SOR*) *scheme* gives an even faster convergence than the Liebmann or Richardson methods in solving the Laplace (or the Poisson) equation. For a rectangular domain of square grids, the successive iteration scheme is given by

$$u_{i,j}^{(n+1)} = u_{i,j}^{(n)} + \frac{\omega}{4}\left[u_{i-1,j}^{(n+1)} + u_{i+1,j}^{(n)} + u_{i,j-1}^{(n+1)} + u_{i,j+1}^{(n)} - 4u_{i,j}^{(n)}\right] \tag{13.4.37}$$

where $\omega$ is called the *acceleration parameter* (or *relaxation factor*) to be determined. In general, $\omega$ lies in the range $1 \leq \omega < 2$. The successive iterations converge fairly rapidly to the desired solution for $1 \leq \omega < 2$. The most rapid rate of convergence is achieved for the optimum value of $\omega$.

EXAMPLE 13.4.6. Obtain the standard five-point formula for the Poisson equation

$$u_{xx} + u_{yy} = -f(x, y) \quad \text{in} \quad D \subset R^2$$

with the prescribed value of $u(x, y)$ on the boundary $\partial D$.

We assume that the domain $D$ is covered by a system of squares with sides of length $h$ parallel to the $x - y$ axes. Using the central difference approximation to the Laplace operator, we obtain

$$\frac{1}{h^2}(u_{i+1,j} - 2u_{i,j} + u_{i-1,j}) + \frac{1}{h^2}(u_{i,j+1} - 2u_{i,j} + u_{i,j-1}) = -f_{i,j}$$

or

$$u_{i,j} = \frac{1}{4}(u_{i+1,j} + u_{i-1,j} + u_{i,j+1} + u_{i,j-1}) + \frac{1}{4}h^2 f_{i,j}$$

where $f_{i,j} = f(ih, jh)$. This is the five-point formula.

EXAMPLE 13.4.7. Find the numerical solution of the torsion problem in a square beam governed by

$$\nabla^2 u = -2 \quad \text{in} \quad D\{0 \leqslant x \leqslant 1, 0 \leqslant y \leqslant 1\}$$

with $u(x, y) = 0$ on $\partial D$.

From the above five-point formula, we obtain

$$u_{i,j} = \frac{1}{4}(u_{i+1,j} + u_{i-1,j} + u_{i,j+1} + u_{i,j-1}) - \frac{1}{2}h^2$$

where $h$ is the side-length of the unit square net.

We choose $h = \frac{1}{2}$, $1/2^2$, $1/2^3$, $1/2^4$ to calculate the corresponding numerical values $u_{i,j} = 0.1250$, $0.1401$, $0.1456$, $0.1469$.

Note that the known exact analytical solution is $0.1474$.

EXAMPLE 13.4.8. Using the explicit finite difference method, find the solution of the Dirichlet problem

$$u_{xx} + u_{yy} = 0 \quad \text{in} \quad 0 < x < 1, \qquad 0 < y < 1$$

$$u(x, 0) = x, \qquad u(x, 1) = 0 \quad \text{on} \quad 0 \leqslant x \leqslant 1$$

$$u(x, y) = 0, \quad \text{on} \quad x = 0, \qquad x = 1 \quad \text{and} \quad 0 \leqslant y \leqslant 1$$

We use four interior grid points (that is, $i = j = 4$) as shown in Fig. 13.4.2 in the $x - y$ plane.

We apply the explicit finite difference formula (13.4.32) to obtain four

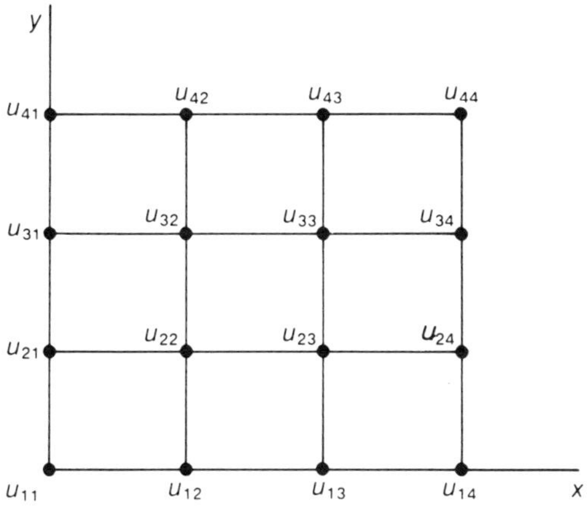


Figure 13.4.2. The square grid system.

algebraic equations

$$-4u_{22} + u_{32} + u_{12} + u_{23} + u_{21} = 0$$
$$-4u_{23} + u_{33} + u_{13} + u_{24} + u_{22} = 0$$
$$-4u_{32} + u_{42} + u_{22} + u_{33} + u_{31} = 0$$
$$-4u_{33} + u_{43} + u_{23} + u_{34} + u_{32} = 0$$

The given boundary conditions imply that $u_{21} = u_{24} = u_{31} = u_{34} = u_{42} = u_{43} = 0$, $u_{12} = \frac{1}{3}$ and $u_{13} = \frac{2}{3}$ so that the above system of equations becomes

$$-4u_{22} + u_{32} + \frac{1}{3} + u_{23} = 0$$
$$-4u_{23} + u_{33} + \frac{2}{3} + u_{22} = 0$$
$$-4u_{32} + u_{22} + u_{33} = 0$$
$$-4u_{33} + u_{23} + u_{32} = 0$$

In matrix notation, this system reads as

$$\begin{bmatrix} -4 & 1 & 1 & 0 \\ 1 & -4 & 0 & 1 \\ 1 & 0 & -4 & 1 \\ 0 & 1 & 1 & -4 \end{bmatrix} \begin{bmatrix} u_{22} \\ u_{23} \\ u_{32} \\ u_{33} \end{bmatrix} = \begin{bmatrix} -\frac{1}{3} \\ -\frac{2}{3} \\ 0 \\ 0 \end{bmatrix}$$

The solutions of this system are

$$u_{22}=\frac{11}{72}, \qquad u_{23}=\frac{16}{72}, \qquad u_{32}=\frac{4}{72}, \qquad u_{33}=\frac{5}{72}.$$

## (D) Simultaneous First Order Equations

We recall the wave equation (13.4.1) in $0<x<1$, $t>0$. Introducing two auxiliary variables $v$ and $w$ by $v=u_t$ and $w=c^2u_x$, the wave equation gives two simultaneous first order equations

$$v_t=w_x, \qquad w_t=c^2v_x \tag{13.4.38ab}$$

The initial values of $v$ and $w$ are given at $t=0$ for all $x$ in $0<x<1$. The boundary condition on $v$ and $w$ is also prescribed on the lines $x=0$ and $x=1$ for $t>0$.

The explicit finite difference method can be used to determine $v$ and $w$ in the triangular domain of dependence bounded by the characteristics $x-ct=0$ and $x+ct=1$.

The finite difference approximations to the differential equations (13.4.38ab) are

$$\frac{1}{k}(v_{i,j+1}-v_{i,j})=\frac{1}{2h}(w_{i+1,j}-w_{i-1,j}) \tag{13.4.39}$$

$$\frac{1}{k}(w_{i,j+1}-w_{i,j})=\frac{c^2}{2h}(v_{i+1,j}-v_{i-1,j}) \tag{13.4.40}$$

where the forward difference for $v_t$ or $w_t$ and the central difference for $v_x$ or $w_x$ are used. However, the central difference approximations to (13.4.38ab) can also be utilized to obtain

$$\frac{1}{2k}(v_{i,j+1}-v_{i,j-1})=\frac{1}{2h}(w_{i+1,j}-w_{i-1,j}) \tag{13.4.41}$$

$$\frac{1}{2k}(w_{i,j+1}-w_{i,j-1})=\frac{c^2}{2h}(v_{i+1,j}-v_{i-1,j}) \tag{13.4.42}$$

We examine the stability of the above two sets of finite difference formulas with $c=1$. The von Neumann stability method is applied by replacing $i$ and $j$ by $r$ and $s$ respectively. The error function $e_{r,s}$ given by (13.4.10) is substituted in (13.4.39)–(13.4.40) to obtain the stability relations

$$A(p-1)=\varepsilon iB\sin\alpha h \tag{13.4.43}$$

$$B(p-1)=\varepsilon iA\sin\alpha h \tag{13.4.44}$$

where the initial perturbations in $v$ and $w$ along $t=0$ are $A\exp(i\alpha rh)$ and $B\exp(i\alpha rh)$ respectively with two different constants $A$ and $B$.

Elimination of $A$ and $B$ from the above relations gives

$$(p-1)^2+\varepsilon^2 \sin^2 \alpha h = 0$$

or

$$p = 1 \pm i\varepsilon \sin \alpha h$$

and

$$|p| = (1+\varepsilon^2 \sin^2 \alpha h)^{\frac{1}{2}} \sim 1 + \frac{1}{2}\varepsilon^2 \sin^2 \alpha h = 1 + O(\varepsilon^2) \quad (13.4.45)$$

Since $|p| > 1 + O(\varepsilon)$, the finite difference scheme for the finite time-step $t = sk$ would be unstable as the grid sizes tend to zero.

A similar stability analysis for (13.4.41)–(13.4.42) leads to the condition

$$\left(p - \frac{1}{p}\right)^2 + 4\varepsilon^2 \sin^2 \alpha h = 0 \quad (13.4.46)$$

This scheme is shown to be stable for $\varepsilon \leqslant 1$.

Another finite difference approximation to the coupled system (13.4.38ab) is

$$\frac{1}{2h}(v_{r+1,s} - v_{r-1,s}) = \frac{1}{k}\left[w_{r,s+1} - \frac{1}{2}(w_{r+1,s} + w_{r-1,s})\right] \quad (13.4.47)$$

$$\frac{1}{2h}(w_{r+1,s} - w_{r-1,s}) = \frac{1}{k}\left[v_{r,s+1} - \frac{1}{2}(v_{r+1,s} + v_{r-1,s})\right] \quad (13.4.48)$$

A similar stability analysis can be carried out for these systems by substituting $v_{r,s} = Ap^s e^{i\alpha rh}$ and $w_{r,s} = Bp^s e^{i\alpha rh}$ into the equations. Elimination of $A/B$ yields the stability equation

$$p = \cos \alpha h \pm \frac{i}{\varepsilon} \sin \alpha h$$

or

$$|p|^2 = \cos^2 \alpha h + \frac{1}{\varepsilon^2} \sin^2 \alpha h \leqslant 1 \quad (13.4.49)$$

Hence the scheme is stable provided

$$\varepsilon \geqslant 1, \quad \text{that is,} \quad k \leqslant h \quad (13.4.50)$$

## 13.5 IMPLICIT FINITE DIFFERENCE METHODS

From a computational point of view, the explicit finite difference algorithm is simple and convenient. However, as shown in § 13.4(B), the major difficulty in the method for solving parabolic partial differential equations is the severe restriction on the time-step imposed by the stability condition $\varepsilon \leqslant \frac{1}{2}$ or $k \leqslant h^2/2\kappa$. This difficulty is also present in the explicit finite difference method for the solution of hyperbolic equations. In this section we develop implicit finite difference schemes for solving partial differential equations.

## (A) Parabolic Equations

One of the successful implicit finite difference schemes is the Crank and Nicolson Method (1947), which is based on six grid points. This method eliminates the major difficulty involved in the explicit scheme. When the Crank–Nicolson implicit scheme is applied to the parabolic equation (13.4.13), $u_{xx}$ is replaced by the mean value of the finite difference values in the $j$th and the $(j+1)$th row so that the finite difference approximation to Eq. (13.4.13) becomes

$$\frac{1}{k}(u_{i,j+1}-u_{i,j}) = \frac{\kappa}{2h^2}[(u_{i+1,j+1}-2u_{i,j+1}+u_{i-1,j+1}) + (u_{i+1,j}-2u_{i,j}+u_{i-1,j})] \qquad (13.5.1)$$

or

$$2(1+\varepsilon)u_{i,j+1} - \varepsilon(u_{i-1,j+1}+u_{i+1,j+1}) = 2(1-\varepsilon)u_{i,j} + \varepsilon(u_{i-1,j}+u_{i+1,j}) \qquad (13.5.2)$$

where $\varepsilon = k\kappa/h^2$ is a parameter.

The left side of (13.5.2) is a linear combination of three unknowns in the $(j+1)$th row, and the right side involves three known values of $u$ in the $j$th row of the grid system in the $x-t$ plane. Equation (13.5.2) is called the *Crank–Nicolson implicit formula*. This formula (or its suitable modification) is widely used for solving parabolic equations. If there are $n$ internal grid points along each $j$th row, then for $j=0$ and $i=1, 2, 3, \ldots, n$ the implicit formula (13.5.2) gives $n$ simultaneous algebraic equations for $n$ unknown values of $u$ along the first $j$th row $(j=0)$ in terms of given boundary and initial data. Similarly, if $j=1$ and $i=1, 2, 3, \ldots, n$, Eq. (13.5.2) represents $n$ unknown values of $u$ along the second $j$th row $(j=1)$ and so on. This means that the method involves the solution of a system of simultaneous algebraic equations. In practice, the Crank–Nicolson scheme is convergent and unconditionally stable for all finite values of $\varepsilon$, and has the advantage of reducing huge amounts of numerical computation.

This implicit scheme can be further generalized by introducing a numerical weight factor $\lambda$ in the modified version of the explicit equation (13.4.16) which is written below by approximating $u_{xx}$ in (13.4.13) in the $(j+1)$th row instead of the $j$th row.

$$\frac{1}{k}(u_{i,j+1}-u_{i,j}) = \frac{\kappa}{h^2}(u_{i+1,j+1}-2u_{i,j+1}+u_{i-1,j+1}) \qquad (13.5.3)$$

or

$$u_{i,j+1} - u_{i,j} = \varepsilon(u_{i+1,j+1} - 2u_{i,j+1} + u_{i-1,j+1}) \tag{13.5.4}$$

Introducing the numerical factor $\lambda$, this can be replaced by a more general difference equation in the form

$$u_{i,j+1} - u_{i,j} = \varepsilon[\lambda\delta_x^2 u_{i,j+1} + (1-\lambda)\delta_x^2 u_{i,j}] \tag{13.5.5}$$

where $0 \leqslant \lambda \leqslant 1$ and $\delta_x^2$ is the difference operator defined by

$$\delta_x^2 u_{i,j} = u_{i+1,j} - 2u_{i,j} + u_{i-1,j} \tag{13.5.6}$$

Another equivalent form of (13.5.5) is

$$(1+2\varepsilon\lambda)u_{i,j+1} - \varepsilon\lambda(u_{i+1,j+1} + u_{i-1,j+1}) = \{1 - 2\varepsilon(1-\lambda)\}u_{i,j} + \varepsilon(1-\lambda)(u_{i+1,j} - u_{i-1,j}) \tag{13.5.7}$$

This is a fairly general implicit formula which reduces to (13.5.4) when $\lambda = 1$. When $\lambda = \frac{1}{2}$, (13.5.7) becomes the Crank–Nicolson formula (13.5.2). Finally, if $\lambda = 0$, this implicit difference equation reduces to the explicit equation (13.4.17).

The Richardson scheme explicit was found to be unconditionally unstable in § 13.4. This undesirable feature of the scheme can be eliminated by considering the corresponding implicit scheme. In terms of $\delta_x^2$, the Richardson equation (13.4.22) can be expressed as

$$u_{i,j+1} = 2\varepsilon\delta_x^2 u_{i,j} + u_{i,j-1} \tag{13.5.8}$$

To obtain the implicit Richardson formula, we replace $\delta_x^2 u_{i,j}$ by $\frac{1}{3}\delta_x^2(u_{i+1,j} + u_{i,j} + u_{i-1,j})$ in (13.5.8) so that

$$\left(1 - \frac{2\varepsilon}{3}\delta_x^2\right)u_{i,j+1} = \frac{2\varepsilon}{3}\delta_x^2 u_{i,j} + \left(1 + \frac{2\varepsilon}{3}\right)u_{i,j-1} \tag{13.5.9}$$

This implicit scheme can be shown to be unconditionally stable. To prove this result, we apply the von Neumann stability method with the error function (13.4.19) to obtain the equation for $p$ as

$$(1+a)p^2 + ap + (a-1) = 0 \tag{13.5.10}$$

where

$$a \equiv \frac{8\varepsilon}{3}\sin^2\frac{\alpha h}{2} \tag{13.5.11}$$

The roots of the quadratic equation are

$$p = \frac{-a \pm (4 - 3a^2)^{\frac{1}{2}}}{2(1+a)} \tag{13.5.12}$$

This gives $|p| \leqslant 1$ for all values of $a$. Hence the result is proved.

EXAMPLE 13.5.1. Obtain the numerical solution of the following parabolic system by using the Crank–Nicolson method:

$$u_t = u_{xx}, \qquad 0 < x < 1, \qquad t > 0$$
$$u(0, t) = u(1, t) = 0, \qquad t \geqslant 0$$
$$u(x, 0) = x(1 - x), \qquad 0 \leqslant x \leqslant 1$$

We recall the Crank–Nicolson equation (13.5.2) and then set $h = 0.2$ and $k = 0.01$ so that $\varepsilon = \frac{1}{4}$. The boundary and initial conditions give $u_{0,0} = u_{5,0} = u_{0,1} = u_{5,1} = 0$ and $u_{i,0} = u(ih, 0) = ih(1 - ih)$, $i = 1, 2, 3, 4$. Consequently, formula (13.5.2) leads to the following system of four equations:

$$-u_{0,1} - u_{2,1} + 10u_{1,1} = u_{0,0} + u_{2,0} + 6u_{1,0}$$
$$-u_{1,1} - u_{3,1} + 10u_{2,1} = u_{1,0} + u_{3,0} + 6u_{2,0}$$
$$-u_{2,1} - u_{4,1} + 10u_{3,1} = u_{2,0} + u_{4,0} + 6u_{3,0}$$
$$-u_{3,1} - u_{5,1} + 10u_{4,1} = u_{3,0} + u_{5,0} + 6u_{4,0}$$

Using the boundary and initial conditions, the above system becomes

$$-u_{2,1} + 10u_{1,1} = 1.20$$
$$-u_{1,1} + 10u_{2,1} - u_{3,1} = 1.84$$
$$-u_{2,1} - u_{4,1} + 10u_{3,1} = 1.84$$
$$-u_{3,1} + 10u_{4,1} = 1.20$$

These equations can be solved by direct elimination to obtain the solutions as $u_{1,1} = 0.1418$, $u_{2,1} = 0.2202$, $u_{3,1} = 0.2202$, $u_{4,1} = 0.1420$.

## (B) Hyperbolic Equations

We consider an implicit finite difference scheme to solve the initial boundary-value problem consisting of the first order hyperbolic equation

$$\frac{\partial u}{\partial t} + c\frac{\partial u}{\partial x} = 0, \qquad (c > 0) \tag{13.5.13}$$

with the initial data $u(x, 0) = U(x)$ and the boundary condition $u(0, t) = V(t)$ where $0 \leqslant x$, $t < \infty$.

The implicit finite difference approximation to (13.5.13) is

$$\frac{1}{k}(u_{i,j+1} - u_{i,j}) + \frac{c}{h}(u_{i,j+1} - u_{i-1,j+1}) = 0$$

or

$$u_{i,j} = (1 + \varepsilon)u_{i,j+1} - \varepsilon u_{i-1,j+1} \tag{13.5.14}$$

where $\varepsilon = ck/h$.

The stability of the scheme can be examined by using the von Neumann method with the error function (13.4.10). It turns out that

$$p = [1 - \varepsilon + \varepsilon \exp(-i\alpha h)]^{-1} \tag{13.5.15}$$

from which it follows that $|p| \leqslant 1$ for all $h$. Hence the implicit scheme is unconditionally stable.

We next solve the wave equation $u_{tt} = c^2 u_{xx}$ by an implicit finite difference scheme. In this case, $u_{tt}$ is replaced by the central difference formula, and $u_{xx}$ by the mean value of the central difference values in the $(j-1)$th and $(j+1)$th rows. Consequently, the implicit difference approximation to the wave equation is

$$u_{i,j+1} - 2u_{i,j} + u_{i,j-1} = \frac{\varepsilon^2}{2}[(u_{i+1,j+1} - 2u_{i,j+1} + u_{i-1,j+1}) + (u_{i+1,j-1} - 2u_{i,j-1} + u_{i-1,j-1})] \tag{13.5.16}$$

where $\varepsilon = ck/h$.

Expressing the solution for the $(j+1)$th step in terms of the two preceding steps gives

$$2(1+\varepsilon^2)u_{i,j+1} - \varepsilon^2(u_{i-1,j+1} + u_{i+1,j+1}) = 4u_{i,j} + \varepsilon^2(u_{i-1,j-1} + u_{i+1,j-1}) - 2(1+\varepsilon^2)u_{i,j-1} \tag{13.5.17}$$

With $N$ grid points along each time step, $j = 0$, $i = 1, 2, 3, \ldots, N$, (13.5.17) along with the finite difference approximation to the boundary condition gives $N$ simultaneous equations for the $N$ unknown values of $u$ along the first time step. This constitutes a tridiagonal system of equations that can be solved by direct or iterative numerical methods.

To investigate the stability of the implicit schemes, we apply the von Neumann stability method with the error function (13.4.10). This leads to the equation

$$p + \frac{1}{p} = 2\left(1 + 2\varepsilon^2 \sin^2 \frac{\alpha h}{2}\right)^{-1}$$

or

$$p^2 - 2bp + 1 = 0 \tag{13.5.18}$$

where $b = (1 + 2\varepsilon^2 \sin^2 \alpha h/2)^{-1}$, which implies that $0 < b \leqslant 1$.

Hence the stability condition is

$$|p| \leqslant 1 \tag{13.5.19}$$

which is always satisfied provided $0 < b \leqslant 1$, that is, $\varepsilon < 1$ for all positive $h$. This confirms the unconditional stability of the scheme.

A more general implicit scheme can be introduced by replacing $u_{xx}$ in the wave equation (13.4.1) with

$$u_{xx} \sim \frac{1}{h^2}[\lambda(\delta_x^2 u_{i,j+1} + \delta_x^2 u_{i,j-1}) + (1-2\lambda)\delta_x^2 u_{i,j}] \tag{13.5.20}$$

where $\lambda$ is a numerical weight (relaxation) factor and the central difference operator $\delta_x^2$ is given by (13.5.6). This general scheme allows us to approximate the wave equation with $c = 1$ to the form

$$\delta_t^2 u_{i,j} = \varepsilon^2[\lambda(\delta_x^2 u_{i,j+1} + \delta_x^2 u_{i,j-1}) + (1 - 2\lambda)\delta_x^2 u_{i,j}] \qquad (13.5.21)$$

where $\varepsilon = k/h$. This equation reduces to (13.5.16) when $\lambda = \frac{1}{2}$, and to the explicit finite difference result when $\lambda = 0$.

It follows from the von Neumann stability analysis that the implicit scheme is unconditionally stable for $\lambda \geq \frac{1}{4}$.

Von Neumann introduced another fairly general finite difference algorithm for the wave equation (13.4.1) in the form

$$\delta_t^2 u_{i,j} = \varepsilon^2 \delta_x^2 u_{i,j} + \frac{\omega}{h^2} \delta_t^2 \delta_x^2 u_{i,j} \qquad (13.5.22)$$

This equation with appropriate boundary conditions can be solved by the tridiagonal method. Von Neumann discussed the question of stability of the implicit scheme, and proved that the scheme is conditionally stable or unconditionally stable accordingly as $\omega \leq \frac{1}{4}$ or $> \frac{1}{4}$.

## 13.6 VARIATIONAL METHODS AND THE EULER–LAGRANGE EQUATIONS

To describe the variational methods and Rayleigh–Ritz approximate method, it is convenient to introduce the concepts of the inner product (pre-Hilbert) and Hilbert spaces. An *inner product* space $X$ consisting of elements $u, v, w, \ldots$ over the complex number field $C$ is a complex linear space with an inner product $(u, v): X \times X \to C$ such that

(i) $(u, v) = \overline{(v, u)}$ where the bar denotes the complex conjugate of $(v, u)$
(ii) $(\alpha u + \beta v, w) = \alpha(u, w) + \beta(v, w)$ for any scalars $\alpha, \beta \in C$
(iii) $(u, v) \geq 0$ and equality holds if and only if $u = 0$

By (i) $(u, u) = \overline{(u, u)}$, and so $(u, u)$ is real. We denote $(u, u)^{\frac{1}{2}} = \|u\|$, which is called the *norm* of $u$. The norm thus obtained is induced by the inner product. Thus every inner product space is a normed linear space under the norm $\|u\| = \sqrt{(u, u)}$.

Let $X$ be an inner product space. A sequence $\{u_n\}$ when $u_n \in X$ for every $n$ is called a *Cauchy Sequence* in $X$ if and only if

$$\|u_n - u_m\| < \varepsilon \quad \text{for all} \quad n, m > N(\varepsilon)$$

The space $X$ is called *complete* if every Cauchy Sequence converges to a point in $X$. A complete normed linear space is called a *Banach Space.* A complete linear inner product space is called a *Hilbert Space* and is usually denoted by $H$.

EXAMPLE 13.6.1. Let $C^n$ be a set of all $n$-tuples of complex numbers. $C^n$ is an $n$-dimensional Hilbert space with the inner product

$$(x, y) = \sum_{k=1}^{n} x_k \overline{y_k}$$

Obviously, $R^n$ (a set of all $n$-tuples of real numbers) is an $n$-dimensional Hilbert space.

EXAMPLE 13.6.2. Let $l_2$ be the set of all sequences such that $\sum_{k=1}^{\infty} |x_k|^2 < \infty$. This forms a Hilbert space with the inner product

$$(x, y) = \sum_{k=1}^{\infty} x_k \overline{y_k}$$

EXAMPLE 13.6.3. Let $L_2(a, b)$ be the set of all square integrable functions in the Lebesgue sense in an interval $(a, b)$. $L_2(a, b)$ is a Hilbert space with the inner product

$$(u, v) = \int_a^b u\bar{v}\, dx$$

We next introduce the notion of an operator in a Hilbert space $H$. An *operator* $A$ is a mapping from $H$ to $H$ (that is $A: H \to H$). It assigns to an element $u$ in $H$ a new element $Au$ in $H$. An operator $A$ is called *linear* if it satisfies the property

$$A(\alpha u + \beta v) = \alpha Au + \beta Av \quad \text{for every} \quad \alpha, \beta \in C$$

An operator is said to be *bounded* if there exists a constant $k$ such that $\|Au\| \leqslant k\, \|u\|$ for all $u \in H$.

We consider a bounded operator $A$ on a Hilbert space $H$. For a fixed element $v$ in $H$, the inner product $(Au, v)$ in $H$ can be regarded as a number $I(u)$ which varies with $u$. Thus $(Au, v) = I(u)$ is a *linear functional* on $H$.

If there exists an operator $A^*$ on a Hilbert space $(A^*: H \to H)$ such that

$$(Au, v) = (u, A^*v) \quad \text{for all} \quad u, v \in H$$

then $A^*$ is called the *adjoint* of $A$. In general, $A \neq A^*$. If $A = A^*$, that is $(Au, v) = (u, Av)$ for all $u$, $v$ in $H$, then $A$ is called *self-adjoint.*

It is important to note that any bounded operator $T$ on a real Hilbert space $(T: H \to H)$ of the form $T = A^*A$ is self-adjoint. This follows from the fact that

$$(Tu, v) = (A^*Au, v) = (Au, Av) = (u, A^*Au) = (u, Tv)$$

A self-adjoint operator $A$ on a Hilbert space $H$ is said to be *positive* if $(Au, u) \geqslant 0$ for all $u$ in $H$, where equality implies $u = 0$ in $H$. Further, if there exists a positive constant $k$ such that $(Au, u) \geqslant k(u, u)$ for all $u$ in $H$, then $A$ is called *positive definite* in $H$.

The rest of this section is essentially concerned with linear operators in a real Hilbert space, which means that the associated scalar field involved is real. Some specific inner products which will be used in the subsequent sections are

$$(u, v) = \int_a^b u(x)v(x)\,dx, \qquad (u, v) = \iint_D u(x, y)v(x, y)\,dx\,dy$$

where $D \subset R^2$.

EXAMPLE 13.6.4. Determine whether the differentiable operators (i) $A = d/dx$, (ii) $A = d^2/dx^2$, and (iii) $A = \nabla^2 = (\partial^2/\partial x^2) + (\partial^2/\partial y^2)$ are self-adjoint for functions that are differentiable in $a \leqslant x \leqslant b$ or in $D \subset R^2$ and vanish on the boundary.

(i)
$$(Au, v) = \int_a^b \left(\frac{du}{dx}\right) v\,dx = \int_a^b u\left(-\frac{dv}{dx}\right) dx + [uv]_a^b$$
$$= (u, A^*v) \quad \text{where} \quad A^* = -\frac{d}{dx} \neq A$$

Hence $A$ is not self-adjoint.

(ii)
$$(Au, v) = \int_a^b \left(\frac{d^2u}{dx^2}\right) v\,dx = \int_a^b \left(\frac{du}{dx}\right)\left(-\frac{dv}{dx}\right) dx + \left[v\frac{du}{dx}\right]_a^b$$
$$= \int_a^b u\left(\frac{d^2v}{dx^2}\right) dx + \left[v\frac{du}{dx} - u\frac{dv}{dx}\right]_a^b$$
$$= \int_a^b u\left(\frac{d^2v}{dx^2}\right) dx = (u, Av)$$

Thus $A$ is a self-adjoint.

(iii)
$$(Au, v) = \iint_D (\nabla^2 u)v\,dx\,dy = \iint_D [\nabla \cdot (\nabla u)v - \nabla u \cdot \nabla v]\,dx\,dy$$
$$= \int_{\partial D} (\hat{\mathbf{n}} \cdot \nabla u)v\,dS - \iint_D \nabla u \cdot \nabla v\,dx\,dy$$
$$= -\iint_D (\nabla u \cdot \nabla v)\,dx\,dy$$

where the divergence theorem is used with the unit outward normal vector $\hat{\mathbf{n}}$.

Noting the symmetry of the right hand side in $u$ and $v$, it follows that

$$(Au, v) = (Av, u) = (u, Av)$$

This means that $A = \nabla^2$ is self-adjoint.

EXAMPLE 13.6.5. Use the inner product $(\mathbf{u}, \mathbf{v}) = \iiint_D (\mathbf{u} \cdot \mathbf{v})\, dV$ and the operator $A = \text{grad}$ to show that $A^* = -\text{div}$ provided the functions vanish on the boundary surface $\partial D$ of $D$.

We use the divergence theorem to obtain

$$(A\phi, \mathbf{v}) = \iiint_D \text{grad}\,(\phi \cdot \mathbf{v})\, dV = \iiint_D [\text{div}\,(\phi\mathbf{v}) - \phi\, \text{div}\,\mathbf{v}]\, dV$$

$$= \iiint_D \phi(-\text{div}\,\mathbf{v})\, dV + \iint_{\partial D} (\hat{\mathbf{n}} \cdot \phi\mathbf{v})\, dS$$

$$= \iiint_D \phi(-\text{div}\,\mathbf{v})\, dV = (\phi, A^* v)$$

In the calculus of variations it is a common practice to use $\delta u$, $\delta^2 u$, etc. to denote the first and second variations of a function $u$. Thus $\delta$ can be regarded as an operator that changes $u$ into $\delta u$, $u_x$ into $\delta(u_x)$, and $u_{xx}$ into $\delta(u_{xx})$ with the meaning, $\delta u = \varepsilon v$, $\delta(u_x) = \varepsilon v_x$, $\delta(u_{xx}) = \varepsilon v_{xx}$ where $\varepsilon$ is a small arbitrary real parameter. The operators $\delta$, $\delta^2$ are called the *first* and *second variational operators* respectively.

Some simple properties of $\delta$ are

$$\frac{\partial}{\partial x}(\delta u) = \frac{\partial}{\partial x}(\varepsilon v) = \varepsilon \frac{\partial v}{\partial x} = \delta\left(\frac{\partial u}{\partial x}\right) \tag{13.6.1}$$

$$\delta\left[\int_a^b u\, dx\right] = \varepsilon \int_a^b v\, dx = \int_a^b \varepsilon v\, dx = \int_a^b (\delta u)\, dx \tag{13.6.2}$$

The variational operator can be interchanged with the differential and integral operators, and proves to be very useful in the calculation of the variation of a functional.

The main task of the calculus of variations is concerned with the problem of minimizing or maximizing a functional involved in mathematical, physical, and engineering problems. The variational principles have their origins in the simplest kind of variational problem, which was first considered by Euler (1744) and Lagrange (1760–61).

The classical Euler–Lagrange variational problem is to determine the extremum value of the functional

$$I(u) = \int_a^b F(x, u, u')\, dx, \qquad u' = \frac{du}{dx} \tag{13.6.3}$$

with the boundary conditions

$$u(a)=\alpha \quad \text{and} \quad u(b)=\beta \tag{13.6.4ab}$$

where $u$ belongs to the class $C^{(2)}[a, b]$ of functions which have continuous derivatives up to second order in $a \leqslant x \leqslant b$, and $F$ has continuous second order derivatives with respect to all of its arguments.

We assume that $I(u)$ has an extremum at some $u \in C^{(2)}$. Then we consider the set of all variations $u+\varepsilon v$ for fixed $u$ where $v$ is an arbitrary function belonging to $C^{(2)}$ such that $v(a)=v(b)=0$. We next consider the increment of the functional

$$\delta I = I(u+\varepsilon v) - I(u) = \int_a^b [F(x, u+\varepsilon v, u'+\varepsilon v') - F(x, u, u')]\, dx \tag{13.6.5}$$

Using the Taylor series expansion

$$F(x, u+\varepsilon v, u'+\varepsilon v') = F(x, u, u') + \varepsilon\left(v\frac{\partial F}{\partial u} + v'\frac{\partial F}{\partial u'}\right) + \frac{\varepsilon^2}{2!}\left(v\frac{\partial F}{\partial u} + v'\frac{\partial F}{\partial u'}\right)^2 + \ldots$$

it follows from (13.6.5) that

$$I(u+\varepsilon v) = I(u) + \varepsilon\delta I + \frac{\varepsilon^2}{2!}\delta^2 I + \ldots \tag{13.6.6}$$

where the first and second variations of $I$ are given by

$$\delta I = \int_a^b \left(v\frac{\partial F}{\partial u} + v'\frac{\partial F}{\partial u'}\right) dx \tag{13.6.7}$$

$$\delta^2 I = \int_a^b \left(v\frac{\partial F}{\partial u} + v'\frac{\partial F}{\partial u'}\right)^2 dx \tag{13.6.8}$$

Thus the necessary condition for the functional $I(u)$ to have an extremum (that is, $I(u)$ is stationary at $u$) is that the first variation becomes zero at $u$ so that

$$0 = \delta I = \int_a^b \left(v\frac{\partial F}{\partial u} + v'\frac{\partial F}{\partial u'}\right) dx \tag{13.6.9}$$

which is, by partial integration of the second integral,

$$= \int_a^b \left[\frac{\partial F}{\partial u} - \frac{d}{dx}\left(\frac{\partial F}{\partial u'}\right)\right] v\, dx + \left[v\frac{\partial F}{\partial u'}\right]_a^b$$

This means that

$$\int_a^b \left[\frac{\partial F}{\partial u} - \frac{d}{dx}\left(\frac{\partial F}{\partial u'}\right)\right] v\, dx = 0 \tag{13.6.10}$$

Since $v$ is arbitrary in $a \leqslant x \leqslant b$, it follows from (13.6.10) that

$$\frac{\partial F}{\partial u} - \frac{d}{dx}\left(\frac{\partial F}{\partial u'}\right) = 0 \tag{13.6.11}$$

which is the famous *Euler-Lagrange equation.* We therefore can state:

***Theorem 13.6.1*** *A necessary condition for the functional $I(u)$ to be stationary at $u$ is that $u$ is the solution of the Euler-Lagrange equation*

$$\frac{\partial F}{\partial u} - \frac{d}{dx}\left(\frac{\partial F}{\partial u'}\right) = 0, \qquad a \leqslant x \leqslant b \tag{13.6.12}$$

*with*

$$u(a) = \alpha, \qquad u(b) = \beta \tag{13.6.13ab}$$

This is the *Euler-Lagrange variational principle.*

After we have determined the function $u$ which makes $I(u)$ stationary, the question arises on the nature of the maximum and the extremum, that is, its minimum, maximum, or saddle point properties. To answer this question, we look at the second variation defined in (13.6.8). If terms of $O(\varepsilon^3)$ can be neglected in (13.6.6), or if they vanish for the case of quadratic $F$, it follows from (13.6.6) that a necessary condition for the functional $I(u)$ to have a minimum $I(u) \geqslant I(u_0)$ at $u = u_0$ or a maximum $I(u) \leqslant I(u_0)$ at $u = u_0$ is that $\delta^2 I \geqslant 0$ or $\leqslant 0$ at $u = u_0$ respectively for all admissible values of $v$. These results enable us to determine the upper or lower bounds for the stationary value $I(u_0)$ of the functional.

EXAMPLE 13.6.6. Find out the shortest distance between given points $A$ and $B$ in the $x - y$ plane.

Suppose $APB$ is any curve in the plane through $A$ and $B$, and $s = \text{arc}\, AP$. Thus the problem is to determine the curve for which the functional

$$I(y) = \int_A^B ds \tag{13.6.14}$$

is a minimum.

Since $ds/dx = (1 + y'^2)^{\frac{1}{2}}$, (13.6.14) becomes

$$I(y) = \int_{x_1}^{x_2} (1 + y'^2)^{\frac{1}{2}}\, dx \tag{13.6.15}$$

In this case, $F = (1 + y'^2)^{\frac{1}{2}}$ which depends on $y'$ only, so $\partial F/\partial y = 0$. Hence the Euler-Lagrange equation is

$$\frac{d}{dx}\left(\frac{\partial F}{\partial y'}\right) = 0$$

or

$$y'' = 0 \tag{13.6.16}$$

This means that the curvature for all points on the curve $AB$ is zero. Hence the path $AB$ is a straight line. It follows from integration of (13.6.16) that $y = mx + c$ is a two-parameter family of lines.

EXAMPLE 13.6.7 (Fermat principle in optics). In an optically homogeneous isotropic medium, light travels from one point $A$ to another point $B$ along a path for which the travel time is minimum.

The velocity of light $v$ is the same at all points of the medium, and hence the minimum time is equivalent to the minimum path length. For simplicity consider a path joining the two points $A$ and $B$ in the $x-y$ plane. The time to travel an elementary arc length $ds$ is $ds/v$. Thus the variational problem is to find the path for which

$$\int_A^B \frac{ds}{v} = \int_{x_1}^{x_2} \frac{(1+y'^2)^{\frac{1}{2}}\,dx}{v} = \int_{x_1}^{x_2} F(y, y')\,dx \tag{13.6.17}$$

is a minimum, where $y' = dy/dx$, and $v = v(y)$.

When $F$ is a function of $y$ and $y'$, then the Euler-Lagrange equation (13.6.1) becomes

$$\frac{d}{dx}(F - y'F_{y'}) = 0 \tag{13.6.18}$$

This follows from the result

$$\begin{aligned}\frac{d}{dx}(F - y'F_{y'}) &= \frac{d}{dx}F(y, y') - y''F_{y'} - y'\frac{d}{dx}F_{y'}\\ &= y'F_y + y''F_y - y''F_y - y'\frac{d}{dx}(F_{y'})\\ &= y'\left[F_y - \frac{d}{dx}(F_{y'})\right] = 0, \text{ by (13.6.1).}\end{aligned}$$

Hence

$$F - y'F_{y'} = \text{constant} \tag{13.6.19}$$

Or

$$\frac{(1+y'^2)^{\frac{1}{2}}}{v} - \frac{y'^2}{v(1+y'^2)^{\frac{1}{2}}} = \text{constant}$$

Or

$$v^{-1}(1+y'^2)^{-\frac{1}{2}} = \text{constant} \tag{13.6.20}$$

In order to give a simple physical interpretation, we rewrite (13.6.20) in terms of the angle $\phi$ made by the tangent to the minimum path with the vertical $y$-axis so that

$$\sin\phi = (1+y'^2)^{-\frac{1}{2}}. \quad \text{Hence}$$

$$\frac{1}{v}\sin\phi = \text{constant} = K \tag{13.6.21}$$

for all points on the minimum curve. For a ray of light $1/v$ must be directly proportional to the refractive index $n$ of the medium through which light is travelling. Equation (13.6.21) is called the *Snell law of refraction of light.* Often this law can be stated as

$$n \sin \phi = \text{constant} \tag{13.6.22}$$

## (A) Hamilton Principle

The difference between the kinetic energy $T$ and the potential energy $V$ of a dynamical system is denoted by $L = T - V$. The quantity $L$ is called the *Lagrangian* of the system. The *Hamilton principle* states that the first variation of the time integral of $L$ is zero, that is

$$\delta \int_{t_1}^{t_2} L\, dt = \delta \int_{t_1}^{t_2} (T - V)\, dt = 0 \tag{13.6.23}$$

This result is supposed to be valid for all dynamic systems whether they are conservative or non-conservative.

For a conservative system the force field $\mathbf{F} = -\nabla V$ and $T + V = C$ where $C$ is a constant, so (13.6.23) gives the *principle of least action*

$$\delta A = 0, \qquad A = \int_{t_i}^{t_2} L\, dt \tag{13.6.24ab}$$

where $A$ is called the *action* of the system.

EXAMPLE 13.6.8. Derive the Newton Second law of motion from the Hamilton principle.

Consider a particle of mass $m$ at the position $\mathbf{r} = (x, y, z)$ which is moving under the action of a field of force $\mathbf{F}$. The kinetic energy of the particle is $T = \frac{1}{2}m\dot{\mathbf{r}}^2$, and the variation of work done is $\delta W = \mathbf{F} \cdot \delta\mathbf{r}$ and $\delta V = -\delta W$. Thus the Hamilton principle for the system is

$$0 = \delta \int_{t_1}^{t_2} (T - V)\, dt = \int_{t_1}^{t_2} (\delta T - \delta V)\, dt = \int_{t_1}^{t_2} (m\dot{\mathbf{r}} \cdot \delta\dot{\mathbf{r}} + \mathbf{F} \cdot \delta\mathbf{r})\, dt$$

Integrating this result by parts and noting that $\delta\mathbf{r}$ vanishes at $t = t_1$ and $t = t_2$, we obtain

$$\int_{t_1}^{t_2} (m\ddot{\mathbf{r}} - \mathbf{F}) \cdot \delta\mathbf{r}\, dt = 0$$

This is true for every virtual displacement $\delta\mathbf{r}$, and hence the integrand must vanish, that is,

$$m\ddot{\mathbf{r}} = \mathbf{F} \tag{13.6.25}$$

This is the Newton second law of motion.

EXAMPLE 13.6.9. Derive the equation for a simple harmonic oscillator in a non-resisting medium from the Hamilton principle.

For a simple harmonic oscillator, $T = \frac{1}{2} m\dot{x}^2$ and $V = \frac{1}{2}m\omega^2 x^2$. According to the Hamilton principle

$$\delta \int_{t_1}^{t_2} \left(\frac{1}{2} m\dot{x}^2 - \frac{1}{2} m\omega^2 x^2\right) dt = \delta \int_{t_1}^{t_2} F(x, \dot{x})\, dt = 0$$

This leads to the Euler–Lagrange equation

$$\frac{\partial F}{\partial x} - \frac{d}{dt}(m\dot{x}) = 0$$

or

$$\ddot{x} + \omega^2 x = 0 \tag{13.6.26}$$

This is the equation for the simple harmonic oscillator.

EXAMPLE 13.6.10. A straight uniform elastic beam of length $l$, line density $\rho$, cross-sectional moment of inertia $I$, and modulus of elasticity $E$ is fixed at each end. The beam performs small transverse oscillations in the horizontal $x - y$ plane. Derive the equation of motion of the beam.

The potential energy of the elastic beam is

$$V = \frac{1}{2} \int_0^l \frac{M^2}{EI}\, dx = \frac{1}{2} \int_0^l EIy''^2\, dx$$

where the bending moment $M$ is proportional to the curvature so that

$$M = EI \frac{y''}{(1 + y'^2)^{\frac{3}{2}}} \sim EIy'' \qquad \text{for small } y'$$

The variational principle gives

$$\delta \int_{t_1}^{t_2} (T - V)\, dt = \delta \int_{t_1}^{t_2} F(y'', \dot{y})\, dt = 0$$

where

$$F(y'', \dot{y}) = \frac{1}{2} \int_0^l (\rho \dot{y}^2 - EIy''^2)\, dx$$

This principle leads to the Euler–Lagrange equation

$$-\int_0^l (\rho \ddot{y} + EIy^{iv})\, dx = 0$$

or

$$\rho \ddot{y} + EIy^{iv} = 0 \tag{13.6.27}$$

This represents the partial differential equation of the transverse vibration of the beam.

## (B) The Generalized Coordinates, Lagrange Equation, and Hamilton Equation

The Euler–Lagrange analysis of a dynamical system can be extended to more complex cases where the configuration of the system is described by *generalized coordinates* $q_1, q_2, \ldots, q_n$. Without loss of generality, we consider a system of three variables where the familiar Cartesian coordinates $x$, $y$, $z$ can be expressed in terms of the generalized coordinates $q_1, q_2, q_3$ as

$$x = x(q_1, q_2, q_3), \qquad y = y(q_1, q_2, q_3), \qquad z = z(q_1, q_2, q_3) \qquad (13.6.28)$$

For example, if $(q_1, q_2, q_3)$ represents the cylinincal polar coordinates $(r, \theta, z)$, the above result becomes

$$x = r\cos\theta, \qquad y = r\sin\theta, \qquad z = z$$

Since the coordinates are functions of time $t$, we obtain the following result by differentiation

$$\dot{x} = \frac{\partial x}{\partial q_1}\dot{q}_1 + \frac{\partial x}{\partial q_2}\dot{q}_2 + \frac{\partial x}{\partial q_3}\dot{q}_3 \qquad (13.6.29)$$

with similar expressions for $\dot{y}$ and $\dot{z}$.

If these results are substituted in $T = \frac{1}{2}m(\dot{x}^2 + \dot{y}^2 + \dot{z}^2)$ and $V = V(x, y, z)$, then both $T$ and $V$ can be written in terms of the generalized coordinates $q_i$ and the generalized velocities $\dot{q}_i$ as

$$T = T(q_1, q_2, q_3; \dot{q}_1, \dot{q}_2, \dot{q}_3), \qquad V = V(q_1, q_2, q_3) \qquad (13.6.30ab)$$

so that the Lagrangian has the form

$$L = T - V = L(q_1, q_2, q_3; \dot{q}_1, \dot{q}_2, \dot{q}_3) \qquad (13.6.31)$$

The Hamilton principle gives

$$\delta \int_{t_1}^{t_2} L(q_1, q_2, q_3; \dot{q}_1, \dot{q}_2, \dot{q}_3)\, dt = 0 \qquad (13.6.32)$$

The simple variation of this integral with fixed end points, interchange of the variation operations and time derivatives for the variation of the generalized velocities, and then integration by parts yield

$$\int_{t_1}^{t_2} \left[ \sum_{i=1}^{3} \left\{ \frac{\partial L}{\partial q_1} - \frac{d}{dt}\left( \frac{\partial L}{\partial \dot{q}_1} \right) \right\} \delta q_i \right] dt = 0 \qquad (13.6.33)$$

where the integrated components vanish by virtue of the conditions $\delta q_i = 0$ $(i = 1, 2, 3)$ at $t = t_1$ and $t = t_2$.

When the generalized coordinates are independent and the variations $\delta q_i$ are independent for all $t$ in $(t_1, t_2)$, the coefficients of the variations $\delta q_i$ vanish independently for arbitrary values of $t_1$ and $t_2$. This means that the integrand in (13.5.33) vanishes, that is,

$$\frac{d}{dt}\left(\frac{\partial L}{\partial \dot{q}_i}\right) - \frac{\partial L}{\partial q_i} = 0, \qquad i = 1, 2, 3 \tag{13.6.34}$$

These are called the *Lagrange equations* of motion.

If a particle of mass $m$ at position $r = (x_1, x_2, x_3)$ moves under the action of a conservative force field $F_i = -\partial V/\partial x_i$, the Lagrangian function is

$$L = T - V = \frac{1}{2}m(\dot{x}_1^2 + \dot{x}_2^2 + \dot{x}_3^2) - V(x_1, x_2, x_3) \tag{13.6.35}$$

Consequently,

$$\frac{\partial L}{\partial \dot{x}_i} = m\dot{x}_i, \qquad \frac{\partial L}{\partial x_i} = -\frac{\partial V}{\partial x_i} = F_i \tag{13.6.36ab}$$

The former represents the momentum of the particle and the latter is the force acting on the particle. In view of (13.6.36ab), the Lagrange equation (13.6.34) gives the Newton second law of motion in the form

$$\frac{d}{dt}(m\dot{x}_i) = F_i \tag{13.6.37}$$

EXAMPLE 13.6.11. Apply the Lagrange equations of motion to derive the equations of motion of a particle under the action of a central force $-mF(r)$ where $r$ is the distance of the particle of mass $m$ from the center of force.

It is convenient to use the polar coordinates $r$ and $\theta$. In terms of the generalized coordinates $q_1 = r$ and $q_2 = \theta$, we write

$$x = r\cos\theta = q_1 \cos q_2, \qquad y = r\sin\theta = q_1 \sin q_2$$

The kinetic energy $T$ is

$$T = \frac{1}{2}m(\dot{x}^2 + \dot{y}^2) = \frac{1}{2}m(\dot{r}^2 + r^2\dot{\theta}^2) = \frac{1}{2}m(\dot{q}_1^2 + q_1^2\dot{q}_2^2) \tag{13.6.38}$$

Since $\mathbf{F} = \nabla V$, the potential is

$$V(r) = \int^r F(r)\,dr = \int^{q_1} F(q_1)\,dq_1 \tag{13.6.39}$$

Then the Lagrangian $L$ is

$$L = T - V = \frac{1}{2}m\left[\dot{q}_1^2 + q_1^2\dot{q}_2^2 - 2\int^{q_1} F(q_1)\,dq_1\right] \tag{13.6.40}$$

Thus the Lagrange equations (13.6.34) with $i = 1, 2, 3$ gives the equations of motion

$$\ddot{q}_1 - q_1\dot{q}_2^2 + F(q_1) = 0, \qquad \frac{d}{dt}(q_1^2\dot{q}_2) = 0 \tag{13.6.41ab}$$

In terms of the polar coordinates, these equations become

$$\ddot{r} - r\dot{\theta}^2 = -F(r), \qquad \frac{d}{dt}(r^2\dot{\theta}) = 0 \tag{13.6.42ab}$$

Equation (13.5.36b) gives immediately

$$r^2\dot{\theta} = h \tag{13.6.43}$$

where $h$ is a constant. In this case, $r\dot{\theta}$ represents the transverse velocity component, and $mr^2\dot{\theta} = mh$ is the constant angular momentum of the particle about the center of force.

Introducing $r = 1/u$, we obtain

$$\dot{r} = \frac{dr}{dt} = -\frac{1}{u^2}\frac{du}{dt} = -\frac{1}{u^2}\frac{du}{d\theta}\cdot\frac{d\theta}{dt} = -h\frac{du}{d\theta}$$

$$\ddot{r} = \frac{d^2r}{dt^2} = -h\frac{d}{dt}\left(\frac{du}{d\theta}\right) = -h\frac{d^2u}{d\theta^2}\frac{d\theta}{dt} = -h^2u^2\frac{d^2u}{d\theta^2}$$

Substituting these into (13.6.42a) gives

$$-h^2u^2\frac{d^2u}{d\theta^2} - h^2u^3 = -F\left(\frac{1}{u}\right)$$

or

$$\frac{d^2u}{d\theta^2} + u = \frac{1}{h^2u^2}F\left(\frac{1}{u}\right) \tag{13.6.44}$$

This is the differential equation of the central orbit and can be solved by standard methods.

In particular, if the law of force is the attractive inverse square $F(r) = \mu/r^2$ so that the potential $V(r) = -\mu/r$, the differential equation (13.6.44) becomes

$$\frac{d^2u}{d\theta^2} + u = \frac{\mu}{h^2} \tag{13.6.45}$$

if the particle is projected initially from distance $a$ with velocity $V$ at an angle $\beta$ that the direction of motion makes with the outward radius vector. Thus the constant $h$ in (13.6.43) is $h = Va\sin\beta$. The angle $\phi$ between the tangent and radius vector of the orbit at any point is given by

$$\cot\phi = \frac{1}{r}\frac{dr}{d\theta} = u\frac{d}{d\theta}\left(\frac{1}{u}\right) = -\frac{1}{u}\frac{du}{d\theta} \tag{13.6.46}$$

At $t = 0$, the initial conditions are

$$u = \frac{1}{a}, \qquad \frac{du}{d\theta} = -\frac{1}{a}\cot\beta \quad \text{when} \quad \theta = 0 \tag{13.6.47}$$

The general solution of (13.6.45) is

$$u = \frac{\mu}{h^2}[1 + e\cos(\theta + \alpha)] \tag{13.6.48}$$

when $e$ and $\alpha$ are constants to be determined from the initial data.

Finally, the solution can be written as

$$\frac{l}{r} = 1 + e\cos(\theta + \alpha) \tag{13.6.49}$$

where

$$l = \frac{h^2}{\mu} = (Va\sin\beta)^2/\mu \tag{13.6.50}$$

This represents a conic section of semi-latus rectum $l$ and eccentricity $e$ with its axis inclined at an angle $\alpha$ to the radius vector at the point of projection.

The initial conditions (13.6.47) lead to

$$\frac{l}{a} = 1 + e\cos\alpha, \qquad -\frac{1}{a}\cot\beta = -e\sin\alpha \tag{13.6.51ab}$$

which give

$$\tan\alpha = \frac{l\cot\beta}{l - a}$$

$$e^2 = \left(\frac{l}{a} - 1\right)^2 + \frac{l^2}{a^2}\cot^2\beta = \frac{l^2}{a^2}\operatorname{cosec}^2\beta - \frac{2l}{a} + 1$$

$$= 1 - \frac{2aV^2\sin^2\beta}{\mu} + \frac{a^2V^4\sin^2\beta}{\mu^2} \tag{13.6.52}$$

Thus the conic is an ellipse, parabola, or hyperbola accordingly as $e <=> 1$ that is, $V^2 <=> 2\mu/a$.

To derive the Hamilton equations, we introduce the concept of generalized momentum, $p_i$ and generalized force, $F_i$ as

$$p_i = \frac{\partial L}{\partial \dot{q}_i}, \qquad F_i = \frac{\partial L}{\partial q_i} \tag{13.6.53ab}$$

Consequently, the Lagrange equations (13.6.34) become

$$\frac{\partial L}{\partial q_i} = \frac{d}{dt}p_i = \dot{p}_i \tag{13.6.54}$$

The Hamiltonian function $H$ is defined by

$$H = \sum_{i=1}^{n} p_i \dot{q}_i - L \tag{13.6.55}$$

In general, $L = L(q_i, \dot{q}_i, t)$ is a function of $q_i$, $\dot{q}_i$ and $t$ where $\dot{q}_i$ enters through the kinetic energy as a quadratic term. Hence Eq. (13.6.53a) will give $p_i$ as a linear function of $\dot{q}_i$. This system of linear equations involving $p_i$ and $\dot{q}_i$ can be solved to determine $\dot{q}_i$ in terms of $p_i$, and then the $\dot{q}_i$'s can in principle be eliminated from (13.6.55). This means that $H$ can always be expressed as a function of $p_i$, $q_i$, and $t$ so that $H = H(p_i, q_i, t)$. Thus

$$dH = \sum \frac{\partial H}{\partial p_i} dp_i + \sum \frac{\partial H}{\partial q_i} dq_i + \frac{\partial H}{\partial t} dt \tag{13.6.56}$$

On the other hand, differentiating $H$ in (13.6.55) with respect to $t$ gives

$$\begin{aligned}\frac{dH}{dt} &= \sum p_i \frac{d}{dt}\dot{q}_i + \sum \dot{q}_i \frac{d}{dt} p_i - \sum \frac{\partial L}{\partial q_i}\frac{d}{dt} q_i \\ &\quad - \sum \frac{\partial L}{\partial \dot{q}_i}\frac{d}{dt}\dot{q}_i - \frac{\partial L}{\partial t}\end{aligned} \tag{13.6.57}$$

Or

$$dH = \sum p_i \, d\dot{q}_i + \sum \dot{q}_i \, dp_i - \sum \frac{\partial L}{\partial q_i} dq_i - \sum \frac{\partial L}{\partial \dot{q}_i} d\dot{q}_i - \frac{\partial L}{\partial t} dt \tag{13.6.58}$$

which becomes, in view of (13.6.53a),

$$dH = \sum \dot{q}_i \, dp_i - \sum \frac{\partial L}{\partial q_i} dq_i - \frac{\partial L}{\partial t} dt \tag{13.6.59}$$

Evidently, two expressions of $dH$ in (13.6.56) and (13.6.59) must be equal so that the coefficients of the corresponding differentials can be equated to obtain

$$\dot{q}_i = \frac{\partial H}{\partial p_i}, \qquad -\frac{\partial L}{\partial q_i} = \frac{\partial H}{\partial q_i}, \qquad -\frac{\partial L}{\partial t} = \frac{\partial H}{\partial t} \tag{13.6.60abc}$$

Using the Lagrange equation (13.6.54), the first two of the above equations become

$$\dot{q}_i = \frac{\partial H}{\partial p_i}, \qquad \dot{p}_i = -\frac{\partial H}{\partial q_i} \tag{13.6.61ab}$$

These are commonly known as the *Hamilton canonical equations* of motion. They play a fundamental role in advanced dynamics.

Finally, the Lagrange–Hamilton theory can be used to derive the law of conservation of energy. In general, the Lagrangian $L$ is independent of time $t$ and hence (13.6.60c) implies that $H = \text{constant}$. Again, $T$ involved in $L = T - V$ is given by

$$T = \frac{1}{2}\sum_{i=1}^{n}\sum_{j=1}^{n} a_{ij}\dot{q}_i\dot{q}_j \tag{13.6.62}$$

where the coefficients $a_{ij}$ are symmetric functions of the generalized coordinates $q_i$, that is, $a_{ij} = a_{ji}$.

On the other hand, $V$ is, in general, independent of $q_i$ and hence

$$p_i = \frac{\partial L}{\partial \dot{q}_i} = \frac{\partial T}{\partial \dot{q}_i} = \sum_{j=1}^{n} a_{ij}\dot{q}_j \tag{13.6.63}$$

Thus the Hamiltonian $H$ becomes

$$H = \sum_{i=1}^{n} p_i\dot{q}_i - L = \sum_{i=1}^{n}\left(\sum_{j=1}^{n} a_{ij}\dot{q}_j\right)\dot{q}_i - L = 2T - L = T + V \tag{13.6.64}$$

Thus $H$ is equal to the total energy. It has already been observed that, if $L$ does not contain $t$ explicitly, $H$ is a constant. This means that the sum of the potential and kinetic energies is constant. This is the *law of the conservation of energy*.

EXAMPLE 13.6.12. Use the Hamiltonian equations to derive the equations of motion for the problem stated in Example 13.6.11.

The Lagrangian $L$ for this problem is given by (13.6.40) with $q_1 = r$ and $q_2 = \theta$. It follows from the definition (13.6.53) of the generalized momentum that

$$p_1 = m\dot{q}_1 = m\dot{r}, \qquad p_2 = mq_1^2\dot{q}_2 = mr^2\dot{\theta} \tag{13.6.65ab}$$

Expressing the results of the kinetic energy (13.6.38) and the potential energy (13.6.39) in terms of $p_1$ and $p_2$, the Hamiltonian $H = T + V$ can be written as

$$H = \frac{1}{2m}\left(p_1^2 + \frac{p_2^2}{q_1^2}\right) + m\int^{q_1} F(q_1)\,dq_1 \tag{13.6.66}$$

Then the Hamilton equations (13.6.61ab) with $q_1 = r$ and $q_2 = \theta$ give

$$p_1 = m\dot{r}, \qquad p_2 = mr^2\dot{\theta} \tag{13.6.67ab}$$

$$\dot{p}_1 = \frac{1}{m}\frac{p_2}{q_1^3} + mF(q_1), \qquad \dot{p}_2 = 0 \tag{13.6.68ab}$$

Clearly, these equations are identical with the equations of motion (13.6.36ab).

EXAMPLE 13.6.13. Derive the equation of a simple pendulum by using (i) the Lagrange equations and (ii) the Hamilton equations.

We consider the motion of a simple pendulum of mass $m$ attached at the end of a rigid massless string of length $l$ that pivots about a fixed point. We suppose that the pendulum makes an angle $\theta$ with its vertical position. The force $F$ acting on the mass $m$ is $F = -mg \sin \theta$, so that the potential $V$ is obtained from $F = -\nabla V$ as $V = mgl\ (1 - \cos \theta)$. The kinetic energy $T = \frac{1}{2} m l^2 \dot{\theta}^2$.

Thus the Lagrangian $L$ is

$$L = T - V = \frac{1}{2} m l^2 \dot{\theta}^2 - mg(1 - \cos \theta) = L(\theta, \dot{\theta}) \qquad (13.6.69)$$

The Lagrange equation is

$$\frac{\partial L}{\partial \theta} - \frac{d}{dt}\left(\frac{\partial L}{\partial \dot{\theta}}\right) = 0 \qquad (13.6.70)$$

or

$$-mgl \sin \theta - \frac{d}{dt}(m l^2 \dot{\theta}) = 0$$

or

$$\ddot{\theta} + \omega^2 \sin \theta = 0, \qquad \omega^2 = g/l \qquad (13.6.71ab)$$

This is the equation of the simple pendulum.

To derive the same equation from the Hamilton equations, we choose $q_1 = l(\dot{q}_1 = 0)$ and $q_2 = \theta$ as the generalized (polar) coordinates. The kinetic and potential energies are

$$T = \frac{1}{2} m l^2 \dot{q}_2^2, \qquad V = mgl(1 - \cos q_2) \qquad (13.6.72ab)$$

Thus $H = T + V$ and $L = T - V$ are given by

$$(H, L) = \frac{1}{2} m l^2 \dot{q}_2^2 \pm mgl(1 - \cos q_2) \qquad (13.6.73ab)$$

From the definition of the generalized momentum, we find

$$p_2 = \frac{\partial L}{\partial \dot{q}_2} = m l^2 \dot{q}_2$$

so that

$$H = \frac{1}{2} \frac{p_2^2}{m l^2} + mgl(1 - \cos q_2)$$

Thus the Hamilton equation (13.6.61ab) gives

$$\ddot{\theta} + \omega^2 \sin \theta = 0 \qquad (13.6.74)$$

The variational methods can be further extended for functional depending on functions or more independent variables in the form

$$I[u(x, y)] = \iint_D F(x, y, u, u_x, u_y)\, dx\, dy \qquad (13.6.75)$$

where the values of the function $u(x, y)$ are prescribed on the boundary $\partial D$ of a finite domain $D$ in the $x-y$ plane. We assume that $F$ is differentiable and the surface $u = u(x, y)$ giving an extremum is also continuously differentiable twice.

The first variation $\delta I$ of $I$ is defined by

$$\delta I[u, \varepsilon] = I(u + \varepsilon) - I(u) \tag{13.6.76}$$

which is, by Taylor's expansion theorem

$$= \iint_D [\varepsilon F_u + \varepsilon_x F_p + \varepsilon_y F_q]\, dx\, dy \tag{13.6.77}$$

where $\varepsilon \equiv \varepsilon(x, y)$ is small and $p = u_x$ and $q = u_y$.

According to the variational principle, $\delta I = 0$ for all admissible values of $\varepsilon$. The partial integration of (13.6.77) combined with $\varepsilon = 0$ on $\partial D$ gives

$$0 = \delta I = \iint_D \left[F_u - \frac{\partial}{\partial x} F_p - \frac{\partial}{\partial y} F_q\right] dx\, dy \tag{13.6.78}$$

This is true for all arbitrary $\varepsilon$, and hence the integrand must vanish, that is,

$$\frac{\partial}{\partial x} F_p + \frac{\partial}{\partial y} F_q - F_u = 0 \tag{13.6.79}$$

This is the Euler-Lagrange equation which is the second order PDE to be satisfied by the extremizing function $u(x, y)$.

EXAMPLE 13.6.14. Derive the equation of motion for the free vibration of an elastic string of length $l$.

The potential energy $V$ of the string is

$$V = \frac{1}{2} T^* \int_0^l u_x^2\, dx \tag{13.6.80}$$

where $u = u(x, t)$ is the displacement of the string from its equilibrium position and $T^*$ is the constant tension of the string.

The kinetic energy $T$ is

$$T = \frac{1}{2} \int_0^l \rho u_t^2\, dx \tag{13.6.81}$$

where $\rho$ is a constant line-density of the string.

According to the Hamilton principle

$$\delta I = \delta \int_{t_1}^{t_2} (T - V)\, dt = \delta \int_{t_1}^{t_2} \int_0^l \frac{1}{2} (\rho u_t^2 - T^* u_x^2)\, dx\, dt = 0 \tag{13.6.82}$$

which has the form

$$\delta \int_{t_1}^{t_2} \int_0^l L(u_t, u_x) = 0 \tag{13.6.83}$$

where

$$L = \frac{1}{2}(\rho u_t^2 - T^* u_x^2)$$

Then the Euler-Langrange equation is

$$\frac{\partial}{\partial t}(\rho u_t) - \frac{\partial}{\partial x}(T^* u_x) = 0 \tag{13.6.84}$$

or

$$u_{tt} - c^2 u_{xx} = 0, \qquad c^2 = T^*/\rho \tag{13.6.85ab}$$

This is the equation of motion of the string.

EXAMPLE 13.6.15. Derive the Laplace equation from the functional

$$I(u) = \iint_D (u_x^2 + u_y^2)\, dx\, dy$$

with a boundary condition $u = f(x, y)$ on $\partial D$.

The variational principle gives

$$\delta I = \delta \iint_D (u_x^2 + u_y^2)\, dx\, dy = 0$$

This leads to the Euler–Langrange equation

$$u_{xx} + u_{yy} = 0 \quad \text{in} \quad D$$

Similarly, the functional

$$I[u(x, y, z)] = \iiint_D (u_x^2 + u_y^2 + u_z^2)\, dx\, dy\, dz$$

will lead to the three-dimensional Laplace equation.

## 13.7 THE RAYLEIGH–RITZ APPROXIMATE METHOD

We consider the boundary-value problem governed by the differential equation

$$Au = f \quad \text{in} \quad D \tag{13.7.1}$$

with the boundary condition

$$B(u) = 0 \quad \text{on} \quad \partial D \tag{13.7.2}$$

where $A$ is a self-adjoint differential operator in a Hilbert space $H$ and $f \in H$.

In general, the determination of the exact solution of the problem is often a difficult task. However, it can be shown that the solution of (13.7.1)–(13.7.2) is equivalent to finding the minimum of a functional $I(u)$ associated with the differential system. In other words, the solution can be characterized as the function which minimizes (or maximizes) the functional $I(u)$. For an approximate solution of the extremum problem, a simple and efficient method was independently formulated by Lord Rayleigh and W. Ritz.

We next prove a fundamental result which states that the solution of the equation (13.7.1) is equivalent to finding the minimum of the quadratic functional

$$I(u) \equiv (Au, u) - 2(f, u) \tag{13.7.3}$$

Suppose $u = u_0$ is the solution of (13.7.1) so that $Au_0 = f$. Consequently,

$$I(u) \equiv (Au, u) - 2(Au_0, u) = (A(u - u_0), u) - (Au_0, u)$$

Since the inner product is symmetrical and $(Au_0, u) = (u_0, Au) = (Au, u_0)$, $I(u)$ can be written as

$$\begin{aligned} I(u) &= (A(u - u_0), u) - (Au, u_0) + (Au_0, u_0) - (Au_0, u_0) \\ &= (A(u - u_0), u - u_0) - (Au_0, u_0) \\ &= (A(u - u_0), u - u_0) + I(u_0) \end{aligned} \tag{13.7.4}$$

Since $A$ is a positive operator, $(A(u - u_0), u - u_0) \geqslant 0$ where equality holds if and only if $u - u_0 = 0$. Thus it follows that

$$I(u) \geqslant I(u_0) \tag{13.7.5}$$

where equality holds if and only if $u = u_0$. We thus conclude from this inequality that $I(u)$ assumes its minimum as the solution $u = u_0$ of equation (13.7.1).

Conversely, the function $u = u_0$ that minimizes $I(u)$ is a solution of Eq. (13.7.1). Clearly, $I(u) \geqslant I(u_0)$, that is, in particular, $I(u_0 + \alpha v) \geqslant I(u_0)$ for any real $\alpha$ and any function $v$. Explicitly,

$$\begin{aligned} I(u_0 + \alpha v) &= (A(u_0 + \alpha v), u_0 + \alpha v) - 2(f, u_0 + \alpha v) \\ &= (Au_0, u_0) + 2\alpha(Au_0, v) + \alpha^2(Av, v) - 2(f, u_0) - 2\alpha(f, v) \end{aligned}$$

This means that $I(u_0 + \alpha v)$ is a quadratic expression in $\alpha$. Since $I(u)$ is minimum at $u = u_0$, then $\delta I(u_0, v) = 0$, that is,

$$\begin{aligned} 0 &= \left[\frac{d}{d\alpha} I(u_0 + \alpha v)\right]_{\alpha=0} \\ &= 2(Au_0, v) - 2(f, v) \\ &= 2(Au_0 - f, v) \end{aligned}$$

This is true for any arbitrary but fixed $v$. Hence $Au_0 - f = 0$. This proves the assertion.

In the Rayleigh–Ritz method an approximate solution of (13.7.1)–(13.7.2) is sought in the form

$$u_n(\mathbf{x}) = \sum_{i=1}^{n} a_i \phi_i(\mathbf{x}) \tag{13.7.6}$$

where $a_1, a_2, \ldots, a_n$ are $n$ unknown coefficients to be determined so that $I(u_n)$ is minimum, and $\phi_1, \phi_2, \ldots, \phi_n$ represent a linearly independent and complete set of arbitrarily chosen functions that satisfy (13.7.2). This set of functions is often called a *trial set.* We substitute (13.7.6) into (13.7.3) to obtain

$$I(u_n) = \left(\sum_{i=1}^{n} a_i A(\phi_i), \sum_{j=1}^{n} a_j \phi_j\right) - 2\left(f, \sum_{i=1}^{n} a_i \phi_i\right)$$

Then the necessary condition for $I$ to obtain a minimum (or maximum) is that

$$\frac{\partial I}{\partial a_j}(a_1, a_2, \ldots, a_n) = 0, \qquad j = 1, 2, \ldots, n \tag{13.7.7}$$

or

$$\frac{\partial}{\partial a_j}\left[\left(\sum_{i=1}^{n} a_i A(\phi_i), \sum_{j=1}^{n} a_j \phi_j\right) - 2\left(f, \sum_{i=1}^{n} a_i \phi_i\right)\right] = 0$$

or

$$\sum_{i=1}^{n} \Big(A(\phi_i), \phi_j\Big) a_i + \sum_{j=1}^{n} (A(\phi_j), \phi_i) a_j - 2(f, \phi_j) = 0$$

or

$$2\sum_{i=1}^{n} (A(\phi_i), \phi_j) a_i = 2(f, \phi_j)$$

Therefore

$$\sum_{i=1}^{n} (A(\phi_i), \phi_j) a_i = (f, \phi_j), \qquad j = 1, 2, \ldots, n \tag{13.7.8}$$

This is a linear system of $n$ equations for the $n$ unknown coefficients $a_j$. Once $a_1, a_2, \ldots, a_n$ are determined, the approximate solution is given by (13.7.6).

In particular, when

$$(A(\phi_i), \phi_j) = \begin{Bmatrix} 0, & i \neq j \\ 1, & i = j \end{Bmatrix} \tag{13.7.9}$$

the equation (13.7.8) gives $a_j$ as

$$a_j = (\phi_j, f) \tag{13.7.10}$$

so that the Rayleigh–Ritz approximate series (13.7.6) becomes

$$u_n(\mathbf{x}) = \sum_{i=1}^{n} (\phi_i, f)\phi_i(\mathbf{x}) \tag{13.7.11}$$

This is similar to the Fourier series solution with known Fourier coefficients $a_i$.

In the limit $n \to \infty$, a limit function can be obtained from (13.7.6) as

$$u(\mathbf{x}) = \lim_{n\to\infty} \sum_{i=1}^{n} a_i\phi_i(\mathbf{x}) = \sum_{i=1}^{\infty} a_i\phi_i(\mathbf{x}) \tag{13.7.12}$$

provided the series converges. Under certain assumptions imposed on the functional $I(u)$ and the trial functions $\phi_1, \phi_2, \ldots, \phi_n$, $u(\mathbf{x})$ represents an exact solution of the problem. In any event, (13.7.6) or (13.7.11) give a reasonable approximate solution.

In the simplest case corresponding to $n = 1$, the Rayleigh–Ritz method gives a simple form of the functional

$$I(u_1) = I(a_1\phi_1) = a_1^2(A\phi_1, \phi_1) - 2a_1(f, \phi_1)$$

where $a_1$ is readily determined from the necessary condition for extremum

$$0 = \frac{\partial}{\partial a_1} I(a_1\phi_1) = 2a_1(A\phi_1, \phi_1) - 2(f, \phi_1)$$

or

$$a_1 = \frac{(f, \phi_1)}{(A\phi_1, \phi_1)} \tag{13.7.13}$$

The corresponding minimum value of the functional is given by

$$I(a_1\phi_1) = -\frac{(f, \phi_1)^2}{(A\phi_1, \phi_1)} \tag{13.7.14}$$

Thus the essence of the Rayleigh–Ritz method can be summarized as follows. For a given boundary-value problem, an approximate series solution is sought so that the trial functions $\phi_i$ satisfy the boundary conditions. We solve the system of algebraic equations (13.7.7) to determine the coefficients $a_i$.

We now illustrate the method by several examples.

EXAMPLE 13.7.1. Find an approximate solution of the Dirichlet problem

$$\nabla^2 u \equiv u_{xx} + u_{yy} = 0 \quad \text{in} \quad D$$

$u = f$ on $\partial D$ where $D \subset R^2$, and $f$ is a given function.

This problem is equivalent to finding the minimum of the associated functional

$$I(u) = \iint_D (u_x^2 + u_y^2)\, dx\, dy$$

We seek an approximate series solution in the form

$$u_2(x, y) = a_1\phi_1 + a_2\phi_2$$

with $a_1 = 1$ so that $u_2$ satisfies the given boundary conditions, that is, $\phi_1 = f$ and $\phi_2 = 0$ on $\partial D$. Substituting $u_2$ into the functional gives

$$I(u_2) = \iint_D \left[\left(\frac{\partial u_2}{\partial x}\right)^2 + \left(\frac{\partial u_2}{\partial y}\right)^2\right] dx\, dy$$

$$= \iint_D (\nabla\phi_1)^2\, dx\, dy + 2a_2 \iint_D (\nabla\phi_1 \cdot \nabla\phi_2)\, dx\, dy + a_2^2 \iint_D |\nabla\phi_2|^2\, dx\, dy$$

The necessary condition for an extremum of $I(u_2)$ is

$$\frac{\partial I}{\partial a_2} = 0$$

or

$$2\iint_D (\nabla\phi_1 \cdot \nabla\phi_2)\, dx\, dy + 2a_2 \iint_D |\nabla\phi_2|^2\, dx\, dy = 0$$

Therefore

$$a_2 = -\frac{\iint (\nabla\phi_1 \cdot \nabla\phi_2)\, dx\, dy}{\iint |\nabla\phi_2|^2\, dx\, dy}$$

This $a_2$ minimizes the functional and hence the approximate solution is obtained.

However, this procedure can be generalized by seeking an approximate solution in the form

$$u_n = \sum_{i=1}^{n} a_i\phi_i \qquad (a_1 = 1)$$

so that $\phi_1 = f$ and $\phi_i = 0$ $(i = 2, 3, \ldots, n)$ on $\partial D$.

The coefficients $a_i$ can be obtained by solving the system (13.7.7) with $j = 2, 3, \ldots, n$.

EXAMPLE 13.7.2. A uniform elastic beam of length $l$ carrying a uniform load $W$ per unit length is freely hinged at $x = 0$ and $x = l$. Find the approximate solution of the boundary-value problem

$$EIy^{\text{iv}}(x) = W$$
$$y = y'' = 0 \quad \text{at} \quad x = 0 \quad \text{and} \quad x = l$$

where $y = y(x)$ is the displacement function.

This problem is equivalent to finding a function $y(x)$ that minimizes the energy functional

$$I(y) = \int_0^l \left(Wy - \frac{EI}{2} y''^2\right) dx$$

We seek an approximate solution

$$y_n(x) = \sum_{r=1}^{n} a_r \sin \frac{r\pi x}{l}$$

which satisfies the boundary conditions.

Substitution of this solution into the energy functional gives

$$I(y_n) = \sum_{r=1}^{n} \left[\int_0^l Wa_r \sin \frac{r\pi x}{l} dx - \frac{EI}{2} \int_0^l \frac{r^4\pi^4}{l^4} a_r^2 \sin^2 \frac{r\pi x}{l} dx\right]$$
$$= \frac{2Wl}{\pi} \sum_{r=1}^{n} \frac{a_r}{r} - \frac{EI\pi^4}{4l^3} \sum_{r=1}^{n} r^4 a_r^2$$

The necessary conditions for extremum are

$$0 = \frac{\partial I}{\partial a_r} = \frac{2Wl}{r\pi} - \frac{EI\pi^4}{4l^3} 2a_r r^4 \qquad r = 1, 2, \ldots, n$$

which gives $a_r$ as

$$a_r = \frac{4Wl^4}{\pi^5 r^5 EI}, \qquad r = 1, 2, \ldots, n$$

Thus the approximate function $y$ is

$$y_n(x) = \frac{4Wl^4}{\pi^5 EI} \sum_{r=1}^{n} \frac{1}{r^5} \sin \frac{r\pi x}{l}$$

The maximum deflection at $x = l/2$ is

$$y_{\max} = \frac{4Wl^4}{\pi^5 EI}\left(1 - \frac{1}{3^5} + \frac{1}{5^4} - \cdots\right)$$

In this case, the first term of the series solution gives a reasonably good approximate solution as

$$y_1(x) \sim \frac{4Wl^4}{EI\pi^5} \sin \frac{\pi x}{l}$$

EXAMPLE 13.7.3. Apply the Rayleigh–Ritz method to investigate the free vibration of a fixed elastic wedge of constant thickness governed by the energy functional

$$I(y)=\int_0^1 (\alpha x^3 y''^2-\omega x y^2)\,dx, \qquad y(1)=y''(1)=0$$

where the free vibration of the wedge is described by the function $u(x,t)=e^{i\omega t}y(x)$, $\omega$ is the frequency.

We seek an approximate solution in the form

$$y_n(x)=\sum_{r=1}^{n} a_r y_r(x)=\sum_{r=1}^{n} a_r (x-1)^2 x^{r-1}$$

which satisfies the given boundary conditions.

We take only the first two terms so that $y_2(x)=a_1y_1+a_2y_2=(x-1)^2(a_1+a_2x)$. Substituting $y_2$ into the functional gives

$$I_2=I(y_2)=\int_0^1 [\alpha x^3(6a_2x+2a_1-4a_2)^2-\omega x(x-1)^4(a_1+a_2x)^2]\,dx$$

$$=\alpha\left[(a_1-2a_2)^2+\frac{24}{5}(a_1-2a_2)a_2+6a_2^2\right]-\frac{\omega}{5}\left[\frac{a_1^2}{6}+\frac{2a_1a_2}{21}+\frac{a_2^2}{56}\right]$$

The necessary conditions for an extremum are

$$\frac{\partial I_2}{\partial a_1}=2a_1\left(\alpha-\frac{\omega}{30}\right)+\frac{2}{5}a_2\left(2\alpha-\frac{\omega}{21}\right)=0$$

$$\frac{\partial I_2}{\partial a_2}=\frac{2a_1}{5}\left(2\alpha-\frac{\omega}{21}\right)+\frac{2a_2}{5}\left(2\alpha-\frac{\omega}{56}\right)=0$$

For non-trivial solutions, the determinant of this algebraic system must be zero, that is,

$$\begin{vmatrix} \alpha-\dfrac{\omega}{30} & \dfrac{1}{5}\left(2\alpha-\dfrac{\omega}{21}\right) \\ 2\alpha-\dfrac{\omega}{21} & 2\alpha-\dfrac{\omega}{56} \end{vmatrix}=0$$

or

$$5\left(\alpha-\frac{\omega}{30}\right)\left(2\alpha-\frac{\omega}{56}\right)-\left(2\alpha-\frac{\omega}{21}\right)^2=0$$

This represents the frequency equation of the vibration which has two roots $\omega_1$ and $\omega_2$. The smaller of these two frequencies gives an approximate value of the fundamental frequency of the vibration of the wedge.

EXAMPLE 13.7.4. An elastic beam of length $l$, density $\rho$, cross-sectional area $A$, and modulus of elasticity $E$ has its end $x = 0$ fixed and the other end connected to a rigid support through a linear elastic spring with spring constant $k$. Apply the Rayleigh–Ritz method to investigate the harmonic axial motion of the beam.

The kinetic energy and the potential energy associated with the axial motion of the beam are

$$T = \int_0^l \frac{\rho A}{2} U_t^2 \, dx, \qquad V = \int_0^l \frac{EA}{2} U_x^2 \, dx + \frac{k}{2} U^2(l)$$

where $U(x, t)$ is the displacement function.

Since the axial motion is simple harmonic, $U(x, t) = u(x)e^{i\omega t}$ where $\omega$ is the frequency of vibration. Consequently, the expressions for $T$ and $V$ can be written in terms of $u(x)$. We then apply the Hamilton variational principle

$$\delta I(u) = \delta \left[ \int_{t_1}^{t_2} \int_0^l \frac{1}{2} (\rho A \omega^2 u^2 - EA u_x^2) \, dx - \frac{k}{2} u^2(l) \right] dt = 0$$

The Euler equation for the variational principle is

$$\frac{d}{dx}\left(EA \frac{du}{dx}\right) + \rho A \omega^2 u = 0, \qquad 0 < x < l$$

$$EA \frac{du}{dx} + ku = 0 \quad \text{at} \quad x = l$$

In terms of non-dimensional variables $(x^*, u^*) = (1/l)(x, u)$ and parameters $\lambda = \omega^2 \rho l^2 / E$ and $\alpha = kl/EA$, this system becomes, dropping the asterisks,

$$u_{xx} + \lambda u = 0, \qquad 0 < x < 1$$

$$u_x + \alpha u = 0 \quad \text{at} \quad x = 1$$

The associated functional for the system is

$$I(u) = \frac{1}{2} \int_0^1 (\lambda u^2 - u_x^2) \, dx - \frac{\alpha}{2} u^2(1)$$

According to the Rayleigh–Ritz method, we seek approximate solution with $\alpha = 1$

$$u_2(x) = a_1 x + a_2 x^2$$

so that $I(u_2)$ is minimum.

We substitute $u_2(x)$ into the functional to obtain

$$I_2 = I(u_2) = \frac{1}{2} \int_0^1 [\lambda(a_1 x + a_2 x)^2 - (a_1 + 2a_2 x)^2] \, dx - \frac{1}{2}(a_1 + a_2)^2$$

The necessary conditions for extremum of the functional are

$$0 = \frac{\partial I_2}{\partial a_1} = a_1\left(\frac{\lambda}{3} - 2\right) + a_2\left(\frac{\lambda}{4} - 2\right)$$

$$0 = \frac{\partial I_2}{\partial a_2} = a_1\left(\frac{\lambda}{4} - 2\right) + a_2\left(\frac{\lambda}{5} - \frac{7}{3}\right)$$

For non-trivial solutions, the determinant of this linear algebraic system must be zero, that is,

$$\begin{vmatrix} \frac{\lambda}{3} - 2 & \frac{\lambda}{4} - 2 \\ \frac{\lambda}{4} - 2 & \frac{\lambda}{5} - \frac{7}{3} \end{vmatrix} = 0$$

or

$$3\lambda^2 - 128\lambda + 480 = 0$$

This equation gives two solutions:

$$\lambda_1 = 4.155, \qquad \lambda_2 = 38.512$$

The corresponding values of the frequency are

$$\omega_1 = 2.038\left(\frac{E}{\rho l^2}\right)^{\frac{1}{2}}, \qquad \omega_2 = 6.206\left(\frac{E}{\rho l^2}\right)^{\frac{1}{2}}$$

The exact solution is determined by the equation

$$\sqrt{\lambda} + \tan\sqrt{\lambda} = 0$$

The first two roots of this equation can be obtained graphically as

$$\omega_{01} \sim 2.0288\left(\frac{E}{\rho l^2}\right)^{\frac{1}{2}}, \qquad \omega_{02} \sim 4.9132\left(\frac{E}{\rho l^2}\right)^{\frac{1}{2}}$$

## 13.8 THE GALERKIN APPROXIMATION METHOD

As an extension of the Rayleigh–Ritz method, Galerkin formulated an ingeneous approximate method which may be applied to a problem in which no simple variational principle exists. The differential operator $A$ in Eq. (13.7.1) need not be linear for the solution of this equation. In order to solve the boundary-value problem (13.7.1)–(13.7.2), we construct an approximate solution $u(\mathbf{x})$ in the form

$$u_n = u_0(\mathbf{x}) + \sum_{i=1}^{n} a_i \phi_i(\mathbf{x}) \tag{13.8.1}$$

where the $\phi_i(\mathbf{x})$ are known functions, $u_0$ is introduced to satisfy the boundary conditions, and the coefficients $a_i$ are to be determined. Substituting (13.8.1) into (13.7.1) gives a non-zero *residual* $R_n$

$$R_n(a_1, a_2, \ldots, a_n, x, y) = A(u_n) = A(u_0) + \sum_{i=1}^{n} a_i A(\phi_i) \quad (13.8.2)$$

In this method the unknown coefficients $a_i$ are determined by solving the following system of equations

$$(R_n, \phi_j) = 0, \qquad j = 1, 2, \ldots, N \quad (13.8.3)$$

Since $A$ is linear, this can be written as

$$\sum_{i=1}^{n} a_i(A(\phi_i), \phi_j) = -(Au_0, \phi_j) \quad (13.8.4)$$

which determines $a_j$'s. Substitution of the $a_j$'s obtained from the solution of (13.8.4) into (13.8.1) gives the required approximate solution $u_n$.

We find an interesting connection between the Galerkin solution and the Fourier representation of the function $u$. We seek a Galerkin solution in the form

$$u_n(\mathbf{x}) = \sum_{i=1}^{n} a_i \phi_i(\mathbf{x}) \quad (13.8.5)$$

with a special restriction on the operator $A$ which satisfies the condition

$$(A\phi_i, \phi_j) = \begin{bmatrix} 0, & i \neq j \\ 1, & i = j \end{bmatrix} \quad (13.8.6)$$

Thus the application of the Galerkin method to (13.7.1) gives

$$\sum_{i=1}^{n} (A(\phi_i), \phi_j) a_i = (f, \phi_j) \quad (13.8.7)$$

which is, by (13.8.6),

$$a_j = (f, \phi_j) \quad (13.8.8)$$

so the Galerkin solution (13.8.5) becomes

$$u_n = \sum_{i=1}^{n} (f, \phi_i)\phi_i(\mathbf{x}) \quad (13.8.9)$$

Evidently, the Galerkin solution (13.8.5) is just the finite Fourier series solution.

Finally, we shall cite an example to show the equivalence of the Galerkin and Rayleigh–Ritz methods. Consider the Poisson equation

$$u_{xx} + u_{yy} = f(x, y) \quad \text{in} \quad D \subset R^2 \quad (13.8.10)$$

with a homogeneous boundary condition $u = 0$ on $\partial D$. The solution of this equation is equivalent to finding the minimum of the functional

$$I(u)=\iint_D (u_x^2+u_y^2+2fu)\,dx\,dy \tag{13.8.11}$$

According to the Rayleigh–Ritz method, we seek a trial solution in the form

$$u_n=\sum_{i=1}^{n} a_i\phi_i(x, y) \tag{13.8.12}$$

where the trial functions $\phi_i$ are chosen so that they satisfy the given boundary condition on $\partial D$.

We substitute the Rayleigh–Ritz solution $u_n$ into $I(u)$ and use $\partial I(u_n)/\partial a_k=0,\ k=1, 2, \ldots, n$ to obtain

$$2\iint_D \left(\frac{\partial u_n}{\partial x}\frac{\partial \phi_k}{\partial x}+\frac{\partial u_n}{\partial y}\frac{\partial \phi_k}{\partial y}+f\phi_k\right) dx\,dy=0 \tag{13.8.13}$$

Application of Green's theorem with the homogeneous boundary condition leads to

$$\iint_D (\nabla^2 u_n-f)\phi_k\,dx\,dy=0 \tag{13.8.14}$$

or

$$(R_n, \phi_k)=0, \qquad R\equiv\nabla^2 u_n-f \tag{13.8.15}$$

This is the Galerkin equation for the undetermined coefficients $a_k$. This establishes the equivalence of the two methods.

EXAMPLE 13.8.1. Use the Galerkin method to find an approximate solution of the Poisson equation

$$\nabla^2 u\equiv u_{xx}+u_{yy}=-1 \quad \text{in} \quad D\{(x, y)\colon |x|<a,\ |y|<b\}$$

with the boundary conditions

$$u=0 \quad \text{on} \quad \partial D\{x=\pm a,\ y=\pm b\}$$

We seek a trial solution in the form

$$u_N(x, y)=\sum_{m,n=1,3,5,\ldots}^{N}\ \sum^{N} a_{mn}\phi_{mn}(x, y)$$

where

$$\phi_{mn}(x, y)=\cos\frac{m\pi x}{2a}\cos\frac{n\pi y}{2b}$$

In this case $A = \nabla^2$ and the residual $R_N$ is

$$R_N = Au_N + 1 = \nabla^2 u_N + 1$$
$$= -\left[\sum_{m=1}^{N} \sum_{n=1}^{N} \left(\frac{m^2\pi^2}{4a^2} + \frac{n^2\pi^2}{4b^2}\right) a_{mn}\phi_{mn}\right] + 1$$

According to the Galerkin method

$$0 = (R_N, \phi_{kl}) = \int_{-a}^{a}\int_{-b}^{b} R_N \cos\frac{k\pi x}{2a} \cos\frac{l\pi y}{2b}\, dx\, dy$$
$$= \frac{ab\pi^2}{4}\left(\frac{k^2}{a^2} + \frac{l^2}{b^2}\right) a_{kl} - \frac{16ab}{\pi^2 kl}(-1)^{(k+l)/2-1}$$

or

$$a_{kl} = \left(\frac{8ab}{\pi^2}\right)^2 \frac{(-1)^{(k+l)/2-1}}{(b^2k^2 + a^2l^2)}$$

Thus the solution of the problem is

$$u_N(x, y) = \left(\frac{8ab}{\pi^2}\right)^2 \sum_{m,n=1,3,5\ldots}^{N} \sum^{N} \frac{(-1)^{(m+n)/2-1}\phi_{mn}}{(b^2m^2 + a^2n^2)}$$

In particular, the solution for $u_N(x, y)$ can be derived for the square domain $D\{(x, y): |x| < 1, |y| < 1\}$. In the limit $N \to \infty$, these solutions are in perfect agreement with those obtained by the double Fourier series.

EXAMPLE 13.8.2. Solve the problem in Example 13.8.1 by using algebraic polynomials as trial functions.

We seek an appropriate solution in the form

$$u_N(x, y) = (x^2 - a^2)(y^2 - b^2)(a_1 + a_2x^2 + a_3y^2 + a_4x^4y^4 + \ldots)$$

Obviously, this satisfies the boundary conditions. In the first approximation, the solution assumes the form

$$u_1(x, y) \equiv a_1\phi_1 = a_1(x^2 - a^2)(y^2 - b^2)$$

where the coefficient $a_1$ is determined by the Galerkin integral

$$\int_{-a}^{a}\int_{-b}^{b} (\nabla^2 u_1 + 1)\phi_1\, dx\, dy = 0$$

or

$$\int_{-a}^{a}\int_{-b}^{b} [2a_1(y^2 - b^2) + 2a_1(x^2 - a^2) + 1](x^2 - a^2)(y^2 - b^2)\, dx\, dy = 0$$

A simple evaluation gives

$$a_1 = \frac{5}{4}(a^2 + b^2)^{-1}$$

Hence the solution is

$$u_1(x, y) = a_1(x^2 - a^2)(y^2 - b^2)$$

## 13.9 THE KANTOROVICH METHOD

In 1932 Kantorovich gave an interesting generalization of the Rayleigh–Ritz method which leads from the solution of a partial differential equation to the solution of a system of algebraic equations in terms of unknown coefficients. The essence of the Kantorovich method is to reduce the problem of the solution of partial differential equations to the solution of ordinary differential equations in terms of undetermined functions.

Consider the boundary-value problem governed by (13.7.1)–(13.7.2). It has been shown in § 13.7 that the solution of the problem is equivalent to finding the minimum of the quadratic functional $I(u)$ given by (13.7.3).

When the Rayleigh–Ritz method is applied to this problem, we seek an approximate solution in the form (13.7.6) where the coefficients $a_k$ are constant. We then determine $a_k$ so as to minimize $I(u_n)$.

In the Kantorovich method, we assume that $a_k$ in (13.7.6) are no longer constants but unknown functions of one of the independent variables $x$ of $\mathbf{x}$ so that the Kantorovich solution has the form

$$u_n(\mathbf{x}) = \sum_{k=1}^{n} a_k(x)\phi_k(\mathbf{x}) \tag{13.9.1}$$

where the product $a_k(x)\phi_k(\mathbf{x})$ satisfies the same boundary conditions as $u$. Thus the problem leads to minimizing the functional

$$I(u_n(\mathbf{x})) = I\left(\sum_{k=1}^{n} a_k(x)\phi_k(\mathbf{x})\right) \tag{13.9.2}$$

Since $\phi_k(\mathbf{x})$ are known functions, we can perform integration with respect to all independent variables except $x$ and obtain a functional $\bar{I}(a_1(x), a_2(x), \ldots, a_n(x))$ depending on $n$ unknown functions $a_k(x)$ of one independent variable $x$. These functions must be so determined that they minimize the functional $\bar{I}(a_1, a_2, \ldots, a_n)$. Finally, under certain conditions, the solution $u_n(\mathbf{x})$ converges to the exact solution $u(\mathbf{x})$ as $n \to \infty$.

In order to describe the method more precisely, we consider the following example in two dimensions:

$$\nabla^2 u = f(x, y) \qquad \text{in } D \tag{13.9.3}$$

$$u(x, y) = 0 \qquad \text{on } \partial D \tag{13.9.4}$$

where $D$ is a closed domain bounded by the curves $y = \alpha(x)$, $y = \beta(x)$ and two vertical lines $x = a$ and $x = b$.

The associated functional for the Poisson equation is

$$I(u)=\iint_D (u_x^2+u_y^2-2u)\,dx\,dy$$

We seek the Kantorovich solution in the first approximation

$$u(x, y)\sim\left(y^2-\frac{x^2}{3}\right)u_1(x)$$

so that $u_1(a)=0$.

Substituting the solution into $I(u)$ gives

$$I(u_1)=\int_0^a dx\int_{-x/\sqrt{3}}^{x/\sqrt{3}}\left[\left\{\left(y^2-\frac{x^2}{3}\right)u_1'-\frac{2xu_1}{3}\right\}^2+4y^2u_1^2(x)-2\left(y^2-\frac{x^2}{3}\right)u_1\right]dy$$

$$=\frac{8}{135\sqrt{3}}\int_0^a (2x^5u_1'^2+10x^4u_1u_1'+30x^3u_1^2+15x^3u_1)\,dx$$

The Euler equation for this functional is

$$4(x^2u_1''+5xu_1'-5u_1)=15$$

This is a nonhomogeneous ordinary differential equation of order two. We seek a solution of the corresponding homogeneous equation in the form $x^r$ where $r$ is determined by the equation $(r-1)(r+5)=0$. The particular integral of the equation is $u_1=-\frac{3}{4}$. Hence the general solution is

$$u_1(x)=Ax+Bx^{-5}-\frac{3}{4}$$

where the constants $A$ and $B$ are determined by the boundary conditions. For the bounded solution, $B\equiv 0$. The condition $u_1(a)=0$ implies $A=-3a/4$. Therefore the final solution is

$$u(x, y)=\frac{3}{4}\left(\frac{x}{a}-1\right)\left(y^2-\frac{x^2}{3}\right)$$

## 13.10 EXERCISES

**1.** Obtain the explicit finite difference solution of the problem

$$u_{tt}-4u_{xx}=0, \qquad 0<x<1, \qquad t>0$$

$$u(0, t)=u(1, t)=0, \qquad t\geq 0$$

$$u(x, 0)=\sin 2\pi x, \qquad u_t(x, 0)=0, \qquad 0\leq x\leq 1$$

Compare the numerical solutions with the analytical solution, $u(x, t)=\cos 4\pi t\sin 2\pi x$ at several points.

**2.** Calculate the explicit finite difference solution of the equation

$$u_{xx} - u_{tt} = 0, \qquad 0 < x < 1, \qquad t > 0$$

satisfying the boundary conditions

$$u(0, t) = u(1, t) = 0, \qquad t \geqslant 0$$

and the initial conditions

$$u(x, 0) = \frac{1}{8} \sin \pi x, \qquad u_t(x, 0) = 0, \qquad 0 \leqslant x \leqslant 1$$

Show that the exact solution of the problem is

$$u(x, t) = \frac{1}{8} \cos \pi t \sin \pi x$$

Compare the two solutions at several points.

**3.** Use the Lax–Wendroff method to find the numerical solution of the problem

$$\begin{aligned} u_x + u_t &= 0, \qquad && x > 0, \qquad t > 0 \\ u(x, 0) &= 2 + x, && x > 0 \\ u(0, t) &= 2 - t, && t > 0 \end{aligned}$$

Show that the exact solution of the problem is

$$u(x, t) = 2 + (x - t)$$

Compare the two solutions at various points.

**4.** Show that the finite difference approximation to the equation

$$au_t + bu_x = f(x, t)$$

is

$$u_{i,j+1} - \frac{1}{2}(u_{i+1,j} + u_{i-1,j}) + \frac{\varepsilon b}{2a}(u_{i+1,j} - u_{i-1,j}) - f_{i,j} = 0$$

where $a$, $b$ are constants and $\varepsilon = k/h$.

**5.** Obtain the finite difference solution of the heat-conduction problem

$$u_t = \kappa u_{xx}, \qquad 0 < x < l, \qquad t > 0$$

with the boundary conditions

$$u(0, t) = u(l, t) = 0, \qquad t > 0$$

and the initial condition

$$u(x, 0) = \frac{4x}{l}(l - x), \qquad 0 \leqslant x \leqslant l$$

**6.** a. Find the explicit finite difference solution of the parabolic system

$$u_t = u_{xx}, \qquad 0<x<1, \qquad t>0$$
$$u(0, t) = u(1, t) = 0, \qquad t>0$$
$$u(x, 0) = \sin \pi x \quad \text{on} \quad 0\leqslant x\leqslant 1$$

Compare the numerical results with the analytic solution

$$u(x, t) = e^{-\pi^2 t} \sin \pi x$$

at $t = 0.5$ and $t = 0.05$.

b. Prove that the Richardson finite difference scheme for the problem (a) is

$$u_{i,j+1} = u_{i,j-1} + 2\varepsilon\delta_x^2 u_{i,j}$$

Hence show that the exact solution of this equation is

$$u_{i,j} = (A_1\alpha_1^j + A_2\alpha_2^j) \sin \pi hi$$

where $\alpha_1$ and $\alpha_2$ are the roots of the quadratic equation

$$x^2 + 8\varepsilon x \sin^2 (\pi h/2) - 1 = 0$$

**7.** Using the four internal grid points, find the explicit finite difference solution of the Dirichlet problem

$$\nabla^2 u \equiv u_{xx} + u_{yy} = 0, \qquad 0<x<1, \qquad 0<y<1$$
$$u(x, 0) = x(1-x), \qquad u(x, 1) = 0 \quad \text{on} \quad 0\leqslant x\leqslant 1$$
$$u(0, y) = u(1, y) = 0 \quad \text{on} \quad 0\leqslant y\leqslant 1$$

Compare the numerical solution with the exact analytical solution

$$u(x, y) = \frac{4}{\pi^3} \sum_{n=0}^{\infty} \frac{2}{(2n+1)^3} \frac{\sin n\pi x \sinh n\pi(1-y)}{\sinh n\pi}$$

at the point $(x, y) = (\frac{1}{3}, \frac{1}{3})$.

**8.** Solve the Dirichlet problem by the explicit finite difference method

$$u_{xx} + u_{yy} = 0, \qquad 0<x<1, \qquad 0<y<1$$
$$u(x, 0) = \sin \pi x, \qquad u(x, 1) = 0 \quad \text{on} \quad 0\leqslant x\leqslant 1$$
$$u(x, y) = 0 \quad \text{on} \quad x = 0, \qquad x = 1 \quad \text{and} \quad 0\leqslant y\leqslant 1$$

**9.** Using a square grid system with $h = \frac{1}{2}$, find the finite difference solution of the Laplace equation on the quarter-disk given by

$$u_{xx} + u_{yy} = 0, \qquad x^2 + y^2 < 1, \qquad y>0$$
$$u(x, 0) = 0, \qquad -1<x<1$$
$$u(x, y) = 10^2, \qquad x^2 + y^2 = 1, \qquad y>0$$

**10.** Find the finite difference solution of the wave problem

$$u_{tt} - u_{xx} = 0, \qquad 0 < x < 1, \qquad t > 0$$
$$u(0, t) = u(1, t) = 0, \qquad t \geqslant 0$$
$$u(x, 0) = \frac{1}{2}x(1 - x), \qquad u_t(x, 0) = 0, \qquad 0 \leqslant x \leqslant 1$$

Compare the numerical results with the exact analytical solution

$$u(x, t) = \frac{2}{\pi^3} \sum_{r=1}^{\infty} \frac{1}{r^3} \{1 - (-1)^r\} \cos \pi rt \sin \pi rx$$

at various points.

**11.** Obtain the finite difference solution of the problem

$$u_{tt} = c^2 u_{xx}, \qquad 0 < x < 1, \qquad t > 0$$
$$u(0, t) = \sin \pi ct, \qquad u(1, t) = 0, \qquad t \geqslant 0$$
$$u(x, 0) = u_t(x, 0) = 0, \qquad 0 \leqslant x \leqslant 1$$

**12.** Show that the transformation $v = \log u$ transforms the nonlinear system

$$v_t = v_{xx} + v_x^2, \qquad 0 < x < 1, \qquad t > 0$$
$$v_x(0, t) = 1, \qquad v(1, t) = 0, \qquad t \geqslant 0$$
$$v(x, 0) = 0, \qquad 0 \leqslant x \leqslant 1$$

into the linear system

$$u_t = u_{xx}, \qquad 0 < x < 1, \qquad t > 0$$
$$u_x(0, t) = u(0, t), \qquad u(1, t) = 1, \qquad t \geqslant 0$$
$$u(x, 0) = 1, \qquad 0 \leqslant x \leqslant 1$$

Solve the linear system by the explicit finite difference method with the derivative boundary condition approximated by the central difference formula.

**13.** Solve the following parabolic system by the Crank–Nicolson method

$$u_t = u_{xx}, \qquad 0 < x < 1, \qquad t > 0$$
$$u(0, t) = u(1, t) = 0, \qquad t \geqslant 0$$

with the initial condition

a. $u(x, 0) = 1, \qquad 0 \leqslant x \leqslant 1$

b. $u(x, 0) = \sin \pi x, \qquad 0 \leqslant x \leqslant 1$

**14.** Use the Crank–Nicolson implicit method with the central difference formula for the boundary condtions to find the numerical solution of the differential system

$$u_t = u_{xx}, \qquad 0 < x < 1, \qquad t > 0$$
$$u_x(0, t) = u, \qquad u_x(1, t) = -u, \qquad t \geq 0$$
$$u(x, 0) = 1, \qquad 0 \leq x \leq 1$$

**15.** Find the numerical solution of the wave equation

$$u_{tt} = c^2 u_{xx}, \qquad 0 < x < l, \qquad t > 0$$

with the boundary and initial conditions

$$u = \frac{l}{20} u_x \quad \text{at} \quad x = 0 \quad \text{and} \quad x = l, \qquad t > 0$$

$$u(x, 0) = 0, \qquad u_t(x, 0) = a \sin \frac{\pi x}{l}, \qquad 0 \leq x \leq l$$

**16.** Determine the function representing a curve which makes the following functional extremum:

a. $I(y(x)) = \int_0^1 (y'^2 + 12\, xy)\, dx, \qquad y(0) = 0, \qquad y(1) = 1$

b. $I(y(x)) = \int_0^{\pi/2} (y'^2 - y^2)\, dx, \qquad y(0) = 0, \qquad y\left(\frac{\pi}{2}\right) = 1$

c. $I(y(x)) = \int_{x_0}^{x_1} \frac{1}{x} (1 + y'^2)^{\frac{1}{2}}\, dx$

**17.** In the problem of tautochroneous motion, find the equation of the curve joining the origin $O$ and a point $A$ in a vertical $x - y$ plane so that a particle sliding freely from $A$ to $O$ under the action of gravity reaches the origin $O$ in the shortest time, friction and resistance of the medium being neglected.

**18.** In the problem of minimum surface of revolution, determine a curve with given boundary points $(x_0, y_0)$ and $(x_1, y_1)$ such that rotation of the curve about the $x$-axis generates a surface of revolution of minimum area.

Hint: $I(y(x)) = 2\pi \int_{x_0}^{x_1} y(1 + y'^2)^{\frac{1}{2}}\, dx.$

**19.** Show that the Euler equation of the variational principle

$$\delta I[u(x, y)] = \delta \iint_D F(x, y, u, p, q, l, m, n)\, dx\, dy = 0$$

is

$$F_u - \frac{\partial}{\partial x} F_p - \frac{\partial}{\partial y} F_q + \frac{\partial^2}{\partial x^2} F_l + \frac{\partial^2}{\partial x\, \partial y} F_m + \frac{\partial^2}{\partial y^2} F_n = 0$$

where

$$p = u_x, \qquad q = u_y, \qquad l = u_{xx}, \qquad m = u_{xy}, \qquad n = u_{yy}$$

**20.** Prove that the Euler–Lagrange equation for the funtional

$$I = \iiint_R F(x, y, z, u, p, q, r, l, m, n, a, b, c)\, dx\, dy\, dz$$

is

$$F_u - \frac{\partial}{\partial x} F_p - \frac{\partial}{\partial y} F_q - \frac{\partial}{\partial z} F_r + \frac{\partial^2}{\partial x^2} F_l + \frac{\partial^2}{\partial y^2} F_m$$
$$+ \frac{\partial^2}{\partial z^2} F_n + \frac{\partial^2}{\partial x\, \partial y} F_a + \frac{\partial^2}{\partial y\, \partial z} F_b + \frac{\partial^2}{\partial z\, \partial x} F_c = 0$$

where $(p, q, r) = (u_x, u_y, u_z)$, $(l, m, n) = (u_{xx}, u_{yy}, u_{zz})$, $(a, b, c) = (u_{xy}, u_{yz}, u_{zx})$.

**21.** In each of the following cases apply the variational principle or its simple extension with appropriate boundary conditions to derive the corresponding equations:

a. $F = u_x^2 + u_y^2 + 2u_{xy}^2$

b. $F = \frac{1}{2}[u_t^2 - \alpha(u_x^2 + u_y^2) - \beta^2 u^2]$

c. $F = \frac{1}{2}(u_t u_x + \alpha u_x^2 - \beta u_{xx}^2)$

d. $F = \frac{1}{2}(u_t^2 - \alpha^2 u_{xx}^2)$

e. $F = p(x)u'^2 + \frac{d}{dx}(q(x)u^2) - [r(x) + \lambda s(x)]u^2$

where $p$, $q$, $r$, and $s$ are given functions of $x$ and $\alpha$, $\beta$ are constants.

**22.** Derive the Schrödinger equation from the variational principle

$$\delta \iiint_R \left[\frac{\hbar^2}{2m}(\psi_x^2 + \psi_y^2 + \psi_z^2) + (V - E)\psi^2\right] dx\, dy\, dz = 0$$

where $h = 2\pi\hbar$ is the Planck constant, $m$ is the mass of a particle moving under the action of a force field described by the potential $V(x, y, z)$ and $E$ is the total energy of the particle.

**23.** Derive the Poisson equation $\nabla^2 u = F(x, y)$ from the variational principle with the functional

$$I(u) = \iint_D [u_x^2 + u_y^2 + 2uF(x, y)]\, dx\, dy$$

where $u = u(x, y)$ is given on the boundary $\partial D$ of $D$.

**24.** Derive the equation of motion of a vibrating string of length $l$ under the action of an external force $F(x, t)$ from the variational principle

$$\delta \int_{t_1}^{t_2} \int_0^l \left[\left(\frac{1}{2}\rho u_t^2 - T^* u_x^2\right) + \rho u F(x, t)\right] dx\, dt = 0$$

where $\rho$ is the line density and $T^*$ is the constant tension of the string.

**25.** The kinetic and potential energies associated with the transverse vibration of a thin elastic plate of constant thickness $h$ are

$$T = \frac{1}{2}\rho \iint_D \dot{u}^2\, dx\, dy$$

$$V = \frac{1}{2}\mu_0 \iint_D [(\nabla u)^2 - 2(1-\sigma)(u_{xx}u_{yy} - u_{xy}^2)]\, dx\, dy$$

where $\rho$ is the surface density and $\mu_0 = 2h^3 E/3(1-\sigma^2)$.
Use the variational principle

$$\delta \int_{t_1}^{t_2} \iint_D [(T - V) + fu]\, dx\, dy\, dt = 0$$

to derive the equation of motion of the plate

$$\rho \ddot{u} + \mu_0 \nabla^4 u = f(x, y, t)$$

where $f$ is the transverse force per unit area acting on the plate.

**26.** The kinetic and potential energies associated with the wave motion in elastic solids are

$$T = \frac{1}{2} \iiint_D \rho(u_t^2 + v_t^2 + w_t^2)\, dx\, dy\, dz$$

$$V = \frac{1}{2} \iiint_D [\lambda(u_x + v_y + w_z)^2 + 2\mu(u_x^2 + v_y^2 + w_z^2) + \mu\{(v_x + u_y)^2 + (w_y + v_z)^2 + (u_z + w_x)^2\}]\, dx\, dy\, dz$$

Use the variational principle

$$\delta \int_{t_1}^{t_2} \iiint_D (T - V)\, dx\, dy\, dz\, dt = 0$$

to derive the equation of wave motion in the elastic medium

$$(\lambda + \mu)\, \text{grad div}\, \mathbf{u} + \mu \nabla^2 \mathbf{u} = \rho \mathbf{u}_{tt}$$

where $\mathbf{u} = (u, v, w)$ is the displacement vector.

**27.** From the variational principle

$$\delta \iint_D L\, d\mathbf{x}\, dt = 0 \quad \text{with} \quad L = -\rho \int_{-h}^{\eta} \left\{ \phi_t + \frac{1}{2} (\nabla \phi)^2 + gz \right\} dz$$

derive the basic equations of water waves

$$\nabla^2 \phi = 0, \qquad -h(x, y) < z < \eta(x, y, t), \qquad t > 0$$

$$\eta_t + \nabla \phi \cdot \nabla \eta - \phi_z = 0 \quad \text{on} \quad z = \eta$$

$$\phi_z + \frac{1}{2} (\nabla \phi)^2 + gz = 0 \quad \text{on} \quad z = \eta$$

$$\phi_z = 0 \qquad \text{on} \quad z = -h$$

where $\phi(x, y, z, t)$ is the velocity potential, and $\eta(x, y, t)$ is the free surface displacement function in a fluid of depth $h$.

**28.** Derive the Boussinesq equation for water waves

$$u_{tt} - c^2 u_{xx} - \mu u_{xxtt} = \frac{1}{2} (u^2)_{xx}$$

from the variational principle

$$\delta \iint L\, dx\, dt = 0$$

where $L \equiv \frac{1}{2}\phi_t^2 - \frac{1}{2}c^2\phi_x^2 + \frac{1}{2}\mu\phi_{xt}^2 - \frac{1}{6}\phi_x^3$ and $\phi$ is the potential for $u (u = \phi_x)$.

**29.** Determine an approximate solution of the problem of finding an extremum of the functional

$$I(y(x)) = \int_0^1 (y'^2 - y^2 - 2xy)\, dx, \qquad y(0) = y(1) = 0$$

Hint: $y_n = x(1 - x) \sum_{r=1}^{n} a_r x^{r-1}$. Work out the solution for $n = 1$ and $n = 2$.

**30.** Find the approximate solution of the torsion problem of a cylinder with an elliptic base so that the domain of integration $D$ is the interior of the ellipse with the major and minor axes $2a$ and $2b$ respectively. The associated functional is

$$I(u(x, y)) = \iint_D \left[ \left( \frac{\partial u}{\partial x} - y \right)^2 + \left( \frac{\partial u}{\partial y} + x \right)^2 \right] dx\, dy$$

Hint: $u_1(x, y) = a_1 xy$.

**31.** Use the Rayleigh–Ritz method to find an approximate solution of the problem

$$\nabla^2 u = 0, \qquad 0 < x < 1, \qquad 0 < y < 1$$
$$u(0, y) = 0 = u(1, y)$$
$$u(x, 0) = x(1 - x)$$

Hint: $u_3 = x(1-x)(1-y) + x(1-x)y(1-y)(a_2 + a_3 y)$.

**32.** Find an approximate solution of the boundary value problem

$$\nabla^2 u = 0, \qquad x > 0, \qquad y > 0, \qquad x + 2y < 2$$
$$u(0, y) = 0, \qquad u(x, 0) = x(2 - x)$$
$$u(2 - 2y, y) = 0$$

Hint: $u_2 = x(2 - x - 2y) + a_2 xy(2 - x - 2y)$.

**33.** In the torsion problem in elasticity, the Prandtl stress function $\Psi(x, y) = \psi(x, y) - \frac{1}{2}(x^2 + y^2)$ satisfies the boundary problem

$$\nabla^2 \Psi = -2 \qquad \text{in } D$$
$$\Psi = 0 \qquad \text{on } \partial D$$

Use the Galerkin method to find the approximate solution of the problem in a rectangular domain $D\{(x, y): -a \leqslant x \leqslant a, \ -b \leqslant y \leqslant b\}$.

Hint: $\Psi_N = \sum_{m,n=1}^{N} a_{mn}\phi_{mn} = \sum_{m,n=1}^{N} a_{mn} \cos \frac{m\pi x}{2a} \cos \frac{n\pi y}{2b}$.

**34.** Apply the Galerkin approximate method to find the first eigenvalue of the problem of a circular membrane of radius $a$ governed by the equation

$$\nabla^2 u \equiv \frac{d^2 u}{dr^2} + \frac{1}{r}\frac{du}{dr} = \lambda u \quad \text{in} \quad 0 < r < a$$
$$u = 0 \quad \text{on} \quad r = a$$

Hint: $\phi_n = \cos (2n - 1)\frac{\pi r}{2a}$.

**35.** Use the Rayleigh–Ritz method to find the solution in the first approximation of the problem of deformation of an elastic plate $(-a \leqslant x \leqslant a, \ -a \leqslant y \leqslant a)$ by a parabolic distribution of tensile forces over its opposite sides at $x = \pm a$. The problem is governed by the differential system

$$\nabla^4 u = \frac{2\alpha}{a^2} \qquad \text{in} \quad |x| < a \quad \text{and} \quad |y| < a$$
$$u = \frac{\partial u}{\partial x} = 0 \quad \text{on} \quad |x| = a$$
$$u = \frac{\partial u}{\partial y} = 0 \quad \text{on} \quad |y| = a$$

where $U = u_0 + u = \frac{1}{2}\alpha y^2(1 - \frac{1}{6}y^2) + u$, and $\nabla^4 U = 0$.

Hint: $I(u) = \int_{-a}^{a}\int_{-a}^{a}\left[(\nabla^2 u)^2 - \frac{4\alpha}{a^2}u\right]dx\,dy = \text{min}$, and

$$u_n = (x^2 - a^2)^2(y^2 - a^2)^2(a_1 + a_2x^2 + a_3y^2 + \ldots).$$

**36.** Show that the Kantorovich solution of the torsion problem in exercise 33 is

$$\Psi_1(x, y) = \frac{1}{2}(b^2 - y^2)\left(1 - \frac{\cosh kx}{\cosh ka}\right), \qquad k = \frac{1}{b}\sqrt{\frac{5}{2}}$$

Hint: $\Psi_1 = (b^2 - y^2)U(x)$.

# Answers to Exercises

## 1.5 EXERCISES

1. a. Linear, nonhomogeneous, second-order; b. quasilinear, first-order; c. nonlinear, first-order; d. linear, homogeneous, fourth-order; e. linear, nonhomogeneous, second-order; f. quasilinear, third-order; g. nonlinear, second-order.
5. $u(x, y) = f(x) \cos y + g(x) \sin y$
6. $u(x, y) = f(x)e^{-y} + g(y)$
7. $u(x, y) = f(x + y) + g(3x + y)$
8. $u(x, y) = f(y + x) + g(y - x)$

## 2.7 EXERCISES

14. (i) $\dfrac{\partial^2}{\partial x^2}\begin{pmatrix} I \\ V \end{pmatrix} = \dfrac{1}{c^2}\dfrac{\partial^2}{\partial t^2}\begin{pmatrix} I \\ V \end{pmatrix}, \qquad c^2 = \dfrac{1}{LC}$

(ii) $\dfrac{\partial}{\partial t}\begin{pmatrix} I \\ V \end{pmatrix} = \kappa \dfrac{\partial^2}{\partial x^2}\begin{pmatrix} I \\ V \end{pmatrix}, \qquad \kappa = \dfrac{1}{RC}$

(iii) $\left(\dfrac{\partial^2}{\partial t^2} + 2k\dfrac{\partial}{\partial t} + k^2\right)\begin{pmatrix} V \\ I \end{pmatrix} = c^2 \dfrac{\partial^2}{\partial x^2}\begin{pmatrix} V \\ I \end{pmatrix}$

## 3.6 EXERCISES

1. a. $x < 0$, hyperbolic;

$$u_{\xi\eta} = \frac{1}{4}\left(\frac{\xi - \eta}{4}\right)^4 - \frac{1}{2}\left(\frac{1}{\xi - \eta}\right)(u_\xi - u_\eta)$$

$x = 0$, parabolic, the given equation is then in canonical form; $x > 0$, elliptic.

$$u_{\alpha\alpha} + u_{\beta\beta} = \frac{\beta^4}{16} + \frac{1}{\beta} u_\beta$$

b. $y = 0$, parabolic; $y \neq 0$, elliptic.

$$u_{\alpha\alpha} + u_{\beta\beta} = u_\alpha + e^\alpha$$

d. Parabolic everywhere

$$u_{\eta\eta} = \frac{2\xi}{\eta^2} u_\xi + \frac{1}{\eta^2} e^{\xi/\eta}$$

f. Elliptic everywhere for finite values of $x$ and $y$

$$u_{\alpha\alpha} + u_{\beta\beta} = u - \frac{1}{\alpha} u_\alpha - \frac{1}{\beta} u_\beta$$

g. Parabolic everywhere

$$u_{\eta\eta} = \frac{1}{1 - e^{2(\eta - \xi)}} [\sin^{-1} e^{\eta - \xi} - u_\xi]$$

h. $y = 0$, parabolic; $y \neq 0$, hyperbolic,

$$u_{\xi\eta} = \frac{1 + \xi - \ln \eta}{\eta} u_\xi + u_\eta + \frac{1}{\eta} u$$

2. i. $u(x, y) = f(y/x) + g(y/x)e^{-y}$; ii. $u(r, t) = (1/r)f(r + ct) + (1/r)g(r - ct)$;
3. a. $\xi = (y - x) + i\sqrt{2}\,x$, $\eta = (y - x) - i\sqrt{2}\,x$, $\alpha = y - x$, $\beta = \sqrt{2}\,x$, $u_{\alpha\alpha} + u_{\beta\beta} = -\frac{1}{2}u_\alpha - 2\sqrt{2}\,u_\beta - \frac{1}{2}u + \frac{1}{2}e^{\beta/\sqrt{2}}$.
   b. $\xi = y + x$, $\eta = y$, $u_{\eta\eta} = -\frac{3}{2}u$.
   c. $\xi = y - x$, $\eta = y - 4x$, $u_{\xi\eta} = \frac{7}{9}(u_\xi + u_\eta) - \frac{1}{9}\sin[(\xi - \eta)/3]$.
   d. $\xi = y + ix$, $\eta = y - ix$, $\alpha = y$, $\beta = x$; the given equation is already in canonical form.
   e. $\xi = x$, $\eta = x - y/2$, $u_{\xi\eta} = 18u_\xi + 17u_\eta - 4$.
   f. $\xi = y + x/6$, $\eta = y$, $u_{\xi\eta} = 6u - 6\eta^2$.
   g. $\xi = x$, $\eta = y$; the given equation is already in canonical form.
   h. $\xi = x$, $\eta = y$; the given equation is already in canonical form.
4. i. $u(x, y) = f(x + cy) + g(x - cy)$; ii. $u(x, y) = f(x + iy) + g(x - iy)$;
   iii. $u(x, y) = (x - iy)f_1(x + iy) + f_2(x + iy) + (x + iy)f_3(x - iy) + f_4(x - iy)$;
   iv. $u(x, y) = f(y + x) + g(y + 2x)$; v. $u(x, y) = f(y) + g(y - x)$;
   vi. $u(x, y) = (-y/128)(y - x)(y - 9x) + f(y - 9x) + g(y - x)$.
5. i. $v_{\xi\eta} = -(1/16)v$; ii. $v_{\xi\eta} = (84/625)v$.

## 4.12 EXERCISES

1. a. $u(x, t) = t$; b. $u(x, t) = \sin x \cos ct + x^2 t + \frac{1}{3}c^2 t^3$; c. $u(x, t) = x^3 + 3c^2 x t^2 + xt$; d. $u(x, t) = \cos x \cos ct + t/e$; e. $u(x, t) = 2t + \frac{1}{2}[\log(1 + x^2 + 2cxt + c^2t^2) + \log(1 + x^2 - 2cxt + c^2t^2)]$; f. $u(x, t) = x + (1/c)\sin x \sin ct$.
2. a. $u(x, t) = 3t + \frac{1}{2}xt^2$; c. $u(x, t) = 5 + x^2 t + \frac{1}{3}c^2 t^3 + (1/2c^2)(e^{x+ct} + e^{x-ct} - 2e^x)$; e. $u(x, t) = \sin x \cos ct + (e^t - 1)(xt + x) - xte^t$; f. $u(x, t) = x^2 + t^2(1 + c^2) + (1/c)\cos x \sin ct$.
3.

$$s(r, t) = \begin{cases} 0, & 0 \leqslant t < r - R \\ \dfrac{s_0(r-t)}{2r}, & r - R < t < r + R \\ 0, & r + R < t < \infty \end{cases}$$

4. $u(x, y) = \frac{1}{4}\sin(y + x) + \frac{3}{4}\sin(-y/3 + x) + y^2/3 + xy$
5. $u(x, t) = \sin x \cos at + xt$, $a$ = physical constant
6. b.

$$u(x, t) = \frac{1}{2}f(xt) + \frac{1}{2}tf\left(\frac{x}{t}\right) + \frac{1}{4}\sqrt{xt}\int_{xt}^{x/t}\frac{f(\tau)}{\tau^{\frac{3}{2}}}\,d\tau - \frac{1}{2}\sqrt{xt}\int_{xt}^{x/t}\frac{g(\tau)}{\tau^{\frac{3}{2}}}\,d\tau$$

19. $$u(x, y) = f\left(\sqrt{\frac{y^2 + x^2 - 8}{2}}\right) + g\left(\sqrt{\frac{x^2 - y^2 + 16}{2}}\right) - f(2)$$

22. $$u(x, y) = f\left[\frac{y - \dfrac{x^2}{2} + 4}{2}\right] + g\left(\frac{y + x^2/2}{2}\right) - f(2)$$

## 5.13 EXERCISES

1. a. $$f(x) = -\frac{\pi}{4} + \frac{h}{2} + \sum_{k=1}^{\infty}\frac{1}{\pi k^2}[1 + (-1)^{k+1}]\cos kx + \frac{1}{\pi k}[h + (h + \pi)(-1)^{k+1}]\sin kx$$

c. $$f(x) = \sin x + \sum_{k=1}^{\infty}\frac{2(-1)^{k+1}}{k}\sin kx$$

e. $$f(x) = \frac{\sinh \pi}{\pi}\left[1 + \sum_{k=1}^{\infty}\frac{2(-1)^k}{1 + k^2}(\cos kx - k\sin kx)\right]$$

2. a. $$f(x) = \sum_{k=1}^{\infty}\frac{2}{k}\sin kx$$

b. $f(x)=\sum_{k=1}^{\infty}\frac{2}{\pi k}\left[1-2(-1)^k+\cos\frac{k\pi}{2}\right]\sin kx$

c. $f(x)=\sum_{k=1}^{\infty}\left[2(-1)^{k+1}\frac{\pi}{k}+\frac{4}{\pi k^3}((-1)^k-1)\right]\sin kx$

d. $f(x)=\sum_{k=2}^{\infty}\frac{2k}{\pi}\left[\frac{1+(-1)^k}{k^2-1}\right]\sin kx$

3. a. $f(x)=\frac{3}{2}\pi+\sum_{k=1}^{\infty}\frac{2}{\pi k^2}[(-1)^k-1]\cos kx$

b. $f(x)=\frac{\pi}{2}+\sum_{k=1}^{\infty}\frac{2}{\pi k^2}[(-1)^k-1]\cos kx$

c. $f(x)=\frac{\pi^2}{3}+\sum_{k=1}^{\infty}\frac{4(-1)^k}{k^2}\cos kx$

d. $f(x)=\frac{2}{3\pi}+\sum_{k=1,2,4,\ldots}^{\infty}\frac{6}{\pi}\left[\frac{1+(-1)^k}{9-k^2}\right]\cos kx, \qquad k\neq 3$

4. b. $f(x)=\sum_{k=1}^{\infty}\frac{2}{k\pi}\sin\frac{k\pi}{2}\cos\frac{k\pi x}{6}$

c. $f(x)=\frac{2}{\pi}+\sum_{k=2}^{\infty}\frac{2}{k\pi}\left[\frac{1+(-1)^k}{1-k^2}\right]\cos\frac{k\pi x}{l}$

f. $f(x)=\sum_{k=1}^{\infty}\frac{k\pi}{1+k^2\pi^2}(-1)^{k+1}(e-e^{-1})\sin k\pi x$

5. a. $f(x)=\sum_{k=-\infty}^{\infty}\frac{1}{\pi}\left(\frac{2+ik}{4+k^2}\right)(-1)^k\sinh 2\pi e^{ikx}$

b. $f(x)=\sum_{k=-\infty}^{\infty}\frac{(-1)^k}{\pi(1+k^2)}\sinh \pi e^{ikx}$

d. $f(x)=\sum_{k=-\infty}^{\infty}(-1)^k\frac{i}{k\pi}e^{ik\pi x}$

8. a. $\sin^2 x=\sum_{k=1,3,4,\ldots}^{\infty}\frac{4(1-\cos k\pi)}{k\pi(4-k^2)}\sin kx$

b. $\cos^2 x=\sum_{k=1,3,4,\ldots}^{\infty}\frac{2}{k\pi}\left(\frac{1-k^2}{4-k^2}\right)(1-\cos k\pi)\sin kx$

c. $\sin x\cos x=\sum_{k=1,3,4,\ldots}^{\infty}\frac{2}{\pi}\left(\frac{1-\cos k\pi}{4-k^2}\right)\cos kx$

9. a. $\frac{x^2}{4}=\frac{\pi^2}{12}-\sum_{k=1}^{\infty}\frac{(-1)^{k+1}}{k^2}\cos kx$

c. $\int_0^{\infty} \ln\left(2\cos\frac{x}{2}\right) dx = \sum_{k=1}^{\infty} (-1)^{k+1} \frac{\sin kx}{k^2}$

e. $\frac{\pi}{2} - \frac{4}{\pi} \sum_{k=1}^{\infty} \frac{\cos(2k-1)x}{(2k-1)^2} = \begin{cases} -x, & -\pi < x < 0 \\ x, & 0 < x < \pi \end{cases}$

6. a. $f(x) = \frac{\pi}{8} + \sum_{k=1}^{\infty} \frac{1}{2\pi k^2}[(-1)^k - 1]\cos kx + \frac{(-1)^{k+1}}{2k}\sin kx$

7. a. $f(x) = \frac{l^2}{3} + \sum_{k=1}^{\infty} 4(-1)^k \left(\frac{1}{k\pi}\right)^2 \cos\frac{k\pi x}{l}$

10. a. $f(x, y) = \frac{16}{\pi^2} \sum_{m=1,3,\ldots}^{\infty} \sum_{n=1,3,\ldots}^{\infty} \frac{1}{mn} \sin mx \sin ny$

c. $$f(x, y) = \frac{\pi^4}{9} + \frac{1}{2}\sum_{m=1}^{\infty} \frac{8}{3}\pi^2 \frac{(-1)^m}{m^2}\cos mx + \frac{1}{2}\sum_{n=1}^{\infty} \frac{8}{3}\pi^2 \frac{(-1)^n}{n^2}\cos ny + \sum_{m=1}^{\infty}\sum_{n=1}^{\infty} \frac{16(-1)^{m+n}}{m^2n^2}\cos mx \cos ny$$

e. $f(x, y) = \sum_{m=1}^{\infty} \frac{2(-1)^{m+1}}{m} \sin mx \sin y$

## 6.8 EXERCISES

1. a. $u(x, t) = \sum_{n=1}^{\infty} \frac{4}{(n\pi)^3}[1 - (-1)^n]\cos n\pi ct \sin n\pi x$

b. $u(x, t) = 3\cos ct \sin x$

2. a. $u(x, t) = \sum_{u=1,3,4,\ldots}^{\infty} \frac{32[(-1)^n - 1]}{\pi c n^2(n^2 - 4)} \sin nct \sin nx$

5. $u(x, t) = \sum_{n=1}^{\infty} a_n T_n(t) \sin\frac{n\pi x}{l}$

where

$$a_n = \frac{2}{l}\int_0^l f(x)\sin\frac{n\pi x}{l}\,dx$$

and

$$T_n(t) = \begin{cases} e^{-at/2}\left(\cosh \alpha t + \dfrac{a}{2\alpha}\sinh \alpha t\right) & \text{for } \alpha^2 > 0 \\ e^{-at/2}\left(1 + \dfrac{at}{2}\right) & \text{for } \alpha = 0 \\ e^{-at/2}\left(\cos \beta t + \dfrac{a}{2\beta}\sin \beta t\right) & \text{for } \alpha^2 < 0 \end{cases}$$

in which

$$\alpha = \frac{1}{2}\left[a^2 - 4\left(b + \frac{n^2\pi^2c^2}{l^2}\right)\right]^{\frac{1}{2}}$$

$$\beta = \frac{1}{2}\left[4\left(b + \frac{n^2\pi^2c^2}{l^2}\right) - a^2\right]^{\frac{1}{2}}$$

6. $u(x, t) = \sum_{n=1}^{\infty} a_n T_n(t) \sin \dfrac{n\pi x}{l}$

where

$$a_n = \frac{2}{l}\int_0^l g(x) \sin \frac{n\pi x}{l}\, dx$$

and

$$T_n(t) = \begin{cases} \dfrac{2e^{-at/2}}{\sqrt{(a^2 - \alpha)}} \sinh \dfrac{\sqrt{(a^2 - \alpha)}}{2} t & \text{for } a^2 > \alpha \\ te^{-at/2} & \text{for } a^2 = \alpha \\ \dfrac{2e^{-at/2}}{\sqrt{(\alpha - a^2)}} \sin \dfrac{\sqrt{(\alpha - a^2)}}{2} t & \text{for } a^2 < \alpha \end{cases}$$

7. $\theta(x, t) = \sum_{n=1}^{\infty} a_n \cos a\alpha_n t \sin(\alpha_n x + \phi_n)$ where

$$a_n = \frac{2(\alpha_n^2 + h^2)}{2h + (\alpha_n^2 + h^2)l}\int_0^l f(x) \sin(\alpha_n x + \phi_n)\, dx$$

and

$\phi_n = \arctan \dfrac{\alpha_n}{h}$; $\alpha_n$ are the roots of the equation

$$\tan \alpha l = \frac{2h\alpha}{\alpha^2 - h^2}$$

11. $u(x, t) = v(x, t) + U(x)$ where

$$v(x, t) = \sum_{n=1}^{\infty} \left[ \frac{-2}{l} \int_0^l U(\tau) \sin \frac{n\pi\tau}{l} \, d\tau \right] \cos \frac{n\pi ct}{l} \sin \frac{n\pi x}{l}$$

and

$$U(x) = -\frac{A}{c^2} \sinh x + \left( \frac{A}{c^2} \sinh l + k - h \right) \frac{x}{l} + h$$

12. $u(x, t) = \dfrac{A}{6c^2} x^2(1 - x) + \displaystyle\sum_{n=1}^{\infty} \frac{12}{(n\pi)^3} (-1)^n \cos n\pi ct \sin n\pi x$

14. $u(x, t) = -\dfrac{hx^2}{2k} + \left( 2u_0 + \dfrac{h}{2k} \right) x - \dfrac{4h}{k\pi} e^{-k\pi^2 t} \sin \pi x$

$$+ \sum_{n=2}^{\infty} a_n e^{-kn^2\pi^2 t} \sin n\pi x$$

where

$$a_n = \frac{2u_0}{n\pi} [1 + (-1)^n] + \frac{2u_0 n}{(n^2 - 1)\pi} [1 - (-1)^n] + \frac{2h}{k\pi^3 n^3} [(-1)^n - 1]$$

15. a. $u(x, t) = \displaystyle\sum_{n=1}^{\infty} \frac{4}{n^3\pi^3} [2(-1)^{n+1} - 1] e^{-4n^2\pi^2 t} \sin n\pi x$

b. $u(x, t) = \displaystyle\sum_{n=1,3,4,\ldots}^{\infty} [(-1)^n - 1] \left[ \frac{n}{\pi(4 - n^2)} - \frac{1}{n\pi} \right] e^{-n^2 kt} \sin nx$

16. $u(x, t) = \displaystyle\sum_{n=1}^{\infty} \frac{2l^2}{n^3\pi^3} [1 - (-1)^n] e^{-(n\pi/l)^2 t} \cos \frac{n\pi x}{l}$

18. $v(x, t) = Ct\left(1 - \dfrac{x}{l}\right) - \dfrac{Cl^2}{6k} \left[ \left(\dfrac{x}{l}\right)^3 - 3\left(\dfrac{x}{l}\right)^2 + 2\left(\dfrac{x}{l}\right) \right]$

$$+ \frac{2Cl^2}{\pi^3 k} \sum_{k=1}^{\infty} \frac{e^{-n^2\pi^2 kt/l^2}}{n^3} \sin \frac{n\pi x}{l}$$

21. $u(x, t) = v(x, t) + w(x)$ where

$$v(x, t) = e^{-kt} \sin x + \sum_{u=1}^{\infty} a_u e^{-u^2 kt} \sin ux$$

and

$$a_u = \frac{-u}{(u^2 + a^2)} [(-1)^u_e{}^{-a\pi} - 1] + \frac{2A}{a^2 k\pi} \left[ \frac{1}{u} \{(-1)^u - 1\} \right.$$

$$\left. + \frac{(-1)^u}{u} [e^{-a\pi} - 1] \right]$$

$$w(x) = \frac{A}{a^2 k} \left[ 1 - e^{-ax} + \frac{x}{\pi} (e^{-a\pi} - 1) \right]$$

## 7.14 EXERCISES

1. a. $\lambda_n = n^2$, $\phi_n(x) = \sin nx$ for $n = 1, 2, 3, \ldots$
   b. $\lambda_n = ((2n-1)/2)^2$, $\phi_n(x) = \sin((2n-1)/2)\pi x$ for $n = 1, 2, 3, \ldots$
   c. $\lambda_n = n^2$, $\phi_n(x) = \cos nx$ for $n = 1, 2, 3, \ldots$
2. a. $\lambda_n = 0,\ n^2\pi^2$, $\phi_n(x) = 1,\ \sin n\pi x,\ \cos n\pi x$ for $n = 1, 2, 3, \ldots$
   b. $\lambda_n = 0,\ n^2$, $\phi_n(x) = 1,\ \sin nx,\ \cos nx$ for $n = 1, 2, 3, \ldots$
   c. $\lambda_n = 0,\ 4n^2$, $\phi_n(x) = 1,\ \sin 2nx,\ \cos 2nx$ for $n = 1, 2, 3, \ldots$
3. a. $\lambda_n = -(3/4 + n^2\pi^2)$, $\phi_n(x) = e^{-x/2}\sin n\pi x$, $n = 1, 2, 3, \ldots$
4. a. $\lambda_n = 1 + n^2\pi^2$, $\phi_n(x) = (1/x)\sin(n\pi \ln x)$, $n = 1, 2, 3, \ldots$
   b. $\lambda_n = \frac{1}{4} + (n\pi/\ln 3)^2$, $\phi_n(x) = [1/(x+2)^{\frac{1}{2}}]\sin[(n\pi/\ln 3)\ln(x+2)]$, $n = 1, 2, 3, \ldots$
   c. $\lambda_n = \frac{1}{12}[1 + (2n\pi/\ln 2)^2]$, $\phi_n(x) = [1/(1+x)^{\frac{1}{2}}]\sin[(n\pi/\ln 2)\ln(1+x)]$, $n = 1, 2, 3, \ldots$

5. a. $\phi(x) = \sin(\sqrt{\lambda}\ln x)$, $\lambda > 0$
   b. $\phi(x) = \sin\sqrt{\lambda}\,x$, $\lambda > 0$

7. $$f(x) \sim \sum_{n=1}^{\infty} \frac{2}{\pi}\left[\frac{(-1)^n - 1}{n^2}\right]\cos nx$$

11. a. $$G(x,\xi) = \begin{cases} x, & x \leqslant \xi \\ \xi, & x > \xi \end{cases}$$

12. a. $$u(x) = -\cos x + \left(\frac{\cos 1 - 1}{\sin 1}\right)\sin x + 1$$
    b. $$u(x) = -\frac{2}{5}\cos 2x - \frac{1}{10}\left(\frac{1 + 2\sin 2}{\cos 2}\right)\sin 2x + \frac{1}{5}e^x$$

16. $$G(x,\xi) = \begin{cases} x^3\xi/2 + x\xi^3/2 - 9x\xi/5 + x & \text{for } 0 \leqslant x < \xi \\ x^3\xi/2 + x\xi^3/2 - 9x\xi/5 + \xi & \text{for } \xi \leqslant x \leqslant 1 \end{cases}$$

## 8.10 EXERCISES

8. a. $$u(r,\theta) = \frac{4}{3}\left(\frac{1}{r} - \frac{r}{4}\right)\sin\theta$$
   c. $$u(r,\theta) = \sum_{n=1}^{\infty} a_n \sinh\left[(n\pi/\ln 3)\left(\theta - \frac{\pi}{2}\right)\right]\sin[(n\pi/\ln 3)\ln r]$$

   where

$$a_n = \frac{2}{\ln 3 \sinh(n\pi^2/2\ln 3)}\left\{\frac{n\pi \ln 3}{n^2\pi^2 + 4(\ln 3)^2}[9(-1)^n - 1] - \frac{4n\pi\ln 3}{n^2\pi^2 + (\ln 3)^2}[3(-1)^n - 1] + \frac{3\ln 3}{n\pi}[(-1)^n - 1]\right\}$$

9. $$u(r, \theta) = \sum_{n=1}^{\infty} a_n(r^{-n\pi/\alpha} - b^{-2n\pi/\alpha} r^{n\pi/\alpha}) \sin \frac{n\pi\theta}{\alpha} + \sum_{n=1}^{\infty} b_n \sinh\left[\frac{n\pi}{\ln(b/a)}(\theta - \alpha)\right] \sin\left[\frac{n\pi}{\ln(b/a)}(\ln r - \ln a)\right]$$

where

$$a_n = \frac{2}{\alpha(a^{-n\pi/\alpha} - b^{-2n\pi/\alpha} a^{n\pi/\alpha})} \int_0^{\alpha} f(\theta) \sin \frac{n\pi\theta}{\alpha}\, d\theta$$

$$b_n = \frac{-2}{\ln(b/a) \sinh[\alpha n\pi/\ln(b/a)]} \cdot \int_a^b f(r) \sin[(n\pi/\ln[b/a])(\ln r + \ln a)] \frac{dr}{r}$$

12. $$u(r, \theta) = \sum_{n=1}^{\infty} \frac{2}{\alpha J_v(a)} \left[\int_0^{\alpha} f(\theta) \sin\left(\frac{n\pi\tau}{\alpha}\right) d\tau\right] J_v(r) \sin \frac{n\pi\theta}{\alpha},$$
$v = n\pi/\alpha$

13. $$u(r, \theta) = \frac{1}{2}(a^2 - r^2)$$

14. a. $$u(r, \theta) = -\frac{1}{3}\left(r + \frac{4}{r}\right) \sin\theta + \text{constant}$$

16. $$u(r, \theta) = \sum_{n=1}^{\infty} \frac{1}{nR^{n-1}} r^n \sin n\theta$$

18. $$u(r, \theta) = \frac{a_0}{2} + \sum_{n=1}^{\infty} r^n(a_n \cos n\theta + b_n \sin n\theta)$$

where

$$a_n = \frac{R^{1-n}}{(n + Rh)\pi} \int_0^{2\pi} f(\theta) \cos n\theta\, d\theta, \quad n = 0, 1, 2, \ldots$$

$$b_n = \frac{R^{1-n}}{(n + Rh)\pi} \int_0^{2\pi} f(\theta) \sin n\theta\, d\theta, \quad n = 1, 2, 3, \ldots$$

20. $$u(r, \theta) = c - \frac{r^4}{12} \sin 2\theta + \frac{1}{6}\left(\frac{r_1^6 - r_2^6}{r_1^4 - r_2^4}\right) r^2 \sin 2\theta + \frac{1}{6}\left(\frac{r_1^2 - r_2^2}{r_1^4 - r_2^4}\right) r_1^4 r_2^4 r^{-2} \sin 2\theta$$

22. a. $$u(x, y) = \sum_{n=1}^{\infty} \frac{4[1 - (-1)^n]}{(n\pi)^3 \sinh n\pi} \sin n\pi x \sinh n\pi(y - 1)$$

c. $u(x, y) = \sum_{n=1}^{\infty} a_n(\sinh n\pi x - \tanh n\pi \cosh n\pi x) \sin n\pi y$ where

$$a_n = \frac{1}{\tanh n\pi}\left[\frac{2n\pi^3}{n^2\pi^4 - 4} + \frac{(-1)^n - 1}{n\pi}\right]$$

23. a. $u(x, y) = c + \sum_{n=1}^{\infty} a_n(\cosh nx - \tanh n\pi \sinh nx) \cos ny$

where $a_n = 2[1 - (-1)^n]/n^3\pi \tanh n\pi$

c. $u(x, y) = -\frac{1}{\tanh \pi}[\cosh y - \tanh \pi \sinh y] \cos x + C$

25. $u(x, y) = xy(1 - x) + \sum_{n=1}^{\infty} \frac{4(-1)^n}{(n\pi)^3 \sinh n\pi} \sin n\pi x \sinh n\pi y$

27. $u(x, y) = c + (x^2/2)\left(\frac{x^2}{3} - y^2\right) + \sum_{n=1}^{\infty} \frac{8a^4(-1)^{n+1}}{(n\pi)^3 \sinh n\pi} \cosh \frac{n\pi x}{a} \cos \frac{n\pi y}{a}$

29. $u(x, y) = x[(x/2) - \pi] + \sum_{n=1}^{\infty} a_n \sin\left(\frac{2n-1}{2}\right)x \cosh\left(\frac{2n-1}{2}\right)y$

where

$$a_n = \frac{2}{A\pi}\int_0^{\pi}\left[f(x) - h\left(\frac{x^2}{2} - \pi x\right)\right] \sin\left(\frac{2n-1}{2}\right)x\, dx$$

with

$$A = \left(\frac{2n-1}{2}\right) \sinh\left(\frac{2n-1}{2}\right)\pi + h \cosh\left(\frac{2n-1}{2}\right)\pi$$

## 9.13 EXERCISES

1. $u(x, y, z) = \frac{\sinh [(\pi/b)^2 + (\pi/c)^2]^{\frac{1}{2}}(a - x)}{\sinh [(\pi/b)^2 + (\pi/c)^2]^{\frac{1}{2}} a} \sin \frac{\pi y}{b} \sin \frac{\pi z}{c}$

2. $u(x, y, z) = \left[\frac{\sinh \sqrt{2}\,\pi z}{\sqrt{2}\,\pi} - \frac{\cosh \sqrt{2}\,\pi z}{\sqrt{2}\,\pi \tanh \sqrt{2}\,\pi}\right] \cos \pi x \cos \pi y$

4. a. $u(r, \theta, z) = \sum_{m=0}^{\infty} \sum_{n=1}^{\infty} (a_{mn} \cos m\theta + b_{mn} \sin m\theta) J_m(\alpha_{mn} r/a)$

$$\cdot \frac{\sinh \alpha_{mn}(l - z)/a}{\sinh \alpha_{mn} l/a}$$

where

$$a_{mn} = \frac{2}{a^2\pi\varepsilon_n[J_{m+1}(\alpha_{mn})]^2}\int_0^{2\pi}\int_0^a f(r,\theta)J_m(\alpha_{mn}r/a)\cos m\theta\, r\,dr\,d\theta$$

$$b_{mn} = \frac{2}{a^2\pi[J_{m+1}(\alpha_{mn})]^2}\int_0^{2\pi}\int_0^a f(r,\theta)J_m(\alpha_{mn}r/a)\sin m\theta\, r\,dr\,d\theta$$

with

$$\varepsilon_n = \begin{cases} 1 & \text{for } m \neq 0 \\ 2 & \text{for } m = 0 \end{cases}$$

and $\alpha_{mn}$ is the $n$th root of the equation $J_m(\alpha_{mn}) = 0$.

5. $u(r, \theta) = \frac{1}{3} + (2/3a^2)r^2P_2(\cos\theta)$

7. $u(r, z) = \sum_{n=1}^{\infty} \frac{a_n \sinh \alpha_n(l - z)/a}{\cosh \alpha_n l/a} J_0(\alpha_n r/a)$

where

$$a_n = \frac{2qu}{k\alpha_n^2 J_x(\alpha_n)}$$

and $\alpha_n$ is the root of the equation $J_0(\alpha_n) = 0$ and $k$ is the coefficient of heat conduction.

8. $u(r, z) = \frac{4u_0}{\pi}\sum_{n=1}^{\infty}\frac{I_0[(2n+1)\pi r/l]}{I_0[(2n+1)\pi a/l]}\frac{\sin(2n+1)\pi z/l}{(2n+1)}$

9. $u(r, \theta) = u_2 + \frac{u_1 - u_2}{2}\sum_{n=1}^{\infty}\left(\frac{2n+1}{n+1}\right)P_{n-1}(0)\left(\frac{r}{a}\right)^n P_n(\cos\theta)$

11. $u(r, \theta, \phi) = C + \sum_{n=1}^{\infty}\sum_{m=0}^{\infty} r^n P_n^m(\cos\theta)[a_{nm}\cos m\phi + b_{nm}\sin n\phi]$

where

$$a_{nm} = \frac{(2n+1)(n-m)!}{2n\pi(n+m)!}\int_0^{2\pi}\int_0^{\pi} f(\theta,\phi)P_n^m(\cos\theta)\cos m\phi\sin\theta\,d\theta\,d\phi$$

$$b_{nm} = \frac{(2n+1)(n-m)!}{2n\pi(n+m)!}\int_0^{2\pi}\int_0^{\pi} f(\theta,\phi)P_n^m(\cos\theta)\sin m\phi\sin\theta\,d\theta\,d\phi$$

$$a_{n0} = \frac{2n+1}{4n\pi}\int_0^{2\pi}\int_0^{\pi} f(\theta,\phi)P_n(\cos\theta)\sin\theta\,d\theta\,d\phi$$

12. $u(x, y, t) = \sum_{n=1,3,4,\ldots}^{\infty}\left(-\frac{4}{\pi}\right)\frac{[1-(-1)^n]}{n(n^2-4)}\cos\sqrt{(n^2+1)}\,\pi ct\,[\sin n\pi x \sin n\pi y]$

13. $u(r, \theta, t) = \sum_{n=0}^{\infty} \sum_{m=1}^{\infty} J_n(\alpha_{mn} r/a) \cos(\alpha_{mn} ct/a)[a_{mn} \cos n\theta + b_{mn} \sin n\theta]$

$$+ \sum_{n=0}^{\infty} \sum_{m=1}^{\infty} J_n(\alpha_{mn} r/a) \sin(\alpha_{mn} ct/a)[c_{mn} \cos n\theta + d_{mn} \sin n\theta]$$

where

$$a_{mn} = \frac{2}{\pi a^2 \varepsilon_n [J_n'(\alpha_{mn})]^2} \int_0^{2\pi} \int_0^a f(r, \theta) J_n(\alpha_{mn} r/a) \cos n\theta \, r \, dr \, d\theta$$

$$b_{mn} = \frac{2}{\pi a^2 [J_n'(\alpha_{mn})]^2} \int_0^{2\pi} \int_0^a f(r, \theta) J_n(\alpha_{mn} r/a) \sin n\theta \, r \, dr \, d\theta$$

$$c_{mn} = \frac{2}{\pi a c \alpha_{mn} \varepsilon_n [J_n'(\alpha_{mn})]^2} \int_0^{2\pi} \int_0^a g(r, \theta) J_n(\alpha_{mn} r/a) \cos n\theta \, r \, dr \, d\theta$$

$$d_{mn} = \frac{2}{\pi a c \alpha_{mn} [J_n'(\alpha_{mn})]^2} \int_0^{2\pi} \int_0^\alpha g(r, \theta) J_n(\alpha_{mn} r/a) \sin n\theta \, r \, dr \, d\theta$$

in which $\alpha_{mn}$ is the root of the equation $J_n(\alpha_{mn}) = 0$ and

$$\varepsilon_n = \begin{cases} 2; & n = 0 \\ 1; & n \neq 0 \end{cases}$$

15. $u(r, \theta, t) = \sum_{n=0}^{\infty} \sum_{m=1}^{\infty} J_n(\alpha_{mn} r) e^{-k\alpha_{mn} t}[a_{nm} \cos n\theta + b_{nm} \sin n\theta]$

where

$$a_{nm} = \frac{2}{\pi \varepsilon_n [J_n'(\alpha_{mn})]^2} \int_0^{2\pi} \int_0^1 f(r, \theta) J_n(\alpha_{mn} r) \cos n\theta \, r \, dr \, d\theta$$

$$b_{nm} = \frac{2}{\pi [J_n'(\alpha_{mn})]^2} \int_0^{2\pi} \int_0^1 f(r, \theta) J_n(\alpha_{mn} r) \sin n\theta \, r \, dr \, d\theta$$

where $\alpha_{mn}$ is the root of $J_n(\alpha_{mn}) = 0$, and

$$\varepsilon_n = \begin{cases} 1 & \text{for } n \neq 0 \\ 2 & \text{for } n = 0 \end{cases}$$

16. $u(x, y, z, t) = \sin \pi x \sin \pi y \sin \pi z \cos \sqrt{3}\, \pi ct$

18. $u(r, \theta, z, t) = \sum_{n=0}^{\infty} \sum_{m=1}^{\infty} \sum_{l=1}^{\infty} J_n(\alpha_{mn} r/a) \sin(m\pi z/l) \cos \omega ct$

$$\cdot [a_{nml} \cos n\theta + b_{nml} \sin n\theta]$$

$$+ \sum_{n=0}^{\infty} \sum_{m=1}^{\infty} \sum_{l=1}^{\infty} J_n(\alpha_{mn} r/a) \sin(m\pi z/l) \sin \omega ct$$

$$\cdot [c_{nml} \cos n\theta + d_{nml} \sin n\theta]$$

where

$$a_{nml}=\frac{4}{\pi a^2 l\varepsilon_n[J_n'(\alpha_{mn})]^2}\int_0^{2\pi}\int_0^a\int_0^l f(r,\theta,z)$$

$$\cdot J_n(\alpha_{mn}r/a)\sin(m\pi z/l)\cos n\theta\, r\,dr\,d\theta\,dz$$

$$b_{nml}=\frac{4}{\pi a^2 l[J_n'(\alpha_{mn})]^2}\int_0^{2\pi}\int_0^a\int_0^l f(r,\theta,z)$$

$$\cdot J_n(\alpha_{mn}r/a)\sin(m\pi z/l)\sin n\theta\, r\,dr\,d\theta\,dz$$

$$c_{nml}=\frac{4\omega^{-1}}{\pi a^2 l\varepsilon_n[J_n'(\alpha_{mn})]^2}\int_0^{2\pi}\int_0^a\int_0^l g(r,\theta,z)$$

$$\cdot J_n(\alpha_{mn}r/a)\sin(m\pi z/l)\cos n\theta\, r\,dr\,d\theta\,dz$$

$$d_{nml}=\frac{4\omega^{-1}}{\pi a^2 l[J_n'(\alpha_{mn})]^2}\int_0^{2\pi}\int_0^a\int_0^l g(r,\theta,z)$$

$$\cdot J_n(\alpha_{mn}r/a)\sin(m\pi z/l)\sin n\theta\, r\,dr\,d\theta\,dz$$

where $\alpha_{mn}$ is the root of the equation $J_n(\alpha_{mn})=0$ and

$$\omega=[(m\pi/l)^2+(\alpha_{mn}/a)^2]^{\frac{1}{2}},\qquad \varepsilon_n=\begin{Bmatrix}1 & \text{for } n\neq 0\\ 2 & \text{for } n=0\end{Bmatrix}$$

20. $$u(r,\theta,z,t)=\sum_{n=0}^{\infty}\sum_{m=1}^{\infty}\sum_{p=1}^{\infty}[a_{nmp}\cos n\theta+b_{nmp}\sin n\theta]$$

$$\cdot J_n(\alpha_{mn}r/a)\sin(p\pi z/l)e^{-\omega t}$$

where

$$a_{nmp}=\frac{4}{\pi a^2 l\varepsilon_n[J_n'(\alpha_{mn})]^2}\int_0^{2\pi}\int_0^a\int_0^l f(r,\theta,z)$$

$$\cdot J_n(\alpha_{mn}r/a)\sin(p\pi z/l)\cos n\theta\, r\,dr\,d\theta\,dz$$

$$b_{nmp}=\frac{4}{\pi a^2 l[J_n'(\alpha_{mn})]^2}\int_0^{2\pi}\int_0^a\int_0^l f(r,\theta,z)$$

$$\cdot J_n(\alpha_{mn}r/a)\sin(p\pi z/l)\sin n\theta\, r\,dr\,d\theta\,dz$$

in which

$$\varepsilon_n=\begin{cases}1 & \text{for } n\neq 0\\ 2 & \text{for } n=0\end{cases}\qquad \text{and } \omega=[(p\pi/l)^2+(\alpha_{mn}/a)^2]$$

23. $u(x, y, t) = \sum_{m=1}^{\infty} \sum_{n=1}^{\infty} u_{mn}(t) \sin mx \sin ny$ where

$$u_{mn}(t) = \frac{4(-1)^{m+n+1}}{mn\alpha_{mn}c} \left\{ \sin \alpha_{mn}ct \left[ \frac{\cos (1-\alpha_{mn}c)t - 1}{2(1-\alpha_{mn}c)} + \frac{\cos (1+\alpha_{mn}c)t - 1}{2(1+\alpha_{mn}c)} \right] \right\}$$
$$+ \cos \alpha_{mn}ct \left[ \frac{\sin (1-\alpha_{mn}c)t}{2(1-\alpha_{mn}c)} + \frac{\sin (1+\alpha_{mn}c)t}{2(1+\alpha_{mn}c)} \right]$$

and $\alpha_{mn} = (m^2 + n^2)^{\frac{1}{2}}$

25. $$u(x, y, t) = \sum_{n=1}^{\infty} \sum_{m=1}^{\infty} \frac{4A}{mn\pi^2} \frac{[(-1)^n - 1][(-1)^n - 1]}{k(n^2+m^2)} [1 - e^{-k(n^2+m^2)t}] \cdot \sin nx(\sin my - m \cos my).$$

27. $$u(x, y, t) = x(x-\pi)\left(1 - \frac{y}{\pi}\right) \sin t + \sum_{n=1}^{\infty} \sum_{m=1}^{\infty} v_{mn}(t) \sin nx \sin my$$
where $\alpha_{mn}^2 = m^2 + n^2$ and

$$v_{mn}(t) = \frac{8e^{-c^2\alpha_{mn}^2 t}[1 - (-1)^n]}{\pi^2 mn(1 + c^4\alpha_{mn}^4)} \left[ \frac{c^2}{n^2} (\alpha_{mn}^2 - n^2)(\cos t\, e^{-c^2\alpha_{mn}^2 t} - 1) + \left( \frac{1}{n^2} + c^4\alpha_{mn}^2 \right) \sin t\, e^{c^2\alpha_{mn}^2 t} \right]$$

30. $$u(x, y, t) = \frac{4qb^4}{\pi^5 D} \sum_{n=1,3,\ldots}^{\infty} \frac{1}{n^5} \left[ 1 - \frac{v_n(x)}{1 + \cosh (n\pi a/b)} \right] \sin (n\pi y/b)$$
where

$$v_n(x) = 2 \cosh \frac{n\pi a}{2b} \cosh \frac{n\pi x}{b} + \frac{n\pi a}{2b} \sinh \frac{n\pi a}{2b} \cosh \frac{n\pi x}{b} - \frac{n\pi x}{b} \sinh \frac{n\pi x}{b} \cosh \frac{n\pi a}{2b}$$

## 10.10 EXERCISES

3. $$u(\rho, \theta) = \frac{1}{2\pi} \int_0^{2\pi} \frac{(\rho^2 - 1) f(\beta)\, d\beta}{[1 - \rho^2 - 2\rho \cos (\beta - \theta)]}$$

7. $$u(x, y) = -\frac{2}{b} \sum_{n=1}^{\infty} \frac{\sin (n\pi y/b)}{\sinh (n\pi a/b)} \left[ \sinh \frac{n\pi}{b}(a-x) \int_0^x f(\xi) \sinh \frac{n\pi\xi}{b}\, d\xi + \sinh \frac{n\pi x}{b} \int_x^a f(\xi) \sinh \frac{n\pi}{b}(a-\xi)\, d\xi \right]$$

8. $u(r, \theta) = -\sum_{n=0}^{\infty} \sum_{k=1}^{\infty} (R/\alpha_{nk})^2 J_n(\alpha_{nk} r/R)(A_{nk} \cos n\theta + B_{nk} \sin n\theta)$

where

$$A_{0k} = \frac{1}{\pi R^2 J_1^2(\alpha_{0k})} \int_0^R \int_0^{2\pi} rf(r, \theta) J_0(\alpha_{0k} r/R)\, dr\, d\theta$$

$$A_{nk} = \frac{2}{\pi R^2 J_{n+1}^2(\alpha_{nk})} \int_0^R \int_0^{2\pi} rf(r, \theta) J_n(\alpha_{nk} r/R) \cos n\theta\, dr\, d\theta$$

$$B_{nk} = \frac{2}{\pi R^2 J_{n+1}^2(\alpha_{nk})} \int_0^R \int_0^{2\pi} rf(r, \theta) J_n(\alpha_{nk} r/R) \sin n\theta\, dr\, d\theta$$

$n = 1, 2, 3, \ldots;\ k = 1, 2, 3, \ldots$ and $\alpha_{nk}$ are the roots of $J(\alpha_{nk}) = 0$.

9. $G(r, r') = \dfrac{e^{ik|r-r'|}}{|r-r'|} - \dfrac{e^{ik|\rho-r'|}}{|\rho-r'|}$

where $r = (\xi, \eta, \zeta)$, $r' = (x, y, z)$, $\rho = (\xi, \eta, -\zeta)$

10. $G(r, r') = \dfrac{e^{ik|r-r'|}}{|r-r'|} + \dfrac{e^{ik|\rho-r'|}}{|\rho-r'|}$

14. $G = -\dfrac{4a}{\pi} \sum_{n=1}^{\infty} \int_0^{\infty} \dfrac{1}{(\alpha^2 a^2 + n^2\pi^2)} \sin \dfrac{n\pi x}{a} \sin \dfrac{n\pi\xi}{a} \sin \alpha y \sin \alpha\eta\, d\alpha$

16. $u(r, z) = \dfrac{2C}{\pi} \int_0^{\infty} \int_0^{\infty} \dfrac{1}{(\kappa^2 - \lambda^2 - \beta^2)} J_0(\beta_r) J_1(\beta a) \cos \lambda z\, d\beta\, d\lambda$

17. $u(r, \theta) = A r^{\frac{1}{2}} \sin(\theta/2)$

18. $G = -\dfrac{2}{a} \sum_{n=1}^{\infty} \dfrac{\sinh \sigma y' \sinh \sigma(y-b)}{\sigma \sinh \sigma b} \sin \dfrac{n\pi x}{a} \sin \dfrac{m\pi x'}{a},$

$\sigma = \sqrt{(\kappa^2 + (n^2\pi^2)/\alpha^2)} \qquad 0 < x' < x < a, \qquad 0 < y' < y < b$

## 11.16 EXERCISES

1. $F(\alpha) = \sqrt{(1/2a)}\, e^{-\alpha^2/4a}$
2. $F(\alpha) = \sqrt{(2/\pi)}(\sin \alpha a)/\alpha$
3. $F(\alpha) = 1/|\alpha|$
4. (a) $F(\alpha) = \sqrt{(1/2)} \sin\left(\dfrac{\alpha^2}{4} + \dfrac{\pi}{4}\right)$

   (b) $F(\alpha) = \sqrt{(1/2)} \cos\left(\dfrac{\alpha^2}{4} - \dfrac{\pi}{4}\right)$

11. $u(x, t) = \dfrac{1}{2}[f(x+ct) + f(x-ct)] + \dfrac{1}{2c} \int_{x-ct}^{x+ct} g(\tau)\, d\tau$

12. $u(x, t) = \dfrac{1}{2\sqrt{t}} \displaystyle\int_0^\infty [e^{-(x-\xi)^2/4t} - e^{-(x+\xi)^2/4t}] f(\xi)\, d\xi$

13. $u(x, t) = \dfrac{1}{\sqrt{2\pi}} \displaystyle\int_{-\infty}^\infty f(\xi) \cos(a\xi^2 t) e^{-i\xi x}\, d\xi$

14. $u(x, t) = \dfrac{1}{\sqrt{\pi}} \displaystyle\int_{x/\sqrt{2at}}^\infty g\left(\dfrac{t - x^2}{2a\xi^2}\right)\left(\sin\dfrac{\xi^2}{2} + \cos\dfrac{\xi^2}{2}\right) d\xi$

15. $u(x, t) = \dfrac{\phi_0}{2\pi} \displaystyle\int_{-\infty}^\infty \dfrac{\sin a\xi}{\xi} \dfrac{e^{i\xi x}}{|\xi|} e^{-|\xi| y}\, d\xi$

16. $u(x, t) = \dfrac{1}{\sqrt{4\pi t}} \displaystyle\int_{-\infty}^\infty e^{-(x-\xi)^2/4t} f(\xi)\, d\xi$

19. $u(x, y) = \dfrac{2y}{\pi} \displaystyle\int_0^\infty \dfrac{(x^2 + \tau^2 + y^2) f(\tau)\, d\tau}{[y^2 + (x-\tau)^2][y^2 + (x+\tau)^2]}$

$$- \frac{1}{2\pi} \int_0^\infty g(\tau) \log \frac{[x^2 + (y+\tau)^2]}{[x^2 + (y-\tau)^2]}\, d\tau$$

21. $u(x, t) = \dfrac{2}{\pi} \displaystyle\int_0^\infty \xi \sin x\xi \int_0^t e^{-\xi^2(t-\tau)} f(\tau)\, d\tau\, d\xi$

22. $u(x, y) = \dfrac{2}{\pi} \displaystyle\int_0^\infty \dfrac{\sin \xi x \sin h\xi(1-y)}{\sin h\xi} \int_0^\infty f(\tau) \sin \xi\tau\, d\tau\, d\xi$

24. a. $\dfrac{1}{3}(\cos t - \cos 2t)$; b. $\dfrac{1}{3}\sin t - \dfrac{1}{6}\sin 2t$; c. $e^{2t} - e^t$; d. $1 - e^{-t} - te^{-t}$; e. $1 - e^{-t}$; f. $t \cos 2t$.

28. $u(x, t) = \begin{cases} \frac{1}{2}[f(ct + x) - f(ct - x)] & \text{for } t > x/c \\ \frac{1}{2}[f(x + ct) + f(x - ct)] & \text{for } t < x/c \end{cases}$

29. $u(x, t) = \begin{cases} 0 & \text{for } t < x/c \\ f(t - x/c) & \text{for } x/c < t \leq (2 - x)/c \end{cases}$

30. $u(x, t) = f_0 + (f_1 - f_0) \operatorname{erfc}[\sqrt{(x^2/4\kappa t)}]$

31. $u(x, t) = x - x \operatorname{erfc}(x/\sqrt{4\kappa t})$

32. $u(x, t) = 2\int_0^t \int_0^\eta \operatorname{erfc}(x/\sqrt{4\kappa\xi})\, d\eta\, d\xi$

33. $u(x, t) = f_0 e^{-ht}[1 - \operatorname{erfc}(x/\sqrt{4\kappa t})]$

34. $u(x, t) = f_0 \operatorname{erfc}(x/\sqrt{4\kappa t})$

35. $u(x, t) = \begin{cases} f_0 t & \text{for } t < x/c \\ f_0 x/c & \text{for } t > x/c \end{cases}$

36. $u(x, t) = \begin{cases} \frac{1}{2}[f(x+ct) - f(ct-x)], & t > x/c \\ \frac{1}{2}[f(x+ct) + f(x-ct)], & t < x/c \end{cases}$

37. $V(x, t) = V_0\left(t - \frac{x}{c}\right)H\left(t - \frac{x}{c}\right)$

(i) $V = V_0 H\left(t - \frac{x}{c}\right)$, (ii) $V = V_0 \cos\left\{\omega\left(t - \frac{x}{c}\right)\right\}H\left(t - \frac{x}{c}\right)$

38. $u(z, t) = Ut\left[(1 + 2\zeta^2)\operatorname{erfc}(\zeta) - \frac{2\zeta}{\sqrt{\pi}}e^{-\zeta^2}\right], \ \zeta = \frac{z}{2\sqrt{\nu t}}$

41. $V(x, t) = V_0 \operatorname{erfc}\left(\frac{x}{2\sqrt{\kappa t}}\right)$

42. 
$$\begin{aligned} q(z, t) = \frac{a}{2}e^{i\omega t}[&e^{-\lambda_1 z}\operatorname{erfc}\{\zeta - [it(2\Omega + \omega)]^{\frac{1}{2}}\} \\ &+ e^{\lambda_1 z}\operatorname{erfc}\{\zeta + [it(2\Omega + \omega)]^{\frac{1}{2}}\}] \\ &+ \frac{b}{2}e^{-i\omega t}[e^{-\lambda_2 z}\operatorname{erfc}\{\zeta - [it(2\Omega - \omega)]^{\frac{1}{2}}\} \\ &+ e^{\lambda_2 z}\operatorname{erfc}\{\zeta + [it(2\Omega - \omega)]^{\frac{1}{2}}\}] \end{aligned}$$

where

$$\lambda_{1,2} = \left\{\frac{i(2\Omega \pm \omega)}{\nu}\right\}^{\frac{1}{2}}.$$

$$q(z, t) \sim a \exp(i\omega t - \lambda_1 z) + b \exp(-i\omega t - \lambda_2 z)$$

$$\delta_{1,2} = \left\{\frac{\nu}{|2\Omega \pm \omega|}\right\}^{\frac{1}{2}}$$

43. $\left(\frac{\nu}{2\Omega}\right)^{\frac{1}{2}}$

45. $f(t) = f(0) + \frac{1}{\Gamma(\alpha)\Gamma(1-\alpha)}\int_0^t g(x)(t-x)^{\alpha-1}\,dx$

46. $x = a(\theta - \sin\theta), \ y = a(1 - \cos\theta)$

55. $u(x, t) = \sum_{n=1}^{\infty}\int_0^t e^{-n^2(t-\tau)}a_n(\tau)\,d\tau \ \sin nx + \sum_{n=1}^{\infty} b_n(0)e^{-n^2 t}\sin nx$

where

$$a_n(t) = \frac{2}{\pi}\int_0^{\pi} g(x, t)\sin nx\,dx$$

$$b_n(0) = \frac{2}{\pi}\int_0^{\pi} f(x)\sin nx\,dx$$

57. $u(x, t) = \sum_{n=1}^{\infty} \frac{2}{\pi} \sin\left(n - \frac{1}{2}\right)x \int_0^t \int_0^{\pi} e^{-(2n-1)^2(t-\tau)/4}$

$$\cdot \sin\left(n - \frac{1}{2}\right)\xi g(\xi, \tau)\, d\xi\, d\tau.$$

61. $u(x, t) = \frac{c}{2}\left[\frac{\sinh \sqrt{1/c}\, x}{\sinh \sqrt{1/c}\, \pi} - \frac{\sin \sqrt{1/c}\, x}{\sin \sqrt{1/c}\, \pi}\right] \sin t$

$$+ \frac{2}{\pi c} \sum_{n=1}^{\infty} (-1)^{n+1} \frac{n}{n^4 - (1/c)^2} \sin n^2 ct \sin nx$$

in which $\sqrt{1/c}$ is not an integer.

65. $f(x) = \int_0^{\infty} g(t)h(xt)\, dt, \qquad h(x) = M^{-1}\left[\frac{1}{K(1-p)}\right]$

## 13.10 EXERCISES

16. (a) $y = x^3$, (b) $y = \sin x$, (c) $x^2 + (y - \beta)^2 = r^2$ where $\beta$ and $r$ are constants.

17. $x = a(\theta - \sin\theta)$, $y = a(1 - \cos\theta)$

18. $x = c_1 t + c_2$, $y = c_1 \cosh t = c_1 \cosh \frac{x - c_2}{c_1}$

(A surface generated by rotation of a cateneory is called a catenoid.)

21. (a) $\nabla^4 u = 0$ (Biharmonic equation)
(b) $u_{tt} - \alpha^2\nabla^2 u + \beta^2 u = 0$ (Klein-Gordon equation)
(c) $\phi_t + \alpha\phi_x + \beta\phi_{xxx} = 0$, $(\phi = u_x)$ (KdV equation)
(d) $u_{tt} + \alpha^2 u_{xxxx} = 0$ (Elastic beam equation)
(e) $\frac{d}{dx}(pu') + (r + \lambda s)u = 0$ (Strum-Liouville equation)

22. $\frac{\hbar^2}{2m}\nabla^2\psi + (E - V)\psi = 0$ (Schrödinger equation)

24. $u_{tt} - c^2 u_{xx} = F(x, t)$, $c^2 = T^*/\rho$

29. $n = 1$: $a_1 = \frac{5}{18}$, $y_1 = a_1 x(1 - x)$

$n = 2$: $a_1 = \frac{71}{369}, \qquad a_2 = \frac{7}{41}, \qquad y_2 = x(1 - x)(a_1 + a_2 x)$

30. $I(u_1) = \frac{\pi ab}{4}[(a_1 + 1)^2 a^2 + (a_1 - 1)^2 b^2], \qquad a_1 = \frac{b^2 - a^2}{b^2 + a^2}$, $u_1 = a_1 xy$

In general,

$$\Gamma(n+1) = n! \qquad \text{for } n = 0, 1, 2, \ldots \tag{4}$$

From Eqs. (3) and (4), we obtain the value of 0!

$$\Gamma(1) = 0! = 1 \tag{5}$$

If we write Eq. (2) in the form

$$\Gamma(x) = (1/x)\Gamma(x+1) \tag{6}$$

we obtain, by repeated application of (6),

$$\Gamma(x) = \frac{\Gamma(x+k)}{x(x+1)(x+2)\ldots(x+k-1)} \tag{7}$$

for $k = 1, 2, 3, \ldots$. We see that $\Gamma(x)$ is infinite for all negative integers.

The values of $\Gamma(x)$ are tabulated for $1 < x < 2$.[21] From these values, we can find, for example,

$$\Gamma(3.5) = (2.5)(1.5)\Gamma(1.5)$$

and

$$\Gamma(-1.4) = \frac{\Gamma(1.6)}{(-1.4)(-0.4)(0.6)}$$

where $\Gamma(1.5) = 0.88623$ and $\Gamma(1.6) = 0.89352$.

## II TABLE OF FOURIER TRANSFORMS

| $f(x)$ | $\mathscr{F}[f(x)] = F(\alpha)$ |
|---|---|
| $c, \quad a \leqslant x \leqslant b$<br>$0, \quad$ otherwise | $\dfrac{ic}{\sqrt{2\pi}\,\alpha}(e^{i\alpha a} - e^{i\alpha b})$ |
| $x^n, \quad 0 \leqslant x \leqslant a$<br>$0, \quad$ otherwise | $\dfrac{1}{\sqrt{2\pi}} n!\,(-i\alpha)^{-(n+1)} - \dfrac{e^{i\alpha a}}{\sqrt{2\pi}} \sum_{k=0}^{n} \dfrac{n!}{k!}(-i\alpha)^{k-(n+1)} a^k$<br>$n = 1, 2, 3, \ldots$ |
| $\sqrt{\dfrac{\pi}{2}}\, e^{-\lvert x\rvert}$ | $1/(1+\alpha^2)$ |

[21] For the table of Gamma function, see H. B. Dwight, *Tables of Integrals and Other Mathematical Data,* Macmillan, 1955.

| | |
|---|---|
| $e^{-ax^2}$ | $e^{-\alpha^2/4a}/\sqrt{2a}$ |
| $1/\lvert x\rvert$ | $1/\lvert\alpha\rvert$ |
| $e^{-a\lvert x\rvert}/\lvert x\rvert^{\frac{1}{2}}$ | $[(a^2+\alpha^2)^{\frac{1}{2}}+a]^{\frac{1}{2}}/(a^2+\alpha^2)^{\frac{1}{2}}$ |
| $(\sin bx)/x$ | $\sqrt{\frac{\pi}{2}}, \quad \lvert\alpha\rvert < b$ <br> $0, \quad \lvert\alpha\rvert > b$ |
| $\sinh ax/\sinh \pi x$ | $\sqrt{\frac{1}{2\pi}}[\sin a/(\cos a + \cosh \alpha)], \; -\pi < a < \pi$ |
| $\cosh ax/\cosh \pi x$ | $\sqrt{\frac{2}{\pi}}\left[\cos\frac{a}{2}\cosh\frac{\alpha}{2}\Big/(\cos a + \cosh \alpha)\right], \; -\pi < a < \pi$ |

| | |
|---|---|
| $(x^2+a^2)^{-1}$ | $\sqrt{\frac{\pi}{2}}\, a^{-1}e^{-a\lvert\alpha\rvert}, \qquad \mathrm{Re}\,(a) > 0$ |
| $x(x^2+a^2)^{-1}$ | $\sqrt{\frac{\pi}{2}}\,\frac{\alpha e^{-a\lvert\alpha\rvert}}{2ia}, \qquad \mathrm{Re}\,(a) > 0$ |
| $x^{-1}$ | $-i\sqrt{\frac{\pi}{2}}\,\mathrm{sgn}\,\alpha$ |
| $x^{-2}$ | $-\lvert\alpha\rvert\sqrt{\frac{\pi}{2}}$ |
| $\begin{Bmatrix} e^{iwx}, & a < x < b \\ 0, & x < a,\; x > b \end{Bmatrix}$ | $\frac{i}{\sqrt{2\pi}}\left[\frac{e^{ia(w+\alpha)} - e^{ib(w+\alpha)}}{\alpha}\right]$ |
| $e^{iwx-ax}H(x)$ | $\frac{i}{\sqrt{2\pi}\,(w+\alpha+ia)}$ |
| $e^{iwx}$ | $\sqrt{2\pi}\,\delta(\alpha+w)$, $w$ is real |
| $e^{-a\lvert x\rvert}$ | $\sqrt{\frac{2}{\pi}}\,\frac{a}{(a^2+\alpha^2)}, \qquad a > 0$ |
| $xe^{-a\lvert x\rvert}$ | $\sqrt{\frac{2}{\pi}}\,\frac{2ia\alpha}{(a^2+\alpha^2)^2}, \qquad a > 0$ |
| $\lvert x\rvert\, e^{-a\lvert x\rvert}$ | $\sqrt{\frac{2}{\pi}}\,\frac{(a^2-\alpha^2)}{(a^2+\alpha^2)^2}, \qquad a > 0$ |

| | |
|---|---|
| $\dfrac{e^{-a\|x\|}}{\sqrt{\|x\|}}$ | $\dfrac{[a+(a^2+\alpha^2)]^{\frac{1}{2}}}{(a^2+\alpha^2)^{\frac{1}{2}}}$ |
| $\dfrac{H(a-\|x\|)}{2(a^2-x^2)^{\frac{1}{2}}}$ | $\sqrt{2\pi}\, J_0(a\alpha)$ |
| $\dfrac{\sin[b(x^2+a^2)^{\frac{1}{2}}]}{(x^2+a^2)^{\frac{1}{2}}}$ | $\sqrt{\dfrac{\pi}{2}}\, J_0(a\sqrt{b^2-\alpha^2})\, H(b-\|\alpha\|)$ |
| $\dfrac{\cos(b\sqrt{a^2-x^2})}{(a^2-x^2)^{\frac{1}{2}}}\, H(a-\|x\|)$ | $\sqrt{\dfrac{\pi}{2}}\, J_0(a\sqrt{\alpha^2+b^2})$ |
| $\dfrac{\cosh(b\sqrt{a^2-x^2})}{(a^2-x^2)^{\frac{1}{2}}}\, H(a-\|x\|)$ | $\sqrt{\dfrac{\pi}{2}}\, J_0(a\sqrt{\alpha^2-b^2})$ |
| $\operatorname{sgn} x$ | $\sqrt{\dfrac{2}{\pi}}\dfrac{1}{(-i\alpha)}$ |
| $H(x)$ | $\sqrt{\dfrac{\pi}{2}}\,\delta(\alpha)-\dfrac{i}{\alpha\sqrt{2\pi}}$ |
| $\delta^{(n)}(x)$ | $\dfrac{(-i\alpha)^n}{\sqrt{2\pi}}$ |

## III TABLE OF LAPLACE TRANSFORMS

| $f(t)$ | $\mathscr{L}[f(t)]=F(s)$ | |
|---|---|---|
| $c$ | $c/s,$ | $c$ is a constant |
| $t^n$ | $n!/s^{n+1},$ | $n=1, 2, 3, \ldots$ |
| $e^{at}$ | $1/(s-a)$ | |
| $te^{at}$ | $1/(s-a)^2$ | |
| $\sin \omega t$ | $\omega/(s^2+\omega^2)$ | |
| $\cos \omega t$ | $s/(s^2+\omega^2)$ | |
| $\sinh \kappa t$ | $\kappa/(s^2-\kappa^2)$ | |
| $\cosh \kappa t$ | $s/(s^2-\kappa^2)$ | |
| $\sqrt{t}$ | $\sqrt{\pi}/2\sqrt{s^3}$ | |
| $1/\sqrt{t}$ | $\sqrt{\pi}/\sqrt{s}$ | |
| $e^{at}\sin \omega t$ | $\omega/[(s-a)^2+\omega^2]$ | |
| $e^{at}\cos \omega t$ | $(s-a)/[(s-a)^2+\omega^2]$ | |
| $\operatorname{erf}(a\sqrt{t})$ | $a/s\sqrt{s+a^2}$ | |
| $\operatorname{erfc}(a/2\sqrt{t})$ | $e^{-a\sqrt{s}}/s$ | |
| $t^{\kappa-1}$ | $\Gamma(\kappa)/s^{\kappa},$ | $\kappa>0$ |
| $J_0(at)$ | $1/\sqrt{s^2+a^2}$ | |

| | | |
|---|---|---|
| $I_0(at)$ | $1/\sqrt{s^2-a^2}$ | |
| $(\sin \omega t)/t$ | $\tan^{-1}(\omega/s)$ | |
| $t \sinh \kappa t$ | $2\kappa s/(s^2-\kappa^2)^2$ | |
| $t \cosh \kappa t$ | $(s^2+\kappa^2)/(s^2-\kappa^2)^2$ | |
| $t \sin \omega t$ | $2\omega s/(s^2+\omega^2)^2$ | |
| $t \cos \omega t$ | $(s^2-\omega^2)/(s^2+\omega^2)^2$ | |
| $\dfrac{e^{at}-e^{bt}}{a-b}$ | $\dfrac{1}{(s-a)(s-b)}$ | $(a \neq b)$ |
| $\dfrac{ae^{at}-be^{bt}}{a-b}$ | $\dfrac{s}{(s-a)(s-b)}$ | $(a \neq b)$ |
| $(1+at)e^{at}$ | $\dfrac{s}{(s-a)^2}$ | |
| $\dfrac{e^{-at}}{\sqrt{\pi t}}$ | $\dfrac{1}{\sqrt{s+a}}$ | |
| $\dfrac{1}{\sqrt{\pi t}}-ae^{a^2 t}\operatorname{erfc}(a\sqrt{t})$ | $\dfrac{1}{\sqrt{s}+a}$ | |
| $\dfrac{4^n n!\, t^{n-\frac{1}{2}}}{(2n)!\sqrt{\pi}}$ | $s^{-(n+\frac{1}{2})}$ | |
| $\dfrac{1}{\sqrt{t}}e^{-a/t}$, $\arg(a) < \dfrac{\pi}{2}$ | $\left(\dfrac{\pi}{s}\right)^{\frac{1}{2}} e^{-2\sqrt{as}}$ | |
| $\dfrac{\sqrt{\pi}}{\Gamma\left(n+\dfrac{1}{2}\right)}\left(\dfrac{t}{2a}\right)^n J_n(at)$ | $(s^2+a^2)^{-n-\frac{1}{2}}$, $\quad \operatorname{Re}(n) > -\dfrac{1}{2}$ | |
| $\dfrac{na^n}{t}J_n(at)$ | $(\sqrt{s^2+a^2}-s)^n$, $\quad \operatorname{Re}(n) > -\dfrac{1}{2}$ | |
| $a^n J_n(at)$ | $\dfrac{(\sqrt{s^2+a^2}-s)^n}{\sqrt{s^2+a^2}}$, $\quad \operatorname{Re}(n) > -1$ | |
| $\dfrac{(e^{bt}-e^{at})}{2t\sqrt{\pi t}}$ | $\sqrt{s-a}-\sqrt{s-b}$ | |
| $J_0(a\sqrt{t})$ | $\dfrac{1}{s}\exp\left(-\dfrac{a^2}{4s}\right)$ | |
| $\left(\dfrac{2}{a}\right)^n t^{n/2} J_n(a\sqrt{t})$ | $s^{-(n+1)}\exp\left(-\dfrac{a^2}{4s}\right)$, $\quad \operatorname{Re}(n) > -1$ | |
| $\dfrac{a}{2\sqrt{\pi}\, t^{\frac{3}{2}}}\exp\left(-\dfrac{a^2}{4t}\right)$ | $e^{-a\sqrt{s}}$, $\quad a > 0$ | |

| | | |
|---|---|---|
| $\frac{1}{\sqrt{\pi t}}\exp\left(-\frac{a^2}{4t}\right)$ | $\frac{1}{\sqrt{s}}e^{-a\sqrt{s}}$, | $a \geqslant 0$ |
| $\exp\left(-\frac{1}{4}a^2t^2\right)$ | $\frac{\sqrt{\pi}}{a}e^{s^2/a^2}\operatorname{erfc}\left(\frac{s}{a}\right)$ | |
| $t^{\alpha-1}e^{-at}$ | $\frac{\Gamma(\alpha)}{(s+a)^{\alpha}}$, | $\operatorname{Re} s > \operatorname{Re} a$ |
| $\frac{H(t-a)}{(t^2-a^2)^{\frac{1}{2}}}$ | $K_0(as)$, | $a > 0$ |
| $\delta(t-a)$ | $e^{-as}$, | $a \geqslant 0$ |
| $H(t-a)$ | $\frac{1}{s}e^{-as}$, | $a \geqslant 0$ |

## IV TABLE OF HANKEL TRANSFORMS

| $f(r)$ | order $n$ | $F_n(k)$ |
|---|---|---|
| $H(a-r)$ | 0 | $\frac{a}{k}J_1(ak)$ |
| $e^{-ar}$ | 0 | $a(k^2+a^2)^{-\frac{3}{2}}$ |
| $\frac{1}{r}e^{-ar}$ | 0 | $(k^2+a^2)^{-\frac{1}{2}}$ |
| $(a^2-r^2)H(a-r)$ | 0 | $\frac{4a}{k^3}J_1(ak)-\frac{2a^2}{k^2}J_0(ak)$ |
| $a(a^2+r^2)^{-\frac{3}{2}}$ | 0 | $e^{-ak}$ |
| $\frac{1}{r}\sin ar$ | 0 | $(a^2-k^2)^{-\frac{1}{2}}H(a-k)$ |
| $\frac{1}{r}\cos ar$ | 0 | $(k^2-a^2)^{-\frac{1}{2}}H(k-a)$ |
| $\frac{1}{r^2}(1-\cos ar)$ | 0 | $\cos h^{-1}\frac{a}{k}H(a-k)$ |
| $\frac{\pi}{2}Y_0(ar)$ | 0 | $(a^2-k^2)^{-1}$ |
| $K_0(ar)$ | 0 | $(k^2+a^2)^{-1}$ |
| $\frac{\delta(r)}{r}$ | 0 | 1 |
| $e^{-ar}$ | 1 | $k(k^2+a^2)^{-\frac{3}{2}}$ |

| | | |
|---|---|---|
| $\frac{1}{r} e^{-ar}$ | 1 | $\frac{1}{k}\left[1-\frac{a}{(k^2+a^2)^{\frac{1}{2}}}\right]$ |
| $\frac{1}{r}\sin ar$ | 1 | $\frac{a}{k}(k^2-a^2)^{-\frac{1}{2}}H(k-a)$ |
| $\frac{1}{r^2} e^{-ar}$ | 1 | $\frac{1}{k}[(k^2+a^2)^{\frac{1}{2}}-a]$ |
| $r^n H(a-r)$ | $n>-1$ | $\frac{1}{k} a^{n+1} J_{n+1}(ak)$ |
| $r^n e^{-ar}$, Re $(a)>0$ | $n>-1$ | $\frac{1}{\sqrt{\pi}}\frac{2^{n+1}\Gamma\left(n+\frac{3}{2}\right)ak^n}{(a^2+k^2)^{n+\frac{3}{2}}}$ |
| $r^n e^{-ar^2}$ | $n>-1$ | $\frac{k^n}{(2a)^{n+1}}\exp\left(-\frac{k^2}{4a}\right)$ |
| $r^{a-1}$ | $n>-1$ | $\frac{2^a\Gamma\left(\frac{1}{2}+\frac{a}{2}+\frac{n}{2}\right)}{k^{n+1}\Gamma\left(\frac{1}{2}-\frac{a}{2}+\frac{n}{2}\right)}$ |
| $r^n(a^2-r^2)^{m-n-1}H(a-r)$ | $n>-1$ | $2^{m-n-1}\Gamma(m-n)a^m k^{n-m} J_m(ak)$ |
| $r^m e^{-r^2/a^2}$ | $n>-1$ | $\frac{k^n a^{m+n+2}}{2^{n+1}\Gamma(n+1)}\Gamma\left(1+\frac{m}{2}+\frac{n}{2}\right)$ $\times F_1\left(1+\frac{m}{2}+\frac{n}{2}; n+1; -\frac{1}{4}a^2k^2\right)$ |
| $\frac{1}{r}J_1(ar)$ | 0 | $\frac{1}{a}H(a-k)$, $a>0$ |
| $\frac{1}{r}\sin ar$ | 1 | $(a/k)(k^2-a^2)^{-\frac{1}{2}}H(k-a)$ |
| $\frac{1}{r}J_{n+1}(ar)$ | $n>-1$ | $k^n a^{-(n+1)}H(a-k)$, $a>0$ |
| $\frac{r^n H(a-r)}{(a^2-r^2)^{\frac{1}{2}}}$ | $n>-1$ | $(\pi/2k)^{\frac{1}{2}}a^{n+1}J_{n+1}(ak)$ |
| $\frac{1}{r^2}J_n(ar)$ | $n>\frac{1}{2}$ | $\begin{cases}\frac{1}{2n}\left(\frac{k}{a}\right)^n, & 0<k<a\\ \frac{1}{2n}\left(\frac{a}{k}\right)^n, & k>a\end{cases}$ |

# Bibliography

## Partial Differential Equations


1. H. Bateman, *Partial Differential Equations of Mathematical Physics,* Cambridge University Press, Cambridge (1959).
2. A. Broman, *Introduction to Partial Differential Equations: from Fourier Series to Boundary Value Problems,* Addison-Wesley, Reading, Massachusetts (1970).
3. P. Berg and J. McGregor, *Elementary Partial Differential Equations,* Holden-Day, New York (1966).
4. M. L. Boas, *Mathematical Methods in the Physical Sciences,* Wiley, New York (1966).
5. H. S. Carslaw and J. C. Jaeger, *Conduction of Heat in Solids,* Oxford University Press, Oxford (1959).
6. R. V. Churchill and J. W. Brown, *Fourier Series and Boundary Value Problems,* McGraw-Hill, New York (1978).
7. R. Courant and D. Hilbert, *Methods of Mathematical Physics, Volume* 2, Interscience, New York (1962).
8. H. K. Crowder and S. W. McCuskey, *Topics in Higher Analysis,* Macmillan, New York (1964).
9. A. F. Danese, *Advanced Calculus, Volume* 2, Allyn and Bacon (1965).
10. R. Dennemeyer, *Partial Differential Equations and Boundary Value Problems,* McGraw-Hill, New York (1968).
11. G. F. D. Duff, *Partial Differential Equations,* University of Toronto, Toronto (1956).
12. G. F. D. Duff and D. Naylor, *Differential Equations of Applied Mathematics,* Wiley, New York (1966).
13. B. Epstein, *Partial Differential Equations,* McGraw-Hill, New York (1962).
14. P. Frank and R. Von Mises, *Die Differential und Integralgleichungen der Mechanik und Physik,* Rosenberg (1943).
15. P. R. Garabedian, *Partial Differential Equations,* John Wiley, New York (1964).

16. M. Greenberg, *Applications of Green's Functions in Science and Engineering,* Prentice-Hall, Englewood Cliffs, New Jersey (1971).
17. G. Grunberg, A new method of solution of certain boundary problems for equations of mathematical physics permitting of a separation of variables, *J. Phys.* **10** 301–320 (1946).
18. J. Hadamard, *Lectures on Cauchy's Problem in Linear Partial Differential Equations,* Dover, New York (1952).
19. G. Hellwig, *Partial Differential Equations,* Blaisdell, Waltham, Massachusetts (1964).
20. J. Irving and N. Mullineux, *Mathematics in Physics and Engineering,* Academic, New York (1959).
21. J. Jeffreys and B. S. Jeffreys, *Methods of Mathematical Physics,* Third Edition, Cambridge University Press, Cambridge (1956).
22. F. John, *Partial Differential Equations,* Lecture Notes, NYU (1953), unpublished.
23. N. S. Koshlyakov, M. M. Smirnov and E. B. Gliner, *Differential Equations of Mathematical Physics,* Holden-Day, New York (1964).
24. D. Kreider, R. Kuller, D. Ostberg, and F. Perkins, *An Introduction to Linear Analysis,* Addison-Wesley, Reading, Massachusetts (1966).
25. E. Kreyszig, *Advanced Engineering Mathematics,* Wiley, New York (1967).
26. P. D. Lax, *Partial Differential Equations,* Lecture Notes NYU (1951), unpublished.
27. S. G. Mikhlin, *Linear Equations of Mathematical Physics,* Holt, Rinehart and Winston, New York (1967).
28. K. S. Miller, *Partial Differential Equations in Engineering Problems,* Prentice-Hall, Englewood Cliffs, New Jersey (1953).
29. P. M. Morse and H. Feshbach, *Methods of Theoretical Physics,* Volume 1 and 2, McGraw-Hill, New York (1953).
30. I. Petrovsky, *Lectures on Partial Differential Equations,* Interscience, New York (1954).
31. L. A. Pipes, *Applied Mathematics for Engineers and Physicists,* McGraw-Hill, New York (1958).
32. H. Sagan, *Boundary and Eigenvalue Problems in Mathematical Physics,* Wiley, New York (1961).
33. R. Seeley, *Introduction to Fourier Series and Integrals,* Benjamin, New York (1966).
34. V. I. Smirnov, *Integral Equations and Partial Differential Equations,* Addison-Wesley, Reading, Massachusetts (1964).
35. M. G. Smith, *Introduction to the Theory of Partial Differential Equations,* Van Nostrand, Princeton, New Jersey (1967).
36. I. N. Sneddon, *Elements of Partial Differential Equations,* McGraw-Hill, New York (1957).
37. I. N. Sneddon, *Fourier Transforms,* McGraw-Hill, New York (1951).
38. I. N. Sneddon, *Mixed Boundary Value Problems in Potential Theory,* North Holland, Amsterdam (1966).
39. S. L. Sobolev *Partial Differential Equations of Mathematical Physics,* Addison-Wesley, Reading, Massachusetts (1964).
40. I. S. Sokolnikoff, *Mathematical Theory of Elasticity,* McGraw-Hill, New York (1956).

41. I. S. Sokolnikoff and R. M. Redheffer, *Mathematics of Physics and Modern Engineering,* McGraw-Hill, New York (1966).
42. A. Sommerfeld, *Partial Differential Equations in Physics,* Academic, New York (1964).
43. I. Stakgold, *Boundary Value Problems of Mathematical Physics,* Macmillan, New York (1967).
44. A. N. Tikhonov and A. A. Samarskii, *Equations of Mathematical Physics,* Macmillan, New York (1963).
45. A. N. Tychonov and A. A. Samarski, *Partial Differential Equations of Mathematical Physics,* Holden-Day, New York (1964).
46. A. G. Webster, *Partial Differential Equations of Mathematical Physics,* Dover, New York (1955).
47. H. Weinberger, *A First Course in Partial Differential Equations,* Blaisdell, Waltham, Massachussetts (1965).
48. E. T. Whittaker and G. N. Watson, *A Course on Modern Analysis,* Cambridge University Press, Cambridge (1952).

## Fourier Series and Fourier Integrals

49. W. E. Byerly, *Fourier Series,* Dover, New York (1959).
50. H. S. Carslaw, *Introduction to the Theory of Fourier Series and Integrals,* Dover, New York (1950).
51. H. F. Davis, *Fourier Series and Orthogonal Functions,* Allyn and Bacon (1963).
52. D. Jackson, *Fourier Series and Orthogonal Polynomials,* Mathematical Association of American Monograph (1941).
53. D. Kreider, R. Kuller, D. Ostberg, and F. Perkins, *An Introduction to Linear Analysis,* Addison-Wesley, Reading, Massachusetts (1966).
54. W. W. Rogosinski, *Fourier Series,* Chelsea (1950).
55. G. Sansone, *Orthogonal Functions,* Interscience, New York (1959).
56. E. C. Titchmarsh, *An Introduction to the Theory of Fourier Integrals,* Oxford University Press, Oxford (1962).
57. G. P. Tolstov, *Fourier Series,* Prentice-Hall, Englewood Cliffs, New Jersey (1962).

## Eigenvalue Problems and Green's Function

58. G. Birkhoff and R-C. Rota, *Ordinary Differential Equations,* Blaisdell, Waltham, Massachusetts (1962).
59. E. A. Coddington and N. Levinson, *Theory of Ordinary Differential Equations,* McGraw-Hill, New York (1955).
60. R. Cole, *Theory of Ordinary Differential Equations,* Appleton-Century-Crofts (1968).
61. L. Collatz, *Eigenwertproblem und ihre Numerische Behandlung,* Academische Verlags Gesellschaft (1945).
62. R. Courant and D. Hilbert, *Methods of Mathematical Physics, Volume* 1, Interscience, New York (1953).
63. F. Hildebrand, *Methods of Applied Mathematics,* Prentice-Hall, Englewood Cliffs, New Jersey (1965).

64. H. Keller, *Numerical Methods for Two Point Boundary Value Problems,* Blaisdell, Waltham, Massachusetts (1968).
65. J. Indriz, *Methods in Analysis,* Macmillan, New York (1963).
66. K. S. Miller, *Linear Differential Equations in the Real Domain,* Norton (1963).
67. T. Myint-U, *Ordinary Differential Equations,* Elsevier North Holland, Inc., New York (1978).
68. H. Sagan, *Boundary and Eigenvalue Problems in Mathematical Physics,* Wiley, New York (1966).
69. E. C. Titchmarsh, *Eigenfunction Expansions Associated with Second-Order Differential Equations,* Oxford University Press, Oxford (1946).

## Integral Transforms and Special Functions

70. G. Doetsch, *Introduction to the Theory and Applications of the Laplace Transformation,* (Translated by Walter Nader), Springer Verlag, New York (1970).
71. H. S. Carslaw and J. C. Jaeger, *Operational Methods in Applied Mathematics,* Oxford University Press, New York (1948).
72. R. V. Churchill, *Operational Mathematics,* (Third Edition) McGraw-Hill, New York (1972).
73. E. T. Copson, *Asymptotic Expansions,* Cambridge University Press, Cambridge (1965).
74. M. Dutta and L. Debnath, *Elements of the Theory of Elliptic and Associated Functions with Applications,* World Press, Calcutta (1965).
75. A. Erdelyi, W. Magnus, F. Oberhettinger and F. Tricomi, *Table of Integral Transforms,* (Vols. I and II), McGraw-Hill, New York (1954).
76. H. Jeffreys, *Operational Methods in Mathematical Physics,* Cambridge University Press, New York (1931).
77. N. W. McLachlan, *Complex Variable and Operational Calculus with Technical Applications,* Cambridge University Press, London (1942).
78. N. W. McLachlan, *Modern Operational Calculus,* Macmillan, London (1948).
79. R. H. Raven, *Mathematics of Engineering Systems,* McGraw-Hill, New York (1966).
80. E. J. Scott, *Transform Calculus with an Introduction to Complex Variables,* Harper, New York (1955).
81. I. N. Sneddon, *Fourier Transforms,* McGraw-Hill, New York (1951).
82. I. N. Sneddon, *The Use of Integral Transforms,* McGraw-Hill, New York (1972).
83. E. C. Titchmarsh, *An Introduction to Theory of Fourier Integrals,* Oxford University Press, New York (1937).
84. C. J. Tranter, *Integral Transforms in Mathematical Physics,* (Third Edition) Wiley, New York (1966).
85. G. N. Watson, *A Treatise on the Theory of Bessel Functions,* Cambridge University Press, Cambridge (1966).
86. D. V. Widder, *The Laplace Transform,* Princeton University Press, Princeton, New Jersey (1941).

## Nonlinear Partial Differential Equations


87. M. J. Ablowitz and H. Segur, *Solitons and Inverse Scattering Transform,* SIAM, Philadelphia (1981).
88. T. B. Benjamin, J. L. Bona and J. J. Mahony, Model Equations for Long Waves in Nonlinear Dispersive Systems, *Phil. Trans. Roy. Soc. A272* (1972) 47–78.
89. F. Calogero and Degasperis, *Soliton and Spectral Transform* I, North-Holland, Amsterdam (1982).
90. J. Crank, *Mathematics of Diffusion,* (Second Edition) Clarendon Press, Oxford (1975).
91. L. Debnath, *Nonlinear Waves,* Cambridge University Press, Cambridge, England (1983).
92. L. Debnath, *Advances in Nonlinear Waves,* Vol. I (1984), Vol. II (1985), Pitman Publishing Company, Boston and London.
93. F. A. Haight, *Mathematical Theory of Traffic Flow,* Academic Press, New York (1963).
94. D. J. Korteweg and G. de Vries, On the Change of Form of Long Waves Advancing in a Rectangular Canal, and on a New Type of Long Stationary Waves, *Phil. Mag.* **39** (1895) 422–443.
95. H. Lamb, *Hydrodynamics,* (Sixth Edition) Cambridge University Press, Cambridge, England (1932).
96. S. Leibovich and A. R. Seebass, *Nonlinear Waves,* Cornell University Press, Ithaca and London (1972).
97. M. J. Lighthill, *Waves in Fluids,* Cambridge University Press, Cambridge (1980).
98. A. C. Newell, *Solitons in Mathematics and Physics,* SIAM, Philadelphia (1985).
99. G. B. Whitham, *Linear and Nonlinear Waves,* John Wiley, New York (1976).
100. N. J. Zabusky and M. D. Kruskal, Interaction of Solitons in a Collisionless Plasma and Recurrence of Initial States, *Phys. Rev. Lett.* **15** (1965) 240–243.
101. V. E. Zakharov and A. B. Shabat, Exact Theory of Two-Dimensional Self-Focusing and One-Dimensional Self-Modulation of Waves in Nonlinear Media, *Sov. Phys. JETP* **34** (1972) 62–69.
102. V. E. Zakharov and A. B. Shabat, Interaction Between Solitons in a Stable Medium, *Sov. Phys. JETP* **37** (1973) 823–828.


## Numerical and Approximation Methods


103. W. F. Ames, *Numerical Methods for Partial Differential Equations* (Second Edition) Academic Press, New York (1977).
104. G. E. Forsythe and W. R. Wasow, *Finite-Difference Methods for Partial Differential Equations,* John Wiley, New York (1967).
105. C. Fox, *An Introduction to the Calculus of Variations,* Oxford University Press, Oxford (1963).
106. I. M. Gelfand and S. V. Fomin, *Calculus of Variations,* (Translated by R. A. Silverman), Prentice-Hall, Englewood Cliffs (1963).

107. L. Kantorovich and V. Krylov, *Approximate Methods in Higher Analysis,* Interscience, New York (1958).
108. S. Mikhlin, *Variational Methods in Mathematical Physics,* Pergamon, Oxford (1964).
109. A. R. Mitchell and D. F. Griffiths, *The Finite Difference Method in Partial Differential Equations,* John Wiley, New York (1980).
110. R. D. Richtmyer and K. W. Morton, *Difference Methods for Initial Value Problems,* Interscience, New York (1967).
111. G. D. Smith, *Numerical Solution of the Partial Differential Equations* (Third Edition) Clarendon Press, Oxford (1985).

## Tables and Formulas

112. G. A. Campbell and R. M. Foster, *Fourier Integrals for Practical Applications,* Van Nostrand, New York (1948).
113. A. Erdelyi, W. Magnus, F. Oberhettinger, and F. G. Tricomi, *Higher Transcendental Functions,* Volumes I, II, and III, McGraw-Hill, New York (1953).
114. E. Jahnke, F. Emde, and F. Losch, *Tables of Higher Functions,* McGraw-Hill, New York (1960).
115. W. Magnus and F. Oberhettinger, *Formulas and Theorems for the Special Functions of Mathematical Physics,* Chelsea (1949).

## Problem Books

116. B. M. Budak, A. A. Samarskii and A. N. Tikhonov, *A Collection of Problems on Mathematical Physics,* Macmillan, New York (1964).
117. N. N. Lebedev, I. P. Skalskaya, and Y. S. Uflyand, *Problems of Mathematical Physics,* Prentice-Hall, Englewood Cliffs, New Jersey (1965).
118. M. M. Smirnov, *Problems on the Equations of Mathematical Physics,* Noordhoff (1967).

# Index

**U**



**V**



**W**